B, PB, B9D.

MEDWAY

WC

This book mu
and may be

12

13 D

2

27

9408620

Exploitation of Microorganisms

Exploitation of Microorganisms

Edited by

Professor D. Gareth Jones

Department of Agricultural Sciences,
University of Wales,
Aberystwyth

CHAPMAN & HALL
London · Glasgow · New York · Tokyo · Melbourne · Madras

Published by Chapman & Hall, 2–6 Boundary Row, London SE1 8HN

Chapman & Hall, 2–6 Boundary Row, London SE1 8HN, UK

Blackie Academic & Professional, Wester Cleddens Road, Bishopbriggs, Glasgow G64 2NZ, UK

Chapman & Hall Inc., 29 West 35th Street, New York NY10001, USA

Chapman & Hall Japan, Thomson Publishing Japan, Hirakawacho Nemoto Building, 6F, 1–7–11 Hirakawa-cho, Chiyoda-ku, Tokyo 102, Japan

Chapman & Hall Australia, Thomas Nelson Australia, 102 Dodds Street, South Melbourne, Victoria 3205, Australia

Chapman & Hall India, R. Seshadri, 32 Second Main Road, CIT East, Madras 600 035, India

First edition 1993

© 1993 Chapman & Hall

Typeset in 10/12pt Palatino by Acorn Bookwork

Printed in Great Britain at the University Press, Cambridge

ISBN 0 412 45740 7

A catalogue record for this book is available from the British Library

Library of Congress Cataloging-in-Publication data available

Contents

Contents

Contributors

M.T. Balba, TreaTek-CRA Company, Technology Center, 2801 Long Road, Grand Island, NY 14072, USA.

D.R. Dent, School of Pure and Applied Biology, University of Wales, Cardiff, PO Box 915, Cardiff CF1 3TL, Wales, UK.

G.M. Evans, Department of Agricultural Sciences, University of Wales, Aberystwyth, Dyfed, SY23 3DD, Wales, UK.

M.J. Jeger, Natural Resources Insitute, Central Avenue, Chatham Maritime, Chatham, Kent, ME4 4TB, UK.

D. Gareth Jones, Department of Agricultural Sciences, University of Wales, Aberystwyth, Dyfed, SY23 3DD, Wales, UK.

K.A. Jones, Natural Resources Institute, Central Avenue, Chatham Maritime, Chatham, Kent, ME4 4TB, UK.

B. Kerry, Entomology and Nematology Department, AFRC-IACR, Rothamsted Experimental Station, Harpenden, Herts, AL5 2JQ, UK.

D.M. Lewis, Department of Agricultural Sciences, University of Wales, Aberystwyth, Dyfed, SY23 3DD, Wales, UK.

M.P. McQuilken, Microbiology and Crop Protection Department, Horticulture Research International, Worthing Road, Littlehampton, West Sussex, BN17 6LP, UK.

D.T. Mitchell, Department of Botany, University College Dublin, Belfield, Dublin 4, Ireland.

K.A. Powell, ICI Agrochemicals, Jealott's Hill Research Station, Bracknell, Berks, RG12 6EY, UK

P.J.A. Reilly, Natural Resources Institute, Central Avenue, Chatham Maritime, Chatham, Kent, ME4 4TB, UK.

D.J. Rhodes, ICI Agrochemicals, Jealott's Hill Research Station, Bracknell, Berks, RG12 6EY, UK

L. Rovesti, Consiglo Nazionale delle Ricerche, Centro di Studio per gli Antiparassitari, 40126 Bologna, Italy.

R.A. Samson, Centraalbureau voor Schimmelcultures, P.O. Box 273, 3740 AG Baarn, The Netherlands.

J.F. Smith, Horticulture Research International, Worthing Road, Littlehampton, West Sussex, BN17 6LP, UK.

D.O. TeBeest, Department of Plant Pathology, University of Arkansas, Fayetteville, Arkansas 72701, USA.

A.H. Varnam, Southern Biological, 47 Hatherley Road, Reading, RG1 5QE, UK.

A. Westby, Natural Resources Institute, Central Avenue, Chatham Maritime, Chatham, Kent, ME4 4TB, UK.

J.M. Whipps, Microbiology and Crop Protection Department, Horticulture Research International, Worthing Road, Littlehampton, West Sussex, BN17 6LP, UK.

B.A. Whitton, Department of Biological Sciences, University of Durham, South Road, Durham, DH1 3LE, UK.

D. Winstanley, Horticulture Research International, Worthing Road, Littlehampton, West Sussex, BN17 6LP, UK.

Preface

Microbiology may be described as one of the younger sciences with its history, as a precise subject, only dating as far back as Pasteur in the mid 1800s and his revelation both of the role of microorganisms in nature and their importance to human welfare. Medical scientists rapidly took up the challenge, with their area of microbiology flourishing and expanding almost in complete isolation from the rest of biology.

We now know, of course, that microorganisms have always played an important, if not essential role, in the biosphere with fermented foods and beverages, plant and animal diseases and nutrient cycling foremost in their sphere of activities. Within the last twenty years, microbiology has received two enormous boosts with the developments in microbial genetics and genetic engineering probably being the most influential, and the greater awareness of pollution and environmental sustainability following a close second.

In 1990, your editor had the privilege and pleasure of being elected as President of The Association of Applied Biologists in the United Kingdom and, as the topic for his three-day Presidential Conference, chose 'The exploitation of microorganisms in applied biology'. This meeting stimulated great interest in a wide range of subject areas, from weed control to nematology, from plant breeding to plant pathology, from mushrooms to mycorrhiza. The proceedings of this meeting were published in *Aspects of Applied Biology, No. 24, 1990*. However, it was quite evident that there were many aspects of the exploitation of microorganisms that were not covered in these Proceedings and your editor, in conjunction with Mr Nigel Balmforth of Chapman & Hall, set out to produce a more comprehensive publication. The present book is an attempt to combine reviews and current positions of the 'state of the art' in 17 different areas of microorganism exploitation. It is hoped that biologists of every denomination will be able to find something of interest in, or near, to their own work field but, perhaps more importantly, they may also derive benefit from the volume of information presented on topics as diverse as nitrogen fixing bacteria, detoxification of industrial waste and the commercial view of inoculant production.

I would like to record my thanks to The Association of Applied Biologists and the many members who have contributed to this book. Also to Mr Nigel Balmforth and his colleagues at Chapman & Hall, to my very patient wife and to my very hard working secretaries, Mrs Dorothy Jones and Mrs Doreen Thomas.

D. Gareth Jones

1

Biological control of weeds with fungal plant pathogens

D.O. TeBeest

1.1 INTRODUCTION

Weeds continue to be a problem in agriculture throughout the world, reducing yield and quality of crops by competing for the water, nutrients and sunlight essential for vigorous crop growth. The importance of weeds in agriculture is reflected by the rapid growth of the use of chemical herbicides to control weeds. The effectiveness of herbicides has dramatically increased yields and also stimulated development of new chemical herbicides. Nevertheless, losses due to weeds continue to be documented. Plant pathogens have recently been suggested as one of several possible means of controlling weeds which remain problematic in otherwise successful weed control programmes in intensive agriculture, or even as an alternative to chemical herbicides. Plant pathogens are also being considered as a potential, cost-effective means of reducing weed populations in crops, forests or rangelands where costs and profit margins prohibit large expenditures for chemical herbicides.

Biological control of weeds with plant pathogens has been reviewed a number of times and is the subject of several books to which the reader is. referred for additional information (Templeton *et al.*, 1979; Charudattan and Walker, 1982; Charudattan, 1990a; TeBeest, 1990). This review will concentrate on selected examples of the strategies that have been developed for using plant pathogens as biological control agents, and on the effectiveness of pathogens in biological control of weeds.

Exploitation of Microorganisms Edited by D.G. Jones
Published in 1993 by Chapman & Hall, London ISBN 0 412 45740 7

1.2 STRATEGIES FOR CONTROLLING WEEDS WITH PLANT PATHOGENS

Research conducted in the past two decades has led directly to the development of several strategies that utilize plant pathogens to control weeds.

The **classical** strategy resulted from research that led to the release of *Puccinia chondrillina* Bubak. & Syd. in Australia for the control of *Chondrilla juncea* L. The name and strategy is similar to the entomological strategy in which insects are released to control weeds or other insects through natural predation or spread. This strategy clearly requires little, if any, manipulation after an initial release, and the amount of disease is a function of natural constraints on the spread of the pathogen, subsequent development of disease, and the populations of host and pathogen. The pathogen is expected to survive from year to year, providing long-term control or population reductions without additional re-introductions.

The **mycoherbicide** strategy resulted from research that led to the commercial use of *Phytophthora palmivora* (Butl.) Butl. and *Colletotrichum gloeosporioides* (Penz.) Sacc. f.sp. *aeschynomene*. This strategy utilizes pathogens in ways consistent with modern chemical herbicide technology. Applications of pathogens are made to specific areas in predetermined volumes and dosages that achieve control of the target weed within an allotted amount of time and before economic losses are incurred. The fungus is not expected to survive or to provide control beyond the growing season in which it is applied. This strategy emphasizes manipulation of fungal pathogens with respect to establishing and controlling the amount of disease caused in a specific population of the host. The fungus usually is applied as a formulated product bearing little resemblance to the usual concept of inoculum.

Technically, to be consistent with accepted labelling and registration practices, the term mycoherbicide should be reserved for fungal plant pathogens registered as commercial products for control of weeds.

1.3 EXAMPLES OF THE CLASSICAL STRATEGY

Several pathogens have been imported into the United States and Australia for the specific purpose of controlling weeds in the classical strategy.

1.3.1 Entyloma compositarum

In 1975, *Cercosporella ageratinae* was introduced into Hawaii from Jamaica to control hamakua pamakani, *Ageratina riparia* K.&R. (= *Eupatorium riparium* Regel) (Trujillo 1976, 1985; Trujillo *et al.*, 1988). *A riparia* was determined to be the most serious weed pest of the Hawaiian range from 800 to 6500ft in elevation. Host range studies, which included 40 species representing 29 plant families, had previously indicated that the fungus was specific to hamakua pamakani. The fungus was released into Hawaii from November 1975 until May 1976 and the ensuing epidemic resulted in the infection and devastation of hamakua pamakani. Weed populations were reduced from 80% to <5% within one year. Although the disease was reported to occur in all areas of Hawaii, control of the weed was most noticeable in the 1500–6000ft elevation range where the most favourable temperatures for the disease coincided with rainfall in spring and summer seasons. More than 50 000 hectares of pastureland have been rehabilitated by removal of the weed by *C. ageratinae*. In 1988, Trujillo *et al.* (1988) tentatively reidentified this fungus as *Entyloma compositarum* Farlow, based on the two distinct types of conidia produced in culture and on host plants.

1.3.2 Puccinia chondrillina

P. chondrillina is an autoecious, macrocylic rust fungus that was released into Australia in 1971 for the control of skeleton weed, *C. juncea*, an extremely serious weed of cultivation in south-east Australia. This release appears to be the first deliberate introduction of a pathogen for weed control in any country (Cullen *et al.*, 1972). Previous studies conducted in Europe had shown that the fungus was specific to the genus *Chondrilla* (Hasan, 1972; Hasan and Wapshere, 1973) and that the fungus consisted of strains. Each strain exhibited specificity for a particular strain, or strains, of *C. juncea*. A strain collected near Vieste, Italy, was highly virulent against the common form of *C. juncea* found in Australia (Hasan and Jenkins, 1972). In Europe, reductions of *C. juncea* populations were associated with heavy infections by *P. chondrillina* and were estimated to be between 16% and 70% (Hasan and Wapshere, 1973). In Australia, although the first successful introduction was made near Wagga, New South Wales, the fungus was found 80, 160 and 320 km from the initial release site by the eighth, tenth and twelfth generations after release, respectively (Cullen *et al.*, 1972). The density of *C. juncea* in Australia, the virulence of the selected strain toward the dominant form of the weed and weather conditions all together were considered to account for the rapid dispersal of the pathogen in Australia (Cullen *et al.*, 1972; Hasan and Jenkins, 1972).

Two strains of *P. chondrillina* from Eboli, Italy, were introduced into the United States in 1975 (Cheney *et al.*, 1981; Emge *et al.*, 1981). The fungus spread rapidly and within two years caused severe infections of plants throughout several populations of skeleton weed in California and Oregon. The fungus is now found in all four states of the Pacific Northwest: California, Idaho, Oregon and Washington (Lee *et al.*, 1981; Lee, 1986). Although three major biotypes of skeleton weed have been identified in Australia (Hull and Groves, 1973), the number and distribution of biotypes found in the western United States are less certain. In the US, variations in flowering and plant height have been observed (Schirman and Robocher, 1967) and various populations of skeleton weed have also been described on the basis of susceptibility to rust strains (Hull and Groves, 1973; Emge *et al.*, 1981). For example, among seven collections of the pathogen, four virulence patterns were expressed on the host populations in work conducted by Emge *et al.* (1981).

The effectiveness of *P. chondrillina* as a biological control agent also requires management and manipulation of initial dispersal of the inoculum (Lee *et al.*, 1981; Lee, 1986). Introductions of the inoculum to weed populations can be made by aircraft applying as little as two grams of inoculum per hectare, and applications are usually made during brief periods conducive to rust infection. Infection of *C. juncea* by *P. chondrillina* in the United States reduces plant size, the number of flowers and the number of viable seed produced by infected plants (Emge *et al.*, 1981; Lee *et al.*, 1981). Heavily infected rosettes may also die before flowering, and plant mortality may exceed 90% in areas where climatic conditions are favourable to infection.

Populations of *C. juncea* were reduced following the introduction of three different biological control organisms, *P. chondrillina* and two insects, into northern California (Supkoff *et al.*, 1988). The reductions of *Chondrilla* at three locations varied from 56% to 87%. The best correlation between change in plant density and the number of plants attacked by a control agent was between the incidence of *P. chondrillina* on rosettes in the spring and the change in rosette number the following autumn (Supkoff *et al.*, 1988). Therefore, of the three agents, *P. chondrillina* appeared to be the most effective organism for biological control of *C. juncea* in California.

The two insects released with *P. chondrillina* had either only a local impact or no significant impact on populations of skeleton weed. The number of *C. juncea* in the test plots in California has now reached levels within the range of normal densities of *C. juncea* found in Mediterranean Europe.

1.4 EXAMPLES OF THE MYCOHERBICIDE STRATEGY

Two fungi currently are commercially available in the United States for biological control of weeds in agriculture.

1.4.1 Phytophthora palmivora

A fungus, originally identified as *Phytophthora citrophthora* (R.E. Sm. and E.H. Sm.) (Burnett *et al.*, 1973, 1974) but later re-named *Phytophthora palmivora* (Feichtenberger *et al.*, 1984), has been used commercially since 1981 to control stranglervine (= milkweed vine), *Morrenia odorata* Lindl., in citrus groves of Florida (Kenney, 1986). The fungus is sold under the tradename DeVine. *M. odorata* was introduced from South America as an ornamental plant and became a serious weed problem in citrus. It competes for water, sunlight and nutrients, girdles tree limbs and interferes with harvesting, pesticide applications and with irrigation. The weed infests approximately 120 000 hectares in Florida. The weed can be controlled by mechanical cultivation and with herbicides at an estimated cost of $124 per hectare per year (Kenney, 1986). In many instances, these control measures were inadequate. *P. palmivora* was initially isolated in 1972 from diseased and dead plants found in Orange County, Florida. Small-scale field tests in which 96% of the vines were killed within 10 weeks after infestation of soil in a grove demonstrated the effectiveness of this fungus (Burnett *et al.*, 1973).

The fungus infects a number of susceptible species in addition to milkweed vine in controlled environments. In greenhouse pre-emergence tests, onion, *Allium cepa* L.; cantaloupe, *Cucumis melo*; watermelon, *Citrullus vulgaris* Schrad.; and citrus, *Citrus sinensis* (L.) Osbeck X *Poncirus trifoliata* (L.) Raf. 'Carrizo' were described as susceptible, while in post-emergence tests English pea, *Pisum sativum* L.; squash, *Cucurbita pepo* L.; and watermelon were susceptible, while potato, *Solanum tuberosum* L., and tomato, *Lycopersicon esculentum* L., were susceptible following foliage inoculations (Ridings, 1986). It was reasoned that although these plants were susceptible under laboratory conditions, this level of susceptibility was not representative of field conditions since the plants would not be exposed to direct application of the fungus under high inoculum concentrations in sterile soil (Ridings, 1986).

The fungus infecting milkweed vine was re-identified *P. palmivora* by Feichtenberger *et al.* (1984) on the basis of comparative morphological, physiological and pathological test results. In original studies, oospores were formed in culture in pairings of the fungus identified as *P. citrophthora* from milkweed vine with isolates of *Phytophthora nicotianae* B. deHaan var. *parasitica* (Dastur) and *P. palmivora* (Ridings, 1986). Feichtenberger *et al.* (1984) confirmed that milkweed vine isolates formed oospores in pairings with isolates of *P. palmivora* and *P. parasitica*. However, ten isolates of *P citrophthora* from citrus did not produce sexual structures in single culture or when paired with A_1 or A_2 mating types of *P. palmivora* or *P. parasitica*. Therefore, the fungus from milkweed vine was re-identified as *P. palmivora*.

Ridings (1986) reasoned that, eventually, as host plants become less numerous, the level of inoculum of the fungus in soil also would be reduced. However, single applications to groves from 1978 until 1980 were still providing 95% to 100% control in 1986. Kenney (1986) suggested that periodic emergence of milkweed vine seedlings resulting from windblown seeds maintained the populations of *P. palmivora* in infested groves.

1.4.2 Colletotrichum gloeosporioides f.sp. aeschynomene

C. gloeosporioides f.sp. *aeschynomene* (Cga) has been used commercially since 1982 to control infestations of northern jointvetch (NJV), *Aeschynomene virginica* (L.) BSP, in rice and soybeans in several states in the US (Bowers, 1986). Although a biological agent, spores of the fungus have been formulated as a wettable powder so that it can be used, handled and applied like any post-emergent chemical herbicide. Northern jointvetch reduces yields of rice and quality of white milled rice by seed contamination during harvesting of rice. NJV densities of 1 to 11 plants per square metre reduced grain yields from 4% to 19%, and 11% of the rice crop was discounted in 1980 due to the presence of weed seeds (Smith *et al.*, 1973). The discount loss in Arkansas was estimated to be $7.6 million dollars (Templeton *et al.*, 1981). Chemicals such as 2,4-D and 2,4,5-T were the standard management practice for NJV. *C. gloeosporioides* f.sp. *aeschynomene* causes an anthracnose on northern jointvetch seedlings, infecting stems, petioles and leaflets. Enlargement and coalescence of stem lesions result in the girdling and death of the plant above the lesions (TeBeest *et al.*, 1978a, b). The fungus sporulates profusely on the lesion surfaces and rainfall contributes to dispersal of the fungus spores on the plant, increasing the severity of infection (TeBeest *et al.*, 1978a; Yang and TeBeest, 1991). Although the fungus is dispersed in water, survival is limited in rice irrigation water (TeBeest, 1982). The fungus overwinters on refuse and infected and infested seed of northern jointvetch (TeBeest and Brumley, 1978).

Small-scale field tests demonstrated that 100% of the seedlings inoculated with the fungus were controlled within a few weeks after treatment (Daniel *et al.*, 1973). In the hands of growers, the commercial formulation of the fungus provides greater than 90% control of NJV when used according to label directions (Bowers, 1986). A major concern with the use of *C. gloeosporioides* f.sp. *aeschynomene* is the integration with fungicides, principally the fungicides including benzimidazoles used for control of various rice diseases. However, proper timing of fungicide treatments with *C. gloeosporioides* f.sp. *aeschynomene* provides for the effective use of both pesticide groups (Klerk *et al.*, 1985). *C. gloeosporioides* f.sp. *aeschynomene* has been integrated into the rice management system in Arkansas (Klerk *et al.*, 1985; Khodayari and Smith, (1988), which utilizes chemical pesticides for weeds, insects and diseases.

The host range of Cga has been studied extensively in greenhouse tests and consists of species within nine genera of the Leguminosae, subfamily Papilionideae, including *Aeschynomene, Cicer, Indigofera, Lathyrus, Lens, Lotus, Lupinus, Vicia* and *Pisum* (TeBeest, 1988; Weidemann *et al.*, 1989). However, the only species killed as a result of infection by *C. gloeosporioides* f.sp. *aeschynomene* in these studies was northern jointvetch. On other susceptible species, symptoms ranged from small necrotic non-sporulating lesions to sporulating lesions several centimetres in length on susceptible cultivars of *P. sativum*. Each of the above genera also contained species or varieties immune to infection. It is uncertain whether Cga would severely infect even susceptible varieties of *P. sativum* under field conditions because such infections resulted only after inoculum was placed in the apical buds of seedlings given 24 hours of dew (TeBeest, unpublished information).

1.5 PATHOGENS UNDER EVALUATION FOR BIOLOGICAL WEED CONTROL

Many fungi not listed above have been investigated or are still being evaluated for commercial development as mycoherbicides, and the reader is directed to recent reviews for additional information (Charudattan, 1990b; TeBeest, 1990). Various species within *Alternaria, Aschochyta, Bipolaris, Cercospora, Colletotrichum, Fusarium, Helminthosporium, Phoma, Phomopsis, Phytophthora, Puccinia* and *Sclerotinia* have been field tested or evaluated in studies to determine their potential effectiveness as biological control agents.

1.5.1 Pathogens under study in the classical strategy

Three imported rust fungi are being evaluated currently in the United States for biological control of leafy spurge, *Euphorbia esula* L.; musk thistle, *Carduus nutans* L. ssp. *leiphyllus* (Petrovic) Stoj. & Stef.; and yellow starthistle, *Centaurea solstitialis* L., in the United States (Bruckart and Dowler, 1986) in containment facilities of the United States Department of Agriculture (Melching *et al.*, 1983).

A study of 18 isolates of *Melampsora euphorbiae* (Schub.) Cast. collected in Austria, Hungary, Romania, Switzerland and Yugoslavia resulted in the establishment of only one compatible combination on two collections of cypress spurge, *Euphorbia cyparissias* L. (Bruckart and Dowler, 1986), although other isolates produced infections that were difficult to maintain on both leafy and cypress spurge (Turner *et al.*, 1983).

One isolate of *Puccinia carduorum* Jacky collected in Turkey was found to be very aggressive on 23 of 27 collections of musk thistle from Canada, France and the US (Bruckart and Dowler, 1986). The rust was pathogenic to 8 of 17 *Cirsium* species tested but was much less aggressive on these species than on *C. nutans*. The susceptibility of musk thistle and related composites has also been determined (Politis *et al.*, 1984). This rust was released for field studies in the US in 1987 after it was determined that it was not aggressive on globe artichoke, *Cynara scolymus* L. (Politis *et al.*, 1984).

Puccinia jaceae Otth. was evaluated for control of yellow starthistle with rust collections from Turkey. Isolates of the rust were very aggressive on yellow starthistle but also infected safflower, *Carthamus tinctorius* L. (Bruckart and Dowler, 1986).

In the United States an endemic rust has also been evaluated for control of a weed utilizing an augmentative approach (Bruckart and Hasan, 1991) rather than a truly classical approach. *Puccinia canaliculata* (Schw). Lagerh. has been evaluated for control of nutsedges, *Cyperus rotundus* L. and *C. esculentus* L., in the US (Phatak *et al.*, 1983; Callaway *et al.*, 1985). When released early in the spring, the rust inhibits flowering and tuber formation (Callaway *et al.* 1985). Phatak *et al.* (1983) had earlier reported that a simple release of the rust into plots resulted in reductions in root growth and fresh and dry weight of shoots. This fungus is capable of rapid dispersal and infection for in 60 days, approximately 78% of the leaf area was rusted in these tests. In one test, rust pustules were found in a previously healthy stand of yellow nutsedge within 12 days after a single pot of rusted seedlings was placed in similar stand of yellow nutsedge 7 km away. An epiphytotic reportedly developed over the entire area. The rust appears to be adapted to a wide range of environmental conditions, since epiphytotics developed over several hectares following releases made throughout the growing season under all conditions where nutsedge was growing.

The effects of the rust, *Puccinia lagenophorae* Cooke, on groundsel, *Senecio vulgaris* L., have been studied under summer and winter conditions in the United Kingdom (Paul and Ayres, 1985, 1986, 1987a, b). The rust was introduced into Great Britain in 1961 (Paul and Ayres, 1987b). Field-grown groundsel, infected as a result of inoculations made in the autumn, showed 70% mortality when measured in the spring whereas mortality was only 40% in controls inoculated during the spring and summer. The higher mortality of plants inoculated in the autumn was attributed to infection of the hypocotyls, which usually killed the host plants within 1 to 2 weeks and compromised the ability of seedlings to withstand water stress. Infection of hypocotyls in the summer, though as severe, did not result in significant mortality although infection had very substantial effects on growth and reproduction of populations during the summer. Infection resulted in a 50% reduction of vegetative growth, a 15% reduction in plant density, and a 24% reduction in floret production. The fungus does not naturally survive at sufficiently high levels to cause significant mortality. In this case, manipulation of the host-pathogen population relationship over time has a large effect on host survival. In glasshouse and field experiments, rust infections of groundsel decreased the competitive ability of groundsel with lettuce, although lettuce yields were not significantly reduced by rust-infected plants until weed densities exceeded 4000 plants per square metre (Paul and Ayres, 1987a). Rust infections reduced the impact of groundsel upon lettuce yield without causing any significant increase in groundsel mortality.

Other pathogens not listed above have been studied throughout the world, and the reader is referred to other reviews (Templeton *et al.*, 1979; Charudattan, 1990a; Watson, 1991) for a more comprehensive discussion of the classical strategy.

1.5.2 Pathogens under study in the mycoherbicide strategy

Many species of fungal plant pathogens have been studied as potential mycoherbicides and extensive reviews have been published (Templeton *et al.*, 1978; Charudattan, 1990b; TeBeest, 1990). Two species of *Colletotrichum* appear to be very near registration and actual commercial use as biological control agents for control of weeds in Australia and Canada at this time.

Colletotrichum gloeosporioides f.sp. *malvae* has been evaluated as a biological control agent for round-leaved mallow, *Malva pusilla* Sm. in Canada (Grant *et al.*, 1988, 1990; Mortenson, 1988). The fungus was isolated from round-leaved mallow but infects several *Malva* species, velvetleaf, *Abutilon theophrasti* Medic., and hollyhock, *Althea rosea* (L.) Cav. Round-leaved mallow plants inoculated with spore suspensions were killed within 17 to 20 days after inoculation. The disease is severe only on *Malva pusilla*. *C. gloeosporioides* f.sp. *malvae*, *Colletotrichum malvarum* (A. Braun and Casp.) and a *Colletotrichum gloeosporioides* (Penz.) Sacc. recently isolated from a *Lavatera* sp. apparently share an overlapping host range within the Malvaceae, each infecting a separate species more seriously than the others (Mortenson, 1991). Only *C. gloeosporioides* f.sp. *malvae* is being developed as a mycoherbicide.

Colletotrichum orbiculare (Berk. et Mont.) v. Arx is being re-evaluated as a biological control agent for Bathurst bur, *Xanthium spinosum* L., in Australia (Auld and Tisdell, 1985; Auld *et al.*, 1990). This fungus was previously identified as *C. xanthii* Halst. and evaluated in 1951 in a more classical approach (Butler, 1951). When applied as a mycoherbicide, the fungus controlled 50% to 100% of the seedlings in field tests conducted from 1987 to 1988. The best results of 98% to 100% control were achieved in a dryland grazing site (Auld *et al.*, 1990). The efficacy of the fungus increases with increasing periods of high humidity. The presence of the extracellular conidial matrix hastened the onset of visual symptoms of the disease and the level of anthracnose development on *X. spinosum* (McRae and Stevens, 1990).

1.6 THE EFFECTIVENESS OF FUNGAL PLANT PATHOGENS AS WEED CONTROL AGENTS

Fungal plant pathogens have been shown repeatedly to be an effective means of controlling weeds when used in the classical, augmentative or mycoherbicide strategies, whether used alone or even in combination with other pathogens, or whether applied to foliage or to soil as mycoherbicides (Table 1.1).

1.6.1 The effectiveness of single plant pathogens

The application or release of a single plant pathogen into weed infestations has been repeatedly shown to be an effective strategy to control weeds (Table 1.1). Fungi used as mycoherbicides have been especially effective in agriculture, controlling weeds at levels approaching 100%. The effectiveness of many of these pathogens is the combined result of susceptibility of the weed, the application of viable and aggressive inoculum, application under conditions which favour infection and the rapid establishment of very high levels of disease at inoculation.

1.6.2 The effectiveness of pathogens used in combination

Several examples have also been reported in which pathogens incapable of causing significant levels of control when used alone were more effective in combination with other pathogens. For example, Dimock and Baker (1951) showed that *Fusarium roseum* Lk. emend. Snyd. & Hans. infected snapdragon, *Antirrhinum majus* L., through lesions caused by the rust fungus, *Puccinia antirrhini* D. & H. Apparently, *F. roseum* infected healthy tissue beyond the rust lesion and caused death of leaves and shoots or even entire plants, whereas, infection by the rust alone seldom caused death. Rust-free plants, even when severely wounded, were not infected by *Fusarium roseum*. Thus it appears that a facultative parasite, incapable of infecting a plant alone, contributed to increased disease severity by invading lesions produced by another pathogen. This phenomenon has been extended to biological control of weeds.

Table 1.1. The reported effectiveness of fungal plant pathogens as biological weed control agents utilized in the augmentative, classical or mycoherbicide strategies. Control reported as percent plants killed by disease

Pathogen	Host	Percentage control	Reference
Augmentative strategy			
Puccinia lagenophorae	Senecio vulgaris	70	Paul and Ayres, 1986
Puccinia canaliculata	Cyperus esculentus	46–66	Phatak et al., 1985
Classical strategy			
Entyloma compositarum	Ageratina riparia	5–80	Trujillo, 1985
Puccinia chondrillina	Chondrilla juncea	56–90	Emge et al., 1981; Lee et al., 1981; Supkoff et al., 1988
Mycoherbicide strategy			
Alternaria cassiae	Cassia obtusifolia	86	Charudattan et al., 1986
Alternaria crassa	Datura stramonium	100	Boyette, 1986
Alternaria macrospora	Anoda cristata	100	Walker and Sciumbato, 1979; Walker, 1981
Colletotrichum coccodes	Abutilon theophrasti	46	Wymore et al., 1988
Colletotrichum gloeosporioides	Aeschynomene virginica	96	Boyette et al., 1979; Bowers, 1986
Colletotrichum gloeosporioides	Hakea sericea	80	Morris, 1983
Colletotrichum gloeosporioides	Hypericum perforatum	72–83	Hildebrand and Jensen, 1991
Colletotrichum gloeosporioides	Jussiaea decurrens	100	Boyette et al., 1979
Colletotrichum gloeosporioides	Malva pusilla	100	Mortenson, 1988
Colletotrichum malvarum	Sida spinosa	84–95	Kirkpatrick et al., 1982
Colletotrichum orbiculare	Xanthium spinosum	50–100	Auld et al., 1990
Fusarium solani	Cucurbita texana	80–99	Boyette et al., 1984; Weidemann, 1988
Phytophthora palmivora	Morrenia odorata	95–100	Burnett et al., 1973
Sclerotinia sclerotiorum	Cirsium arvense	20–80	Brosten and Sands, 1986
Sclerotinia sclerotiorum	Taraxacum officinale	81	Riddle et al., 1991

Alternaria macrospora Zimm. has been investigated as a potential myco-herbicide to control spurred anoda, *Anoda cristata* (L.) Schlecht. in the United States (Walker and Sciumbato 1979; Walker, 1981; Crawley *et al.*, 1985). The susceptibility of spurred noda to *A. macrospora* is highly correlated with plant age. Approximately 100% of seedlings inoculated at the cotyledonary stage were killed by infection, but less than 10% were killed at the 3- to 4-leaf-stage (Walker and Sciumbato, 1979). However, 100% of plants inoculated at the 4- to 5-leaf-stage were killed by the interaction of *A. macrospora* and *Fusarium lateritium* Nees ex Fr. *F. lateritium* is a weak pathogen of spurred anoda and causes less than 20% mortality when inoculated to seedlings alone, although *Fusarium later-itium* usually killed wound inoculated spurred anoda seedlings (Crawley *et al.*, 1985). The highest mortality occurred when *Alternaria* was inocu-lated five days before *Fusarium*. When *Fusarium* was inoculated five days before *Alternaria* only 11% of the seedlings were killed. One suggested explanation for the *Alternaria/Fusarium* interaction was that *F. lateritium* penetrated and infected through the lesions induced by *Alternaria*. These results indicate that sequential applications of both fungi were more effective that either fungus used alone for control of spurred anoda.

Premature mortality of groundsel infected by *P. lagenophorae* has been attributed to invasion of rust lesions by *Botrytis cinerea* Pers. (Hallett *et al.*, 1990a,b). Inoculation of healthy groundsel with *B. cinerea* caused only 10% mortality, and only 40% mortality of abiotically wounded plants. However, all plants previously infected by *P. lagenophorae* died after inoculation with *Botrytis*. Death of plants was attributed to growth of *Botrytis* into stem bases. The time necessary to kill plants was dependent upon several factors including the inoculum concentration of *Botrytis* and on initial pustule numbers of the rust (Hallett *et al.*, 1990).

These examples illustrate a possible effective alternative which would utilize pathogens that have only a very limited potential as biological control agents due to lack of virulence or aggressiveness.

1.7 SUMMARY

This brief survey of the literature shows that fungal plant pathogens can effectively control weeds in agricultural systems. The successful intro-duction of a plant pathogen into Australia and the US and the commer-cialization of two pathogens fuelled additional research on biological control of weeds. Though not specifically discussed above, remarkable advances have been made in the technical aspects associated with biological control of weeds and high among these advances is the development of inoculum that had an extended shelf-life, for it permitted the extension of research from laboratory and small field trials to commercial application. Considerable work must still be done, however, to develop formulations that will protect inoculum from harsh environmental conditions subsequent to inoculation.

Pathogens capable of controlling economically important weeds in agriculture, forests, rangelands and turf must be found and developed. Economical methods for producing large quantities of inoculum of fungi such as *Alternaria* must be found before these organisms can be fully utilized. The development of effective methods to screen and/or develop effective strains of pathogens now considered to be ineffective may provide additional tools for weed control.

Continued progress in biological control of weeds with plant pathogens will depend upon the continued interest by industry in developing products and on the continuation of active research on some of the more basic questions regarding the use of plant pathogens for biological control.

REFERENCES

Auld, B.A., Say, M.M., Ridings, H.I., *et al.* (1990) Field applications of *Colletotrichum orbiculare* to control *Xanthium spinosum*. *Agriculture, Ecosystems and Environment*, **32**, 315–23.

Auld, B.A. and Tisdell, C.A. (1985) Biological weed control-equilibria model. *Agriculture, Ecosystems and Environment*, **13**, 1–8.

Bowers, R.C. (1986) Commercialization of Collego – An Industrialist's view. *Weed Science*, **34** (Suppl 1), 24–5.

Boyette, C.D. (1986) Evaluation of *Alternaria crassa* for biological control of jimsonweed: host range and virulence. *Plant Science*, **45**, 223–8.

Boyette, C.D., Templeton, G.E. and Oliver, L.R. (1984) Texas gourd (*Cucurbita texana*) control with *Fusarium solani* f.sp. *cucurbitae*. *Weed Science*, **32**, 649–55.

Boyette, C.D., Templeton, G.E. and Smith, R.J. Jr. (1979) Control of winged waterprimrose (*Jussiaea decurrens*) and northern jointvetch (*Aeschynomene virginica*) with fungal pathogens. *Weed Science*, **27**, 497–501.

Brosten, B.S. and Sands, D.C. (1986) Field trials of *Sclerotinia sclerotiorum* to control canada thistle (*Cirsium arvense*). *Weed Science*, **34**, 377–80.

Bruckart, W.L. and Dowler, W.M. (1986) Evaluation of exotic rust fungi in the United States for classical biological control of weeds. *Weed Science*, **34**, (Suppl. 1), 11–14.

Bruckart, W.L. and Hasan, S. (1991) Options with plant pathogens intended for classical control of range and pasture weeds, in *Microbial Control of Weeds*, (ed. D.O. TeBeest), Chapman & Hall New York, pp.69–79.

Burnett, H.C., Tucker, D.P.H., Patterson, M.E., *et al.* (1973) Biological control of milkweed vine with a race of *Phytophthora citrophthora*. *Proceedings Florida Horticultural Society*, **86**, 111–15.

Burnett, H.C., Tucker, D.P.H. and Ridings, W.H. (1974) *Phytophthora* root and stem rot of milkweed vine. *Plant Disease Reporter*, **58**, 355–7.

Butler, F.C. (1951) Anthracnose and seedling blight of bathurst burr caused by *Colletotrichum xanthii* Halst. *Australian Journal Agricultural Research*, **2**, 401–11.

Callaway, M.B., Phatak, S.C. and Wells, H.D. (1985) Studies on alternate hosts of the rust *Puccinia canaliculata*, a potential biological control agent for nutsedges. *Plant Disease*, **69**, 924–6.

Charudattan, R. (1990a) Pathogens with potential for weed control, in *Microbes and Microbial Products as Herbicides*, (ed R.E. Hoagland), American Chemical Society, pp.132–54.

Charudattan, R. (1990b) Assessment of efficacy of mycoherbicide candidates, in *Proceedings VII International Symposium on Biological Control of Weeds*, (ed E.S. Delfosse), Rome, Italy, pp.455–64.

Charudattan, R. and Walker, H.L. (1982) *Biological Control of Weeds with Plant Pathogens* J. Wiley & Sons, New York.

Charudattan, R., Walker, H.L. Boyette, C.D., *et al.* (1986) Evaluation of *Alternaria cassiae* as a mycoherbicide for sicklepod (*Cassia obtusifolia*) in regional field tests. *Southern Regional Cooperative Series Bulletin* 317. Alabama Agricultural Experiment. Station, Auburn University.

Cheney, T.M., Piper, G.L., Lee, G.A., *et al.* (1981) Rush skeleton weed, biology and control in the Pacific Northwest. *Idaho Current Information Series*, No. 585.

Crawley, D.K., Walker, H.L., and Riley, J.A. 1985 Interaction of *Alternaria macrospora* and *Fusarium lateritium* on spurred anoda. *Plant Disease*, **69**, 977–9.

Cullen, J.M., Kable, P.F. and Catt, M. (1972) Epidemic spread of a rust imported for biological control. *Nature*, **244**, 462–4.

Daniel, J.T., Templeton, G.E., Smith, R.J., Jr., *et al.* (1973) Biological control of northern jointvetch in rice with an endemic fungal disease. *Weed Science*, **21**, 303–7.

Dimock, A.W. and Baker, K.F. (1951) Effect of climate on disease development, injuriousness, and fungicidal control, as exemplified by snapdragon rust. *Phytopathology*, **41**, 536–52.

Emge, R.G., Melching, J.S. and Kingsolver, C.H. (1981) Epidemiology of *Puccinia chondrillina*, a rust pathogen for the biological control of rush skeleton weed in the United States. *Phytopathology*, **71**, 839–43.

Feichtenberger, E., Zentmeyer, G.A. and Menge, J.A. (1984) Identity of *Phytophthora* isolated from milkweed vine. *Phytopathology*, **74**, 50–55.

Grant, N.T., Prusinkiewicz, E., Makowski, R.M.D., *et al.* (1990) Effect of selected pesticides on survival of *Colletotrichum gloeosporioides* f.sp. *malvae*, a bioherbicide for round-leaved mallow (*Malva pusilla*). *Weed Technology*, **4**, 701–15.

Grant, N.T., Prusinkiewicz, E., Mortenson, K., *et al.* (1988) Herbicide interactions with *Colletotrichum gloeosporiodes* f.sp. *malvae*: a bioherbicide for round-leaved mallow (*Malva pusilla*) control. *Weed Technology*, **4**, 716–23.

Hallett, S.G., Hutchinson, P., Paul, N.D., *et al.* (1990a) Conidial germination of *Botrytis cinerea* in relation to aeciospores and aecia of groundsel rust (*Puccinia lagenophorae*). *Mycological Research*, **94**, 603–6.

Hallett, S.G., Paul, N.D. and Ayres, P.G. (1990b) *Botrytis cinerea* kills groundsel (*Senecio vulgaris*) infected by rust (*Puccinia lagenophorae*). *New Phytologist*, **114**, 105–9.

Hasan, S. (1972) Specificity and host specialization of *Puccinia chondrillina*. *Annals of Applied Biology*, **72**, 257–63.

Hasan, S. and Jenkins, P.T. (1972) The effect of some climatic factors on infectivity of the skeleton weed rust, *Puccinia chondrillina*. *Plant Disease Reporter*, **56**, 858–60.

Hasan, S. and Wapshere, A.J. (1973) The biology of *Puccinia chondrillina*, a potential biological control agent of skeleton weed. *Annals of Applied Biology*, **74**, 325–32.

Hildebrand, P.D. and Jensen, K.I.N. (1991) Potential for the biological control of St. John's-wort (*Hypericum perforatum*) with an endemic strain of *Colletotrichum gloeosporioides. Canadian Journal of Plant Pathology*, **13**, 6–70.

Hull, V.J. and Groves, R.H. (1973) Variation in *Chondrilla juncea* L. in southeastern Australia. *Australian Journal of Botany*, **21**, 113–15.

Kenney, D.S. (1986) DeVine – The way it was developed – an industrialist's view. *Weed Science*, **34**, (Suppl. 1), 15–16.

Khodayari, K. and Smith, R.J., Jr. (1988) A mycoherbicide integrated with fungicides in rice, *Oryza sativa. Weed Technology*, **2**, 282–5.

Kirkpatrick, T.L., Templeton, G.E., TeBeest, D.O., *et al.* (1982) Potential of *Colletotrichum malvarum* for biological control of prickly sida. *Plant Disease*, **66**, 323–5.

Klerk, R.A., Smith, R.J., Jr. and TeBeest, D.O. (1985) Integration of a microbial herbicide into weed and pest control programs in rice (*Oryza sativa*). *Weed Science*, **33**, 95–9.

Lee, G.A. (1986) Integrated control of rush skeleton weed (*Chondrilla juncea*) in the Western United States. *Weed Science*, **34**, (Suppl. 1), 2–6.

Lee, G.A., Cheney, T.M. and Thill, D.C. (1981) The influence of *Puccinia chondrillina* on the flowering, seed production and seed variability of rush skeleton weed (*Chondrilla juncea*). *93rd Proceedings Western Society Weed Science.*

McRae, C.F. and Stevens, G.R. (1990) Role of conidial matrix of *Colletotrichum orbiculare* in pathogenesis of *Xanthium spinosum, Mycological Research*, **94**, 890–6.

Melching, J.S., Bromfield, K.R. and Kingsolver, C.H. (1983) The plant pathogen containment facility at Frederick, MD. *Plant Disease*, **67**, 717–22.

Morris, M.J. (1983) Evaluation of field trials with *Colletotrichum gloeosporioides* for the biological control of *Hakea sericea. Phytophylactica*, **15**, 13–16.

Mortenson, K. (1988) The potential of an endemic fungus, *Colletotrichum gloeosporioides* f.sp. *malvae*, for biological control of round-leaved mallow (*Malva pusilla*) and velvetleaf, (*Abutilon theophrasti*). *Weed Science*, **36**, 473–8.

Mortenson, K. (1991), *Collectotrichum gloeosporioides* causing anthracnose of *Lavatera* sp. *Canadian Plant Disease Survey*, **71**, 155–9.

Paul, N.D. and Ayres. P.G. (1985) Seasonal effects of rust disease (*Puccinia lagenophorae*) of *Senecio vulgaris. Symbiosis*, **2**, 165–73.

Paul, N.D. and Ayres, P.G. (1985) The impact of a pathogen (*Puccinia lagenophorae*) on populations of groundsel (*Senecio vulgaris*) overwintering in the field. *Journal of Ecology*, **74**, 1069–84.

Paul, N.D. and Ayres, P.G. (1987a) Effects of rust infection of *Senecio vulgaris* on competition with lettuce. *Weed Research*, **27**, 431–41.

Paul, N.D. and Ayres, P.G. (1987b) Survival, growth and reproduction of groundsel (*Senecio vulgaris*) infected by rust (*Puccinia lagenophorae*) in the field during summer. *Journal of Ecology*, **75**, 61–71.

Phatak, S.C., Sumner, D.R., Wells, H.D., *et al.* (1983) Biological control of yellow nutsedge with the indigenous rust fungus, *Puccinia canaliculata. Science*, **219**, 1446–7.

Phatak, S.C., Callaway, M.B. and Vavrina, C.S. (1985) Biological control and its integration in weed managment systems for purple and yellow nutsedge (*Cyperus rotundus* and *C. esculentus*). *Weed Technology*, **1**, 84–91.

Politis, D.J., Watson, A.K. and Bruckart, W.L. (1984) Susceptibility of musk thistle and related composites to *Puccinia carduorum. Phytopathology*, **74**, 687–91.

Riddle, G.E., Burpee, L.L. and Boland, G.J (1991) Virulence of *Sclerotinia sclerotiorum* and *S. minor* on dandelion (*Taraxacum officinale*). *Weed Science*, **39**, 109–18.

Ridings, W.H. (1986) Biological control of stranglervine in citrus – A researcher's view. *Weed Science*, **34**, (Suppl. 1), 31–2.

Schirman, R. and Robocher, W.C. (1967) Rush skeleton weed-threat to dryland agriculture. *Weeds*, **15**, 310–12.

Smith, R.J., Jr., Daniel, J.T., Fox, W.T., *et al.* (1973) Distribution in Arkansas of a fungus disease used for bio-control of northern jointvetch in rice. *Plant Disease Reporter*, **57**, 695–7.

Supkoff, D.M., Joley, D.B., and Marois, J.J. (1988) Effect of introduced biological control organisms on the density of *Chondrilla juncea* in California. *Journal of Applied Ecology*, **25**, 1089–95.

TeBeest, D.O. (1982) Survival of *Colletotrichum gloeosporioides* f.sp. *aeschynomene* in rice irrigation water and soil. *Plant Disease*, **66**, 469–72.

TeBeest, D.O. (1988) Additions to the host range of *Colletotrichum gloeosporioides* f.sp. *aeschynomene*. *Plant Disease*, **72**, 16–18.

TeBeest, D.O. (1990) Ecology and epidemiology of fungal plant pathogens studied as biological control agents of weeds, in *Microbial Control of Weeds*, (ed D.O. TeBeest), Chapman & Hall, New York, pp. 97–114.

TeBeest, D.O. and Brumley, J.M. (1978) *Colletotrichum gloeosporioides* f.sp. *aeschynomene* borne within the seed of *Aeschynomene virginica*. *Plant Disease Reporter*, **62**, 675–8.

TeBeest, D.O., Templeton, G.E. and Smith, R.J., Jr. (1978a) Temperature and moisture requirements for development of anthracnose on northern jointvetch. *Phytopathology*, **68**, 389–93.

TeBeest, D.O., Templeton, G.E. and Smith, R.J. Jr. (1978b) Histopathology of *Colletotrichum gloeosporioides* f.sp. *aeschynomene* on northern jointvetch. *Phytopathology*, **68**, 1271–5.

Templeton, G.E., Smith, R.J., Jr., TeBeest, D.O., *et al.* (1981) Field evaluation of dried fungus spores for biocontrol of curly indigo in rice and soybeans. *Arkansas Farm Research*, **30**, 8.

Templeton, G.E., TeBeest, D.O. and Smith, R.J., Jr. (1979) Biological weed control with mycoherbicides. *Annual Review of Phytopathology*, **17**, 301–10.

Trujillo, E.E. (1976) Biological control of hamakua pamakani with plant pathogens. *Proceedings American Phytopathology Society*, **3**, 298 (Abstr).

Trujillo, E.E. (1985) Biological control of hamakua pamakani with *Cercosporella* sp. in Hawaii, in *Proceedings VI International Symposium Biological Control of Weeds*, (ed E.S. Delfosse), pp.661–5.

Trujillo, E.E., Aragaki, M. and Shoemaker, R.A. (1988) Infection, disease development and axenic culture of *Entyloma compositarum*, the cause of hamakua pamakani blight in Hawaii. *Plant Disease*, **72**, 355–7.

Turner, S.K., Bruckart, W.L. and Fay, P.K. (1983) European rust fungi pathogenic to collections of leafy spurge from the United States. *Phytopathology*, **73**, 969 (Abstr.).

Walker, H.L. (1981) Factors affecting biological control of spurred anoda (*Anoda cristata*) with *Alternaria macrospora*. *Weed Science*, **29**, 505–7.

Walker, H.L. and Sciumbato, G.L. (1979) Evaluation of *Alternaria macrospora* as a potential biocontrol agent for spurred anoda (*Anoda cristata*): Host range studies. *Weed Science*, **27**, 612–14.

Watson, A.K. (1991) The classical approach with plant pathogens, in *Microbial Control of Weeds*, (ed D.O. TeBeest), Chapman & Hall, New York, pp.3–23.

Weidemann, G.J. (1988) Effects of nutritional amendments on conidial production of *Fusarium solani* f.sp. *cucurbitae* on sodium alginate granules and on control of texas gourd. *Plant Disease*, **72**, 757–9.

Weidemann, G.J., TeBeest, D.O. and Cartwright, R.D. (1989) Host specificity of *Colletotrichum gloeosporioides* f.sp. *aeschynomene* and *C. truncatum* in the Leguminosae. *Phytopathology*, **72**, 986–90.

Wymore, L.A., Poirier, C., Watson, A.K., *et al.* (1988) *Colletotrichum coccoides*, a potential bioherbicide for control of velvetleaf (*Abutilon theophrasti*). *Plant Disease*, **72**, 534–8.

Yang, X.B. and TeBeest, D.O. (1991) Rain dispersal of *Colletotrichum gloeosporioides* under simulated rice field conditions. *Phytopathology*, **81**, 815 (Abstr.).

2

The use of *Bacillus thuringiensis* as an insecticide

David R. Dent

2.1 INTRODUCTION

Bacillus thuringiensis (*Bt*) is a Gram-positive soil bacterium characterized by its ability to produce crystalline inclusions during sporulation. These inclusions consist of protein which exhibits a highly specific insecticidal activity (Höfte and Whiteley, 1989). The use of this bacterium as a microbial insecticide dates back nearly 50 years to when the first commercial formulation 'Sporiene' was made available in France (Deacon, 1983). Since that time the exploitation and development of *Bt* as an insecticide has been influenced and shaped by the changing fortunes of the agrochemical industry, the growth in interest in integrated pest management and the advances made in genetic manipulation, as well as the continued effort of invertebrate pathologists promoting *Bt* as a control agent.

Pest control practice since the mid-1940s has been dominated by chemical pesticides, firstly by the organochlorines, then the organophosphates, the carbamates and the pyrethroids. This chemical-based approach, especially in the earlier years, emphasized the use of highly toxic, broad spectrum, persistent insecticides that could be used against a wide variety of pests in diverse agroecosystems. *Bt*, having relatively lower toxicity, high host specificity and a low persistence was considered suitable only for very specific and restricted markets. However,

Exploitation of Microorganisms Edited by D.G. Jones
Published in 1993 by Chapman & Hall, London ISBN 0 412 45740 7

the increasing problems of chemical insecticide use (insecticide resistance and secondary pest outbreaks), increasing public concern over the environment, the concomitant development of integrated pest management (IPM), combined with the availability of a more toxic strain of *Bt* (HD-1) (having a broad spectrum of activity against lepidoptera) meant that by the early 1970s *Bt* was beginning to establish a small but sure foothold in the pest control market place. *Bt* underwent substantial development during the 1970s (Burgess, 1981), largely under the umbrella of IPM, where the advantages of *Bt* (specificity, non-toxic to the environment or beneficial organisms) could be more fully exploited in tandem with other pest control methods. Larger scale field research programmes and commercial use became firmly established in the 1980s. During this same period an increase in the study of the molecular genetics of *Bt* started to yield important results. The realization that δ-endotoxin production is controlled by single genes and that these can be manipulated, opened the way to a new understanding of *Bt* mode of action and the mechanisms involved in host specificity.

The balance of the two approaches, the traditional one based on isolation and selection of appropriate *Bt* strains, and the more recent approach of genetic manipulation has brought the use of *Bt* as an insecticide to a new crossroads. The following account considers this and other, more recent, advances that have been made, and their implication for the future exploitation of *Bt* as an insecticide.

2.2 MODE OF ACTION AND SPECIFICITY

A number of insecticidal metabolites of *Bt* have been identified but it is only the β-exotoxin and the δ-endotoxin that have as yet been studied in any detail. The β-exotoxin acts by inhibiting protein synthesis through interference of DNA-dependent RNA polymerase by structurally mimicking ATP and competing for the binding site (Sebasta *et al.*, 1981). This disruption of RNA synthesis is manifested during critical stages of development (moulting and pupation) causing mortality or prolonged development of immature stages (Bond *et al.*, 1971). Adult insects may be infertile or have reduced fecundity (Sebasta *et al.*, 1981). The pattern of occurrence of the β-exotoxin production suggests that a single gene might be involved (Levinson *et al.*, 1990) which, in the same way as for the δ-endotoxin, has clear implications for genetic manipulation. In the case of the β-exotoxin, genetic manipulation could be used where there is the need to remove the ability of certain serotypes to produce this toxin during the production process.

The confirmation that the crystal protoxins of the δ-endotoxin are single gene products (Kronstad *et al.*, 1983) has simplified the transfer of toxin production to other biological organisms (e.g. *E. coli*) and thereby enhanced mode of action studies and interest, at the molecular level, in differences in the host specificity of this toxin.

The crystal protoxin is ingested by the insect larva, after which a combination of the alkaline pH and proteolytic activity of the gut fluids dissolve the crystal protein, thus activating the toxic polypeptide fragment (the δ-endotoxin). The activated toxin penetrates the peritrophic membrane (Mathavan *et al.*, 1989) targeting the midgut epithelium and inducing the formation of small non-specific pores (0.5–1.0 nm) (although possibly K^+ specific; Sacchi *et al.*, 1986) in the membrane of susceptible cells, resulting in a net influx of ions and an accompanying inflow of water (Knowles and Ellar, 1987). As a consequence of this, the cells swell and lyse, and ultimately, massive epithelial disruption and death of the insect occurs (Endo and Nishiitsutsuji-Uwo, 1980; Percy and Fast, 1983; Reisner *et al.*, 1989).

2.2.1 Mechanisms of action

Elucidation of the general mechanism of action of the δ-endotoxin has been important in identifying key processes that could be responsible for observed levels of specificity of some single crystal proteins (Höfte and Whiteley, 1989). The most obvious factors that could influence host range of a crystal protein are: (i) differences in the larval gut affecting the solubilization and/or processing efficiency of the protoxin and (ii) the presence of specific toxin-binding sites (receptors) in the gut of different insects.

Since the alkaline pH and proteolytic activity of the fluid contained in the midgut of susceptible larvae are required for the solubilization and activation of toxic protein from intact crystal (Whiteley and Schnepf, 1986), it would seem reasonable that differential effects of these processes in the insect's gut could account for the levels of specificity encountered. Both the pH (Jaquet *et al.*, 1987) and proteases (Haider *et al.*, 1986; Jaquet *et al.*, 1987; Haider and Ellar, 1989) in the insect gut lumen have been suggested as important factors, although Johnson *et al.* (1990) working to determine the mechanism for *Bt* resistance in *Plodia interpunctella*, demonstrated that midguts from susceptible and resistant strains of the larvae were similar both in their ability to activate *Bt* protoxin and other proteolytic activity. This suggests that proteases may not therefore be involved in the determination of toxicity and hence the observed specificity of the proteins.

Obviously further work is required on this subject but evidence is now accumulating to suggest that the second mechanism may play a more predominant role, i.e. the presence of specific toxic receptor sites in the midgut epithelium. Research has shown that heterogeneity of midgut cell receptors, at least among different species of lepidoptera, is a major determinant of toxin specificity (Hoffmann *et al.*, 1988; Van Rie *et al.*, 1989; Van Rie *et al.*, 1990a; Van Rie *et al.*, 1990b). Together these studies indicate that a species can possess more than one type of receptor and that these receptors can differ in their concentration and affinity for different toxins (van Frankenhuyzen *et al.*, 1991). However, van Frank-enhuyzen *et al.* (1991) warn that such specificity data need to be considered within the constraints imposed by the procedures that were used for production and assay of the toxins. Arvidson *et al.* (1989) using protoxin proteins produced in *E. coli*, found that the expression of the *cry1A(c)* toxin gene in *E. coli* altered in toxicity towards *Heliothis virescens* but not *Trichoplusia ni*, although other studies have reported no differences between native and *E. coli*-produced Cry1A(c) proteins (MacIntosh *et al.*, 1990; van Frankhuyzen *et al.*, 1991).

One of the fascinating aspects that has arisen from the study of host specificity has been the discovery that very similar gene products can have different host specificities. Widner and Whiteley (1989) cloned and sequenced two genes (designated *cryIIA* and *cryIIB*) encoding crystal proteins, from *Bt kurstaki* HD-1, of 633 amino acids and a molecular mass *ca.* 71 kDa. Despite the fact that these two proteins display 87% identity in amino acid sequence, and differ by only 18 amino acids (Widner and Whiteley, 1990), they exhibit different toxic specificities. The *cryIIA* gene product is toxic to both diptera (*Aedes aegyptii*) and lepidoptera (*Manduca sexta*) larvae, the *cryIIB* gene product is toxic only to the latter (Widner and Whiteley, 1989; 1990). Von Tersch *et al.* (1991) sequenced *cry1A(c)* from *Bt kenyae* and compared it with *cry1A(c)* from *Bt kurstaki* to find that the two genes are more than 99% identical (only seven amino acid differences among the predicted sequences of 1177 amino acids), yet this was sufficient to reduce host specificity of *Bt kenyae*. A difference of three amino acids is sufficient to account for differences in specificity between the larvicidal protein of *Bt aizawai* ICI that is toxic to both lepidoptera and diptera and the monospecific lepidopteran toxin from *Bt berliner* (Haider and Ellar, 1989). Such similarity lends support to the notion that most of these strains are evolutionarily related (Widner and Whiteley, 1989; Höfte and Whiteley, 1989).

A number of the same crystal proteins occur in *Bt* strains of different subspecies (Table 2.1) and many strains produce several crystal proteins simultaneously (Höfte and Whiteley, 1989). The use of gene-specific probes led to the discovery that various subspecies of *Bt* contained one, two or three closely related genes (Kronstad *et al.*, 1983; Kronstad and Whiteley, 1986). For example, *Bt kurstaki* HD-1 contains all three genes *cry1A(a)*, *cry1A(b)* and *cry1A(c)*, while HD-1 Dipel has only the *cry1A(a)* and *cry1A(c)* genes. More recently work by Pang and Mathieson (1991) has shown that in 9 out of 16 wild *Bt* strains tested, only one of the *cry1A* genes was expressed; in six strains, one or two genes were

Table 2.1. Distribution of three crystal protein types among different *B. thuringiensis* strains as determined by reaction with monoclonal antibodies the *Cry1A* subtype (a or c) is indicated when known. (After Höfte and Whiteley, 1989.)

Flagellar serotype	B. thuringiensis subsp.	Strain	Presence of protein:		
			Cry1A	Cry1B	Cry1C
1	*thuringiensis*	HD2, Berliner 1715	+	+	−
		HD-14	+(a)	−	−
		4412	−	+	−
3a	*alesti*	HD-4	+	−	−
3a3b	*kurstaki*	HD-1	+	−	−
		HD-73	+(c)	−	−
4a4b	*dendrolimus*	HD-7	+	−	−
		HD-37	+(a)	−	−
4a4c	*kenyae*	HD-5, HD-64	+(c)	−	−
		HD-136			
5a5b	*galleriae*	HD-8, HD-129	+	?	?
6	*entomocidus*	HD-110, 4448	+(a)	+	+
6	*subtoxicus*	HD-10	+	+	−
7	*aizawai*	HD-11, HD-68	+	−	−
		HD-127, HD-854, HD-229, HD-133, HD-137	+	−	+
		HD-272	+	+	−
		HD-847	+	−	+
8a8b	*morrisoni*	HD-12	+	?	−
9	*tolworthi*	HD-121	+	−	−

(a) Proteins that react with only some of the monoclonal antibodies that specify a crystal protein type are designated by ?. All Cry1A-specific antibodies recognized Cry1A(b); two did not bind to Cry1A(a); two other monoclonal antibodies did not react with Cry1A(c). Hence, the last two crystal proteins could be distinguished, provided no other Cry1A proteins were present in the crystals.

identified with the presence of an additional gene possible, but masked because of the overlap of peptide bonds; in one strain none of the genes were expressed. Such results would suggest that in multigene strains, toxicity of the crystal proteins may not be determined simply by the properties of a single toxin protein, but by the activity, relative abundance and potential interaction of each of the proteins present (Pang and Mathieson, 1991). This provides the opportunity for manipulation and development of more complex *Bt* insecticides having a broad spectrum of activity.

Research over the last 10 years into the mode of action of *Bt*, its host specificity and studies of mechanisms of resistance, have started to provide the basic background information that is required to exploit the use of this organism in the full development of its potential as a microbial insecticide. More work is obviously required to complete the picture, but sufficient is now known to enable the concerted research effort that will produce a whole new generation of *Bt* insecticides.

2.3 SELECTION, MANIPULATION AND EVALUATION OF STRAINS

New strains and isolates of *Bt* are being continually found and evaluated for their use as microbial pesticides against a range of insect groups and species (Ferro and Gelernter, 1989; Cidaria *et al.*, 1991; Whitlock *et al.*, 1991). It is important that the search for new isolates is not neglected, because these will form the basis for the growing need for sources of genetic material for the future exploitation of *Bt*.

Traditionally, strains of *Bt* have been selected for use on the basis of potency for particular pest species or range of species. Strain selection has rarely been used beyond this initial screening procedure, although recent work (Smyth, 1992) would suggest it is possible to use selection as a means of producing *Bt* isolates with favourable characteristics, e.g. UV-light resistance. The main thrust of the development of *Bt* in the last decade has been through using the technique of gene manipulation. Three methods are available for vectoring genetic material into *Bt*, namely protoplast transformation, transduction and conjugation (Faulkner and Boucias, 1985). These methods provide opportunities not only for elucidating mechanisms of action (and resistance) but also for improving formulation (e.g. Evans, 1990; Section 4), and host range and toxicity (e.g. Honée *et al.*, 1990; Bar *et al.*, 1991). The conjugation-like mating system of *Bt* involving growing strains in mixed culture, has been shown to be a highly effective means of transferring plasmids between *Bt* strains (Faulkner and Boucias, 1985), and is also seen by some as a more natural way to improve *Bt* activity and spectrum than the techniques of protoplast transformation and transduction (Jutsum *et al.*, 1989). Such an approach has been utilized to transfer lepidopteran-active crystal genes into coleopteran-active *Bt* strains to generate new hybrid clones active against species in both orders (Jutsum *et al.*, 1989). A similar approach has produced a *Bt* strain active against both lepidoptera and diptera (Klier *et al.*, 1983), and a *Bt* strain with greatly increased activity against *Lymantria dispar* (Evans, 1990).

One drawback of these manipulative approaches that concentrate purely on the role of the endotoxin is the neglect of the role played by the spore in the efficacy of the insecticide. The presence of the spore is of obvious importance in some pest interactions (Mohd-Salleh and Lewis, 1982; Samasanti *et al.*, 1982; Moar *et al.*, 1989, 1990), and the present over-emphasis on the importance of the δ-endotoxin could result in valuable levels of spore-related activity being overlooked. However, regardless of the approach taken to improve the insecticidal activity of *Bt* strains, the most severe limitation to *Bt* improvement is the lack of a reliable analytical method for potency assessment (Jutsum *et al.*, 1989). Recently the use of cell lines and the Lawn Assay have been used as a means of evaluating the activity of *Bt* endotoxins (Gringorten *et al.*, 1990), but for the large part the insect bioassay provides the primary means for assessing the efficacy of *Bt* preparations.

To allow comparisons to be made on the relative efficacy between *Bt* strains and trials, it has been necessary to use a standard strain against which comparison can be made. The first generally accepted standard was 'E61' obtained from a fermentation of H-type *thuringiensis* (Burgess, 1967). This was later superseded in the USA by the HD-1 strain of H-type *kurstaki* when this was selected for commercial production in that country (Dulmage *et al.*, 1981). HD-1-S-80 is now generally established as the microbial standard for insects susceptible to *Bt kurstaki* strains (Navon and Klein, 1990). WHO (1987) recommended IPS-78 as the standard for assaying *Bt* for the control of mosquitos and blackflies.

The standard preparation is compared with the various strains under test, using bioassays usually based on incorporation of *Bt* into larval diet, or surface contamination of the diet (Moar *et al.*, 1986; Moar and Trumble, 1987; Gardner, 1988; McGaughey and Beeman, 1988; Morris 1988; MacIntosh *et al.*, 1990) while a few studies have utilized treated plant material (Zehnder and Gelernter 1989; Ferro and Gelernter, 1989; Dent, 1990). Although the fundamental role and significance of such methods in the development of more effective preparations is well recognized, the techniques nevertheless have a number of fundamental shortcomings (Klein, 1978). For instance, different strains of the same species of insect may have differing susceptibilities to the same preparation of *Bt* (Afify, 1969; Subramanyam and Cutkomp, 1985; Smyth, 1992); hence to provide a truly valid comparison between *Bt* strains carried out in different regions or countries there is a requirement for insect 'tester strains' having a known genetic background (Nolan, 1981). The potency expressed has also been shown to be a function of the substrate used in the bioassay (Subramanyam and Cutkomp, 1985; Pourmirza, 1989) even to the extent that small changes in diet pH can affect insect mortality when treated with *Bt* (Poumirza, 1989). Lewis *et al.* (1984) found that variation occurred in the estimation of potencies of *Bt* preparations even by competent and experienced personnel using the same protocols. This sort of variation indicates the critical need that exists for a standard bioassay with very detailed and uniform protocols if valid comparisons between strains are to be made. Slight variations in dilution techniques, replication, test-larval age and weight, rearing conditions and many other factors can, and do, have considerable influence on bioassay results (Lewis *et al.*, 1984).

One other major drawback with these bioassays is that it is unknown how much *Bt* is ingested by each individual larva. This has meant an emphasis on studies calculating LC_{50} (lethal consumption) as opposed to LD_{50} (lethal dose). Some workers have got round the problem by providing larvae with a small amount of food material containing a known dose of *Bt* which they are sure the insect will consume in a short space of time (Poumirza, 1989; Johnson *et al.*, 1991). Such bioassays allow for more accurate evaluations of dose effects, and for the calculation of an LD_{50}.

Overall, more emphasis needs to be placed on the development of the bioassay from an entomological perspective, where care is taken to ensure insect material is cultured in appropriate ways (Dent, 1990) and the standardization of diets, taking into consideration both the nutritional requirements of the larva and the procedures of the microbial bioassay (Navon and Klein, 1990).

2.4 FERMENTATION, FORMULATION AND FIELD APPLICATION

The isolation or genetic manipulation of a particularly potent strain does not guarantee the development of a *Bt* insecticide, rather it is only the start to a process of evaluation and development. A particularly useful strain must be amenable to large scale production and the consequent need for high yields, not least to enhance its rapid transition to large scale field trials (Nolan, 1981). It will also need to be appropriately formulated and to perform well in tests carried out in the field.

2.4.1 Fermentation

For the fermentation of *Bt*, conventional microbial techniques can be used. A number of review articles are available for information about general production methods (Dulmage, *et al.*, 1981; Kreig and Miltenburger, 1984). The technology of *Bt* production is now well advanced but small improvements can invariably still be made. For instance, with changes to media (Burgess, 1981) or levels of aeration (Lüthy, 1986) although improvements in formulation of only 2× or at most 5× can be expected (Burgess, 1981). Changes to media are commonly attempted in the search for cheaper, locally available ingredients (Salama *et al.*, 1983; Obeta and Okafor, 1984; Mummigatti and Raghunathan, 1990). However, major changes are unlikely to drastically reduce production costs unless the use of genetic engineering manages to produce completely new, reliable and cheap methods of production perhaps, for example, by the insertion of genes coding for easy-to-grow organisms such as *Pseudomonas* spp. (Jutsum, 1988).

2.4.2 Formulation

The *Bt* active ingredient has been formulated as wettable powders, dusts and granular preparations since the early sixties (Briggs, 1963). Wettable powders were most commonly available but they have in the past caused some problems for spraying, such as poor suspendability and sediment formation. During the 1980s, flowable products have come onto the market that can be used without these problems during application (Lüthy, 1986).

A formulation for a *Bt* insecticide may include the use of additives. These may be added to improve persistence, through reducing degradation of the crystal protein by contaminating proteases, screening the spore from inactivating UV-light, and preventing spore germination in the presence of leaf solutes; other additives may include feeding stimulants or materials to improve application and retention on, or in, appropriate transfer to surfaces.

In the past, the persistence of *Bt* preparations has been relatively short, e.g. <4 days (Pinnock *et al.*, 1974), but new insecticides are being developed constantly with improved persistence. Although results vary with dosage applied, half-lives can now be anything up to 10 days (Kirschbaum, 1985; Beckwith and Stelzer, 1987; Payne, 1988). The persistence of commercially-applied *Bt* at levels insufficient to cause insect mortality have been recorded for much longer periods, e.g. one year in treated plots of Balsam fir (Reardon and Haissig, 1983) and eight months in soil (Petras and Casida, 1985) but, in general, the need to provide improved persistence through formulation remains a major task in *Bt* commercialization.

The inactivation of the bacterial spore by ultraviolet light continues to be one of the main factors influencing persistence that needs to be addressed during formulation. Work with NPVs (nuclear polyhedrosis viruses) have shown congo red, folic acid and riboflavin to be useful protectants (Shapiro, 1985, 1989) and similarly with *Bt*, Dunkle and Shasha (1989) have shown that congo red, folic acid and para-aminobenzoic acid are suitable agents as UV-protectants. Work by Cohen *et al.* (1991) indicated that adsorbtion of cationic-chromophores such as acriflavin, methyl green and rhodamine B are suitable agents for use as UV-light protectants. Protectants may be added directly to a *Bt* preparation or spray mix (Smith *et al.*, 1980; Luttrell *et al.*, 1982; Morris, 1983) or more innovative methods may be utilized. Dunkle and Shasha (1989) used formulations of *Bt* spores and crystals encapsulated with a starch matrix containing the UV-screens.

Novel methods of formulating *Bt* are also being used to improve the application and targeting of the *Bt* insecticides. This is particularly true for *Bt isrealensis* (*Bti*) in the control of mosquito larvae in water. Cheung and Hammock (1985) describe the use of a micro-lipid-droplet encapsulation of *Bti* that affects the buoyancy of the insecticide, thereby making it available to surface-feeding *Anopheles* species. With the recent advances in liposome technology, liposomes having different buoyancies can be engineered to carry *Bti* toxin and thereby target mosquito larvae of all species. The technology would also allow for the inclusion of sunlight protectants and antimicrobial agents (Cheung and Hammock, 1985).

Other *Bt* formulations have included adjuvants, such as sorbitol (Smirnoff, 1985), to improve stability and evaporation, and some attempts have been made to improve uptake of *Bt* by the insects through the addition of feeding stimulants. However, use of commercial stimulants such as Coax, Entice, Tenac and Gustol, while tending to increase consumption of the targeted surface, has had variable and limited success with enhanced mortality (Luttrell *et al.*, 1982; Lutwama and Matanmi, 1988; Hough-Goldstein *et al.*, 1991) and the addition of potassium carbonate to Dipel significantly affected larval mortality (Salama *et al.*, 1990), possibly through its solubilizing action of the crystal, rendering the insect susceptible to the action of proteolytic enzymes (Salama *et al.*, 1985, 1986). It appears that the potential for the biggest advances in solving the problems of formulation, at least those for UV resistance, will come once again from genetic manipulation. It is well known that the *Bt* toxins are single gene products and that the genes are located in many *Bt* strains on the plasmids (Kronstad *et al.*, 1983). However, it appears that the *Bt* plasmids increase the UV-sensitivity of the spores (Benoit *et al.*, 1990). If it is assumed that loss of the cryptic plasmids would not have a detrimental effect on the growth and survival of *Bt*, then moving the protoxin genes to the chromosome of a plasmid-cured strain might add UV-resistance to the spores and cells, without any loss of the capacity to produce the toxin inclusion (Benoit *et al.*, 1990). Such an advance would have a major impact on the potential of *Bt* as an insecticide, although clearly the level of persistence achieved could itself create problems for the environment, and hence registration. In general though, the potential to influence some of the factors affecting formulation are certainly within the grasp of genetic manipulation technology to control.

2.4.3 Application

Results from field application studies have been variable but largely positive with successes or partial control reported for *Plodia interpunctella* in stored grain (McGaughey, 1985a), *Choristoneura fumiferana* in spruce-fir forests (Lewis *et al.*, 1984; Smirnoff, 1985), *Spodoptera littoralis* in cotton (Broza *et al.*, 1984), and soybean (Salama *et al.*, 1990), *Helicoverpa armigera* in tomato (Lutwama and Matanmi, 1988), *Lepidoptera decimlineata* in potato (Turnbull *et al.*, 1988) and the blackflies *Simulium* spp. (White and Morris, 1985), *Datana integerrima* and *Hyphantri cunea* on pecan (Tedders and Gottwald, 1986). However, control has been reported to be inadequate or unsuccessful against *Tribolium confusum* in wheat grain storage bins (Kramer *et al.*, 1985) and against *Prosimulium* spp. nuisance flies of humans (White and Morris, 1985). These field results, published over the last 10 years, are certainly encouraging for the future of *Bt* as a microbial insecticide since satisfactory levels of control are very often achieved, although where it is not, or where only partial control is obtained, then a diverse range of factors (other than formulation) tend to be blamed: inadequate spray coverage, contact rate of larvae (Bryant and Yendol, 1988; Yendol, *et al.* 1990), larval consumption rates (Tedders and Gottwald, 1986), variability in tank mix potency and spore deposit density (Lewis *et al.*, 1984), appropriate droplet size (Cheung and Hammock, 1985; Baldwin, 1986), effect of leaf characteristics on deposit retention (Pinnock *et al.*, 1975), and loss of toxin crystals during application (Smith *et al.*, 1977). All of these factors affecting efficacy are under the control of spray application technology, of insect behaviour and pharmacokinetics; these subjects are now being addressed by researchers working with chemical pesticides (Munthali and Scopes, 1982; Ford and Salt, 1987; Adams and Hall, 1989, 1990; Hoy *et al.*, 1990) and are also subjects that need to be addressed by those working with *Bt* insecticides. A better understanding of spray technology and *Bt* foliage deposition has been one important factor in the success in forestry (Jutsum *et al.*, 1989); similar approaches are needed in other systems.

Bt is applied by a variety of means from knapsack sprayers (Lutwama and Matanmi, 1988), hand-applied baits (Lampert and Southern, 1987), raking dusts into stored grain (McGaughey, 1985b), dust blowers (Salama *et al.*, 1990), and ULV hand application in cotton (Patti and Carner 1974), to aircraft (Lewis *et al.*, 1984; Smirnoff, 1985; Yendol *et al.*, 1990). Despite the diversity of application equipment and concomitant effects on spray deposition, it is somewhat surprising that so little work has considered the influence that application may have on *Bt* insecticide efficacy. All the work, as one would expect, states the rate at which the *Bt* was applied, but few seem to consider spray droplet size or deposition characteristics on target surfaces. Studies which have been made (i.e. on droplet size, etc.), other than those of Smith *et al.*, (1977) on *T. ni* and *H. zea*, have been in forestry looking at control of *Lymantria dispar* (Bryant and Yendol, 1988; Yendol *et al.*, 1990). Preliminary observations suggested that once impacted on the leaf surface, droplet density and sizes of *Bt* deposits may interact with the foraging and specific feeding characteristics of the gypsy moth larvae (Bryant and Yendol, 1988). It was thought that this may result in changes in mortality due to droplet density and size characteristics of the dose. Results of the experiments to determine the effect of spray droplet size and density on the efficacy of a commercial *Bt* preparation indicated that efficacy could be increased if for a given dose relatively high densities of small droplets (50–150 μm) of *Bt* were produced compared with larger droplets (>150 μm) present at low densities. Also, for a given level of mortality a lower dose will be needed if applied in droplets between 10–150 μm, than in droplets >150 μm, i.e. there is an increased dose requirement with increased droplet size. Similar results were found by Munthali and Scopes (1982) using the chemical dicofol against *Tetranychus urticae*.

Of course it is not only the distribution and dose of the *Bt* insecticide deposit that will affect subsequent efficacy but also the encounter rate and behaviour of the insect (Ford and Salt, 1987; Dent, 1991). The feeding behaviour of the insect in relation to droplet deposition of the *Bt* insecticide will have a significant effect on efficacy. Feeding behaviour patterns of insects may vary with the type of host, insect stage, development stage of host, presence of other individuals and also in the presence of the insecticide itself. The effects of the presence of chemical insecticides on insect behaviour have been recorded (Hall, 1987) but there is also some evidence from *Bt*-treated leaf choice studies, of a larval preference for non-treated surfaces (Wahundeniya, personal communication). This type of behavioural response clearly requires further investigation but, in general, it needs to be acknowledged that an understanding of insect behaviour is imperative if insect targeting and uptake is to be improved. If this knowledge is combined with studies of optimal placement, density and droplet sizes, then the efficacy of *Bt* insecticides will be greatly enhanced.

2.5 COMMERCIALIZATION OF *Bt* AS A MICROBIAL INSECTICIDE

Commercial exploitation of *Bt* is not only dependent on the selection of appropriate strains or manipulation and incorporation of suitable characters but also on the ability of industry to develop – and **market** – *Bt* as a microbial insecticide. If the insecticide does not have a large enough market and the costs of production are relatively high, then the insecticide will not prove commercially viable.

The size of the market is dependent on the insecticide having a broad spectrum of activity in a wide range of systems. At present, the market is quite small with the sales of all biological control agents and selective chemicals accounting for only 1% of the world market of crop protection

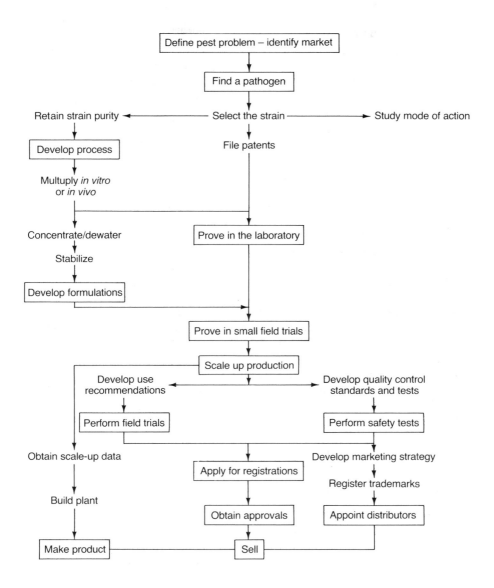

Figure 2.1 Stages involved in the commercialization of microbial pesticides. (After Baldwin, 1986.)

products (although within the insecticide sector *Bt* accounts for a large proportion of the 2.5% share of the market (Jutsum, 1988; Jutsum *et al.*, 1989)). In 1985, *Bt* products for forestry, agriculture and public health were worth $30 million, with Abbott Laboratories, Sandoz Ltd and Biochemicals Ltd sharing the *Bt* market in the USA and south-east Asia (Lüthy, 1986). *Bt* insecticides are competing reasonably well with chemical insecticides at present because the microbial insecticides are much cheaper to develop and commercialize (Figure 2.1). Hence the size of the target market can be much smaller than the minimum needed for commercializing a chemical insecticide (Baldwin, 1986). This approach of exploiting specialist markets is sound in the short term but, in the medium term, production costs will have to be maintained at similarly low levels if the market is to expand.

Commercialization costs of *Bt* are relatively low compared with chemicals, partly because registration requirements for this entomopathogen are less stringent. Hence, they can be obtained more quickly and at a lower cost. Herein lies the paradox of the future exploitation of *Bt* as a microbial pesticide. The most significant advances in extension of *Bt* host range and pathogenicity are going to occur through use of genetic manipulation. It seems unlikely that any organism that has been genetically manipulated will not have imposed upon it a series of costly (time and money) registration requirements. Such high registration costs will certainly reduce the attraction of *Bt* to commercial companies unless the potential gains in sales through improved host range and pathogenicity well exceed these extra costs. This may be possible if the genetically-manipulated *Bt* can be patented and thus provide the company with some guarantee of profitable sales for the life of the patent, but overall, although there is a great deal of interest by industry, *Bt* and similar agents are really not expected to replace chemical insecticides (Jutsum, 1988).

2.6 IPM AND THE FUTURE OF *Bt*

The advent of IPM produced a niche for the development of *Bt* insecticides, away from the direct competition with chemical insecticides, because it allowed for the use of a number of techniques, which although not effective on their own, when in combination with others provided adequate levels of control. However, this niche has never been fully exploited and *Bt* insecticides have continued to be directly compared with chemical insecticides. This has been a disappointing feature of *Bt* development, but it has meant there has been an interest in evaluating compatability with chemical and other microbial agents. Since the mid-to-late 1970s, there has been greater interest shown in use of *Bt* in sequence with chemicals (Karel and Schoonhoven, 1986), in chemical mixtures (Morris, 1975; Hamilton and Attia, 1977; Mohammed *et al.*, 1983a, b; Salama *et al.*, 1984) with insect growth regulators (Mohammed *et al.*, 1983b; Saleh and Wright, 1989) and with other pathogens; NPV (Luttrell *et al.*, 1982; Richter and Fuxa, 1984; Bell and Romine, 1986; Woods *et al.*, 1988); nematodes (Bari and Kaya, 1984; Richter and Fuxa, 1984; Bauer and Nordin, 1989).

Despite this profusion of work there seem to be few general principles that can be drawn from these studies. With chemical mixtures, variations in responses occur within and between the chemical groups in their interactions with *Bt*. The effects seem to vary with the specific insecticide, concentration, method of treatment and duration of exposure (Mohammed *et al.*, 1983b), and hence it would seem advisable for tests to be carried out between *Bt* and other insecticides before their joint utilization on a crop. The addition of *Bt* with NPV generally seems to reduce NPV mortality, both in the laboratory (Bell and Romine, 1986) and in the field (Woods *et al.*, 1988), although synergistic effects have been found in the laboratory at normally sub-lethal concentrations (Bell and Romine, 1986). The advantages of reducing chemical insecticide input, either through use of *Bt* alone, through reduced application of chemicals with chemical/*Bt* mixtures or in sequence application, has been the improvement gained in control through natural mortality factors, particularly natural enemies (Sneh and Gross, 1983; Sneh *et al.*, 1983; Broza *et al.*, 1984). *Bt* is generally considered to be non-harmful to these insects.

Much of the work carried out to evaluate *Bt* effects on natural enemies has been done with one particular commercial product Dipel (Vinson, 1989), and has been conducted with parasitoids rather than predators in the laboratory rather than the field (Croft and Flexner, 1991). The problem with such restrictive approaches is that they may produce misleading generalizations concerning the true value of *Bt* as a biorational insecticide. Only 31% of studies are carried out with predators, in contrast to chemical insecticides tests where this group has commonly been the focus of side effects testing. Hardly a recipe for fair comparison, if different types of insects are being targeted for evaluation. Only 25% of tests are field-based (from Table 11.2, Croft and Flexner, 1991); hence the majority of tests are carried out in the laboratory, thus ignoring the realities of ecological selectivity and the complex interactions that occur in the field. Although there is now substantial evidence to indicate that natural enemies are relatively safe with *Bt* insecticides compared with chemicals, it is clear that the lepidopteran-specific varieties of *Bt* do exhibit some toxic effects to a number of invertebrate non-target organisms (Melin and Cozzi, 1991) and if the magnitude of such effects are to be fully understood and quantified, then there must be a shift away from limited laboratory study of parasitoids, to the more realistic evaluation of the problem in the field and the full range of non-target organisms.

2.6.1 Resistance to *Bt*

It has for a long time been believed that resistance to microbial pathogens such as *Bt* would be unlikely to occur. There were a number of reasons for this belief: in theory, *Bt* will have co-evolved with its hosts to a degree that makes them most fitted to survive; in addition, *Bt* produces several toxins, hence it will affect a number of different targets in the hosts; lastly, because bacteria grow much faster than any host tissue they damage, any bacteria which can grow rapidly in a resistant insect, will kill the insect simply by outgrowing it (Boman, 1981). In practice, significant resistance to *Bt* had not been reported in over three decades of *Bt* use prior to 1985 (Van Rie *et al.*, 1990b). However, 1985 saw the first report of laboratory experiments where high levels of resistance had been obtained (McGaughey, 1985b). This report came from *Bt* studies with the stored product pest *Plodia interpunctella* following a 1984 conference which reported that, in theory, resistance to toxic-forming bacteria could develop along the same lines as chemical insecticide resistance (Briese, 1986). Briese (1986) considered there was sufficient evidence to show that the prerequisites for the development of such resistance exist: (i) there is considerable variability in response to pathogens within a species; (ii) part of this variability is under genetic control and hence is heritable; (iii) selection pressure through exposure to the pathogen can lead to changes in the mean level of response shown by a population.

Subsequent to this, resistance to *Bt* has been demonstrated in several populations of *Plodia interpunctella* (McGaughey 1985a, b), the almond moth *Cadra cautella* (McGaughey and Beeman, 1988) and to a genetically-engineered *Pseudomonas fluorescens* containing the δ-endotoxin of *Bt* in *Heliothis virescens* (Stone *et al.*, 1989).

Resistance to the *Bt* toxin could potentially develop at two different phases in the process. Firstly, at the phase where, after ingestion, the gut pH and the proteolytic activity of the gut fluids solubilize and activate the toxic protein; secondly, at the point where the toxin interacts with the membrane of the midgut epithelial cells (Section 2). Work by Johnson *et al.* (1990) in *Plodia interpunctella* has shown that the first phase is unlikely to be responsible for a resistance mechanism, since they demonstrated that midguts from susceptible and resistant strains of this species were similar both in their ability to activate *Bt* protoxin and in their proteolytic activity. The second phase of the pathogenetic process involves binding of the toxin with the gut membrane. Receptor binding studies carried out by Van Rie *et al.* (1990a) have shown that the resistance of laboratory selected *Plodia interpunctella* to an insecticidal crystal protein is correlated with a 50-fold reduction in affinity to the membrane receptor for this protein, but that the strain is sensitive to a second type of crystal protein that apparently recognizes a different receptor. This would explain the results of McGaughey and Johnson (1987) in which the responses of a wide range of *Bt* strains from several different serotypes were determined, all active on *P. interpunctella*, to identify those that would control a *Bt*-resistant strain. Twenty-one of the 57 isolates (representing 5 of the 8 serotypes) were active against the *Bt*-resistant insects, indicating that resistance is specific rather than general towards all *Bt* crystal types. Hence, when different crystal proteins are available against the same insect species, resistance to one crystal protein does not imply resistance to others (Van Rie *et al.*, 1990a). This is good news for the continued future exploitation of *Bt* as a microbial insecticide, since it means that crystal proteins with different binding properties could be used either in combination, or sequentially, for greater effectiveness and to delay resistance (Van Rie *et al.*, 1990b). However, the question must then be asked, how likely is it that resistance will develop in field populations? After all, most of the practical evidence for resistance comes from laboratory study, or from the stored product environment where the conditions are ideal for the development of resistance. (*Bt* is stable in stored grain, there is little disturbance for long periods, permitting the insects to breed for successive generations in contact with the bacterial spores and toxins (McGaughey, 1985b).) The reasons why so little evidence of resistance of *Bt* was available for over three decades concerned the specificity of *Bt*, its variable potency in field application and its limited geographical use. However, all this could change if genetic manipulation were to permit the development of *Bt* with increased levels of toxicity and with a greater host-pest range; effectively turning a very specific pesticide into one of broad spectrum. Under these circumstances the commercial application and use of *Bt* insecticides would increase rapidly and the chances of field populations developing resistance, just as they have to chemical insecticides, would almost certainly occur. The problem will be further accentuated with the widespread introduction of transgenic plants containing the δ-endotoxin for pest control. Engineered plants, unlike the use of chemical insecticides and microbial insecticides, will not provide temporal refuges from insecticide pressure. Under such circumstances resistance to δ-endotoxins is likely to develop at a very fast rate.

2.6.2 Registration considerations

The development of genetically manipulated *Bt* products either as sprayable insecticides or as part of genetically modified plants will have a dramatic effect on how *Bt* is viewed by the registration authorities. At present *Bt* insecticides have been shown to be highly specific and having negligible toxicity to non-target species including vertebrates and man. The regulatory procedures in the USA and UK, involving a tier system of evaluation, operate in such a way that if the first 'tier' of tests are negative, further testing may be unnecessary (Baldwin, 1986). *Bt* has been cleared with the first tier, hence registration costs are kept low. However, the situation with genetically-manipulated *Bt* may not be so clear-cut. These *Bt* insecticides will probably have an enhanced host range, potency and persistence, all of which may justifiably require further levels of evaluation before release. A most important aspect for regulating such organisms must be that the *Bt* have clearly defined host ranges and that these ranges cannot be altered accidentally after genetic manipulation (Beringer, 1988). Since recent work has shown that plasmid transfer between *Bt* strains can occur under natural conditions, and that *Bt* can be produced in this condition having novel endotoxins (Jarrett and Stephenson, 1990), necessary guarantees of maintenance of host range may not now be made.

The future exploitation of *Bt* as a microbial insecticide is therefore confronted with a number of dilemmas. Firstly, if research continues only along the lines of isolation and selection of natural strains, then *Bt* will remain an insecticide that has a very small market potential restricted to a few very specific pest/crop systems. The *Bt* would, however, keep low registration costs and would probably continue to have some considerable successes in this rather limited way. If environmental concern over chemical pesticide use continues to dominate political and public thinking, then a gradual increase in the market share of insecticides by *Bt* could be envisaged. If, however, as is likely to be the case, genetic manipulation of *Bt* produces new products having wider host ranges, greater potency and persistence, then *Bt* could potentially take a very much larger share of the insecticide market. The genetic manipulation could bring about problems of registration, which could greatly increase development costs and thereby the need by the industry to make a larger return on their investment. This could be achieved by increased widespread use which, as noted above, could lead to the development of insecticide resistance. Such a development would be disastrous for the commercial insecticide producers and for pest control in general. 'We have been there before' – it would seem pointless to carry through with such fundamentally flawed strategies yet again! If genetically-manipulated *Bt* insecticides are not just going to provide a short term economic gain for the commercial producers, then *Bt* insecticide development and use must be carried out within the framework of IPM. At least this way a more sustainable use of *Bt* will be possible, even if the economic returns have to be made over a longer time period. There have been many lessons learned from the past misuse of valuable pest control agents; it would be a shame indeed if the use of genetically-manipulated *Bt* could not be improved and thereby benefit from these past failures. Development within the framework of IPM should at least ensure a sustained and profitable future for those who wish to exploit *Bt* as a microbial insecticide.

REFERENCES

Adams, A.J. and Hall, F.R. (1989) Influence of bifenthrin spray deposit quality on the mortality of *Trichoplusia ni* on cabbage. *Crop Protection*, **8**, 206–11.

Adams, A.J. and Hall, F.R. (1990) Initial behavioural responses of *Aphis gossypii* to defined deposits of bifenthrin on crysanthemum. *Crop Protection*, **9**, 39–43.

Afify, A.M. (1969) Effect of storage of Bactospeine under varying room conditions on the viability and virulence of the spores. *Entomophaga*, **14**, 215–19.

Arvidson, H., Dunn, P.E., Strnad, S. and Aronson, A.I. (1989) Specificity of *Bacillus thuringiensis* for lepidoptera larvae: factors involved *in vivo* and in the structure of a purified protoxin. *Molecular Microbiology*, **3**, 1533–43.

Baldwin, B. (1986) Commercialisation of microbially produced pesticides, in *World Biotech Report 1986 Vol. 1. Proceedings of Biotech 1986 Europe*, London, May 1986, pp. 39–49.

Bar, E., Lieman-Hurwitz, J., Rahamim, E., Keynan, A. and Sandler, N. (1991) Cloning and expression of *Bacillus thuringiensis israelesis* δ-endotoxin DNA in *B. sphaericus. Journal of Invertebrate Pathology*, **57**, 149–58.

Bari, M.A. and Kaya, H.K. (1984) Evaluation of the entomogenous nematode *Neoaplectana carpocapsae* and the bacterium *Bacillus thuringiensis* var. *kurstaki* for suppression of the artichoke plume moth. *Journal of Economic Entomology*, **77**, 225–9.

Bauer, L.S. and Nordin, G.L. (1989) Response of spruce budworm infected with *Nosema fumiferanae* to *Bacillus thuringiensis* treatment. *Environmental Entomology*, **18**, (5), 816–21.

Beckwith, R.C. and Stelzer, M.J. (1987) Persistence of *Bacillus thuringiensis* in two formulations, applied by helicopter against the western spruce budworm in North Central Oregon. *Journal of Economic Entomology*, **80**, 204–7.

Bell, M.R. and Romine, C.L. (1986) *Heliothis virescens* and *H. zea* dosage effects of feeding mixtures of *Bacillus thuringiensis* and a nuclear polyhedrosis virus on mortality and growth. *Environmental Entomology*, **15**, 1161–5.

Benoit, T.G., Wilson, G.R., Bull, D.L. and Aronson, A.I. (1990) Plasmid-associated sensitivity of *Bacillus thuringiensis* to UV light. *Applied and Environmental Microbiology*, **56**, (8), 2282–6.

Beringer, J.E. (1988) Regulating the release of genetically manipulated pest control organsims in the UK, in *British Crop Protection Council Conference – Pests and Diseases*. British Crop Protection Council, Vol. 2, pp.583–6.

Boman, H.G. (1981) Insect responses to microbial infections, in *Microbial Control of Pests and Plant Diseases 1970–1980*, (ed H.D. Burgess), Academic Press, London, pp.769–784.

Bond, R.P.M., Boyce, C.B.C, Rogoff, M.H. and Sheih, T.R. (1971) The thermostable exotoxin of *Bacillus thuringiensis*, in *Microbial Control of Insects and Mites*, (eds H.D. Burgess and N.W. Hussey). Academic Press, New York, pp.275–303.

Briese, D.T. (1986) Host resistance to microbial control agents, in *Biological Plant and Health Protection*, (ed J.M. Franz), Progress in Zoology, Vol. 32, Gustav Fischer Verlag, Stuttgart, pp.233–78.

Briggs, J.D. (1963) Commercial production of insect pathogens, in *Insect Pathology: An Advanced Treatise*, Vol. 2, (ed E.A. Steinhaus), Academic Press, New York, pp. 519–48.

Broza, M., Sneh, B., Yawetz, A., Oron, U. and Honigman, A. (1984) Commercial application of *Bacillus thuringiensis* var. *entomocidus* to cotton fields for the control of *Spodoptera littoralis*. *Journal of Economic Entomology*, **77**, 1530–3.

Bryant, J.E. and Yendol, W.G. (1988) Evaluation of the influence of droplet size and density of *Bacillus thuringiensis* against gypsy moth larvae. *Journal of Economic Entomology*, **81**(1), 130–4.

Burgess, H.D. (1967) The standardisation of *Bacillus thuringiensis*; tests on three candidate reference materials, in *Insect Pathology and Microbial Control*, (ed P.A. van Loan), North-Holland Publ. Co., Amsterdam, pp.314–37.

Burgess, H.D. (ed) (1981) *Microbial Control of Pests and Plant Diseases 1970–1980*. Academic Press, London.

Cheung, P.Y.K. and Hammock, B.D. (1985) Micro-lipid-droplet encapsulation of *Bacillus thuringiensis* subsp. *israelensis* δ-endotoxin for control of mosquito larvae. *Applied and Environmental Microbiology*, **50**(4), 984–8.

Cidaria, D., Cappai, A., Vallesi, A., Caprioli, V. and Pirali, G. (1991) A novel strain of *Bacillus thuringiensis* (NCIMB 40152) active against coleopteran insects. *FEMS Microbiology Letters*, **81**, 129–34.

Cohen, E., Rozen, H., Joseph, T., Braun, S. and Margulies, L. (1991) Photoprotection of *Bacillus thuringiensis kurstaki* from ultraviolet irradiation. *Journal of Invertebrate Pathology*, **57**, 343–51.

Croft, B.A. and Flexner, J.L. (1991) Microbial pesticides, in *Arthropod Biological Control Agents and Pesticides*, (ed B.A. Croft), John Wiley and Sons, New York, pp.269–303.

Deacon, D.W. (1983) *Microbial Control of Plant Pests and Diseases*. Aspects of Microbiology 7, Van Nostrand Reinhold, Wokingham.

Dent, D.R. (1990) *Bacillus thuringiensis* for the control of *Heliothis armigera*: bridging the gap between the laboratory and the field. *Aspects of Applied Biology*, **24**, pp.179–85.

Dent, D.R. (1991) *Insect Pest Management*. CAB International, Wallingford.

Dulmage, H.T., *et al.* (1981) Insecticidal activity of isolates of *Bacillus thuringiensis* and their potential for pest control, in *Microbial Control of Pests and Plant Diseases 1970–1980*, (ed H.D. Burgess), Academic Press, London, pp.193–222.

Dunkle, R.L. and Shasha B.S. (1989) Response of starch-encapsulated *Bacillus thuringiensis* containing ultraviolet screens to sunlight. *Environmental Entomology*, **18**, 1035–41.

Endo, Y. and Nishiitsutsuji-Uwo, J. (1980) Mode of action of *Bacillus thuringiensis* δ-endotoxin: Histopathological changes in silkworm midgut. *Journal of Invertebrate Pathology*, **36**, 90–103.

Evans, H.F. (1990) The use of bacterial and viral control agents in British Forestry (1990). *Aspects of Applied Biology*, **24**, 195–203.

Faulkner, P. and Boucias, D.G. (1985) Genetic improvement of insect pathogens: emphasis on the use of Baculoviruses, in *Biological Control in Agricultural IPM Systems*, (eds M.A. Hoy and D.C. Herzog), Academic Press Inc., Orlando, pp.263–81.

Ferro, D.N. and Gelernter, W.D. (1989) Toxicity of a new strain of *Bacillus thuringiensis* to Colorado Potato Beetle. *Journal of Economic Entomology*, **82** (3), 750–5.

Ford, M.G. and Salt, D.W. (1987) Behaviour of insecticide deposits and their transfer from plant to insect surfaces, in *Pesticides on Plant Surfaces*, (ed H.J. Cottrell), Critical Reports on Applied Chemistry, Vol. 18, John Wiley and Sons, Chichester, pp.26–81.

Frankenhuyzen, van, H., Gringorten J.L., *et al.*, (1991) Specificity of activated Cry1A protiens from *Bacillus thuringiensis* subsp. *kurstaki* HD-1 for defoliating forest lepidoptera. *Applied Environmental Microbiology*, 57 (6), 1650–5.

Gardner, W.A. (1988) Enhanced toxicity of selected combinations of *Bacillus thuringiensis* and Beta-exotoxin against fall armyworm larvae. *Journal of Economic Entomology*, 52, 1115–57.

Gringorten, J.L., Witt D.P., *et al.*, (1990) An *in vitro* system for testing *Bacillus thuringiensis* toxins: The Lawn Assay. *Journal of Invertebrate Pathology*, 56, 237–42.

Haider, M.Z. and Ellar, D.J. (1989) Functional mapping of an entomocidal δ-endotoxin. Single amino acid changes produced by site-directed mutagensis influence toxicity and specificity of the protein. *Journal of Molecular Biology*, 208, 183–94.

Haider, M.Z., Knowles, B.H. and Ellar, D.J. (1986) Specificity of *Bacillus thuringiensis* var. *colmeri* insecticidal delta-endotoxin is determined by differential proteolytic processing of the protoxin by larval gut proteases. *European Journal of Biochemistry*, 156, 531–40.

Hall, F.R. (1987) Studies on dose transfer: effects of drop size, pattern and formulation on pest behaviour. *Aspects of Applied Biology*, 14, 245–56.

Hamilton, J.T. and Attia, J.T. (1977) Effects of mixtures of *Bacillus thuringiensis* and pesticides on *Plutella xylostella* and the parasite *Thyraeela collaris*. *Journal of Economic Entomology*, 70, 146–8.

Hoffman, C., Vanderbruggen, H., *et al.* (1988) Specificity of *Bacillus thuringiensis* δ-endotoxins is correlated with the presence of high-affinity binding sites in the brush border membrane of target insect midguts. *Proceedings of National Academy of Sciences USA*, 85, 7844–8.

Höfte, H. and Whiteley, H.R. (1989) Insecticidal crystal proteins of *Bacillus thuringiensis*. *Microbiological Reviews*, 53, (2), 242–55.

Honée, G., Vriezen, W. and Visser, B. (1990) A translation fusion product of two different insecticidal crystal protein genes of *Bacillus thuringiensis* exhibits an enlarged insecticidal spectrum. *Applied and Environmental Microbiology*, 56, (3), 823–5.

Hough-Goldstein, J., Tisler, A.M., *et al.*, (1991) Colorado potato beetle consumption of foliage treated with *Bacillus thuringiensis* var. *san diego* and various feeding stimulants. *Journal of Economic Entomology*, 84, (1), 87–93.

Hoy, C.W., Adams, A.J. and Hall, F.R. (1990) Behavioural response of *Plutella xylostella* populations to permethrin deposits. *Journal of Economic Entomology*, 83, (4), 1216–21.

Jaquet, F., Hütter, R. and Lüthy, P. (1987) Specificity of *Bacillus thuringiensis* delta-endotoxin. *Applied and Environmental Microbiology*, 53, (3), 500–4.

Jarrett, P. and Stephenson, M. (1990) Plasmid transfer between strains of *Bacillus thuringiensis* infecting *Galleria mellonella* and *Spodoptera littoralis*. *Applied and Environmental Microbiology*, 56, (6), 1608–14.

Johnson, D.E., Brookhart, G.L., *et al.*, (1990) Resistance to *Bacillus thuringiensis* by the Indian moth *Plodia interpuncctella*: Comparison of midgut proteinases from susceptible and resistant larvae. *Journal of Invertebrate Pathology*, 55, 235–44.

Johnson, D.E., McGaughey, W.H. and Barnett, B.D. (1991) Small scale bioassay for the determination of *Bacillus thuringiensis* toxicity toward *Plodia interpunctella*. *Journal of Invertebrate Pathology*, **57**, 159–65.

Jutsum, A.R. (1988) Commercial application of biological control: status and prospects. *Philosophical Transactions of the Royal Society, London*, Series B, **318**, 357–73.

Jutsum, A.R., Poole, N.J., *et al.*, (1989) Insect control using microbial toxins: status and prospects, in *Progress and Prospects in Insect Control*, BCPC Monograph No. 43, pp.131–44.

Karel, A.K. and Schoonhoven, A.V. (1986) Use of chemical and microbial insecticides against pests of common beans. *Journal of Economic Entomology*, **79**, 1692–6.

Kirschbaum, J.B. (1985) Potential implication of genetic engineering and other biotechnologies to insect control. *Annual Review of Entomology*, **30**, 51–70.

Klein, M. (1978) An improved personal administration technique for bioassay of nucleopolyhedrosis viruses against Egyptian cotton worm, *Spodoptera littoralis*. *Journal of Invertebrate Pathology*, **31**, 134–6.

Klier, A., Bourgouin, C. and Rapoport, G. (1983) Mating between *Bacillus subtilis* and *Bacillus thuringiensis* and transfer of cloned crystal genes. *Molecular and General Genetics*, **191**, 257–62.

Knowles, B.H. and Ellar, D.J. (1987) Colloid-osmotic lysis is a general feature of the mechanism of action of *Bacillus thuringiensis* δ-endotoxins with different insect specificities. *Biochimica et Biophysica Acta*, **924**, 509–18.

Kramer, K.J., Hendricks, L.H., Wojciak, J.H. and Fyler, J. (1985) Evaluation of Fenoxycarb, *Bacillus thuringiensis* and Malathion as grain protectants in small bins. *Journal of Economic Entomology*, **78**, 632–6.

Kreig, A. and Miltenburger, H.G. (1984) Bioinsecticides: 1. *Bacillus thuringiensis*, in *Biotechnological Processes*, (eds A. Mizrahi, A.L. van Wezel), Alan R. Liss Inc., New York, pp.273–90.

Kronstad, J.W. and Whiteley, H.R. (1986) Three classes of homogenous *Bacillus thuringiensis* crystal protein genes. *Gene*, **43**, 29–40.

Kronstad, J.W., Schnepf, H.E. and Whiteley, H.R. (1983) Diversity of locations for *Bacillus thuringiensis* crystal protein genes. *Journal of Bacteriology*, **154**, 419–28.

Lampert, E.P. and Southern, P.S. (1987) Evaluation of pesticide application methods for control of tobacco budworms on flue-cured tobacco. *Journal of Economic Entomology*, **80**, 961–7.

Levinson, B.L., Kasyan, K.J., *et al.*, (1990) Identification of β-exotoxin production, plasmids encoding β-exotoxin, and a new exotoxin of *Bacillus thuringiensis* by using high-performance liquid chromatography. *Journal of Bacteriology*, **172**(6), 3172–9.

Lewis, F.B., Walton, G.S., *et al.*, (1984) Aerial application of *Bacillus thuringiensis* against spruce budworm: 1982 *Bacillus thuringiensis* cooperative field tests – combined summary. *Journal of Economic Entomology*, **77**, 999–1003.

Lüthy, P. (1986) Insect pathogenic bacteria as pest control agents, in *Biological Plant and Health Protection*, (ed J.M. Franz), Progress in Zoology, Vol. 32, Gustav Fischer Verlag, Stuttgart, pp.201–16.

Luttrell, R.G., Young, S.G., *et al.*, (1982) Evaluation of *Bacillus thuringiensis*-spray adjuvant-viral insecticide combinations against *Heliothis* spp.*Environmental Entomology*, **11**, 783–7.

Lutwama, J.J. and Matanmi, B.A. (1988) Efficacy of *Bacillus thuringiensis* subsp. *kurstaki* and *Baculovirus heliothis* foliar applications for suppression of *Helicoverpa armigera* and other lepidopterous larvae on tomato in south-western Nigeria. *Bulletin of Entomological Research*, **78**, 173–9.

MacIntosh, S.C., Stone, T.B., et al., (1990) Specificity and efficacy of purified *Bacillus thuringiensis* proteins against agronomically important insects. *Journal of Invertebrate Pathology*, **56**, 258–66.

Mathavan, S., Sudha, P.M. and Pechimuthu, S.M. (1989) Effect of *Bacillus thuringiensis israelensis* on the midget cells of *Bombyx mori* larvae: A histopathologial and histochemical study. *Journal of Invertebrate Pathology*, **53**, 217–27.

McGaughey, W.H. (1985a) Evaluation of *Bacillus thuringiensis* for controlling Indianmeal moths in farm grain bins and elevator silos. *Journal of Economic Entomology*, **78**, 1089–94.

McGaughey, W.H. (1985b) Insect resistance to the biological insecticide *Bacillus thuringiensis*. *Science*, **229**, 193–5.

McGaughey, W.H. and Johnson, D.E. (1987) Toxicity of different serotypes and toxins of *Bacillus thuringiensis* to resistant and susceptible Indianmeal moths. *Journal of Economic Entomology*, **80**,(6), 1122–6.

McGaughey, W.H. and Beeman, R.W. (1988) Resistance to *Bacillus thuringiensis* in colonies of Indian meal moth and almond moth. *Journal of Economic Entomology*, **81**, 28–33.

Melin, B.E. and Cozzi, E.M. (1989) Safety to non-target invertebrates of Lepidopteran strains of *Bacillus thuringiensis* and their β-endotoxins, in *Safety of Microbial Pesticides*, (eds M. Laird, L.A. Lacey and E.W. Davidson), CRC Press, Boca Raton, FL, pp.149–67.

Moar, W.J., Osbrink, W.L.A. and Trumble, J.T. (1986) Potentiation of *Bacillus thuringiensis* var. *kurstaki* with *thuringiensin* on beet armyworm. *Journal of Economic Entomology*, **79**, 1443–6.

Moar, W.J. and Trumble, J.T. (1987) Toxicity, joint action and meantime of mortality of Dipel 2x, avermectin B1, neem and thuringiensin against beet armyworms. *Journal of Economic Entomology*, **80**, 588–92.

Moar, W.J., Trumble, J.T. and Federici, B.A. (1989) Comparative toxicity of spores and crystals from NRD-12 and HD-1 strains of *Bacillus thuringiensis* subsp. *kurstaki* to neonate beet armyworm. *Journal of Economic Entomology*, **82** (6), 1593–603.

Moar, W.J., Masson, L., et al., (1990) Toxicity to *Spodoptera exigua* and *Trichoplusia ni* of individual Pl protoxins and sporulated cultures of *Bacillus thuringiensis* subsp. *kurstaki* HD-1 and NRD-12, *Applied and Environmental Microbiology*, **56**, 2480–3

Mohammed, A.I., Young, S.Y. and Yearian, W.C. (1983a) Effects of microbial agent-chemical mixtures on *Heliothis virescens*. *Environmental Entomology*, **12**, 478–81.

Mohammed, A.I., Young, S.Y. and Yearian, W.C. (1983b) Susceptibility of *Heliothis virescens* larvae to microbial agent-chemical pesticide mixtures on cotton foliage. *Environmental Entomology*, **12**, 1403–5.

Mohd-Salleh, M.B. and Lewis, L.C. (1982) Toxic effects of spore/crystal ratios of *Bacillus thuringiensis* on European Corn Borer larvae. *Journal of Invertebrate Pathology*, **39**, 290–7.

Morris, O.N. (1975) Susceptibility of the spruce budworm and the white marked tussock moth to *Bacillus thuringiensis*: chemical insecticide combination. *Journal of Invertebrate Pathology*, **26**, 193–8.

Morris, O.N. (1983) Protection of *Bacillus thuringiensis* from inactivation by sunlight. *Canadian Entomologist*, **115**, 1215–27.

Morris, O.N. (1988) Comparative toxicity of delta endotoxin and thuringiensin of *Bacillus thuringiensis* and mixtures of the two for the control of the Bertha armyworm. *Journal of Economic Entomology*, **81**, 135–41.

Mummigatti, S.G. and Raghunathan, A.N. (1990) Influence of media composition on the production of δ-endotoxin by *Bacillus thuringiensis* var. *thuringiensis*. *Journal of Invertebrate Pathology*, **55**, 147–51.

Munthali, D.C. and Scopes, N.E.A. (1982) A technique for studying the biological efficiency of small droplets of pesticide solutions and a consideration of the implications. *Pesticide Science*, **13**, 60–2.

Navon, A. and Klein, M. (1990) *Bacillus thuringiensis* potency bioassays against *Heliothis armigera, Earias insulana* and *Spodoptera litoralis* larvae based on standardised diets. *Journal of Invertebrate Pathology*, **55**, 387–93.

Nolan, R.A. (1981) Mass production of pathogens, in *Blackflies the Future for Biological Methods in Integrated Control* (ed. M. Laird), Academic Press, London, pp.319–24.

Obeta, J.A. and Okafor, N. (1984) Medium for the production of primary powder of *Bacillus thuringiensis* subsp. *israelensis*. *Applied Environmental Microbiology*, **47**, 863–7.

Pang, A.S.D. and Mathieson, B. (1991) Peptide mapping of different *Bacillus thuringiensis* toxin gene products by CNBr cleavage in SDS-PAGE gels. *Journal Invertebrate Pathology*, **57**, 82–93.

Patti, J.H. and Carner, G.R. (1974) *Bacillus thuringiensis* investigations for control of *Heliothis spp.* in cotton. *Journal of Economic Entomology*, **67**, 415–18.

Payne, C.C. (1988) Pathogens for the control of insects: where next? *Philosophical Transactions of the Royal Society, London* B, **318**, 225–48.

Percy, J. and Fast, P.G. (1983) *Bacillus thuringiensis* crystal toxin: Ultra-structural studies of its effects on silkworm midgut cells. *Journal of Invertebrate Pathology*, **41**, 86–98.

Petras, S.F. and Casida, L.E. (1985) Survival of *Bacillus thuringiensis* spores in soil. *Applied and Environmental Microbiology*, **50**, (6), 1496–501.

Pinnock, D.E., Brand, R.J., *et al.*, (1974) The field persistence of *Bacillus thuringiensis* spores on *Cercis occidentalis* leaves. *Journal of Invertebrate Pathology*, **23**, 341–16.

Pinnock, D.E., Brand, R.J., *et al.* (1975) Effect of tree species on the coverage and field persistence of *Bacillus thuringiensis* spores. *Journal of Invertebrate Pathology*, **25**, 209–14.

Pourmirza, A.A. (1989) Studies on the susceptibility of *Heliothis armigera* to *Bacillus thuringiensis* and alpha-cypermethrin and the effect of hostplant on susceptibility. Ph.D. thesis, University of Wales, College of Cardiff.

Reardon, R.C. and Haissig, K. (1983) Spruce budworm larval populations and field persistence of *Bacillus thuringiensis* after treatment in Winconsin. *Journal of Economic Entomology*, **76**, 1139–43.

Reisner, W.M., Feir, D.J. *et al.*, (1989) Effect of *Bacillus thuringiensis kurstaki* δ-endotoxin on insects malpighian tubule structure and function. *Journal of Invertebrate Pathology*, **54**, 175–90.

Richter, A.R. and Fuxa, J.R. (1984) Pathogen–pathogen and pathogen–insecticide interactions in the velvetbean caterpillar. *Journal of Economic Entomology*, **77**, 1559–64.

Sacchi, V.F., Parenti P., *et al.*, (1986) *Bacillus thuringiensis* toxin inhibits K^+-gradient-dependent amino acid transport across the brush-border membrane of *Pieris brassicae* midgut cells. *FEBS Letters*, **204**, 213–18.

Salama, H.S., Foda, M.S., *et al.*, (1983) Novel fermentation media for production of δ-endotoxins from *Bacillus thuringiensis*. *Journal of Invertebrate Pathology*, **41**, 8–19.

Salama, H.S., Foda, M.S., *et al.*, (1984) Potency of combinations of *Bacillus thuringiensis* and chemical insecticides on *Spodoptera littoralis*. *Journal of Economic Entomology*, **77**, 885–90.

Salama, H.S., Foda, M.S., *et al.*, (1985) Potential of some chemicals to increase the effectiveness of *Bacillus thuringiensis* against *Spodoptera littoralis*. *Zeitschrift für Angewandte Entomolgie*, **100**, 425–33.

Salama, H.S., Foda, M.S. and Sharaby, A. (1986) Possible extension of the activity spectrum of *Bacillus thuringiensis* strains through chemical additives. *Journal of Applied Entomology*, **101**, 304–13.

Salama, H.S., Saleh, M.R., *et al.*, (1990) Spray and dust applications of *Bacillus thuringiensis* and lannate against *Spodoptera littoralis* on soybean in Egypt. *Journal of Applied Entomology*, **109**, 194–9.

Saleh, M.S. and Wright, R.E. (1989) Effects of the IGR cryomazine and the pathogen *Bacillus thuringiensis* var. *israelensis* on the mosquito *Aedes epacticus*. *Journal of Applied Entomology*, **108**, 381–5.

Samasanti, W., Pantuwatana, S. and Bhumiratana, A. (1982) Role of the parasporal body in causing toxicity of *Bacillus thuringiensis* toward *Aedes aegypti* larvae. *Journal of Invertebrate Pathology*, **39**, 41–8.

Sebasta, K., Farkas, J., *et al.*, (1981) Thuringiensin, the beta-exotoxin of *Bacillus thuringiensis*, in *Microbial Control of Pests and Plant Diseases 1970–1980*, (ed H.D. Burges), Academic Press, New York, pp.249–81.

Shapiro, M. (1985) Effectiveness of B vitamins as UV screens for the gypsy moth nuclear polyhedrosis virus. *Environmental Entomology*, **14**, 705–8.

Shapiro, M. (1989) Congo red as an ultraviolet protectant for the gyspy moth nuclear polyhedrosis virus. *Journal of Economic Entomology*, **82**, (2), 548–50.

Smirnoff, W.A. (1985) Field tests of a highly concentrated formulation of *Bacillus thuringiensis* against spruce budworm *Choristoneura fumiferana*. *Canadian Entomologist*, **117**, 877–81.

Smith, D.B., Hostetter, D.L. and Pinnell, R.E. (1980) Laboratory formulation comparisons for a bacterial (*Bacillus thuringiensis*) and a viral (*Baculovirus heliothis*) insecticide. *Journal of Economic Entomology*, **73**, 18–21.

Smith, D.G., Hostetter, D.L. and Ignoffo, C.M. (1977) Ground spray equipment for applying *Bacillus thuringiensis* suspension on soybeans. *Journal of Economic Entomology*, **70**, 663–7.

Smyth, K. (1992) Isolation and characterization of *Bacillus thuringiensis* strains for control of *Helicoverpa armigera*. Unpublished Ph.D. thesis, University of Wales, College of Cardiff

Sneh, B. and Gross, S. (1983) Biological control of the Egyptian cotton leafworm *Spodoptera littoralis* in cotton and alfalfa fields using a preparation of *Bacillus thuringiensis* supplemented with adjuvants. *Zeitschrift für Anglewandte Entomologie*, **95**, 418–24.

Sneh, B., Schuster, S. and Gross, S. (1983) Improvement of the insecticidal activity of *Bacillus thuringiensis* var. *entomocidus* on larvae of *Spodoptera littoralis* by addition of chitinolytic bacteria, a phagostimulant and a UV protectant. *Zeitschrift für Angelwandte Entomologie*, **96**, 77–83.

Stone, T.B., Sims, S.R. and Marrone, P.G. (1989) Selection of tobacco budworm for resistance to a genetically engineered *Pseudomonas fluorescens* containing the δ-endotoxin of *Bacillus thuringiensis* subsp, *kurstaki*. *Journal of Invertebrate Pathology*, **53**, 228–34.

Subramanyam, Bh. and Cutkomp, L.K. (1985) Moth control in stored grain and the role of *Bacillus thuringiensis*: an overview. *Residue Reviews*, **94**, 1–47.

Tedders, W.L. and Gottwald, T.R. (1986) Evaluation of an insect collecting system and as an ultra-low spray system on a remotely piloted vehicle. *Journal of Economic Entomology*, **79**, 709–13.

Turnbull, S.A., Tolman, J.H. and Harris, C.R. (1988) Colorado potato beetle resistance to insecticides in Ontario, Canada. *British Crop Protection Control*

Conference, Pests and Diseases, Vol.1, British Crop Protection Council, pp.457–60.

Van Rie, J., Jansens, S., *et al.*, (1989) Specificity of *Bacillus thuringiensis* delta-endotoxins. Importance of specific receptors on the brush border membrane of the midgut of target insects. *European Journal of Biochemistry*, **186**, 239–47.

Van Rie, J., Jansens, S. *et al.*, (1990a) Receptors on the brush border membrane of the insect midgut as determinants of the specificity of *Bacillus thuringiensis* delta-endotoxins. *Applied Environmental Microbiology*, **56**, 1378–85.

Van Rie, J., McGaughey, W.H., *et al.*, (1990b) Mechanism of resistance to the microbial insecticide *Bacillus thuringiensis*. *Science*, **247**, 72–4.

Vinson, S.B. (1989) Potential impact of microbial insecticides on beneficial arthropods in the terrestial environment, in *Safety of Microbial Pesticides* (eds M. Laird, L.A. Lacey and E.W. Davidson), CRC Press, Boca Raton, FL.

Von Tersch, M.A., Robbins, H.L., Jany, C.S. and Johnson, T.B. (1991) Insecticidal toxins from *Bacillus thuringiensis* subsp. *kenyae*: Gene cloning and characterization and comparison with *B. thuringiensis* subsp. *kurstaki* CrylA(c) toxins. *Applied and Environmental Microbiology*, **57**, 349–58.

White, D.J. and Morris, C.D. (1985) Seasonal abundance of anthropophilic Simuliidae from the Adirondack Mountains of New York State and effectiveness of an experimental treatment program using *Bacillus thuringiensis* var. *israelensis*. *Environmental Entomology*, **14**, 464–9.

Whiteley, H.R. and Schnepf, H.E. (1986) The molecular biology of parasporal crystal body formation in *Bacillus thuringiensis*. *Annual Review of Microbiology*, **40**, 549–76.

Whitlock, V.H., Lo, M.C., *et al.*, (1991) Two new isolates of *Bacillus thuringiensis* pathogenic to *Spodoptera litura*. *Journal of Invertebrate Pathology*, **58**, 33–9.

WHO (1987) World Health Organisation Report of an informal consultation on 'The detection, isolation, identification and ecology of biocontrol agents of disease vectors', Geneva, 2–4 September, pp.1–41.

Widner, W.R. and Whiteley, H.R. (1989) Two highly related insecticidal crystal proteins of *Bacillus thuringiensis* subsp. *kurstaki* possess different host range spec   *Journal of Bacteriology*, **171**, 965–74.

Widr   Whiteley, H.R. (1990) Location of the Dipteran specificity
re   – Dipteran crystal protein from *Bacillus thuringiensis*.
Journ.   2826–32.

Woods,   nd Shapiro, M. (1988) Effects of *Bacillus thuringiensis*
treatme   ce of nuclear polyhedrosis virus in Gypsy moth
popu   *mic Entomology*, **81**, (6), 1706–14.

Yendol,   nd McManus, M.L. (1990) Penetration of oak
canop   reparation of *Bacillus thuringiensis* applied by air.
Journal   *y*, **83**, (1), 173–9.

Zehnder,   W.D. (1989) Activity of the M-ONE formulation
of a new   *of Bacillus thuringiensis* against the Colorado potato beetle: Relationships between susceptibility and insect life stage. *Journal of Economic Entomology*, **82**, (3), 756–61.

3

Aspects of biocontrol of fungal plant pathogens

John M. Whipps and Mark P. McQuilken

3.1 INTRODUCTION

In recent years there have been considerable changes in attitude towards the use of chemicals in agriculture. Increasing public awareness concerning the quantities and types of chemicals used and their potential impact on the environment has led to more stringent regulations on their use and, in some cases, removal from the market. Consequently, interest has focused on alternatives to chemicals, particularly for pest and disease control. The use of biological disease control measures is one of the strategies available and much experimental work is being carried out to assess its commercial applicability.

This review will examine the direct use of specific antagonists. In addition, commercial and environmental issues associated with adoption of biological control measures will be discussed. By necessity this review will be selective, as many aspects of disease biocontrol have been thoroughly reviewed before (e.g. Cook and Baker, 1983; Burge, 1988; Wood and Way, 1988; Whipps and Lumsden, 1989, 1991; Baker and Dunn, 1990; Hornby, 1990; Beemster *et al.*, 1991; Chet *et al.*, 1991; Fravel and Keinath, 1991; Harman, 1991; Lewis and Papavizas, 1991a; Wilson *et al.*, 1991).

3.2 DIRECT APPLICATION OF ANTAGONISTS

The greatest research effort in biological disease control has been concerned with the direct use of specific antagonists. In various ways

Exploitation of Microorganisms Edited by D.G. Jones
Published in 1993 by Chapman & Hall, London ISBN 0 412 45740 7

viruses, bacteria, fungi and microfauna have all been observed to give some level of disease control, but the greatest interest has been with the use of bacteria and fungi to control fungal pathogens. What follows illustrates the ways in which biocontrol agents can be used and their modes of action in a range of biomes (*sensu* Whipps *et al.*, 1988).

3.2.1 Viruses

Numerous fungi have now been shown to contain viruses and, in many cases, no direct physiological effect of their presence has been observed. However, in some fungi, including isolates of the pathogens *Endothia* (*Cryphonectria*) *parasitica*, *Helminthosporium victoriae*, *Gaeumannomyces graminis* var. *tritici* and *Rhizoctonia solani*, infection by virus-like dsRNA can reduce growth rate, sporulation and pathogenicity, the latter effect often termed hypovirulence (Buck, 1988). In addition, DNA plasmids or defective mitochondrial DNA may also be involved in the expression of these phenotypes but, in those cases which have been critically assessed, the presence of specific types of dsRNA appears to be the most important virus-like material controlling hypovirulence (Hashiba, 1987; Koltin *et al.*, 1987).

In terms of practical biological disease control, only the use of hypovirulent strains of *E. parasitica* on chestnut have proven successful. Chestnut blight, caused by virulent strains of *E. parasitica*, spread rapidly throughout the chestnut growing region of Italy from its first appearance in 1938. However, the decline in its severity since 1950 is associated with the appearance of hypovirulent isolates which still infected trees but formed only restricted cankers. Fewer conidia were produced in the cankers and sexual reproduction was also suppressed. After anastomosis, virulent strains were converted into hypovirulent strains by the acquisition of dsRNA. As most conidia produced by hypovirulent strains were also hypovirulent and the sexual stage (which generally leads to the production of virulent ascospores) was suppressed, the hypovirulent state was maintained (van Alfen and Hansen, 1984; Newhouse *et al.*, 1990). Biocontrol has been promoted by inoculating hypovirulent isolates onto existing virulent cankers (Grente, 1965; Grente and Sauret, 1969), and was particularly successful in Europe where there were relatively few vegetative incompatibility groups present. However, this procedure was not successful in the USA where greater numbers of vegetative incompatibility groups exist (Buck, 1988).

3.2.2 Actinomycetes and bacteria

Aerial microbiomes

There seem to be no reports of the use of actinomycetes to control fungal foliar diseases. In contrast, direct application of bacterial antagonists to aerial plant parts has been shown to decrease a number of diseases caused by fungal plant pathogens. Particular emphasis has been placed on the treatment of foliage with *Pseudomonas* and *Bacillus* spp. For example, when *Pseudomonas* spp. isolated from beetroot leaves were reapplied to beetroot leaves, they inhibited spore germination of *Botrytis cinerea*, *Cladosporium herbarum* and *Phoma betae* by competing for amino acids on the leaf surface (Blakeman and Brodie, 1977). Similarly, *Pseudomonas cepacia* isolated from maize suppressed Alternaria leaf spot on tobacco and Cercospora leaf spot on peanut but, in these cases, production of antibiotics seemed to be responsible for control (Spurr, 1981). In the glasshouse, rust (*Uromyces phaseoli*) infection of bean was reduced by treatments with *Bacillus subtilis*, *B. cereus mycoides*, *B. thuringiensis* and *Erwinia ananas* pv. *uredovora* (Baker *et al.*, 1983). One isolate of *B. subtilis* was extremely effective, giving 95–98% reduction in the number of rust pustules when suspensions were sprayed onto leaves before inoculation with *U. phaseoli*. A peptide-containing metabolite produced by the bacterium was apparently responsible for a reduction in uredospore germination, prevention of normal germ-tube development and cytoplasmic abnormalities in the uredospores (Baker *et al.*, 1983). In later field experiments, in which *B. subtilis* was applied three times a week, severity of bean rust was reduced by at least 75% in two successive years (Baker *et al.*, 1985).

Bacteria have also been sprayed onto the foliage of some tropical crops to control diseases. Spraying leaves of jute with a bacterial suspension of *Bacillus megaterium* before inoculation with *Colletotrichum corchori* reduced the number of anthracnose lesions (Purkayastha and Bhattacharyya, 1982). Similarly, spraying rice in the field with a suspension of *B. megaterium* at regular intervals until the grain was mature reduced the incidence and severity of brown spot disease (*Drechslera oryzae*) (Islam and Nandi, 1985), but the mechanisms of disease control are unknown.

Treatment of pruning wounds of shrubs and trees with bacteria have also been tested. For example, inoculation of *B. subtilis* onto fresh pruning wounds of apple trees gave protection from apple canker (*Nectria galligena*) equivalent to treatment with the fungicide benomyl (Corke and Hunter, 1979). Inoculation also led to the release of 96% fewer conidia of *N. galligena* than uninoculated controls during the following year, but this does not appear to have been followed up commercially.

Bacteria also have potential for controlling diseases on flowers, possibly by competing for exogenous nutrients. For instance, when an *Erwinia* sp. and a coryneform bacterium isolated from rose petals were reapplied, they reduced the number of rose blight lesions caused by *Botrytis cinerea* (Redmond *et al.*, 1987).

Another area of growing interest is the use of bacteria to control post-harvest diseases of fruits. Application of *B. subtilis* to nectarines, peaches and plums has controlled brown fruit rot caused by *Monilinia fructicola* (Pusey *et al.*, 1986). Eleven other isolates of *B. subtilis* have also protected cherry fruits from post-harvest diseases caused by Alternaria rot (*Alternaria alternata*) and brown rot (*M. fructicola*) (Utkhede and Scholberg, 1986). Such protection is due to the production of antifungal antibiotics (Pusey, 1989). In addition, control of post-harvest diseases has also been achieved by use of *Enterobacter cloacae* against Rhizopus rot (*Rhizopus stolonifera*) of peach (Wilson *et al.*, 1987) and *Pseudomonas cepacia* against blue mould (*Penicillium expansum*) on apple and pear (Janisiewicz and Roitman, 1988). *In vitro* studies indicated that nutrient competition, perhaps facilitated by bacterial attachment, may play a key role in the control of *R. stolonifera* by *E. cloacae* (Wisniewski *et al.*, 1989).

Microbiomes below ground

A large number of actinomycete and bacterial antagonists have been coated onto seeds using various stickers and powders, or incorporated directly into soil or other soil-less growing media to give biocontrol of fungal plant pathogens (Table 3.1). However, the actinomycetes have been investigated on only a limited scale in comparison with the bacteria and so emphasis must be placed on the latter group. Nevertheless, a strain of *Streptomyces griseoviridis* isolated from light Finnish peat has been developed for commercial use against several soilborne pathogens (Lahdenperä *et al.*, 1991) and indicates its potential.

Bacteria have been shown to provide biocontrol by various different mechanisms, including direct parasitism, production of antibiotics, organic acids and siderophores (Défago and Haas, 1990). These mechanisms are often complex and generally cannot be explained by a single factor. Because of the huge literature involved in this area, only specific examples involving the use of *Enterobacter cloacae* and *Pseudomonas* spp. in the rhizosphere and spermosphere are mentioned to highlight some of the diverse mechanisms involved in biocontrol by bacteria in general.

Enterobacter cloacae
Damping-off of cotton caused by *Pythium ultimum* has been controlled by applications of *Enterobacter cloacae* to seed (Nelson *et al.*, 1986). The

bacterium binds firmly to the hyphae of *P. ultimum*, probably mediated by a lectin-type interaction. Seeds exuding large concentrations of sugars were not protected by the bacterium and treatment of seeds with sugars that interfered with binding eliminated their biocontrol activity. Furthermore, some biocontrol strains of *E. cloacae* are known to produce ammonia that can inhibit growth of *P. ultimum in vitro* (Howell *et al.*, 1988). Since ammonia production is also inhibited by high sugar levels, it may be an additional part of the mechanism by which the bacterium controls Pythium damping-off.

Recently, monthly top dressing applications of *Enterobacter cloacae* strains to golf course turf significantly reduced dollar spot disease caused by *Sclerotinia homoeocarpa* but the mechanisms involved are unknown (Nelson and Craft, 1991).

Pseudomonas spp

Fluorescent pseudomonads are the most extensively studied bacterial biocontrol agents of a diverse range of soil-borne diseases (Table 3.1). Recently, molecular biological approaches have proved useful in identifying the key mechanisms by which pseudomonads achieve control. For instance, when strains of *Pseudomonas fluorescens* (2–79 and 13–79) isolated from take-all decline soils were applied to wheat seeds they suppressed take-all caused by *Gaeumannomyces graminis* var. *tritici* in field-grown winter and spring wheat (Weller and Cook, 1983). Genetic analysis of *P. fluorescens* 2–79 has established a linkage between production of the antibiotic **phenazine-1-carboxylic acid** (PCA) and disease suppression (Thomashow and Weller, 1988). Mutants generated by transposon (Tn5) mutagenesis that were defective in PCA synthesis were less suppressive of take-all than the parental strain, and complementation of the mutants with cosmids from a genomic library of 2–79 restored both phenazine production and suppressiveness of take-all. In another example, *P. fluorescens* Pf-5 produced an antibiotic, pyoluteorin, which was inhibitory to *Pythium ultimum in vitro* (Howell and Stipanovic, 1980). Although the antibiotic is inactivated by soil colloids, application of the antibiotic or bacterial cells to cotton seed at planting increased seedling survival from 30% to over 65% in both treatments in *P. ultimum* infested soil. Similarly, *P. fluorescens* HV 37 produced the antibiotic oomycin and mutation in the afuE locus required for synthesis of the antibiotic, resulted in reduced suppression of damping-off in cotton caused by *P. ultimum* when the mutant was applied to seed (Howie and Suslow, 1986; Gutterson *et al.*, 1990).

Cyanic acid production has been shown to be important in the potential biocontrol of black root-rot of tobacco caused by *Thielaviopsis basicola* and take-all caused by *Gaeumannomyces graminis* by *P. fluorescens* strain CHA0 (Ahl *et al.*, 1986; Défago and Haas, 1990). Using a gene replacement technique, a cyanide-negative mutant (hcn⁻) CHA5 has

Table 3.1. Examples of actinomycete and bacterial antagonists reported to control fungal diseases in root and soil microbiomes

Antagonist	Pathogen	Host	Application method	Reference
Actinomycetes				
Actinoplanes and Micromonospora species	Phytophthora megasperma f.sp. glycinea	Soya bean	Seed	Filonow and Lockwood, 1985
Neocosmospora and Acrophialophora species	Rhizoctonia solani	Pepper	Incorporation	Turhan and Turhan, 1989
Streptomyces species	Macrophomina phaseolina	Sunflower and mungbean	Seed	Hussain et al., 1990
	Pythium aphanidermatum	Poinsettia	Incorporation	Bolton, 1980
	P. debaryanum	Sugar beet	Surface spray	Tahvonen, 1982
	P. splendens	Geranium	Incorporation	Bolton, 1978
	P. ultimum	Pepper	Incorporation	Turhan and Turhan, 1989
	R. solani	Pepper	Incorporation	Turhan and Turhan, 1989
Bacteria				
Arthrobacter species	P. debaryanum	Tomato	Seed	Mitchell and Hurwitz, 1965
Bacillus species	P. ultimum	Snapdragon	Incorporation	Broadbent et al., 1971
	Sclerotium cepivorum	Onion	Seed	Utkhede and Rahe, 1980, 1983
Enterobacter cloacae	P. ultimum	Cotton	Seed	Nelson, 1988
		Cucumber	Seed	Hadar et al., 1983
		Pea	Seed	Nelson et al., 1986
		Rye	Seed	Nelson et al., 1986
	Pythium	Cucumber	Seed	Paulitz and Loper, 1991
	Sclerotinia homeocarpa	Golf course turf	Top dressing (cornmeal sand mixture)	Nelson and Craft, 1991

Erwinia herbicola		Cotton	Seed	Nelson, 1988
Pseudomonas[a] species	*P. ultimum*	Pea	Seed	Parke et al., 1991
	Aphanomyces euteiches			
	Fusarium oxysporum f.sp. *dianthi*	Carnation	Incorporation/stem inoculation	van Peer and Schippers, 1991
	Gaeumannomyces graminis	Barley/wheat	Seed/incorporation	Weller and Cook, 1983; Cook and Weller, 1987; Thomashow et al., 1990
	P. aphanidermatum	Cucumber	Incorporation	Elad and Chet, 1987
	P. ultimum	Chickpea	Seed	Kaiser et al., 1989
		Cotton	Incorporation	Hagedorn et al., 1989
		Cucumber	Seed	Howell and Stipanovic, 1980
		Maize	Seed	Elad and Chet, 1987
		Sugar beet	Seed	Callan et al., 1990
				Osburn et al., 1989; Sikora et al., 1990a,b
		Wheat	Seed	Weller and Cook, 1986; Becker and Cook, 1988
	Pythium species	Chickpea	Seed	Trapero-Casas et al., 1990
		Cucumber	Seed	Zhang et al., 1990
		Pea	Seed	Parke et al., 1991
	Rhizoctonia solani	Cotton	Seed	Howell and Stipanovic, 1980
	Thielaviopsis basicola	Cotton	Seed/incorporation	Ahl et al., 1986; Voisard et al., 1989

[a]Includes *P. fluorescens*, *P. fluorescens-putida*, *P. putida* and other *Pseudomonas* species

been constructed that gave less biocontrol activity against black root-rot than the wild type in a gnotobiotic system (Voisard *et al.*, 1989). Complementation of CHA5 by the cloned wild-type hcn$^+$ genes restored its ability to suppress the disease. Although these results suggest clearly that cyanide contributed significantly to suppression of black root-rot, the antibiotic 2,4-diacetyl-phloroglucinol (Ph1) was also shown to be involved in disease suppression by this strain (Keel *et al.*, 1989).

Competition for **iron** is another mechanism by which fluorescent pseudomonads may control fungal plant pathogens. Under iron-limiting conditions, pigmented compounds termed pyoverdines are produced and these siderophores serve as vehicles for the transport of iron [Fe (III)] into microbial cells (Lynch, 1990). Clear evidence for the involvement of siderophores in biocontrol again came from experiments in which mutants were used. Seed treatment of cotton with pyoverdine-minus mutants of *P. fluorescens* neither suppressed pre-emergence damping-off induced by *P. ultimum* nor improved growth of wheat in soil artificially infested with *Pythium ultimum* var. *sporangiforum* (Becker and Cook, 1988). It was also shown that siderophore production could only partially account for the biocontrol activity against *P. ultimum* (Loper, 1988).

3.2.3 Fungi

Fungi are by far the most extensively researched group of biocontrol agents and they have been used in both aerial, root and soil microbiomes. Interest in their use for the control of aerial plant pathogens has developed more slowly than in the case of root pathogens and this probably reflects the relative dearth of information on the ecology of micro-organisms on aerial surfaces in comparison with those in the soil. Nevertheless, many preparations have been used directly to suppress aerial and soil-borne plant pathogens in the glasshouse and field, and the emphasis here is placed on examples in which control has been shown under near normal growing conditions.

Aerial microbiomes

Foliage
One approach to biocontrol of foliar diseases is the induction of host resistance or cross protection. Usually, an avirulent strain of a pathogenic fungus is inoculated onto the host and may sometimes invade the tissues. The host is not damaged, but resistant reactions may be induced in the plant which inhibit a pathogen which subsequently invades. In addition, the possibility exists in some cases that the avirulent strain may act directly against the pathogen within the tissues. For example,

when a non-pathogenic strain of *Alternaria alternata* was applied as a
conidial suspension to tobacco leaves and a pathogenic strain of the
same species was inoculated three days later, leaf spotting was reduced
by 60% in the laboratory and by 65% under field conditions (Spurr,
1977). Similarly, application of a spore suspension of an avirulent strain
of *Colletotrichum lagenarium* to the cotyledons or the first leaf of cucumber
made plants resistant to subsequent challenge by *C. lagenarium* and
anthracnose development was suppressed (Descalzo *et al.*, 1990).

Naturally occurring saprophytic fungi have also been used in attempts
to control leaf-infecting pathogens by competition. For example, when
spore suspensions of *Sporobolomyces roseus* and *Cryptococcus laurentii* var.
flavescens were sprayed together with nutrients onto wheat flag leaves
they increased in numbers compared with controls sprayed only with
water (Fokkema *et al.*, 1979). When leaves were subsequently inoculated
with the pathogens, *Septoria nodorum* or *Cochliobolus sativus*, infection by
both pathogens was markedly less than on leaves sprayed with water.
This effect lasted for three weeks, but later disappeared as the population of inoculated yeasts declined. Similarly, suspensions of *Chaetomium
globosum* sprayed onto young apple leaves at regular intervals reduced
the incidence of apple-scab (*Venturia inaequalis*) by more than 20% under
field conditions (Cullen *et al.*, 1984), but again populations of the applied
saprophyte declined between applications. The reduction in the population of the applied saprophytes reflects the high stress environment of
the leaf and is consequently a very difficult one in which to introduce
antagonists, and have them survive and multiply. Competition for
nutrients, limiting growth of the pathogen and preventing subsequent
infection probably occurs here.

Mycoparasites have been shown to infect biotrophic foliar pathogens
such as powdery mildews and rusts. The yeasts *Tilletiopsis minor*, *Stephanoascus flocculosus* and *S. rugulosus* have been found to colonize and
restrict the growth of cucumber powdery mildew (*Sphaerotheca fuliginea*)
in the glasshouse and growth chamber (Hijwegen, 1986; Jarvis *et al.*,
1989). All of these mycoparasites required high humidity for optimum
activity and, therefore, their use in the field may be restricted. Nevertheless, strains of *T. minor* resistant to the fungicides dimethirimol and
fenarimol were found and these have been successfully used in a
scheme of integrated control. Similarly, a strain of *Ampelomyces quisqualis*
(Sztejnberg *et al.*, 1989) resistant to the fungicide triforine has been
found and was used as a mycoparasite in an integrated control programme against cucumber powdery mildew, but again, high relative
humidities were required for effective control (Sundheim and Amundsen, 1982).

One mycoparasite that appears to have a great potential for future use
is *Verticillium lecanii*. This fungus has been used successfully to control

aphid and whitefly infestations when applied in a commercial spray formulation in the glasshouse where humidities are high (Hall, 1982; Kanangaratnam *et al.*, 1982). As well as being entomogenous, it also has the ability to destroy urediniospores of carnation rust (*Uromyces dianthi*), and teliosori of white rust (*Puccinia horiana*) of chrysanthemum (Spencer and Atkey, 1981; Srivastava *et al.*, 1985). The mycoparasite may also be of valuable use in the field as it has the ability to attack urediniospores of the leek rust *Puccinia allii* (Uma and Taylor, 1987).

Another approach to biocontrol is the application of antagonists to crop debris to prevent disease carry-over. For example, when the antagonist *Athelia bombacina* was applied to fallen apple leaves infected with the apple-scab pathogen, *Venturia inaequalis*, the leaves were rapidly colonized (Young and Andrews, 1990). The pseudothecia of the pathogen did not mature to produce asci and consequently the normal spring infection of leaves by ascospores was prevented. Similarly, application of pycnidiospores of the sclerotial mycoparasite *Coniothyrium minitans* may also act to decrease inoculum potential of the pathogen, *Sclerotinia sclerotiorum*, during fallow periods (Trutmann *et al.*, 1982).

Flowers and fruits
A number of fungi have been used to control diseases both on flowers and fruits. For instance, conidial sprays of a fungicide-resistant strain of *Trichoderma harzianum* onto apple blossom controlled *Botrytis cinerea* dry eye rot of fruit better than the fungicide-sensitive parent strain both alone and together with reduced dosages of the fungicide vinclozolin (Tronsmo, 1991). Hyphal interactions and production of non-volatile inhibitors by *Trichoderma* are likely to be important here. Trials in French vineyards have also shown that *Trichoderma* spp. have considerable potential for controlling *B. cinerea* rot of grapes. Sprays of *T. viride* onto grape vines controlled *B. cinerea* rot of fruit to a level similar to that achieved with dichlofluanid sprays applied at the same time (Dubos *et al.*, 1982). Recently, *Penicillium frequentans* has been used to control twig blight on peaches caused by *Monilinia laxa* and to give control equivalent to that obtained with captan (De Cal *et al.*, 1990).

Some attention has also been given to controlling aerial infection by *Sclerotinia sclerotiorum*. Ascospores are the primary source of inocula for causing stem rot in oilseed rape and white mould of beans (Abawi and Grogan, 1975; Morrall and Dueck, 1982). Application of a range of naturally occurring saprotrophs to bean or rape petals decreased ascospore infection and depended upon the antagonist occupying the petals before the pathogen (Inglis and Boland, 1990).

Woody stems
Biocontrol using fungi has been most effective on pruning wounds and cut stumps of woody perennials leading to commercial products in some

cases. For example, the pathogen *Chondrostereum purpureum* often causes silver-leaf and branch die-back of fruit and ornamental trees after its basidiospores have entered wounds caused by pruning. Fresh pruning wounds on plum trees have been protected from infection by applying conidial suspensions of *Trichoderma harzianum* before inoculation with the pathogen. The protective effect of *T. viride* results from its production of antibiotics in the wood vessels and the formation of gum barriers by the host in response to the presence of the fungus (Grosclaude, 1974). A curative treatment has also been developed. Application of *T. viride* to trees severely affected by *C. purpureum* decreased disease symptoms during the following three years compared with equivalent untreated trees (Grosclaude, 1974). A commercial preparation of *Trichoderma*, BINAB-T, is currently sold for use against this and other pathogens.

Another pathogen of woody stems, *Eutypa armeniacae*, causes gummosis and branch die-back of apricots in Australia following infection of pruning wounds. The wounds are naturally colonized by *Fusarium lateritium* which produces a non-volatile, water-soluble antibiotic in culture that inhibits spore germination and mycelial growth of *E. armeniacae* (Carter and Price, 1974). Application of spore suspensions of *F. lateritium* to pruning wounds one day before inoculation of the pathogen gave effective protection. In addition, *F. lateritium* was found to be ten times more tolerant than *E. armeniacae* to benzimidazole fungicides. Integrated control of *F. lateritium* and benomyl sprays gave better protection than either the fungicide or the biocontrol agent alone. A recommended application in Australian orchards was 10^6 conidia of *F. lateritium* ml^{-1} plus 200 μg ml^{-1} of benomyl (Carter, 1983). A simple pneumatic-powered control device has been produced that delivers the biocide to wounds made during pruning by a standard pneumatic secateur (Carter and Perrin, 1985).

One biocontrol agent commercially available in Britain is *Phlebia gigantea* for the control of butt-rot (*Heterobasidion annosum*) on pine trees (Rishbeth, 1963). Commercially-produced inocula containing spores of *P. gigantea* are applied to freshly cut stumps; the fungus competes successfully with *H. annosum* and enters the lateral roots thus preventing advance of the pathogen. Hyphal interference is thought to be involved in this interaction.

Microbiomes below ground

Seeds, tubers, cuttings and transplants
There are many examples of fungi applied to seeds and tubers in a variety of ways for the control of soil-borne diseases (Table 3.2). Spore suspensions have been used most frequently. For example, a conidial suspension of *Verticillium bigutattum* applied to potato seed tubers either as a dip or a spray has been used on a field-scale for the control of collar

Table 3.2. Examples of fungal antagonists applied to seeds for biocontrol of soil-borne disease

Antagonist	Pathogen	Host	Reference
Chaetomium globosum	*Pythium debaryanum*	Squash	Harman *et al.*, 1978
	P. ultimum	Squash	Harman *et al.*, 1978
Gliocladium virens	*P. ultimum*	Cotton	Howell, 1982; 1991
	Rhizoctoria solani	Potato	Beagle-Ristaino and Papavizas, 1985
Penicillium oxalicum	*Pythium* spp.	Chickpea	Trapero-Casas *et al.*, 1990
Pythium oligandrum	*Aphanomyces cochlioides*	Sugar beet	McQuilken *et al.*, 1990b
	P. ultimum	Cress	Lutchmeah and Cooke, 1985; McQuilken *et al.*, 1990b
		Sugar beet	Martin and Hancock, 1987; McQuilken *et al.*, 1990b
	Pythium spp.	Chickpea	Trapero-Casas *et al.*, 1990
		Cress	McQuilken *et al.*, 1990b
Trichoderma hamatum	*Pythium* spp.	Pea	Harman *et al.*, 1980
		Radish	Harman *et al.*, 1980
T. harzianum	*P. ultimum*	Cotton	Harman *et al.*, 1989
		Cucumber	Paulitz *et al.*, 1990
		Wheat	Harman *et al.*, 1989
	Pythium spp.	Cucumber	Taylor *et al.*, 1991
T. koningii	*Pythium* spp.	Pea	Lifschitz *et al.*, 1986
T. viride	*P. ultimum*	Pea	Papavizas and Lewis, 1983
Verticillium biguttatum	*R. solani*	Potato	Jager and Velvis, 1985

rot and black scurf (*Rhizoctonia solani*) (Jager and Velvis, 1985). The antagonist was able to spread from the coated tubers to sprouts and stolons, protecting them from infection by *R. solani* and providing a 60% reduction in sclerotium production by the pathogen. Similarly, *Pythium*-induced damping-off and seed rots of pea have been controlled when conidial suspensions of either *Trichoderma koningii* or *T. harzianum* were coated onto seeds using the adhesive polyvinyl alcohol (Hadar *et al.*, 1984). In addition, incorporation of conidia of *Trichoderma* spp. into fluid-drilling gels has been used to provide control of *R. solani* on radish (Mihuta-Grimm and Rowe, 1986).

A number of seed treatments have been refined to improve biocontrol efficacy of the antagonist. Combining biocontrol agents with seed priming systems has been one approach. For example, osmopriming sugar beet seeds led to greater numbers of bacterial numbers on the seed surface and was related to control of root rots caused by *Pythium* spp. (Osburn and Schroth, 1989). Similarly, incorporating *Trichoderma* spp. or *Enterobacter* spp. into solid matrix priming systems of cucumber and tomato resulted in increased stands and decreased damping-off caused by *Pythium* (Harman and Taylor, 1988; Harman *et al.*, 1989). Another approach concerns the methods of application to seeds or the inclusion of additives. For example, oospores of the biocontrol fungus *Pythium oligandrum* have been incorporated into thin film coating or seed pellets of sugar beet for the control of *Pythium ultimum* and *Aphanomyces cochlioides* using commercial processes (McQuilken *et al.*, 1990b). A double coating procedure designed to slow infection of seeds by *Pythium* spp. and still provide an active environment for *Trichoderma* has also been devised (Harman, 1991). Here, *Trichoderma* conidia are applied to the seed followed by a slurry of a sticker and particulate matter such as lignaceous shale. The shale provides a conducive low pH environment for the *Trichoderma* and physically separates it from the competitive soil microflora. *Trichoderma* activity on seeds has also been improved by addition of specific food bases such as organic acids, polysaccharides and polyhydroxy alcohols into seed coatings (Nelson *et al.*, 1988).

Fungi applied to seeds and other below-ground microbiomes have been shown to provide biocontrol by various different mechanisms, including competition for nutrients or space on the rhizosphere or spermosphere, mycoparasitism or antibiosis (Whipps *et al.*, 1988; Taylor and Harman, 1990). None of these mechanisms is mutually exclusive, though one may appear to dominate. For example, when coated onto seeds, *Pythium oligandrum* may compete for nutrients with *P. ultimum* or exhibit mycoparasitism (Martin and Hancock, 1987). Although the mycoparasitic capabilities of *P. oligandrum* in dual culture with *P. ultimum* are known (Lewis *et al.*, 1989), mycoparasitism is probably not the primary mechanism of disease control because of the rapid germination time (1.5 h), fast growth rate (330 μm h^{-1}) and short seed infection time (24 h) of the pathogenic *Pythium* species.

More evidence for mycoparasitism of plant pathogens comes from *in vitro* studies. For example, *Trichoderma hamatum* may grow towards host hyphae and, after contact, form coils and appressoria-like structures from which penetration may take place (Chet *et al.*, 1981). When hyphae of *Trichoderma* were removed from host hyphae, pits and holes were observed, indicating the action of cell-wall degrading enzymes such as proteases, glucanases, chitinases and cellulases. Similar interactions have also been observed on treated seeds in soil (Hubbard *et al.*, 1983). A role for antibiotic production in biocontrol by fungi has similarly been implied from *in vitro* studies, but the use of antibiotic-minus mutants have strengthened the evidence that antibiotics are important *in vivo*. For instance, a strain of *Gliocladium virens* that suppressed *Pythium*-induced damping-off in cotton seedlings was shown to produce the antibiotic gliovirin in culture (Howell, 1991). Antibiotic-minus mutants failed to give biocontrol and were overgrown by *P. ultimum in vitro*.

Recently, attempts have been made to use antibiotics isolated from culture filtrates directly in seed-coatings (Thompson and Burns, 1989). When metabolites from *Penicillium claviforme* were incorporated into sugar beet pellets, damping-off in *P. ultimum*-infested compost was markedly reduced. Unfortunately, the general use of these metabolites will probably be restricted because some are known carcinogens.

Biocontrol fungi have also been applied to cuttings or transplants. For example, control of Verticillium wilt (*Verticillium dahliae*) of tomato has been obtained in glasshouse trials by dipping roots in a suspension of an avirulent strain of *V. albo-atrum* prior to planting (Matta and Garibaldi, 1977). Similarly, cuttings of sweet potato have been cross-protected against Fusarium wilt (*Fusarium oxysporum* f. sp. *batatas*) by dipping into suspensions of bud cells of non-pathogenic isolates of *F. oxysporum* (Ogawa and Komada, 1985). Some of the non-pathogenic isolates have been distributed to growers by the Japanese government and are available commercially.

Soil and growing media
Numerous preparations of biocontrol fungi derived from both semi-solid and liquid culture systems have been incorporated into soils and growing media either before, or at the time of planting, for control of a wide range of fungal soil-borne pathogens (Table 3.3). Large, bulk semi-solid fungal preparations may be suitable for greenhouse and nursery application but, in the field, they are very difficult to distribute evenly, and the amounts to be applied for efficacy are often agriculturally unrealistic. Moreover, they may require costly specialized equipment for

their handling and application. Nevertheless, a number of such preparations have given effective biocontrol when applied both in the field and
glasshouse. For instance, a series of novel solid-substrate preparations
that contain young actively growing hyphae of *Gliocladium virens* or
Trichoderma spp. have been developed by researchers at USDA Beltsville, USA. The rationale for their use is that actively growing hyphae
(germlings) of the biocontrol fungi are more effective than conidia
against pathogens, which may remain quiescent when introduced into
soil (Lewis and Papavizas, 1987a). One of these preparations has been
produced by inoculating moistened sterile wheat bran with a conidial
suspension of the biocontrol fungus followed by inoculation for only
2–3 days, instead of the usual period of 2–3 weeks. When added to
glasshouse soils, such germling/bran preparations of *G. virens* or *T.
harzianum* significantly reduced damping-off and blight of snapbean
caused by *Sclerotium rolfsii* (Papavizas and Lewis, 1989). Similarly,
germling/bran preparations of *G. virens* or *T. hamatum* significantly
reduced damping-off of cotton caused by *Rhizoctonia solani* in field plots
artificially infested with the pathogen (Lewis and Papavizas, 1991b). In
another preparation, a vermiculite-bran mixture was amended with a
wet or dry biomass preparation of *G. virens* or a *Trichoderma* sp. moistened with dilute acid and incubated for two days to induce the formation
of germlings (Lewis *et al.*, 1991). The vermiculite-bran/germling preparation was then used successfully both in soil and soil-less mixes to reduce
the survival of *Rhizoctonia solani*.

A number of other preparations, including dusts and granular formulations, have been developed using dried biomass produced in liquid
culture. In general, these preparations obtain better spatial distribution
and close contact with pathogen propagules than those prepared from
semi-solid culture systems. Dusts prepared from milled biomass of
several isolates of *G. virens* or *Trichoderma* spp., and mixed with a
pyrophyllite clay carrier, have been shown to reduce Rhizoctonia
diseases of cotton and potato when applied in the field (Beagle-Ristaino
and Papavizas, 1985). Similarly, isolates of *G. virens* or *T. harzianum*,
applied to glasshouse soil as a clay-fermenter biomass mixture, effectively suppressed damping-off and blight of snap bean caused by
Sclerotium rolfsii and reduced sclerotial germination of the pathogen
(Papavizas and Lewis, 1989).

An innovative approach to formulation and delivery of biocontrol
fungi involves the encapsulation of biomass within alginate gel pellets.
Briefly, this method consists of mixing biomass and a carrier (bulking
agent) with a sodium alginate solution. The mixture is then dipped into
calcium chloride solution to form insoluble gelatinous beads which dry
into stable, hard pellets of uniform size. A further refinement of this

Table 3.3. Examples of fungal antagonists incorporated either into soil or growing media for biocontrol of soil-borne diseases

Antagonist	Pathogen	Host	Reference
Acrophialospora levis	Pythium ultimum	Pepper	Turhan and Turhan, 1989
Coniothyrium minitans	Sclerotium cepivorum	Onion	Ahmed and Tribe, 1977
	Sclerotinia sclerotiorum	Bean	Trutmann et al., 1980
		Celery	Budge and Whipps, 1991
		Lettuce	Budge and Whipps, 1991
		Sunflower	Huang, 1980
Gliocladium catenulatum	P. ultimum	Pea	Reyes and Dirks, 1985
G. roseum	Phomopsis sclerotioides	Cucumber	Moody and Gindrat, 1977
G. virens	Lanzia sp.	Bermuda grass	Haygood and Mazur, 1990
	P. ultimum	Cabbage, Cotton	Lumsden and Locke, 1989
	Rhizoctonia solani	Cotton	Lewis and Papavizas, 1991b
	Sclerotium rolfsii	Bean	Papavizas and Lewis, 1989
Laetisaria arvalis	P. ultimum	Sugar beet	Martin et al., 1986
	R. solani	Sugar beet	Herr, 1988

Biocontrol agent	Pathogen	Crop	Reference
Pythium nunn	*P. ultimum*	Cucumber	Paulitz *et al.*, 1990
P. oligandrum	*P. ultimum*	Cress	McQuilken *et al.*, 1992b
		Sugar beet	Martin and Hancock, 1987
Sporidesmium sclerotivorum	*Sclerotinia minor*	Lettuce	Adams *et al.*, 1984
Taloromyces flavus	*Verticillium dahliae*	Eggplant	Marois *et al.*, 1982
Trichoderma hamatum	*P. ultimum*	Pea	Reyes and Dirks, 1985
	R. solani	Cotton	Lewis and Papavizas, 1991b
	S. rolfsii	Bean	Papavizas and Lewis, 1989
T. harzianum	*P. aphanidermatum*	Cucumber, Gypsophila, Pepper, Tomato	Sivan *et al.*, 1984
	P. ultimum	Lettuce	Lumsden *et al.*, 1990
	R. solani	Lettuce	Maplestone *et al.*, 1991
	Sclerotium cepivorum	Onion	Abd-El Moity *et al.*, 1982
	S. rolfsii	Peanut	Backman and Rodriguez-Kabana, 1975
T. viride	*Sclerotinia sclerotiorum*	Beans	Trutmann and Keane, 1990
	R. solani	Lettuce	Coley-Smith *et al.*, 1991

technique is the incorporation in the pellets of appropriate food bases (e.g. wheat bran) for the biocontrol fungus. The pellets are easily handled and can be applied to soil with conventional fertilizer equipment. Alginate pellets containing biomass of *Gliocladium virens* have suppressed damping-off caused by *Pythium ultimum* as well as *Rhizoctonia solani* of ornamental plants (Lumsden and Locke, 1989; Lumsden *et al.*, 1990) and cotton and sugar beet (Lewis and Papavizas, 1987b). An isolate of *G. virens* has recently been registered by the US Environmental Protection Agency (USEPA) for commercial use in specific markets by W.R. Grace and Co., Columbia, MD.

A final major area in which biocontrol fungi have been used is against sclerotia-forming pathogens, especially species of *Sclerotinia* and *Sclerotium*. The general approach has been to reduce sclerotial populations in soil by applying specific sclerotium-destroying mycoparasites including *Coniothyrium minitans*, *Sporidesmium sclerotivorum* and *Trichoderma* spp. (Whipps, 1991). For example, in the glasshouse, preplanting applications of a maizemeal-perlite preparation of *C. minitans* gave control of bottom rot of lettuce caused by *Sclerotinia sclerotiorum* (Budge and Whipps, 1991). The mycoparasite survived and spread in soil for at least a year and continued to degrade sclerotia, indicating that it can exhibit a degree of self-maintenance under normal horticultural growing conditions. *C. minitans* grown on a barley-rye-sunflower seed mixture applied to sunflower seed rows reduced *Sclerotinia* wilt by 30% over a two-year period (Huang, 1980). The mycoparasite releases β 1–3 glucanase and chitinase into culture media when grown on cell walls of sclerotia of *S. sclerotiorum* as sole carbon source and it is likely that such hydrolytic enzymes may play a key role in sclerotial destruction (Jones *et al.*, 1974). *C. minitans* has also been used successfully to decrease white mould of bean caused by *S. sclerotiorum* (Trutmann *et al.*, 1980) and white rot (*Sclerotium cepivorum*) on onion seedlings in the glasshouse (Ahmed and Tribe, 1977).

The mycoparasite *Sporidesmium sclerotivorum* has been shown to be effective against *Sclerotinia minor* on lettuce in field trials. A single application of the antagonist, grown on non-sterile sand containing 1% w/v live sclerotia of *Sclerotinia minor*, to plots at rates of 10^2 and 10^3 macro-conidia g^{-1} of soil reduced lettuce drop by 40–80% in four successive crops over a two-year period (Adams and Ayers, 1982). Spores of the mycoparasite germinated, invaded sclerotia and, after colonization, its hyphae grew from sclerotium to sclerotium through the soil (Ayers and Adams, 1981). Although there are problems in producing large quantities of inoculum for commercial use, W.R. Grace and Co., USA are registering the fungus for use as a biocontrol agent with the USEPA (G.C. Papavizas, personal communication).

3.2.4 Fauna

Fauna may be involved in natural biological disease control of soil-borne pathogens either directly, by parasitizing hyphae, spores and sclerotia, or indirectly, by transferring propagules of biocontrol agents through the soil. Amoebae, nematodes, mites, collembola and sciarid larvae have all been demonstrated to attack a wide range of pathogens with different groups or species showing various feeding preferences (Table 3.4). In some instances, soil fauna may be important factors in the maintenance of disease-suppressive soils (Curl and Harper, 1990) and can have a controlling influence on the populations of fungi in various soil horizons (Visser, 1985).

Most experimental work with mycophagus fauna have been concerned with population studies in the field or feeding preference tests carried out *in vitro*. However, there are a few instances in which fauna have been added to soil and this has resulted in disease control. For example, in greenhouse pot tests, additions of the nematode *Aphelenchoides hamatus* reduced damage to wheat caused by *Fusarium culmorum* (Rossner and Nagel, 1984) and when populations of the collembolans *Proisotoma minuta* and *Onychiurus encarpus* were added to sterilized soil artificially infested with *Rhizoctonia solani*, root disease of cotton was significantly reduced (Curl, 1979). Optimum control was obtained with high numbers of collembola and low levels of infestation with *R. solani*.

Table 3.4. Examples of fungal pathogens attacked by fauna

Fauna	*Pathogen*	*Reference*
Amoebae		
	Cochliobolus sativus	Anderson and Patrick, 1978
	Gaeumannomyces graminis	Chakraborty and Warcup, 1985
	Thielaviopsis basicola	Anderson and Patrick, 1978
Nematodes		
	Fusarium culmorum	Rossner and Nagel, 1984
	F. solani	Barnes *et al.*, 1981
	Pythium arrhenomanes	Rhoades and Linford, 1959
	Rhizoctonia solani	Barnes *et al.*, 1981
Arthropods		
	Botrytis cinerea	Hanlon, 1981
	Fusarium oxysporum	Curl *et al.*, 1985
	Gnomonia leptostyla	Kessler, 1990
	Pythium ultimum	El-Titi and Ulber, 1991
	Rhizoctonia solani	Curl and Harper, 1990; Bollen *et al.*, 1991
	Sclerotinia sclerotiorum	Anas *et al.*, 1989

Further, it was shown that addition of any one of the three biocontrol fungi, *Gliocladium virens*, *Trichoderma harzianum* and *Laetisaria arvalis* with the collembola provided an additional disease control benefit (Lartey *et al.*, 1986, 1991). These fungi are not favoured food sources for the two collembolan species and allow the possibility of an integrated control approach (Lartey *et al.*, 1989).

Interestingly, *Trichoderma viride* may also be involved in the reduction of viability of sclerotia of *Sclerotinia sclerotiorum* following attack by larvae of the fungus gnat *Bradysia coprophyla* (Anas *et al.*, 1989); direct feeding by the insect and the action of chitinase present in its saliva may make the sclerotia vulnerable to attack by the mycoparasite. Sclerotia of *S. sclerotiorum* infected with *Coniothyrium minitans* can also be attacked by larvae of fungus gnats as well as by slugs and may result in localized dispersal of the biocontrol fungus (Turner and Tribe, 1976; Trutmann *et al.*, 1980). Similarly, *Trichoderma* spp. are carried through soil by collembola and earthworms (Wiggins and Curl, 1979; Visser, 1985) and VA mycorrhizas may be spread by a range of soil fauna (McIlveen and Cole, 1976; Rabatin and Stinner, 1988).

Although these observations indicate a possible use of animals for biocontrol of fungal plant pathogens, some caution must be applied until detailed ecological studies of the specific animals of interest are carried out. There is little control over the movement of animals once they are released, large numbers may be required to be effective and many of the species discussed are known to act as minor pests in their own right or to act as vectors of pathogens. For example, the collembolan *Folsomia candida* interfered with VA mycorrhizal establishment in soybean and fed on leek root mycorrhizas (Warnock *et al.*, 1982; Kaiser and Lussenhop, 1991), many mites and collembola carry *Fusarium* spores on their exoskeletons (Visser, 1985) and larvae of the fungus gnat *Bradysia impatiens* transferred oospores of a *Pythium* sp. via their guts (Gardiner *et al.*, 1990). Nevertheless, as so little is known about the ecology and dynamics of these systems in general, the interactions between animals, plants and fungi must be a very important area for future study.

3.3 COMMERCIAL BIOCONTROL AGENTS OF FUNGAL PATHOGENS

The number of commercial biocontrol agents available for use against fungal pathogens is small (Table 3.5). Binab-T and P.g. suspension have been on the market for many years and are fairly widely available, but many of the others have limited markets, supported by the government in the country of use. Mycostop has recently become commercially available in Finland, aimed at small specific niche markets although sales are hoped to be extended elsewhere and for other diseases if tests prove successful and registration is allowed (Lahdenperä *et al.*, 1991). Three fungi, *Gliocladium virens*, *Sporidesmium sclerotivorum* and *Trichoderma harzianum* are undergoing registration with the US Environmental Protection Agency and a *Fusarium* sp. is undergoing registration in France for use against Fusarium wilt (C. Alabouvette, personal communication).

Many others, such as Dagger-G, marketed first in 1988 for the control of damping-off in cotton in the USA, have already been withdrawn. At first sight this would appear strange, particularly in view of the substantial volume of research demonstrating its efficacy. However, it is really an indication of all the hurdles that must be overcome to translate observations made in small scale experiments in the laboratory or glasshouse to widespread commercial field use.

For instance, any biological control agent must be able to be grown in large quantities fairly easily to give material that can survive and remain active in a range of environmental conditions both prior to, and after, application. Such material needs to be genetically stable and non-toxic to animals, plants and the environment and yet still provide good control of the target pathogen. In addition, the product should at least equal the efficacy and ease of use of existing chemical and cultural control methods at the same price, although some slight extra cost may be considered if it is seen as a 'green' product. Given this set of criteria it is surprising that any biological disease control agent reaches the market and remains there. However, there are several positive aspects associated with their development and use. Registration costs may be cheaper (although this may soon change with altered legislation (Lynch, 1992)); chemicals continue to be lost from the market; there are still no suitable chemical or other control measures for some diseases and they may be viewed as more environmentally friendly and acceptable in sustainable agriculture.

In order to optimize the chances of obtaining a successful commercial biocontrol agent, it may be necessary to adopt a much broader based selection and development programme than has been used in the past. From the very beginning the disease aetiology, the crops and cultivation methods, current control measures, and the environment and position of release should be examined. Simultaneously, selection of antagonists must be related to these factors as well as to inoculum production, formulation, storage and application procedures. Toxicological data need to be gathered as well as information on the mode of action and the whole development related to an appropriate cost–benefit analysis. In this way it should prove possible to identify realistic candidate antagonists at an early stage and, following registration, bring them rapidly to commercialization.

Table 3.5. Examples of antagonists available or in the process of registration for use in commercial preparations for biocontrol of fungal plant pathogens

Antagonist	Target pathogen	Disease/host	Product name and source
Bacteria and Actinomycetes			
Streptomyces griseoviridis	Alternaria brassicicola	Damping-off of crucifers	Mycostop (Kemira Oy, Finland)
	Fusarium oxysporum f.sp. dianthi	Carnation wilt	
Fungi			
Coniothyrium minitans	Sclerotinia sclerotiorum	Sunflower disease	Conyothyrin (Russian government)
Fusarium oxysporum (non-pathogenic)	Fusarium oxysporum f.sp. batatas	Fusarium wilt of sweet potato	Japanese government
Gliocladium virens	Pythium ultimum and Rhizoctonia solani	Damping-off of a range of plants	W.R. Grace & Co, USA[a]
Peniophora (Phlebia) gigantea	Heterobasidion annosum	Stem and root rot of pine	P.g. suspension (Ecological Laboratories Ltd, UK)

Biocontrol agent	Pathogen	Disease	Product
Pythium oligandrum	*Pythium ultimum*	Damping-off of sugar beet	Polygandron (Vyzkummy ustov, Czechoslovakia)
Sporidesmium sclerotivorum	*Sclerotinia* spp.	Vegetables	W.R. Grace & Co, USA[b]
Trichoderma spp.	*Botrytis, Pythium, Sclerotinia* and *Verticillium* spp.	Fruit and vegetables	Trichodermin (Bulgarian and Soviet Governments)
	Armillaria mellea	Honey fungus of trees	BINAB-T (Bio-Innovation AB, Töreboda, Sweden)
	Ceratocystis ulmi	Dutch elm disease	
	Chondrostereum purpureum	Silver leaf disease of trees	
	Endothia parasitica	Chestnut wilt	
	Heterobasidion annosum	Stem and root rot of pine	
Trichoderma harzianum	*Pythium* sp.	Damping-off	F-Stop (Eastman Kodak Co)[c]

[a] Refer to Whipps and Lumsden (1991)
[b] G.C. Papavizas (personal communication)
[c] Refer to Cates (1990)

3.4 FUTURE PROSPECTS

Research on biological disease control is likely to continue at an increased pace with reduced pesticide use and concern over the environment. General approaches that may lead to greater numbers of commercial biocontrol agents have already been mentioned, but there are several specific techniques and procedures which have been developed recently that appear to have considerable potential for enhancing the process. For instance, having obtained a useful agent, it is possible to direct further screening to obtain more isolates of the same organism with further attributes of interest. Selection for cold tolerance or rhizosphere competence (Tronsmo and Dennis, 1977; Ahmad and Baker, 1987; Sivan and Harman, 1991) could allow an increased range of use, while screening for fungicide tolerance would enable integrated control with reduced chemical inputs. Such variants may also be obtained by mutagenesis or by using protoplasting techniques (Abd-El Moity *et al.*, 1982; Harman *et al.*, 1989; Pe'er and Chet, 1990). Indeed, F-stop, the *Trichoderma harzianum* isolate being commercialized in the USA was one of the progeny derived from a protoplast fusion of two other biocontrol isolates. Further, if the mode of action is known, genetic engineering could be used to increase copy number and, if expressed, activity of genes of interest. Transformation of bacteria and fungi with chitinase and β 1–3 glucanase genes have already been attempted (Sundheim *et al.*, 1988; Shapira *et al.*, 1989; Goldman *et al.*, 1991). Modes of action can also be used as the basis for further targeted screens, thereby selecting different organisms with required attributes.

Other areas where advances have been made and seem likely to continue are concerned with inoculum production, formulation and application systems. Although solid substrate fermentations may be the only realistic way to produce some antagonists such as *Coniothyrium minitans* or *Sporidesmium sclerotivorum*, liquid fermentation systems involving the use of waste material such as molasses have provided a simple, low cost inoculum production procedure for several biocontrol fungi (Lumsden and Lewis, 1989; McQuilken *et al.*, 1990a; Lewis and Papavizas, 1991a). Interestingly, decreasing the water potential of the culture medium used to produce inoculum of *Trichoderma* resulted in a high yield of conidia with increased desiccation tolerance, a great advantage for subsequent storage and use. Detailed studies of the physiology of biocontrol agents in relation to both inoculum production and their environment of use is also an approach used at Littlehampton (Jackson *et al.*, 1991a, b, c; McQuilken *et al.*, 1992a). In addition, some of the greatest advances have been made in the area of seed coating and priming technologies described earlier and improvements in biocontrol due to this type of work are likely to continue.

In the future, at every research and development stage there will need to be checkpoints where progress towards a commercial biocontrol product is assessed. Information should then be fed back into the network of concomitant screening, testing and production procedures to identify the best way forward. If a particular line of research appears to come to a halt, alternative strategies overcoming or circumventing the problem could then be rapidly invoked. Providing the resources are present, such a multidisciplinary approach should yield a greater number of successful biocontrol agents than are currently available.

REFERENCES

Abawi, G.S. and Grogan, R.G. (1975) Source of primary inoculum and effects of temperature and moisture on infection of beans by *Whetzelinia sclerotiorum*. *Phytopathology*, **65**, 300–9.

Abd-El Moity, T.H., Papavizas, G.C. and Shatla, M.N. (1982) Induction of new isolates of *Trichoderma harzianum* tolerant to fungicides and their experimental use for control of white rot of onion. *Phytopathology*, **72**, 396–400.

Adams, P.B. and Ayers, W.A. (1982) Biological control of *Sclerotinia* lettuce drop in the field by *Sporidesmium sclerotivorum*. *Phytopathology*, **72**, 485–8.

Adams, P.B. Marois, J.J. and Ayers, W.A. (1984) Population dynamics of the mycoparasite *Sporidesmium sclerotivorum*, and its host, *Sclerotinia minor*, in soil. *Soil Biology and Biochemistry*, **16**, 627–33.

Ahl, P., Voisard, C. and Défago, G. (1986) Iron-bound siderophores, cyanide, and antibiotics involved in suppression of *Thielaviopsis basicola* by a *Pseudomonas fluorescens* strain. *Journal of Phytopathology*, **116**, 121–34.

Ahmad, J.S. and Baker, R. (1987) Rhizosphere competence of *Trichoderma harzianum*. *Phytopathology*, **77**, 182–9.

Ahmed, A.H.M. and Tribe, H.T. (1977) Biological control of white rot of onion (*Sclerotium cepivorum*) by *Coniothyrium minitans*. *Plant Pathology*, **26**, 75–8.

Anas, O., Alli, I. and Reeleder, R.D. (1989) Inhibition of germination of *Sclerotinia sclerotiorum* by salivary gland secretions of *Bradysia coprophila*. *Soil Biology and Biochemistry*, **21**, 47–57.

Anderson, T.R. and Patrick, Z.A. (1978) Mycophagous amoeboid organisms from soil that perforate spores of *Thielaviopsis basicola* and *Cochliobolus sativus*. *Phytopathology*, **68**, 1618–26.

Ayers, W.A. and Adams, P.B. (1981) Mycoparasitism and its application to biological control of plant diseases, in *Biological Control in Crop Protection*, (ed G.C. Papavizas), Allanheld and Osmun, Totowa, NJ, pp.91–103.

Backman, P.A. and Rodriguez-Kabana, R. (1975) A system for the growth and delivery of biological control agents to the soil. *Phytopathology*, **65**, 819–21.

Baker, R. and Dunn, P.R. (eds) (1990) *New Directions in Biological Control*, Alan R. Liss, Inc., New York.

Baker, C.J., Stavely, J.R., Thomas, C.A. *et al.* (1983) Inhibitory effect of *Bacillus subtilis* on *Uromyces phaseoli* and on development of rust pustules on bean leaves. *Phytopathology*, **73**, 1148–52.

Baker, C.J., Stavely, J.R. and Mock, N. (1985) Biocontrol of bean rust by *Bacillus subtilis* under field conditions. *Plant Disease*, **69**, 770–2.

Barnes, G.L., Russell, C.C., Foster, W.D. *et al.* (1981) *Aphelenchus avenae*, a potential biological control agent for root rot fungi. *Plant Disease*, **65**, 423–4.

Beagle-Ristaino, J.E. and Papavizas, G.C. (1985) Biological control of Rhizoctonia stem canker and black scurf of potato. *Phytopathology*, **75**, 560–4.

Becker, J.O. and Cook, R.J. (1988) Role of siderophores in suppression of *Pythium* species and production of increased-growth response of wheat by fluorescent pseudomonads. *Phytopathology*, **78**, 778–82.

Beemster, A.B.R., Bollen, G.J., Gerlagh, M. *et al.* (eds) (1991) *Biotic Interactions and Soil-Borne Diseases*, Elsevier, Amsterdam.

Blakeman, J.P. and Brodie, I.D.S. (1977) Competition for nutrients between epiphytic micro-organisms and germination of spores of plant pathogens on beetroot leaves. *Physiological Plant Pathology*, **10**, 29–42.

Bollen, G.J., Middelkoop, J. and Hofman, T.W. (1991) Effects of soil fauna on infection of potato sprouts by *Rhizoctonia solani*, in *Biotic Interactions and Soil-Borne Diseases*, (eds A.B.R. Beemster, G.J. Bollen, M. Gerlagh, M.A. Ruissen, B. Schippers and A. Tempel), Elsevier, Amsterdam, pp.27–34.

Bolton, A.T. (1978) Effects of amending soilless growing medium with soil containing antagonistic organisms on root rot and blackleg of geranium (*Pelargonium hortorum*) caused by *Pythium splendens*. *Canadian Journal of Plant Science*, **58**, 379–83.

Bolton, A.T. (1980) Control of *Pythium aphanidermatum* in poinsettia in a soilless culture by *Trichoderma viride* and a *Streptomyces* sp. *Canadian Journal of Plant Pathology*, **2**, 93–5.

Broadbent, P., Baker, K.F. and Waterworth, Y. (1971) Bacteria and actinomycetes antagonistic to fungal root pathogens in Australian soils. *Australian Journal of Biological Sciences*, **24**, 925–44.

Buck, K.W. (1988) Control of plant pathogens with viruses and related agents. *Philosophical Transactions of the Royal Society, Series B*, **318**, 295–317.

Budge, S.P. and Whipps, J.M. (1991) Glasshouse trials of *Coniothyrium minitans* and *Trichoderma* species for the biological control of *Sclerotinia sclerotiorum* in celery and lettuce. *Plant Pathology*, **40**, 59–66.

Burge, M.N. (ed) (1988) *Fungi in Biological Control Systems*, Manchester University Press, Manchester.

Callan, N.W., Mathre, D.E. and Miller, J.B. (1990) Bio-priming seed treatment for biological control of *Pythium ultimum* pre-emergence damping-off in Sh2 sweet corn. *Plant Disease*, **74**, 368–72.

Carter, M.V. (1983) Biological control of *Eutypa armeniacae*. 5. Guidelines for establishing routine wound protection in commercial apricot orchards. *Australian Journal of Experimental Agriculture and Animal Husbandry*, **23**, 429–36.

Carter, M.V. and Perrin, E. (1985) A pneumatic-powered spraying secateur for use in commercial orchards and vineyards. *Australian Journal of Experimental Agriculture and Animal Husbandry*, **25**, 939–42.

Carter, M.V. and Price, T.V. (1974) Biological control of *Eutypa armeniacae*. III. A comparison of chemical, biological and integrated control. *Australian Journal of Agricultural Research*, **25**, 105–19.

Cates, D. (1990) Biological fungicide closer to market. *Agricultural Consultant*, August, p. 11.

Chakraborty, S. and Warcup, J.H. (1985) Reduction of take-all by mycophagous amoebas in pot bioassays, in *Ecology and Management of Soilborne Plant Pathogens*, (eds C.A. Parker, A.D. Rovira, K.J. Moore, P.T.W. Wong and J.F. Kollmorgen), American Phytopathological Society, St Paul, MN, pp.107–9.

Chet, I., Harman, G.E. and Baker, R. (1981) *Trichoderma hamatum*: Its hyphal interactions with *Rhizoctonia solani* and *Pythium* spp. *Microbial Ecology*, **7**, 29–38.

Chet, I., Ordentlich, A., Shapira, R. *et al.* (1991) Mechanisms of biocontrol of soil-borne plant pathogens by rhizobacteria, in *The Rhizosphere and Plant Growth*, (eds D.L. Keister and P.B. Cregan), Kluwer Academic Publishers, Netherlands, pp.229–36.

Coley-Smith, J.R., Ridout, C.J., Mitchell, C.M. *et al.* (1991) Control of bottom rot disease of lettuce (*Rhizoctonia solani*) using preparations of *Trichoderma viride, T. harzianum* or tolclofos-methyl. *Plant Pathology*, **40**, 359–66.

Cook, R.J. and Baker, K.F. (1983) *The Nature and Practice of Biological Control of Plant Pathogens*, The American Phytopathological Society, St Paul, MN, USA.

Cook, R.J. and Weller, D.M. (1987) Management of take-all in consecutive crops of wheat or barley, in *Innovative Approaches to Plant Disease Control*, (ed I. Chet), John Wiley and Sons, New York, pp.41–66.

Corke, A.T.K. and Hunter, T. (1979) Biocontrol of *Nectria galligena* infection of pruning wounds on apple shoots. *Journal of Horticultural Science*, **54**, 47–55.

Cullen, D., Berbee, F.M. and Andrews, J.H. (1984) *Chaetomium globosum* antagonizes the apple scab pathogen, *Venturia inaequalis*, under field conditions. *Canadian Journal of Botany*, **62**, 1814–18.

Curl, E.A. (1979) Effects of mycophagous Collembola on *Rhizoctonia solani* and cotton seedling disease, in *Soil-Borne Plant Pathogens*, (eds B. Schippers and W. Gams), Academic Press, New York, pp.253–69.

Curl, E.A., Gudauskas, R.T., Harper, J.D. *et al.* (1985) Effects of soil insects on populations and germination of fungal propagules, in *Ecology and Management of Soilborne Plant Pathogens*, (eds C.A. Parker, A.D. Rovira, K.J. Moore, P.T.W. Wong and J.F. Kollmorgen), American Phytopathological Society, St Paul, Minn, pp.20–33.

Curl, E.A. and Harper, J.D. (1990) Fauna-microflora interactions, in *The Rhizosphere*, (ed J.M. Lynch), John Wiley and Sons, Chichester, pp.369–88.

De Cal, A. and Sagasta, E.M.- and Melgarejo, P. (1990) Biological control of peach twig blight (*Monilinia laxa*) with *Penicillium frequentans*. *Plant Pathology*, **39**, 612–18.

Défago, G. and Haas, D. (1990) Pseudomonads as antagonists of soilborne plant pathogens: mode of action and genetic analysis. *Soil Biochemistry*, **6**, 249–91.

Descalzo, R.C., Rahe, J.E. and Mauza, B. (1990) Comparative efficacy of induced resistance for selected diseases of greenhouse cucumber. *Canadian Journal of Plant Pathology*, **12**, 16–24.

Dubos, B., Jailloux, F. and Bulit, J. (1982) L'antagonisme microbien dans la lutte contre la pourriture grise de la vigne. *Bulletin OEPP*, **12**, 171–5.

Elad, Y. and Chet, I. (1987) Possible role of competition for nutrients in biocontrol of Pythium damping-off by bacteria. *Phytopathology*, **77**, 190–5.

El-Titi, A. and Ulber, B. (1991) Significance of biotic interactions between soil fauna and microflora in integrated arable farming, in *Biotic Interactions and Soil-Borne Diseases*, (eds A.B.R. Beemster, G.J. Bollen, M. Gerlagh, M.A. Ruissen, B. Schippers and A. Tempel), Elsevier, Amsterdam, pp.1–19.

Filonow, A.B. and Lockwood, J.L. (1985) Evaluation of several actinomycetes and the fungus *Hyphochytrium catenoides* as biocontrol agents for Phytophthora root rot of soybean. *Plant Disease*, **69**, 1033–6.

Fokkema, N.J., den Houter, J.G., Kosterman, Y.J.C. *et al.* (1979) Manipulation of yeasts on field-grown wheat leaves and their antagonistic effect on *Cochliobolus sativus* and *Septoria nodorum*. *Transactions of the British Mycological Society*, **72**, 19–29.

Fravel, D.R. and Keinath, A.P. (1991) Biocontrol of soilborne plant pathogens with fungi, in *The Rhizosphere and Plant Growth*, (eds D.L. Keister and P.B. Cregan), Kluwer Academic Publishers, Netherlands, pp.237–43.

Gardiner, R.B., Jarvis, W.R. and Shipp, J.L. (1990) Ingestion of *Pythium* spp.by larvae of the fungus gnat *Bradysia impatiens* (Diptera: Sciaridae). *Annals of Applied Biology*, **116**, 205–12.

Goldman, G.H., Geremia, R., Van Montagu, M. *et al.* (1991) Molecular genetics of the biocontrol agents *Trichoderma* spp., in *Biotic Interactions and Soil-Borne*

Diseases, (eds A.B.R. Beemster, G.J. Bollen, M. Gerlagh, M.A. Ruissen, B. Schippers and A. Tempel), Elsevier, Amsterdam, pp.175–80.

Grente, J. (1965) Les formes hypovirulents d'*Endothia parasitica* et les espoirs de lutte contre le chancre du chataigner. *Comptes Rendus Hebdomadaires des Séances de l'Academie d'Agriculture de France*, **51**, 1033–7.

Grente, J. and Sauret, S. (1969) L'hypovirulence exclusive phenomene original en pathologie vegetale. *Comptes Rendus Hebdomadaires des Séances de l'Academie des Sciences Paris*, **D268**, 2347–50.

Grosclaude, C. (1974) Le pénétration des spores de champignons dans les blessures de taille des arbes fruitiers. Application au cas de la protection biologique vis-à-vis du *Stereum purpureum*. *Revue de Zoologie Agricole et de Pathologie Vegetale*, **73**, 1–21.

Gutterson, N.I., Howie, W. and Suslow, T. (1990) Enhancing efficacies of biocontrol agents by use of biotechnology, in *New Directions in Biological Control*, (eds R. Baker and P. Dunn), A.R. Liss Inc., New York, pp.749–65.

Hadar, Y., Harman, G.E., Taylor, A.G. *et al.* (1983) Effects of pregermination of pea and cucumber seeds and of seed treatment with *Enterobacter cloacae* on rots caused by *Pythium* spp. *Phytopathology*, **73**, 1322–5.

Hadar, Y., Harman, G.E. and Taylor, A.G. (1984) Evaluation of *Trichoderma koningii* and *T. harzianum* from New York soils for biological control of seed rot caused by *Pythium* spp. *Phytopathology*, **74**, 106–10.

Hagedorn, C., Gould, W.D. and Bardinelli, T.R. (1989) Rhizobacteria of cotton and their repression of seedling disease pathogens. *Applied and Environmental Microbiology*, **55**, 2793–7.

Hall, R.A. (1982) Control of whitefly *Trialeurodes vaporariorum* and cotton aphid, *Aphis gossypii*, in glasshouses by two isolates of the fungus, *Verticillium lecanii*. *Annals of Applied Biology*, **101**, 1–11.

Hanlon, R.D.G. (1981) Influence of grazing by Collembola on the activity of senescent fungal colonies grown on media of different nutrient concentration. *Oikos*, **36**, 362–7.

Harman, G.E. (1991) Seed treatments for biological control of plant disease. *Crop Protection*, **10**, 166–71.

Harman, G.E., Chet I. and Baker, R. (1980) *Trichoderma hamatum* effects on seed and seedling disease induced in radish and pea by *Pythium* spp.or *Rhizoctonia solani*. *Phytopathology*, **70**, 1167–72.

Harman, G.E., Eckenrode, C.J. and Webb, D.R. (1978) Alteration of spermosphere ecosystems affecting oviposition by the bean seed fly and attack by soilborne fungi on germinating seeds. *Annals of Applied Biology*, **90**, 1–6.

Harman, G.E. and Taylor, A.G. (1988) Improved seedling performance by integration of biological control agents at favourable pH levels with solid matrix priming. *Phytopathology*, **78**, 520–5.

Harman, G.E., Taylor, A.G. and Stasz, T.E. (1989) Combining effective strains of *Trichoderma harzianum* and solid matrix priming to improve biological seed treatment. *Plant Disease*, **73**, 631–7.

Hashiba, T. (1987) An improved system for biological control of damping-off by using plasmids in fungi, in *Innovative Approaches to Plant Disease Control*, (ed I. Chet), Wiley Interscience, New York, pp.337–51.

Haygood, R.A. and Mazur, A.R. (1990) Evaluation of *Gliocladium virens* as a biocontrol agent of dollar spot bermudagrass. *Phytopathology*, **80**, 435 (Abstr.).

Herr, L.J. (1988) Biocontrol of Rhizoctonia crown rot and root rot of sugar beet by binucleate *Rhizoctonia* spp.and *Laetisaria arvalis*. *Annals of Applied Biology*, **113**, 107–18.

Hijwegen, T. (1986) Biological control of cucumber powdery mildew by *Tilletiopsis minor*. *Netherlands Journal of Plant Pathology*, **92**, 93–5.

Hornby, D. (ed) (1990) *Biological Control of Soil-borne Plant Pathogens*. CAB International, Wallingford.

Howell, C.R. (1982) Effect of *Gliocladium virens* on *Pythium ultimum, Rhizoctonia solani*, and damping-off of cotton seedlings. *Phytopathology*, **72**, 496–8.

Howell, C.R. (1991) Biological control of Pythium damping-off with seed-coating preparations of *Gliocladium virens*. *Phytopathology*, **81**, 738–41.

Howell, C.R., Beier, R.C. and Stipanovic, R.D. (1988) Production of ammonia by *Enterobacter cloacae* and its possible role in the biological control of Pythium pre-emergence damping-off by the bacterium. *Phytopathology*, **78**, 1075–8.

Howell, C.R. and Stipanovic, R.D. (1980) Suppression of *Pythium ultimum* induced damping-off of cotton seedlings by *Pseudomonas fluorescens* and its antibiotic pyoluteorin. *Phytopathology*, **70**, 712–15.

Howie, W. and Suslow, T. (1986) Effect of antifungal compound biosynthesis on cotton root colonization and *Pythium* suppression by a strain of *Pseudomonas fluorescens* and its antifungal minus isogenic mutant. *Phytopathology*, **76**, 1069 (Abstr.).

Huang, H.C. (1980) Control of *Sclerotinia* wilt of sunflower by hyperparasites. *Canadian Journal of Plant Pathology*, **7**, 26–32.

Hubbard, J.P., Harman, G.E. and Hadar, Y. (1983) Effect of soilborne *Pseudomonas* spp. on the biological control agent, *Trichoderma hamatum*, on pea seeds. *Phytopathology*, **73**, 655–9.

Hussain, S., Ghaffar, A. and Aslam, M. (1990) Biological control of *Macrophomina phaseolina* charcoal rot of sunflower and mung bean. *Journal of Phytopathology*, **130**, 157–60.

Inglis, G.D. and Boland, G.J. (1990) The microflora of bean and rapeseed petals and the influence of the microflora of bean petals on white mold. *Canadian Journal of Plant Pathology*, **12**, 129–34.

Islam, K.Z. and Nandi, B. (1985) Control of brown spot of rice by *Bacillus megaterium. Zeitschrift für pflanzenkrankheiten und pflanzenshutz*, **92**, 241–6.

Jackson, A.M., Whipps, J.M. and Lynch, J.M. (1991a) Nutritional studies of four fungi with disease biocontrol potential. *Enzyme and Microbial Technology*, **13**, 456–61.

Jackson, A.M., Whipps, J.M. and Lynch, J.M. (1991b) Effects of temperature, pH and water potential on growth of four fungi with disease biocontrol potential. *World Journal of Microbiology and Biotechnology*, **7**, 494–501.

Jackson, A.M., Whipps, J.M. and Lynch, J.M. (1991c) Production, delivery systems, and survival in soil of four fungi with disease biocontrol potential. *Enzyme and Microbial Technology*, **13**, 636–42.

Jager, G. and Velvis, H. (1985) Biological control of *Rhizoctonia solani* on potatoes by antagonists. 4. Inoculation of seed tubers with *Verticillium biguttatum* and other antagonists in field experiments. *Netherlands Journal of Plant Pathology*, **91**, 49–63.

Janisiewicz, W.J. and Roitman, J. (1988) Biological control of blue and gray mold on apple and pear with *Pseudomonas cepacia. Phytopathology*, **78**, 1697–1700.

Jarvis, W.R., Shaw, L.A. and Traquair, J.A. (1989) Factors affecting antagonism of cucumber powdery mildew by *Stephanoascus flocculosus* and *S. rugulosus. Mycological Research*, **92**, 162–5.

Jones, D., Gordon, A.H. and Bacon, J.S.D. (1974) Co-operative action by endo- and exo-β-1 → 3-glucanases from parasitic fungi in the degradation of cell-wall glucans of *Sclerotinia sclerotiorum. Biochemical Journal*, **140**, 47–55.

Kaiser, P.A. and Lussenhop, P.J. (1991) Collembolan effects on establishment of vesicular-arbuscular mycorrhizae in soybean (*Glycine max*). *Soil Biology and Biochemistry*, **23**, 307–8.

Kaiser, W.J., Hannan, R.M. and Weller, D.M. (1989) Biological control of seed rot and preemergence damping-off of chickpea with fluorescent pseudomonads. *Soil Biology and Biochemistry*, **21**, 269–73.

Kanangaratnam, P., Hall, R.A. and Burges, H.D. (1982) Control of glasshouse whitefly, *Trialeurodes vaporariorum* by an 'aphid' strain of the fungus *Verticillium lecanii*. *Annals of Applied Biology*, **100**, 213–19.

Keel, C., Voisard, C., Berling, H. *et al.* (1989) Iron sufficiency, a prerequisite for suppression of tobacco black root rot by *Pseudomonas fluorescens* strain CHAO under gnotobiotic conditions. *Phytopathology*, **79**, 584–9.

Kessler, K.J. (1990) Destruction of *Gnomia leptostyla* perithecia on *Juglans nigra* leaves by microarthropods associated with *Elacagnus umbellata* litter. *Mycologia*, **82**, 387–90.

Koltin, Y., Finkler, A. and Ben-Zvi, B.-S. (1987) Double-stranded RNA viruses of pathogenic fungi: virulence and plant protection, in *Fungal Infection of Plants*, (eds G.F. Pegg and P.G. Ayres), Cambridge University Press, Cambridge, pp.334–48.

Lahdenperä, M.-L., Simon, E. and Uoti, J. (1991) Mycostop – a novel biofungicide based on *Streptomyces* bacteria, in *Biotic Interactions and Soil-Borne Diseases*, (eds A.B.R. Beemster, G.J. Bollen, M. Gerlagh, M.A. Ruissen, B. Schippers and A. Tempel), Elsevier, Amsterdam, pp.258–63.

Lartey, R.T., Curl, E.A. and Peterson, C.M. (1986) Compared biological control of *Rhizoctonia solani* by fungal agents and mycophagous Collembola. *Phytopathology*, **76**, 1104.

Lartey, R.T., Curl, E.A., Peterson, C.M. *et al.* (1989) Mycophagous grazing and food preference of *Proisotoma minuta* (Collembola: Isotomidae) and *Onychiurus encarpatus* (Collembola: Onychiuridae). *Environmental Entomology*, **18**, 334–7.

Lartey, R.T., Curl, E.A., Peterson, C.M. *et al.* (1991) Control of *Rhizoctonia solani* and cotton seedling disease by *Laetisaria arvalis* and a mycophagous insect *Proisotoma minuta* (Collembola). *Journal of Phytopathology*, **133**, 89–98.

Lewis, J.A. and Papavizas, G.C. (1987a) Reduction of inoculum of *Rhizoctonia solani* in soil by germlings of *Trichoderma hamatum*. *Soil Biology and Biochemistry*, **19**, 195–201.

Lewis, J.A. and Papavizas, G.C. (1987b) Application of *Trichoderma* and *Gliocladium* in alginate pellets for control of Rhizoctonia damping-off. *Plant Pathology*, **36**, 438–46.

Lewis, J.A. and Papavizas, G.C. (1991a) Biocontrol of plant diseases: the approach for tomorrow. *Crop Protection*, **10**, 95–105.

Lewis, J.A. and Papavizas, G.C. (1991b) Biocontrol of cotton damping-off caused by *Rhizoctonia solani* in the field with formulations of *Trichoderma* spp.and *Gliocladium virens*. *Crop Protection*, **10**, 396–402.

Lewis, J.A., Papavizas, G.C. and Lumsden, R.D. (1991) A new formulation system for the application of biocontrol fungi to soil. *Biocontrol Science and Technology*, **1**, 59–69.

Lewis, K., Whipps, J.M. and Cooke, R.C. (1989) Mechanisms of biological disease control with special reference to the case study of *Pythium oligandrum* as an antagonist, in *Biotechnology of Fungi for Improving Plant Growth*, (eds J.M. Whipps and R.D. Lumsden), Cambridge University Press, Cambridge, pp.191–217.

Lifschitz, R., Windham, M.T. and Baker, R. (1986) Mechanism of biological control of pre-emergence damping-off of pea by seed treatment with *Trichoderma* spp.*Phytopathology*, **76**, 720–5.

Loper, J.E. (1988) Role of fluorescent siderophore production in biological control of *Pythium ultimum* by a *Pseudomonas fluorescens* strain. *Phytopathology*, **78**, 166–72.

Lumsden, R.D. and Lewis, J.A. (1989) Biological control of soilborne plant pathogens, problems and progress, in *Biotechnology of Fungi for Improving Plant Growth*, (eds J.M. Whipps and R.D. Lumsden), Cambridge University Press, Cambridge, pp.171–90.

Lumsden, R.D. and Locke, J.C. (1989) Biological control of damping-off caused by *Pythium ultimum* and *Rhizoctonia solani* with *Gliocladium virens* in soilless mix. *Phytopathology*, **79**, 361–6.

Lumsden, R.D., Locke, J.C., Lewis, J.A. *et al.* (1990) Evaluation of *Gliocladium virens* for biocontrol of Pythium and Rhizoctonia damping-off of bedding plants. *Biological and Cultural Tests*, **5**, 90.

Lutchmeah, R.S. and Cooke, R.C. (1985) Pelleting of seed with the antagonist *Pythium oligandrum* for biological control of damping-off. *Plant Pathology*, **34**, 528–31.

Lynch, J.M. (1990) Microbial metabolites, in *The Rhizosphere*, (ed J.M. Lynch), John Wiley and Sons, Chichester, pp.177–206.

Lynch, J.M. (1992) Environmental implications for the release of biocontrol agents, in *Biological Control of Plant Diseases: Progress and Challenges for the Future*, (eds E.C. Tjamos, G.C. Papavizas and R.J. Cook), Plenum, New York, pp.389–97.

Maplestone, P.A., Whipps, J.M. and Lynch, J.M. (1991) Effect of peat-bran inoculum of *Trichoderma* species on biological control of *Rhizoctonia solani*. *Plant and Soil*, **136**, 257–63.

Marois, J.J., Johnston, S.A., Dunn, M.T. *et al.* (1982) Biological control of Verticillium wilt of eggplant in the field. *Plant Disease*, **66**, 1166–8.

Martin, S.B., Abawi, G.S. and Hoch, H.C. (1986) The relation of population densities of the antagonist, *Laetisaria arvalis*, to seedling diseases of table beet incited by *Pythium ultimum*. *Canadian Journal of Microbiology*, **32**, 156–9.

Martin, F.N. and Hancock, J.G. (1987) The use of *Pythium oligandrum* for biological control of preemergence damping-off caused by *P. ultimum*. *Phytopathology*, **77**, 1013–20.

Matta, A. and Garibaldi, A. (1977) Control of *Verticillium* wilt of tomato by preinoculation with avirulent fungi. *Netherlands Journal of Plant Pathology*, **83**, (Suppl. 1), 457–62.

McIlveen, W.D. and Cole, H. (1976) Spore dispersal of Endogonaceae by worms, ants, wasps and birds. *Canadian Journal of Botany*, **54**, 1486–9.

McQuilken, M.P., Whipps, J.M. and Cooke, R.C. (1990a) Oospores of the biocontrol agent *Pythium oligandrum* bulk-produced in liquid culture. *Mycological Research*, **94**, 613–16.

McQuilken, M.P., Whipps, J.M. and Cooke, R.C. (1990b) Control of damping-off in cress and sugar beet by commercial seed coating with *Pythium oligandrum*. *Plant Pathology*, **39**, 452–62.

McQuilken, M.P., Whipps, J.M. and Cooke, R.C. (1992a) Nutritional and environmental factors affecting biomass and oospore production of the biocontrol agent *Pythium oligandrum*. *Enzyme and Microbial Technology*, **14**, 106–111.

McQuilken, M.P., Whipps, J.M. and Cooke, R.C. (1992b) Use of oospore formulations of *Pythium oligandrum* for biological control of *Pythium* damping-off in cress. *Journal of Phytopathology*, **135**, 125–34.

Mihuta-Grimm, L. and Rowe, R.C. (1986) *Trichoderma* spp. as biocontrol agents of *Rhizoctonia* damping-off of radish in organic soil and comparison of four delivery systems. *Phytopathology*, **76**, 306–12.

Mitchell, R. and Hurwitz, E. (1965) Suppression of *Pythium debaryanum* by lytic rhizosphere bacteria. *Phytopathology*, **55**, 156–8.

Moody, A.R. and Gindrat, D. (1977) Biological control of cucumber black root rot by *Gliocladium roseum*. *Phytopathology*, **67**, 1159–62.

Morrall, R.A.A. and Dueck, J. (1982) Epidemiology of Sclerotinia stem rot of rapeseed in Saskatchewan. *Canadian Journal of Plant Pathology*, **4**, 161–8.

Nelson, E.B. (1988) Biological control of *Pythium* seed rot and pre-emergence damping-off of cotton with *Enterobacter cloacae* and *Erwinia herbicola* applied as seed treatments. *Plant Disease*, **72**, 140–2.

Nelson, E.B., Chao, W.-L., Norton, J.M. *et al.* (1986) Attachment of *Enterobacter cloacae* to hyphae of *Pythium ultimum*: possible role in biological control of *Pythium* pre-emergence damping-off. *Phytopathology*, **76**, 327–35.

Nelson, E.B. and Craft, C.M. (1991) Introduction and establishment of strains of *Enterobacter cloacae* in golf course turf for the biological control of dollar spot. *Plant Disease*, **75**, 510–14.

Nelson, E.B., Harman, G.E. and Nash, G.T. (1988) Enhancement of *Trichoderma*-induced biological control of *Pythium* seed rot and pre-emergence damping-off of peas. *Soil Biology and Biochemistry*, **20**, 145–50.

Newhouse, J.R., MacDonald, W.L. and Hoch, H.C. (1990) Virus-like particles in hyphae and conidia of European hypovirulent (dsRNA-containing) strains of *Cryphonectria parasitica*. *Canadian Journal of Botany*, **68**, 90–101.

Ogawa, K. and Komada, H. (1985) Biological control of *Fusarium* wilt of sweet potato with cross-protection by non-pathogenic *Fusarium oxysporum*, in *Ecology and Management of Soilborne Plant Pathogens*, (eds C.A. Parker, A.D. Rovira, K.J. Moore, P.T.W. Wong and J.F. Kollmorgen), American Phytopathological Society, St Paul, – Minn, pp.121–3.

Osburn, R.M. and Schroth, M.N. (1989) Effect of osmopriming sugar beet seed on germination rate and incidence of *Pythium ultimum* damping-off. *Plant Disease*, **73**, 21–4.

Osburn, R.M., Schroth, M.N., Hancock, J.G. *et al.* (1989) Dynamics of sugar beet colonization by *Pythium ultimum* and *Pseudomonas* species: Effects on seed rot and damping-off. *Phytopathology*, **79**, 709–16.

Papavizas, G.C. and Lewis, J.A. (1983) Physiological and biocontrol characteristics of stable mutants of *Trichoderma viride* resistant to MBC fungicides. *Phytopathology*, **73**, 407–11.

Papavizas, G.C. and Lewis, J.A. (1989) Effect of *Gliocladium* and *Trichoderma* on damping-off and blight of snapbean caused by *Sclerotium rolfsii*. *Plant Pathology*, **38**, 227–86.

Parke, J.L., Rand, R.E., Joy, A.E. *et al.* (1991) Biological control of Pythium damping-off and Aphanomyces root rot of peas by application of *Pseudomonas cepacia* or *P. fluorescens* to seed. *Plant Disease*, **75**, 987–92.

Paulitz, T.C., Ahmad, J.S. and Baker, R. (1990) Integration of *Pythium nunn* and *Trichoderma harzianum* isolate T-95 for the biological control of Pythium damping-off of cucumber. *Plant and Soil*, **121**, 243–50.

Paulitz, T.C. and Loper, J.E. (1991) Lack of a role for fluorescent siderophore production in the biological control of Pythium damping-off of cucumber by a strain of *Pseudomonas putida*. *Phytopathology*, **81**, 930–5.

Pe'er, S. and Chet, I. (1990) *Trichoderma* protoplast fusion: a tool for improving biocontrol agents. *Canadian Journal of Microbiology*, **36**, 6–9.

Purkayastha, R.P. and Bhattacharyya, B. (1982) Antagonism of microorganisms from jute phylloplane towards *Colletotrichum corchori*. *Transactions of the British Mycological Society*, **78**, 509–13.

Pusey, P.L. (1989) Use of *Bacillus subtilis* and related organisms as biofungicides. *Pesticide Science*, **27**, 133–40.

Pusey, P.L., Wilson, C.L., Hotchkiss, M.W. *et al.* (1986) Compatibility of *Bacillus subtilis* for postharvest control of peach brown rot with commercial fruit waxes, dicloran and cold-storage conditions. *Plant Disease*, **70**, 587–90.

Rabatin, S.C. and Stinner, B.R. (1988) Indirect effects of interactions between VAM fungi and soil-inhabiting invertebrates on plant processes. *Agriculture, Ecosystems and Environment*, **24**, 135–46.

Redmond, J.C., Marois, J.J. and MacDonald, J.D. (1987) Biological control of *Botrytis cinerea* on roses with epiphytic microorganisms. *Plant Disease*, **71**, 799 –802.

Reyes, A.A. and Dirks, V.A. (1985) Suppression of *Fusarium* and *Pythium* pea rot by antagonistic microorganisms. *Phytoprotection*, **66**, 23–9.

Rhoades, H.L. and Linford, M.B. (1959) Control of *Pythium* root rot by the nematode *Aphelenchus avenae*. *Plant Disease Reporter*, **43**, 323–8.

Rishbeth, J. (1963) Stump protection against *Fomes annosus*. III. Inoculation with *Peniophora gigantea*. *Annals of Applied Biology*, **52**, 63–77.

Rossner, J. and Nagel, S. (1984) Untersuchungen zur Okologie und vermehrung des mycohagen nematoden *Aphelenchoides hamatus*. *Nematologica*, **30**, 90–8.

Shapira, R., Ordentlich, A., Chet, I. *et al.* (1989) Control of plant diseases by chitinase expressed from cloned DNA in *Escherichia coli*. *Phytopathology*, **79**, 1246–9.

Sikora, R.A., Bedenstein, F. and Nicolay, R. (1990a) Einfluß der Behandlung von Rübensaatgut mit Rhizosphärebakterien auf den Befall durch Pilze der Gattung *Pythium*. I. Antagonische Wirkung verschiedener Bakterienisolate gegenüber *Pythium* spp.*Journal of Phytopathology*, **129**, 111–20.

Sikora, R.A., Bedenstein, F. and Nicolay, R. (1990b) Einfluß der Behandlung von Rübensaatgut mit Rhizosphärebakterien auf den Befall durch Pilze der Gattung *Pythium*. II. Untersuchungen zum Wirkungsmechanismus. *Journal of Phytopathology*, **129**, 121–32.

Sivan, A., Elad, Y. and Chet, I. (1984) Biological control effects of a new isolate of *Trichoderma harzianum* on *Pythium aphanidermatum*. *Phytopathology*, **74**, 498– 501.

Sivan, A. and Harman, G.E. (1991) Improved rhizosphere competence in a protoplast fusion progeny of *Trichoderma harzianum*. *Journal of General Micro-biology*, **137**, 23–9.

Spencer, D.M. and Atkey, P.T. (1981) Parasitic effects of *Verticillium lecanii* on two rust fungi. *Transactions of the British Mycological Society*, **77**, 535–42.

Spurr, H.W., Jr. (1977) Protective applications of conidia of non pathogenic *Alternaria* sp. isolates for control of tobacco brown spot disease. *Phytopathology*, **67**, 128–32.

Spurr, H.W. (1981) Experiments on foliar disease control using bacterial antagonists, in *Microbial Ecology of the Phylloplane*, (ed. J.P. Blakeman), Academic Press, London, pp.369–81.

Srivastava, A.K., Défago, G. and Kern, H. (1985) Hyperparasitism of *Puccinia horiana* and other microcylic rusts. *Phytopathologische Zeitschrift*, **114**, 73–8.

Sundheim, L. and Amundsen, T. (1982) Fungicide tolerance in the hyperparasite *Ampelomyces quisqualis* and integrated control of cucumber powdery mildew. *Acta Agriculturae Scandinavica*, **32**, 349–55.

Sundheim, L., Poplawsky, A.R. and Ellingboe, A.H. (1988) Molecular cloning of two chitinase genes from *Serratia marcescens* and their expression in *Pseudomonas* species. *Physiological and Molecular Plant Pathology*, **33**, 483–91.

Sztejnberg, A., Galper, S., Mazar, S. *et al.* (1989) *Ampelomyces quisqualis* for biological and integrated control of powdery mildews in Israel. *Journal of Phytopathology*, **124**, 285–95.

Tahvonen, R. (1982) Preliminary experiments into the use of *Streptomyces* spp.

isolated from peat in the biological control of soil and seed-borne diseases in peat culture. *Journal of the Scientific Agricultural Society of Finland*, **54**, 357–69.

Taylor, A.G. and Harman, G.E. (1990) Concepts and technologies of selected seed treatments. *Annual Review of Phytopathology*, **28**, 321–39.

Taylor, A.G., Min, T.-G., Harman, G.E. *et al.* (1991) Liquid coating formulation for the application of biological seed treatments of *Trichoderma harzianum*. *Biological Control*, **1**, 16–22.

Thomashow, L.S. and Weller, D.M. (1988) Role of a phenazine antibiotic from *Pseudomonas fluorescens* in biological control of *Gaeumannomyces graminis* var. *tritici*. *Journal of Bacteriology*, **170**, 3499–508.

Thomashow, L.S., Weller, D.M., Bonsall, R.F. *et al.* (1990) Production of the antibiotic phenazine-1-carboxylic acid by fluorescent *Pseudomonas* species in the rhizosphere of wheat. *Applied and Environmental Microbiology*, **56**, 908–12.

Thompson, R.J. and Burns, R.G. (1989) Control of *Pythium ultimum* with antagonistic fungal metabolites incorporated into sugar beet seed pellets. *Soil Biology and Biochemistry*, **21**, 745–8.

Trapero-Casas, A., Kaiser, W.J. and Ingram, D.M. (1990) Control of Pythium seed rot and pre-emergence damping-off of chickpea in the US Pacific Northwest and Spain. *Plant Disease*, **74**, 563–9.

Tronsmo, A. (1991) Biological and integrated controls of *Botrytis cinerea* on apple with *Trichoderma harzianum*. *Biological Control*, **1**, 59–62.

Tronsmo, A. and Dennis, C. (1977) The use of *Trichoderma* species to control strawberry fruit rots. *Netherlands Journal of Plant Pathology*, **83**, (Supp.1), 449–55.

Trutmann, P. and Keane, P.J. (1990) *Trichoderma koningii* as a biological control agent for *Sclerotinia sclerotiorum* in Southern Australia. *Soil Biology and Biochemistry*, **22**, 43–50.

Trutmann, P., Keane, P.J. and Merriman, P.R. (1980) Reduction of sclerotial inoculum of *Sclerotinia sclerotiorum* with *Coniothyrium minitans*. *Soil Biology and Biochemistry*, **12**, 461–5.

Trutmann, P., Keane, P.J. and Merriman, P.R. (1982) Biological control of *Sclerotinia sclerotiorum* on aerial parts of plants by the hyperparasite *Coniothyrium minitans*. *Transactions of the British Mycological Society*, **78**, 521–9.

Turhan, G. and Turhan, K. (1989) Suppression of damping-off on pepper caused by *Pythium ultimum* Trow and *Rhizoctonia solani* Kühn by some new antagonists in comparison with *Trichoderma harzianum* Rifai. *Journal of Phytopathology*, **126**, 175–82.

Turner, G.J. and Tribe, H.T. (1976) On *Coniothyrium minitans* and its parasitism of *Sclerotinia* species. *Transactions of the British Mycological Society*, **66**, 97–104.

Uma, N.V. and Taylor, G.S. (1987) Parasitism of leek rust urediniospores by four fungi. *Transactions of the British Mycological Society*, **88**, 335–40.

Utkhede, R.S. and Rahe, J.E. (1980) Biological control of onion white rot. *Soil Biology and Biochemistry*, **12**, 101–4.

Utkhede, R.S. and Rahe, J.E. (1983) Interactions of antagonist and pathogen in biological control of onion white rot. *Phytopathology*, **73**, 890–3.

Utkhede, R.S. and Scholberg, P.L. (1986) *In vitro* inhibition of plant pathogens by *Bacillus subtilis* and *Enterobacter aerogenes* and *in vivo* control of two postharvest cherry diseases. *Canadian Journal of Microbiology*, **32**, 963–7.

van Alfen, N.K. and Hansen, D.R. (1984) Hypovirulence, in *Plant Microbe Interactions. Molecular and Genetic Perspectives*, Volume 1, (eds T. Kosuge and E.W. Nester), MacMillan Publishing Co., New York, pp.400–19.

Van Peer, R. and Schippers, B. (1991) Biocontrol of Fusarium wilt by *Pseudomonas* sp. strain WCS417r: induced resistance and phytoalexin accumulation, in *Biotic Interactions and Soil-Borne Diseases*, (eds A.B.R. Beemster, G.J. Bollen, M. Gerlagh, M.A. Ruissen, B. Schippers and A. Tempel), Elsevier, Amsterdam, pp.274–80.

Visser, S. (1985) Role of soil invertebrates in determining the composition of soil microbial communities, in *Ecological Interactions in Soil: Plants, Microbes and Animals*, (eds A.H. Fitter, D. Atkinson, D.J. Read and M.B. Usher), Blackwell Scientific Publications, Oxford, pp.297–317.

Voisard, C., Keel, C., Haas, D. *et al.* (1989) Cyanide production by *Pseudomonas fluorescens* helps suppress black root rot of tobacco under gnotobiotic conditions. *EMBO Journal*, **8**, 351–8.

Warnock, A.J., Fitter, A.H. and Usher, M.B. (1982) The influence of a springtail *Folsomia candida* (Insecta, Collembola) on the mycorrhizal association of leek *Allium porrum* and the vesicular-arbuscular mycorrhizal endophyte *Glomus fasciculatum*. *New Phytologist*, **90**, 285–92.

Weller, D.M. and Cook, R.J. (1983) Suppression of take-all of wheat by seed treatments with fluorescent pseudomonads. *Phytopathology*, **73**, 463–9.

Weller, D.M. and Cook, R.J. (1986) Increased growth of wheat by seed treatments with fluorescent pseudomonads, and implications of *Pythium* control. *Canadian Journal of Plant Pathology*, **8**, 328–34.

Whipps, J.M. (1991) Effects of mycoparasites on sclerotia-forming fungi, in *Biotic Interactions and Soil-Borne Diseases*, (eds A.B.R. Beemster, G.J. Bollen, M. Gerlagh, M.A. Ruissen, B. Schippers and A. Tempel), Elsevier, Amsterdam, pp.129–40.

Whipps, J.M., Lewis, K. and Cooke, R.C. (1988) Mycoparasitism and plant disease control, in *Fungi in Biological Control Systems*, (ed. M.N. Burge), Manchester University Press, pp.161–87.

Whipps, J.M. and Lumsden, R.D. (eds) (1989). *Biotechnology of Fungi for Improving Plant Growth*. Cambridge University Press, Cambridge.

Whipps, J.M. and Lumsden, R.D. (1991) Biological control of *Pythium* species. *Biocontrol Science and Technology*, **1**, 75–90.

Wiggins, E.A. and Curl, E.A. (1979) Interactions of Collembola and microflora of cotton rhizosphere. *Phytopathology*, **69**, 244–9.

Wilson, C.L., Franklin, J.D. and Pusey, P.L. (1987) Biological control of Rhizopus rot of peach with *Enterobacter cloacae*. *Phytopathology*, **77**, 303–5.

Wilson, C.L., Wisniewski, M.E., Biles, C.L. *et al.* (1991) Biological control of post-harvest diseases of fruits and vegetables: alternatives to synthetic fungicides. *Crop Protection*, **10**, 172–7.

Wisniewski, M., Wilson, C. and Hershberger, W. (1989) Characterization of inhibition of *Rhizopus stolonifera* germination and growth by *Enterobacter cloacae*. *Canadian Journal of Botany*, **67**, 2317–23.

Wood, R.K.S. and Way, M.J. (eds) (1988) *Biological Control of Pests, Pathogens and Weeds: Developments and Prospects*. The Royal Society, London.

Young, C.S. and Andrews, J.H. (1990) Inhibition of pseudothecial development of *Venturia inaequalis* by the basidiomycete *Athelia bombacina* in apple leaf litter. *Phytopathology*, **80**, 536–42.

Zhang, B.X., Ge, Q.X., Chen, D.H. *et al.*, (1990) Biological and chemical control of rot diseases on vegetable seedlings in Zhejiang province, China, in *Biological Control of Soil-borne Plant Pathogens*, (ed D. Hornby), CAB International, Wallingford, pp. 181–96.

4

The use of microbial agents for the biological control of plant parasitic nematodes

Brian Kerry

4.1 INTRODUCTION

Most populations of nematodes are probably under 'natural' control and, in several soils, microbial agents have been shown to provide long-term control of pest species (Kerry, 1987; Stirling, 1991). However, nematologists are still some way from manipulating these organisms with treatments that are practical for a grower. Despite much research on the biological control of many soil-borne bacterial and fungal diseases, only two agents have proved successful in practice and none has been developed that will protect the root system throughout a growing season (Deacon, 1991). However, more agents with different modes of action are currently being studied by more research workers than ever before and significant progress is being made. Research on the biological control of nematodes has moved from an empirical observational discipline to one in which quantitative studies on the interrelationships between pest, agent and plant are undertaken and detailed studies on the mode of action of various agents are made. As a consequence, there are more grounds for optimism than in the past (Sayre and Walter, 1991).

4.2 HISTORICAL BACKGROUND

Most research on microbial agents that attack nematodes in soil has concerned fungi, especially those that form traps to ensnare their prey.

Exploitation of Microorganisms Edited by D.G. Jones
Published in 1993 by Chapman & Hall, London ISBN 0 412 45740 7

These fungi form characteristic traps on their mycelium; there are four principal types of trap – namely, adhesive knobs, adhesive two- or three-dimensional networks, non-constricting rings and constricting rings (Barron, 1977). The first nematode-trapping fungus, the ubiquitous *Arthrobotrys oligospora*, was described by Fresenius in 1852 but it was Zopf (1888) who observed that the fungus could capture motile nematodes and parasitize them.

The potential of nematode-trapping fungi as biological control agents (BCAs) was first examined in Hawaii for control of nematodes on pineapples (Linford *et al.*, 1938). Research developed over the next 20 years in the UK and the USSR (Tribe, 1980). Usually, the application of fungi to soil for nematode control was accompanied by the addition of large quantities of organic matter; in general, there is only circumstantial evidence that any reductions in nematode populations resulted from the activities of the nematode trapping fungi and, in most cases, the levels of control were unpredictable (Stirling, 1991). After the Second World War, a number of cheap chemicals became available which, when applied to soil, gave high levels of nematode control in a wide range of conditions. As Stirling (1991) pointed out, such results contrasted sharply with the experiences achieved with biological control agents and resulted in a declining interest in the subject. The soil fumigants D-D (1,3-dichloropropane and 1,2-dichloropropene), EDB (1,2-dibromoethane) and DBCP (1,2-dibromo-3-chloropropane) were discovered in 1943, 1945 and 1954 respectively and were used extensively (Thomason, 1987). In the 1950s and 1960s, a new range of organophosphorus and carbamate nematicides was developed which was non-phytotoxic and could be applied as granules at the time of planting. Such developments led to an intensification of crop production with a shortening of rotations on nematode infested land. In the mid 1970s it became apparent that the nematicides or their degradation products were longer lasting in soil than anticipated and some were found as contaminants of drinking water. The first nematicide to be withdrawn from the market was DBCP, which was observed to cause a reduction in sperm counts of workers in manufacturing plants. Other chemicals have since been withdrawn and there is now much public concern over health and environmental problems associated with the use of pesticides in general, especially those applied to soil. As a consequence, there is an urgent need to develop alternative strategies for nematode control.

Applied plant nematology grew rapidly as the use of nematicides continued to highlight the benefits which could be achieved through control of these pests. However, research on biological control was limited to a few individual scientists (<10) worldwide and, even in the 1980s, more than 60% of nematological research papers still concerned work on chemical control (Kerry, 1981). Several developments in biological control in the mid-1970s coincided with the increased concern over the use of nematicides. Soils that suppressed nematode multiplication were identified for the first time in the early 1970s (Kerry, 1987; Stirling, 1991). At the same time, in France, two commercial agents, Royal 300 and Royal 350, were sold for control of *Ditylenchus myceliophagus* on mushrooms and *Meloidogyne* spp. on vegetables. Both products contained nematode-trapping fungi (Cayrol *et al.*, 1978; Cayrol and Frankowski, 1979) which were applied fresh at large rates, but have not been widely used. More recently, the facultative parasite, *Paecilomyces lilacinus* has been produced in the Philippines as 'Biocon' which can be applied for the control of several nematode species including root-knot nematodes. This fungus has been tested widely throughout the tropics and subtropics. In general, few tests have adequately assessed the biological control potential of the fungus and too often control has been variable (Kerry, 1990). However, some success has been achieved and it remains to be seen how widely this product will be used.

In line with the theme of this volume, this chapter concentrates on the exploitation of microbial agents that may be applied to soil for the control of plant-parasitic nematodes. It does not include a discussion of nematode-suppressive soils which, although providing effective nematode control, have proved difficult to manipulate and has been discussed elsewhere (Kerry, 1990; Sayre and Walter, 1991; Stirling, 1991). The history of research on the biological control of nematodes has demonstrated that the successful use of microbial agents will require a deep knowledge of the interrelationships between agent, pest and plant and the environmental factors which influence them. Nematologists continue to debate whether manipulation of natural enemies in soil is likely to be more successful than the addition of selected agents which are used to inundate pests. At this stage in our knowledge, both approaches appear valid and much can be learnt from either. In this chapter the major considerations in the development of microbial agents for the biological control of plant-parasitic nematodes are addressed.

4.3 POTENTIAL BIOLOGICAL CONTROL AGENTS

Four types of organism have been considered as potential biological control agents for plant parasitic nematodes (Table 4.1). The life cycles and descriptions of most of these agents have appeared in numerous reviews (Sayre, 1971; Mankau, 1980; Tribe, 1980; Kerry, 1987; Stirling, 1991; Sikora, 1992) and will not be discussed. Comments are therefore

Table 4.1. Different groups of organisms examined for their potential as biological control agents for nematodes

Group	Mode of infection	Reference
Obligate parasites		
Pasteuria penetrans	adhesive spores	Sayre and Starr, 1985
Nematophthora gynophila	zoospores	Kerry and Crump, 1980
Hirsutella rhossiliensis	adhesive spores	Jaffee and Zehr, 1982
Facultative parasites		
Nematode-trapping fungi	adhesive/non-adhesive traps	Barron, 1977
Verticillium chlamydosporium	hyphal penetration	de Leij and Kerry, 1991
Paecilomyces lilacinus	hyphal penetration	Gomez Garneiro and Cayrol, 1991
Rhizosphere bacteria	non-parasitic: modification of root exudates or toxin production	Oostendorp and Sikora, 1989; Becker *et al.*, 1988
Competitors		
Mycorrhizae,	competition for root space, modification of root exudates	Hussey and Roncadori, 1982; Stiles and Glawe, 1989
Endophytic fungi	colonization of nematode feeding cells	

confined to a comparison of the likely effectiveness and application of the different types.

Obligate parasites such as the bacterium, *Pasteuria penetrans*, and the fungus, *Nematophthora gynophila*, survive in soil as resistant spores. As these agents do not grow outside their hosts, they must be applied in large numbers and thoroughly incorporated in soil if a large proportion of the nematodes are to be infected. Proliferation and spread of such agents may be too slow to be practical, particularly if they are dependent on the presence of large nematode infestations. Hence, obligate parasites may only be effective as inundative treatments, which may restrict their use to protected or horticultural crops of high value and in which relatively small volumes of soil require treatment. Both *P. penetrans* and *N. gynophila* require complex media, and only limited *in vitro* growth has been achieved (Graff and Madelin, 1989; Williams *et al.*, 1989; Bishop and Ellar, 1991). These organisms also have limited host ranges, which would further restrict their use, but both produce resistant spores to enable them to survive in the absence of hosts and this makes them easy to handle and store.

Facultative parasites, particularly those that colonize the rhizosphere, can be added to soil and increased considerably in the presence or absence of nematodes. Thus a ten-fold reduction in the application rate of chlamydospores of *Verticillium chlamydosporium* had little effect on the amount of fungus developing in the rhizosphere of tomato plants after one and two generations of root-knot nematodes, although substantial differences in soil colonization remained throughout the experiment (Figure 4.1). Usually, facultative parasites are easily cultured *in vitro*,

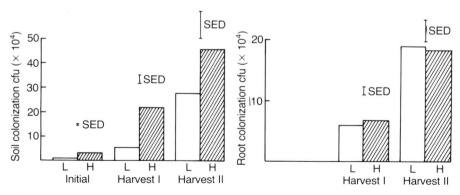

Figure 4.1 Colonization of the soil and rhizosphere of tomato plants by *Verticillium chlamydosporium* added to the soil at rates of 1000 (L) and 10 000 (H) chlamydospores per gram soil for control of root-knot nematodes. (From de Leij *et al.*, 1992a, with permission.)

have wide host ranges and, as a consequence, are more attractive for commercial development than are obligate parasites. As they are not dependent on nematodes for their nutrition, they may be added to soil at relatively small nematode densities to prevent damaging populations developing. Facultative parasites which produce resistant chlamydospores are more easily handled than those (such as some nematode-trapping fungi) that do not and have to be added to soil as fresh mycelium which would have to be transported under refrigeration in order to remain active.

Rhizosphere bacteria are not parasitic but appear to reduce nematode invasion of roots by modifying root exudates or by producing toxins that kill the nematodes before they invade or feed (Oostendorp and Sikora, 1990; Sikora, 1991; Racke and Sikora, 1992). These bacteria have been introduced to soil as seed treatments (Oostendorp and Sikora, 1989) which greatly decreases the amounts of inoculum required and might enable these bacteria to be used on field crops.

Some **fungi**, including vesicular-arbuscular mycorrhizae, may compete for space in plant roots and prevent the invasion and development of some endoparasitic nematodes (Hussey and Roncadori, 1982; Sikora, 1992). Fungi in the root cortex may colonize the nematode's feeding cells and prevent further feeding and development (Crump, 1989; Stiles and Glawe, 1989). Some fungi which colonize roots occur commonly in females and eggs of cyst nematodes, but do not appear significantly to affect plant growth.

Nematicidal metabolites or **enzymes** derived from soil organisms are being tested by a number of commercial crop protection organizations. At least one compound has been tested in the field (Devidas *et al.*, 1990). Such products can be screened and developed in much the same way as a nematicide and will not be discussed further.

4.4 ECOLOGICAL CONSIDERATIONS

The soil environment affects the activities of nematodes and their natural enemies and is likely to have a considerable impact on the efficacy of potential biological control agents in soil (Stirling, 1991). However, little research has been done on the influence of soil conditions on the activities of nematophagous fungi and bacteria. There is a dearth of quantitative methods for studying these organisms in soil although, in recent years, selective media have been developed for the estimation of population densities of some facultative parasites, such as *V. chlamydosporium* and *P. lilacinus* (Mitchell *et al.*, 1987; de Leij and Kerry, 1991). Deacon (1991) considered that an ecological approach is essential for the development of sustainable methods of biological control. In this section, the key factors that affect the complex interactions between the nematode, microbial control agents, and the plant are briefly discussed.

4.4.1 Soil environment

Four key factors have been identified that may influence the efficacy of microbial agents in soil (Table 4.2).

Soil temperature

Soil temperature has a direct effect on the growth of microorganisms and on the activity and rate of development of nematode pests. Too often experiments to test the efficacy of selected agents are performed in the glasshouse at temperatures of *ca.* 25°C when, in temperate climates, soil temperatures at depths of 10 cm rarely exceed 18°C and the agent is likely to be introduced and expected to survive in soil at temperatures considerably below this. The efficacy of the obligate parasite, *P. penetrans*, is greatly reduced at temperatures below 20°C because the root-knot nematode hosts develop more rapidly than the bacterium

Table 4.2. Key soil factors that affect the efficacy of microbial agents introduced into soil for control of plant-parasitic nematodes

Factor	Influence on efficacy of microbial agent
Temperature	Direct effect on the growth and sporulation of the agent and on the rate of development of the nematode target. Agents must be selected for rapid growth at the temperatures at which they are expected to be effective. Greatest control should be achieved at temperatures which are optimal for the agent but not the nematode.
Moisture	Rarely limiting for microbial growth. However, nematode activity and the spread of microorganisms, especially those that infect by zoospores, are directly affected.
Texture/structure	Influences nematode activity and growth and spread of microorganisms. The large bulk of soil makes broadcast application difficult.
Soil microflora	Competes with introduced agent for scarce energy sources. Fungistasis affects establishment and causes variation in efficacy of agent even when latter is introduced with a food source.

at such temperatures and considerable numbers of viable eggs are produced before the parasite kills the females (Stirling, 1981). Temperatures above 25°C reduced the proportion of eggs of root-knot nematodes parasitized by *V. chlamydosporium*; at these temperatures many of the eggs in the egg masses completed their embryonic development and hatched before they were colonized by the fungus (de Leij *et al.*, 1992b). Similarly, *Dactylella oviparasitica* colonized *Meloidogyne* egg masses more slowly at 15°C than at 27°C, but killed more nematode eggs at the lower temperature because nematode development was slowed to a greater extent (Stirling, 1979). Solarization increased the numbers of root-knot nematodes encumbered with spores of *P. penetrans*. Non-lethal temperatures at depth presumably increased the activity of the nematodes in soil and so increased the chance of contact with the infective spores of the bacterium (Walker and Wachtel, 1988).

Soil moisture

Temperature will only increase nematode activity in soil as long as moisture is not limiting. Nematodes are aquatic animals and in soil are confined to films of water on the surface of soil particles in the macropores (>60μm). Nematodes do not create their own living space in soil and are most active when the macropores drain and become air-filled, which occurs when soils are at, or close to, field capacity (Jones, 1975). Even at the permanent wilting point for plants, the relative humidity in the soil pores is approximately 98% and moisture is rarely limiting for the growth of microorganisms in soil except in the surface layers. The major effect of soil moisture levels is therefore likely to be an indirect one, i.e. the influence it has on nematode movement in soil and the chance of making contact with natural enemies that survive in soil as infective spores. However, water movement during drying and wetting cycles is important for the spread of spores and bacteria in soil. Also, fungi that produce zoospores that are most active in water-filled pores are likely to be directly affected by soil moisture levels. Hence, infection of females of the cereal cyst nematode, *Heterodera avenae*, by the fungus, *N. gynophila*, was dependent on rainfall in June, July and August when the females on the roots were exposed to parasitism (Kerry *et al.*, 1982).

Soil bulk

A major problem for the introduction of biological control agents into soil is its large bulk, approximately 2500 t ha^{-1} of mineral matter to plough depth (20 cm). To ensure adequate distribution of an inundative treatment throughout such a mass is likely to be impractical and uneconomic for field crops. Inundative broadcast treatments may be practical in container grown and horticultural crops or in subsistence agriculture in which small areas of land are to be treated. In field crops, the successful development of a BCA is likely to depend on the use of row treatments or the ability of the organism to grow and survive in the rhizosphere so that it can be applied with the seed or other planting material. The great buffering capacity provided by the bulk of soil protects organisms in all but the surface layers from the extremes associated with the aerial environment. In general, organisms are slow to build up in soil but once established are difficult to remove. Hence, the build up of natural enemies to levels that control nematode pests may take 4–5 years in a monoculture of an annual crop (Kerry, 1987) and considerably longer in a perennial one (Stirling and White, 1982) but, once established, these organisms provided long-term control. The effect of soil texture on nematophagous microorganisms has been little studied but it will affect the pore spaces available in soil and thereby moisture retention. Coarse sandy soils tend to have low organic matter contents and as a consequence microbial activity is also low. This is important because the indigenous soil microflora can have a significant impact on the establishment and survival of introduced BCAs (Kerry, 1989).

Soil microflora

The soil microflora will compete for limited energy sources in soil which may affect the establishment of BCAs that grow in soil (Papavizas and Lewis, 1981). Even those organisms that remain dormant until contacted by a nematode host may also be subject to mycostasis which is associated with the soil microflora and may result in the lysis of spores and hyphae or a failure of spores to germinate (Lockwood, 1977). To help overcome such difficulties, facultative parasites such as the nematode-trapping fungi and those that attack nematode eggs have been introduced into soil in a colonized energy source such as wheat bran, rice or wheat grains and crop wastes (Stirling, 1991). In general, the rates of application have been large and, too frequently, little attempt has been made to measure the direct effect of the organic matter on the nematode target. Amending soil with organic matter increases soil microbial activity, which leads to the production of nematicidal breakdown products. It has been assumed that facultative parasites of nematodes attack their hosts when saprophytic competition in soil is great and that this gives nematophagous organisms a competitive advantage. When aggressive soil saprophytes such as *Trichoderma* or *Paecilomyces* were added to soil containing nematode trapping fungi, trapping activity was greater in the presence of the introduced competitors than in their absence (Quinn, 1987). Factors which cause facultative parasites of nematodes to switch from feeding saprophytically to parasitically are not known, but they may be very important in determining the efficacy of an agent.

4.4.2 Nematode target

Nematode parasites may be ectoparasitic and spend all of their life in soil or endoparasitic, with some stages completing their development inside the plant. The latter group are divided into migratory and sedentary endoparasites. The major nematode pests affecting world agriculture are sedentary endoparasites and include cyst (*Globodera* and *Heterodera* spp.) and root-knot (*Meloidogyne* spp.) nematodes. Because of their economic importance, most research on the biological control of plant parasitic nematodes has concerned pests in these genera. The infective stage in the life cycle of these nematodes is the second stage juvenile, which emerges from eggs in soil and migrates towards the roots of a suitable host. Once inside the root, the juveniles become sedentary and complete their development to the adult. Inside the root, the nematodes are protected from natural enemies in the soil and rhizosphere but they are vulnerable to fungi that colonize the root cortex and the feeding cells on which the nematode depends for its nutrition. Adult females of both cyst and root-knot nematodes are saccate; cyst nematode females rupture the root cortex as they enlarge and are exposed on the root surface, whereas infection by root-knot species usually results in the formation of a gall in which the adult female remains embedded so that only the egg mass is exposed in the rhizosphere. These differences in development have important consequences for the efficacy of biological control agents. Female cyst nematodes may take several weeks to complete their development on the root surface, during which time they continue to feed and produce eggs. As a consequence, they are exposed to parasitism by rhizosphere fungi such as *V. chlamydosporium* which may have different effects on the nematode depending on the time at which the female becomes parasitized. The fungus may reduce the number of females that mature, decrease their size and fecundity, and parasitize eggs; most control is achieved when young females are parasitized before they have laid many eggs (Kerry, 1990). Eggs within egg masses of root-knot nematodes complete their development and second stage juveniles hatch within about 10 days in favourable conditions. Thus, *V. chlamydosporium* has much less opportunity to destroy eggs of these nematodes; as discussed, temperature affects the rate of embryonic development and the size of the egg mass and can have an important effect on the numbers of eggs killed.

In general, damage to crop yields caused by nematodes results from the infection of young plants and changes in the morphology of the root system. Nematode feeding and invasion by subsequent generations is considered to have a much smaller effect. Hence, nematode management strategies are aimed at reducing numbers in soil before, or at the time, the crop is planted.

Reducing the invasion of plants by second stage juveniles of cyst and root-knot nematodes has proved difficult. Most research has been on the use of nematode-trapping fungi, but the results have been inconsistent because it has proved difficult to make the period of trapping activity coincide with the relatively short period of nematode migration in soil. Applying the fungus four weeks before planting enabled *Arthrobotrys irregularis* to establish a trap network which significantly reduced root-knot nemtaode invasion of tomato plants (Cayrol and Frankowski, 1979) but control was only significant at relatively small nematode infestations and when rates of application of the fungus were large (1.4 t ha^{-1}). Although significant reductions in nematode invasion and improvements in plant growth have been achieved with applications of rhizosphere bacteria, such treatments have not controlled nematode multiplication. Only agents that infect the reproducing adult female and her eggs have resulted in the natural control of cyst and root-knot nematodes (Kerry, 1987). However, populations of ectoparasites, such as *Criconomella xenoplax*, may be significantly decreased by agents that attack juvenile and adult stages in soil (Jaffee and Zehr, 1982). Clearly, the level of control required to prevent nematode populations from increasing will depend on the number of generations in a growing season, the susceptibility of the crop and on the nematode's fecundity. These factors are greatly affected by plant cultivar.

4.4.3 Host plant

The significance of the host plant in the interaction between nematodes and their natural enemies has largely been ignored in studies on biological control. Kloepper *et al.* (1991) have observed that plants with properties antagonistic to plant parasitic nematodes have rhizosphere microfloras distinct from those of host plants; greater numbers of microorganisms in the rhizospheres of antagonistic plants were chitinolytic, but the authors did not demonstrate that these organisms were responsible for the antagonistic effect. The egg parasitic fungus, *D. oviparasitica*, was unable to control root-knot nematodes on vines in California on which the fecundity of the nematode was three times greater than on peach trees in adjacent orchards, where the nematode was effectively controlled by the fungus (Stirling *et al.*, 1979). Colonization of the rhizosphere is essential for the control of *Meloidogyne* spp. by *V. chlamydosporium* (de Leij and Kerry, 1991) and the extent of colonization depends both on the fungal isolate and on the plant host (Kerry and de Leij, 1992). Also, because this fungus does not colonize

Table 4.3. Gall size, final nematode populations, and the control of three species of root-knot nematodes in soil infested with 1000 second-stage juveniles per plant and treated with 2000 chlamydospores $\cdot g^{-1}$ soil of *V. chlamydosporium* and planted with tomato plants grown at 20°C and 25°C

	M. arenaria		M. incognita		M. javanica		SED
	20°C	25°C	20°C	25°C	20°C	25°C	
Weight of gall (mg) (mean of 15)	2.26	1.82	1.46	1.39	1.45	1.09	0.16
Number of galls/root system	678	7.95	340	455	669	781	142.7
Number of exposed egg masses/g root	209	327	195	481	311	457	59.0
*Final population × 10^3 (juveniles/plant)	186	231	116	322	129	299	22.4
†Control (%)	72	85	57	93	85	90	

*Estimated from plants grown in soil not treated with the fungus

†Calculated from the difference between post crop populations in soil treated with *V. chlamydosporium* and untreated soil (After de Leij *et al.*, 1992b).

the root cortex it will not kill eggs that remain embedded within large galls. The size of the galls produced is related to the numbers of nematodes present and the susceptibility of the plant. Most control will be achieved on plants that support extensive colonization by the fungus in the rhizosphere and that support only small nematode galls so that most egg masses will be exposed to infection on the root surface. These interactions are complex and may be influenced by environmental factors such as temperature (Table 4.3). Few other observations have been published comparing biological control on different crops.

4.5 SELECTION OF POTENTIAL AGENTS

The successful development of a BCA relies much on the careful selection of suitable microbial isolates. Selection should be made bearing in mind the nematode target for control and the cropping system to which the agent will be applied. If the agent is required to control active stages in soil then it is essential that it produces either traps or adhesive spores. Agents that kill sedentary stages must be able to survive in the rhizosphere, as the nematode targets are concentrated close to the root surface. There are several stages in the selection of potential biological control agents (Table 4.4). Initially, surveys are required to isolate potential agents and it is likely that targeted surveys in known nematode suppressive soils or in areas where the pest species is indigenous are more likely to yield useful isolates (Cook and Baker, 1983).

4.5.1 Methods for the isolation of potential agents

The methods of isolation used and the time of sampling also determine the types of organisms found (Crump, 1987; Gintis *et al.*, 1983). Methods for the isolation and the enumeration of nematophagous microbes have

Table 4.4. Stages in the selection procedure of potential biological control agents

Stage	Requirements
1. Identify candidate microorganisms	Targeted surveys
2. Establish *in vitro* or *in vivo* cultures	Basic laboratory studies
3. Initial selection of potential agents	Simple laboratory-based screen
4. Testing efficacy of selected isolates	Pot experiments in non-sterilized soil
5. Testing in field plots	Mass culturing and formulation methods

been reviewed recently (Bailey and Gray, 1989; Kerry and Crump, 1991; Stirling, 1991). Obligate parasites may be isolated from diseased nematodes extracted from soil (Jaffee *et al.*, 1988) and spores or other resting structures can be extracted directly from soil (Crump and Kerry, 1981; Oostendorp *et al.*, 1991a). It is important to use a physical method of extraction which does not rely on the activity of the nematode, as this may be much reduced if the nematode is parasitized.

Baiting techniques have been used to isolate this type of parasite, especially *P. penetrans*. Usually, large numbers of susceptible nematodes from a laboratory culture are added to soil and then extracted after a period of incubation and assessed for infection (Stirling and White, 1982). Such baiting techniques may lead to the isolation of a limited range of variation in the parasite present in soil; 'probing' the same soil with different nematode species may lead to the isolation of more useful isolates as several different populations of *P. penetrans* may occur in the same soil sample (Giblin-Davies *et al.*, 1990); also, attempts have been made to quantify the numbers of fungal spores in soil using baiting techniques (McInnis and Jaffee, 1989). Facultative parasites are more easily recovered from soil on sprinkle plates (Bailey and Gray, 1989) or, in some cases, selective media. Those parasites attacking eggs and females of cyst and root-knot nematodes can be readily isolated from diseased nematodes extracted by routine techniques (Kerry, 1988a).

A number of different observation chambers have been used to study the maturation of cyst nematode females on roots and their infection by fungi, from which isolations can be made (Crump, 1987). Three main principles should be followed when isolating parasites from these nematodes: (i) spread the contents of the female, cyst or egg mass over the surface of the agar; (ii) use a weak agar medium, preferably tapwater agar; (iii) isolate from young females and immature eggs. These principles should assist in the isolation of the slow growing parasitic species and reduce the problem of them being overgrown by fast growing saprophytes that have no, or very little, potential as biological control agents. Methods to quantify the effects of parasites of females and eggs in soil have been developed (Kerry *et al.*, 1982; Nicolay and Sikora, 1989).

Rhizobacteria have been isolated from the soil adhering to root pieces using well-established techniques (Racke and Sikora, 1992). Heating soil suspensions at 90°C for 10 min enabled the selective isolation of spore-forming bacteria. All methods of isolation are very selective, and this is reflected in the estimate that 90% of soil microorganisms have not been cultured *in vitro* (MacDonald, 1986). It is important to ensure that the organism is isolated from the stage of nematode, or in the environmental niche in which it will be expected to be effective as a biological control agent.

4.5.2 Laboratory culture and screening

Culture

Once an organism has been identified from a number of nematodes in which it is causing disease, it is necessary to obtain the pathogen or parasite in pure culture so that it can be increased and its efficacy tested. Most facultative and some obligate parasites such as *Hirsutella rhossiliensis* grow readily on standard laboratory media. *Pasteuria penetrans* and other obligate parasites are best maintained in their nematode hosts. Stirling and Wachtel (1980) developed a method for culturing *P. penetrans* in root-knot nematodes in which the roots containing infected females were harvested and dried. The roots were then milled to a powder which may contain upto 10^8 spores of the bacterium per gram of root; the resistant spores retained their viability for many months in this state and the method is widely used to produce inoculum for experimentation and introductions into soil.

Once established in culture, it is important to store the original isolate in a dried or frozen state (Smith and Onions, 1983) in case its characteristics change through repeated subculture. Thereafter, a range of simple laboratory experiments are necessary to determine the minimal media required for growth and sporulation and the conditions such as pH and temperature necessary to optimize these.

Screening

To assess the virulence of the large numbers of isolates that may be collected in a routine survey can be very time consuming, particularly as individual organisms may affect nematode populations in a number of ways (Kerry, 1990). The influence of parasites or pathogens on the multiplication of cyst and root-knot nematodes may take 12–16 weeks to assess in pot experiments and there is a need to precede such tests with a simple laboratory-based screen to select the most promising agents for further study. Considerable variation has been recognized between isolates of *P. penetrans* (Stirling, 1985), *P. lilacinus* (Villanueva and Davide, 1984), *Fusarium oxysporum* and *Acremonium strictum* (Nigh *et al.*, 1980) and *V. chlamydosporium* (Irving and Kerry, 1986), and this underlines the need for routine screening as part of the development of a biological control agent. This need is further highlighted by the observation that less than 12% of rhizosphere bacteria isolated in several studies have had any activity against plant parasitic nematodes (Racke and Sikora, 1992). However, few screening methods have been developed and the lack of information on the mode of action, epidemiology, and growth and survival of most potential agents means that the development of such screens is based largely on intuition rather than fact. Exposing eggs of cyst and root-knot nematodes to a range of fungi on water agar has detected a considerable range of virulence between isolates of the same species of fungus (Stirling and Mankau, 1978; Nigh *et al.*, 1980; Irving and Kerry, 1986). Such simple screens give no indication of those isolates that will perform well in soil, but those that infect few eggs when they are in intimate contact on agar are unlikely to be useful and can be rejected. Racke and Sikora (1992) screened more than 150 isolates of rhizosphere bacteria by applying them in suspension to potato tuber pieces which were then planted in non-sterile soil to which eggs of the potato cyst nematode were added. Effects on root growth and nematode invasion were examined after three weeks; such a screen did not separate toxic effects on the nematode from indirect effects on root exudation which may have affected the hatch of the nematode eggs or made the roots less attractive to the infective juveniles. In more detailed studies with fewer isolates, the rhizobacteria appeared not to produce toxins or make the root less attractive to *Heterodera schachtii*, but egg hatch and juvenile invasion of the roots were decreased by the bacteria (Oostendorp and Sikora, 1990). To demonstrate the toxic effects of these bacteria, the nematode *Panagrellus redivivus* was placed on agar plates or added to suspensions containing the bacteria and its activity and reproduction recorded. Although the spores from different isolates of *P. penetrans* have been mixed in suspension with a range of nematode species to observe differences in the numbers that attach, such tests do not detect differences in host range or virulence unless the spores are given a chance to germinate and the nematodes to be colonized.

Isolates may also be selected for other characteristics, such as spore production, growth at low temperatures, and rhizosphere colonization, which are considered essential for their role as BCAs. Isolates of *V. chlamydosporium* and rhizobacteria differ markedly in their ability to grow and survive in the rhizosphere (Oostendorp and Sikora, 1990; de Leij and Kerry, 1991; Kerry and de Leij, 1992). Rhizobacteria which are required to protect the roots of seedlings from nematode attack over a period of only a few weeks may be successfully applied as seed treatments (Oostendorp and Sikora, 1989). However, sufficient colonization is unlikely from a seed treatment with *V. chlamydosporium*, which must survive to infect females and eggs developing at the end of the nematode's life cycle.

4.5.3 Tests to determine biological control potential in soil

After the preliminary screen in the laboratory, the efficacy of selected isolates should be assessed in non-sterile soil in pot experiments to identify the key factors affecting their efficacy as biological control agents. This is a critical stage before field testing can start and much published research that has purported to demonstrate biological control has been inadequate; consequently there is a need for more exacting experimentation (Stirling, 1988; Kerry, 1990). Tests which are done in sterilized soil favour the establishment and survival of the introduced agent because competition from the indigenous microflora has been much reduced. To help overcome such competition in non-sterilized soils, fungal parasites of nematodes have often been added to soil in a colonized food base (Papavizas and Lewis, 1981), with applications ranging from 0.1–2.5% w/v soil which is equivalent to 2.5–62.5 t ha^{-1} applied as a broadcast treatment to the top 20 cm of soil. Clearly, such treatments have an important amendment affect and yet the effects of the food base alone or of the colonized and autoclaved food base have been rarely estimated (Kerry, 1990). Nematicidal breakdown products, such as ammonia or fatty acids, are produced by the soil microflora when organic amendments are added to soil and may result in a significant reduction in nematode populations. Hence, the wheat inoculum used to add *P. lilacinus* to soil significantly reduced *M. incognita* populations and the presence of the fungus had no additional effect (Cabanillas and Barker, 1989). As the commercial development of a BCA is likely to be dependent upon much smaller rates of application than are often used in experiments, the amendment effect of these treatments is likely to be small and it is essential during the screening process that the effect of the food base is separated from that of the fungus. Some parasites can be added to soil and will survive without an energy base; *V. chlamydosporium* requires an energy base if it is produced in liquid culture (Kerry, 1988b) but not if it is cultured on solid media (de Leij and Kerry, 1991). Application without an energy base makes screening simpler as fewer control treatments are required.

Stirling (1988) noted that in tests on the efficacy of BCAs, few attempts were made to monitor the rates of nematode infection and that the control observed was not supported by estimates of the parasitism caused by the agent. Also, the establishment and survival of the agent over the period for which it was intended to be active were rarely assessed. Such observations are difficult to make without suitable methods for re-isolation from soil and a detailed knowledge of the biology of the agent applied (Kerry, 1990). As a consequence, there is only circumstantial evidence of the biological control activity of many of the organisms tested and results from such experiments fail to attract the investment necessary for commercial development. Nematologists involved in research on biological control must improve their experimental methods and provide details of the epidemiology of the organisms under test. Some information is available for *P. penetrans* (Stirling,

1984), *H. rhossiliensis* (Jaffee *et al*, 1988), *V. chlamydosporium* (de Leij and Kerry, 1991) and *P. lilacinus* (Gomes Carneiro and Cayrol, 1991). The observations suggest that inoculum densities of approximately 10^4 propagules g^{-1} soil are required for *P. penetrans* and *V. chlamydosporium* to be effective but that *P. lilacinus* required 10^6 propagules g^{-1} soil and Gomes Carneiro and Cayrol suggest that, even at these high densities, *P. lilacinus* is unlikely to be an effective biological control agent. Until such information is available for a wider range of organisms, it is difficult to compare their potential for biological control. The densities of *V. chlamydosporium* required to control root-knot nematodes are similar to those estimated to be present in soils suppressive to cyst nematodes (Kerry, unpublished data).

Field trials

Studies on the dose rate, nematode density, effect of soil type and host plant can be made in pots and so provide information on the key factors involved in the interactions (de Leij, 1992). However, there is no substitute for testing the efficacy of selected organisms in the less well controlled situation in the field. Field trials require methods of mass production, formulation and application. Most tests have involved simple formulations on sand/wheat bran mixtures or on colonized cereal grains such as rice, millet or wheat on which the fungus is applied to soil in an active state. Some attempts have been made at producing dried formulations including alginate granules (Kerry, 1988b; Cabanillas *et al.*, 1989; Schuster and Sikora, 1992). Most agents have been produced on media that could not be used in large scale production because of their complexity or cost. Interest in the application of agents for the biological control of a wide range of soil-borne diseases is increasing but expertise in industry is essential before large scale field testing is possible. So far, tests with nematodes have been done in small plots in which the agent has been intimately mixed as a broadcast treatment. Such treatments will not be possible in large scale field trials and there is a need to study the spread of agents in soil in order to develop practical methods of application.

4.6 BIOLOGICAL CONTROL STRATEGIES

In general, biological control is slower acting, more specific, less predictable and often less effective than chemical control. Experience suggests that some of these criticisms apply to those agents that have been considered for the control of nematode populations. However, in combination with other control methods BCAs could have an important role in the management of nematode pests and contribute to predictable and effective management strategies which rely less on the use of nematicides. At present there is a lack of information and this section can only draw attention to the possibilities which warrant more research.

Those agents, such as trapping fungi, mycorrhizae, rhizobacteria and some endoparasites that attack active nematodes in soil or reduce their invasion of, or feeding on, plant roots may only need to protect the plant for a relatively short period; the length of protection required will depend upon the crop and the duration of the nematode life cycle. These agents could be applied at high densities on seed or planting material to provide reductions in damage, although they are unlikely to prevent nematode populations from increasing (Oostendorp and Sikora, 1989). The parasites that kill the adult females and eggs have an effect similar to a resistant cultivar, in that damage may not be prevented but the final nematode infestation is significantly reduced. Agents of this type should be applied at relatively small nematode population densities or in combination with a nematicide or a tolerant host so that the initial damage is prevented or minimized. Once it has been applied to the soil, crop rotation (Oostendorp *et al.*, 1991b), soil amendments (Rodriguez-Kabana *et al.*, 1987; Spiegel *et al.*, 1987; Schlang *et al.*, 1988), and other factors can be used to try to maximize the survival and activity of the agent. However, it seems likely that most agents will be used as microbial pesticides that require regular application to maintain effective densities in soil. Management strategies such as partial soil sterilization (B'Chir *et al.*, 1983), bare fallowing and solarization (Walker and Wachtel, 1988), that reduce nematode pest populations and competition from the residual soil microflora, should enhance the establishment and activity of applied agents and provide more effective control.

Nematicides usually do not prevent nematode populations from increasing if the pest species is able to complete more than one generation on the crop. Biological control agents such as *P. penetrans* (Brown and Nordmeyer, 1985) and *V. chlamydosporium* (Kerry and de Leij, 1992), applied with a nematicide can provide longer lasting control of such pests. Also, agents that kill females that develop on the roots of resistant plants could be used to slow the development of virulent populations and extend the useful life of the resistant cultivar. In general, the effective use of biological control agents would seem to depend on their introduction on specific crops and at known nematode densities and, as a consequence, their use will require the support of expert advice. It remains to be seen whether changes in the availability of nematicides and public pressure for the development of sustainable methods of crop protection make these more complex management strategies acceptable to growers. Widespread acceptibility will be achieved only if biological control agents can be demonstrated to provide cost effective and reliable protection from nematode pests (Powell and Faull, 1989).

4.7 RESEARCH NEEDS AND PROSPECTS

At present, a wide range of organisms has been identified that affect nematode damage to crops and population build up. It is debatable whether more effective agents would be found through intensive surveys. Therefore research effort should concentrate on identifying the advantages and limitations of the known organisms, and their potential as biological control agents critically evaluated. Research needs have been reviewed recently (Kerry, 1990; Stirling, 1991) and highlight a lack of information on the epidemiology, mode of action and general biology of most agents. The possibilities offered by molecular biology to track organisms in soil and to enhance their performance is dependent on a basic understanding of their biology and ecology. Natural enemies may also provide a useful source of genes that could be used to develop plants with novel forms of resistance, as has already been demonstrated for some insect pests (Feitelson *et al.*, 1992).

The demonstration that some important nematode pests can be controlled naturally in soils has done much to change attitudes towards the potential use of BCAs for nematode control. The challenge now is to see whether applications of selected agents can speed the development of, and provide similar levels to, natural control, so that they can be exploited by growers.

REFERENCES

Bailey, F. and Gray, N.F. (1989) The comparison of isolation techniques for nematophagous fungi from soil. *Annals of Applied Biology*, **114**, 125–32.

Barron, G.L. (1977) *The Nematode Destroying Fungi*. Canadian Biological Publications Ltd., Guelph.

B'Chir, M.M., Horrigue, N. and Verlodt, H. (1983) Mise au point d'une methode de lutte integrée associant un agent biologique et un substance chimique, pour combattre les *Meloidogyne* sous-abris plastiques en Tunisie. *Mededelingen van de Faculteit Landbouwwetenschappen Rijksuniversiteit, Gent*, **48**, 421–32.

Becker, J.O., Zavaleta-Mejia, E., Colbert, S.F. *et al.* (1988) Effect of rhizobacteria on root-knot nematodes and gall formation. *Phytopathology*, **78**, 1466–9.

Bishop, A.H. and Ellar, D.J. (1991) Attempts to culture *Pasteuria penetrans in vitro*. *Biocontrol Science and Technology*, **1**, 101–14.

Brown, S.M. and Nordmeyer, D. (1985) Synergistic reduction in root galling by *Meloidogyne javanica* with *Pasteuria penetrans* and nematicides. *Revue de Nématologie*, **8**, 285–6.

Cabanillas, E. and Barker, K.R. (1989) Impact of *Paecilomyces lilacinus* inoculum level and application time on control of *Meloidogyne incognita* on tomato. *Journal of Nematology*, **21**, 115–20.

Cabanillas, E., Barker, K.R. and Nelson, L.A. (1989) Survival of *Paecilomyces lilacinus* in selected carriers and related effects on *Meloidogyne incognita* on tomato. *Journal of Nematology*, **21**, 121–30.

Cayrol, J.C. and Frankowski, J.P. (1979) Une methode de lutte biologique contre les nematodes a galles des racines appartenant au genre *Meloidogyne*. *Pepinieristes, Horticulteurs, Maraichers, Revue Horticole*, **193**, 15–23.

Cayrol, J.C., Frankowski, J.P., Laniece, A. *et al.* (1978) Contre les nematodes en champigonniere. Mise au point d'une methode de lutte biologique a l'aide d'un Hyphomycete predateur: *Arthrobotrys robusta* souche 'antipolis' (Royal 300). *Pepinieristes, Horticulteurs, Maraichers, Revue Horticole*, **184**, 23–30.

Cook, R.J. and Baker, K.F. (1983) *The Nature and Practice of Biological Control of Plant Pathogens*. American Phytopathological Society, St. Paul, pp.539.

Crump, D.H. (1987) A method for assessing the natural control of cyst nematode populations. *Nematologica*, **33**, 232–43.

Crump, D.H. (1989) Interaction of cyst nematodes with their natural antagonists. *Aspects of Applied Biology*, **22**, 135–40.

Crump, D.H. and Kerry, B.R. (1981) A quantitative method for extracting resting spores of two nematode parasitic fungi, *Nematophthora gynophila* and *Verticillium chlamydosporium*, from soil. *Nematologica*, **27**, 330–9.

Deacon, J.W. (1991) Significance of ecology in the development of biocontrol agents against soil-borne plant pathogens. *Biocontrol Science and Technology*, **1**, 5–20.

de Leij, F.A.A.M. (1992) Significance of ecology in the development of *Verticillium chlamydosporium* as a biological control agent against root-knot nematodes (*Meloidogyne* spp.). *PhD Thesis, University of Wageningen*, pp.140.

de Leij, F.A.A.M., Davies, K.G. and Kerry, B.R. (1992a) The use of *Verticillium chlamydosporium* and *Pasteuria penetrans* alone and in combination to control *Meloidogyne incognita* on tomato plants. *Fundamental and Applied Nematology*, **15**, 235–42.

de Leij, F.A.A.M., Dennehy, J.A. and Kerry, B.R. (1992b) The effect of temperature and nematode species on interactions between the nematophagous fungus *Verticillium chlamydosporium* and root-knot nematodes (*Meloidogyne* spp.). *Nematologica*, **38**, 65–79.

de Leij, F.A.A.M. and Kerry, B.R. (1991) The nematophagous fungus, *Verticillium chlamydosporium*, as a potential biological control agent for *Meloidogyne arenaria*. *Revue de Nématologie*, **14**, 157–64.

Devidas, P., Rehberger, L.A. and Crovetti, A.J. (1990) AARC # 0255 – a nematicidal composition of microbial origin – discovery and evaluation. *Nematologica*, **36**, 344–5.

Feitelson, J.S., Payne, J. and Kim, L. (1992) *Bacillus thuringiensis*: insects and beyond. *Biotechnology*, **10**, 271–5.

Giblin-Davis, R.M., McDaniel, L.L. and Bilz, F.G. (1990) Isolates of the *Pasteuria penetrans* group from phytoparasitic nematodes in bermudagrass turf. *Journal of Nematology*, **22**, 750–62.

Gintis, B.O., Morgan-Jones, G. and Rodriguez-Kabana, R. (1983) Fungi associated with several developmental stages of *Heterodera glycines* from an Alabama soybean field soil. *Nematropica*, **13**, 181–200.

Gomes Carneiro, R.M.D. and Cayrol, J.C. (1991) Relationship between inoculum density of the nematophagous fungus *Paecilomyces lilacinus* and control of *Meloidogyne arenaria* on tomato. *Revue de Nématologie*, **14**, 629–34.

Graff, N.J. and Madelin, M.F. (1989) Axenic culture of the cyst-nematode parasitizing fungus, *Nematophthora gynophila*. *Journal of Invertebrate Pathology*, **53**, 301–6.

Hussey, R.S. and Roncadori, R.W. (1982) Vesicular-arbuscular mycorrhizae may limit nematode activity and improve plant growth. *Plant Disease*, **66**, 9–14.

Irving, F. and Kerry, B.R. (1986) Variation between strains of the nematophagous fungus, *Verticillium chlamydosporium* Goddard.II. Factors affecting parasitism of cyst nematode eggs. *Nematologica*, **32**, 474–85.

Jaffee, B.A. and Zehr, E.I. (1982) Parasitism of the nematode *Criconemella xenoplax* by the fungus *Hirsutella rhossiliensis*. *Phytopathology*, **72**, 1378–81.

Jaffee, B.A., Gaspard, J.T., Ferris, H. and Muldoon, A.E. (1988) Quantification of parasitism of the soil-borne nematode *Criconemella xenoplax* by the nematophagous fungus *Hirsutella rhossiliensis*. *Soil Biology and Biochemistry*, **20**, 631 –6.

Jones, F.G.W. (1975) The soil as an environment for plant parasitic nematodes. *Annals of Applied Biology*, **79**, 113–39.

Kerry, B.R. (1981) Progress in the use of biological agents for control of nematodes, in *Biological Control in Crop Production*, (ed G.C.Papavizas), Allanheld, Osmum, Totowa, pp.79–90.

Kerry, B.R. (1987) Biological control, in *Principles and Practice of Nematode Control in Crops*, (eds R.H.Brown and B.R.Kerry), Academic Press, New York, pp.233 –63.

Kerry, B.R. (1988a) Fungal parasites of cyst nematodes. *Agriculture, Ecosystems and Environment*, **24**, 293–305.

Kerry, B.R. (1988b) Two microorganisms for the biological control of plant parasitic nematodes. *Proceedings of the Brighton Crop Protection Conference – 1988*, **2**, 603–7.

Kerry, B.R. (1989) Fungi as biological control agents for plant parasitic nematodes, in *Biotechnology of Fungi for Improving Plant Growth*, (eds J.M. Whipps and R.D. Lumsden), Cambridge University Press, Cambridge, pp.153–70.

Kerry, B.R. (1990) An assessment of progress toward microbial control of plant parasitic nematodes. *Journal of Nematology*, **22**, 621–31.

Kerry, B.R. and Crump, D.H. (1980) Two fungi parasitic on females of cyst-nematodes (*Heterodera* spp.). *Transactions of the British Mycological Society*, **74**, 119–25.

Kerry, B.R. and Crump, D.H. (1991) (eds) *Methods for Studying nematophagous fungi*. IOBC/WPRS Bulletin XIV/2, pp.64.

Kerry, B.R., Crump, D.H. and Mullen, L.A. (1982) Studies of the cereal cyst nematode, *Heterodera avenae* under continuous cereals, 1975–1978. II. Fungal parasitism of nematode eggs and females. *Annals of Applied Biology*, **100**, 489 –99.

Kerry, B.R. and de Leij, F.A.A.M. (1992) Key factors in the development of fungal agents for the control of cyst and root-knot nematodes, in *Biological control of Plant Diseases*, (eds E.C. Tjamos, G.C. Papavizas and R.J. Cook), Plenum, New York, pp.139–44.

Kloepper, J.W., Rodriguez-Kabana, R., McInroy, J.A. and Collins, D.J. (1991) Analysis of populations and physiological characterisation of microorganisms in rhizospheres of plants with antagonistic properties to phytopathogenic nematodes. *Plant and Soil*, **136**, 95–102.

Linford, M.B., Yap, F. and Oliveira, J.M. (1938) Reduction of soil populations of the root-knot nematode during decomposition of organic matter. *Soil Science*, **45**, 127–41.

Lockwood, J.L. (1977) Fungistasis in soils. *Biological Reviews*, **52**, 1–43.

Macdonald, R.M. (1986) Extraction of microorganisms from soil. *Biological Agriculture and Horticulture*, **3**, 361–5.

McInnis, T.M. and Jaffee, B.A. (1989) An assay for *Hirsutella rhossiliensis* spores and the importance of phialides for nematode inoculation. *Journal of Nematology*, **21**, 229–34.

Mankau, R. (1980) Biological control of nematode pests by natural enemies. *Annual Review of Phytopathology*, **18**, 415–40.

Mitchell, D.J., Kannwischer-Mitchell, M.E. and Dickson, D.W. (1987) A semi-selective medium for the isolation of *Paecilomyces lilacinus* from soil. *Journal of Nematology*, **19**, 255–6.

Nicolay, R. and Sikora, R.A. (1989) Techniques to determine the activity of fungal egg parasites of *Heterodera schachtii* in field soil. *Revue de Nématologie*, **12**, 97–102.

Nigh, E.A., Thomason, I.J. and van Gundy, S.D. (1980) Identification and distribution of fungal parasites of *Heterodera schachtii* eggs in California. *Phytopathology*, **70**, 884–9.

Oostendorp, M., Hewlett, T.E., Dickson, D.W. and Mitchell, D.J. (1991a) Specific gravity of spores of *Pasteuria penetrans* and extraction of spore-filled nematodes from soil. *Supplement to Journal of Nematology*, **23**, 729–32.

Oostendorp, M. and Sikora, R.A. (1989) Seed treatment with antagonistic rhizobacteria for the suppression of *Heterodera schachtii* early root infection of sugar beet. *Revue de Nématologie*, **12**, 77–83.

Oostendorp, M. and Sikora, R.A. (1990) *In vitro* interrelationships between rhizosphere bacteria and *Heterodera schachtii*. *Revue de Nématologie*, **13**, 269–74.

Oostendorp, M., Dickson, D.W. and Mitchell, D.J. (1991b) Population development of *Pasteuria penetrans* on *Meloidogyne arenaria*. *Journal of Nematology*, **23**, 58–64.

Papavizas, G.C. and Lewis, J.A. (1981) Introduction and augmentation of microbial antagonists for the control of soilborne plant pathogens, in *Biological Control in Crop Production (B.A.R.C Symposium No. 5)*, (ed G.C. Papavizas), Allenheld and Osmum, Totowa, pp.305–22.

Powell, K.A. and Faull, J.L. (1989) Commercial approaches to the use of biological control agents, in *Biotechnology of Fungi for Improving Plant Growth*, (eds J.M. Whipps and R.D. Lumsden), Cambridge University Press, Cambridge, pp.259–75.

Quinn, M.A. (1987) The influence of saprophytic competition on nematode predation by nematode-trapping fungi. *Journal of Invertebrate Pathology*, **49**, 170–4.

Racke, J. and Sikora, R.A. (1992) Isolation, formulation and antagonistic activity of rhizobacteria toward the potato cyst nematode *Globodera pallida*. *Soil Biology and Biochemistry*, **24**, 521–6.

Rodriguez-Kabana, R., Morgan-Jones, G. and Chet, I. (1987) Biological control of nematodes: soil amendments and microbial antagonists. *Plant and Soil*, **100**, 237–47.

Sayre, R.M. (1971) Biotic influence in soil environment, in *Plant Parasitic Nematodes* (eds B.M. Zuckerman, W.F. Mai and R.A. Rhode), Academic Press, New York, pp.235–56.

Sayre, R.M. and Starr, M.P. (1985) *Pasteuria penetrans* (ex Thorne 1940) nom.rev., comb.n., sp.n., a mycelial and endospore-forming bacterium parasitic in plant parasitic nematodes. *Proceedings of the Helminthological Society of Washington*, **52**, 149–65.

Sayre, R.M. and Walter, D.E. (1991) Factors affecting the efficacy of natural enemies of nematodes. *Annual Review of Phytopathology*, **29**, 149–66.

Schlang, J., Steudel, W. and Muller, J. (1988) Influence of resistant green manure crops on the population dynamics of *Heterodera schachtii* and its fungal egg parasites. *Nematologica*, **34**, 193.

Schuster, R.-P. and Sikora, R.A. (1992) Persistence and growth of an egg pathogenic fungus applied in alginate granules to field soil and its pathogenicity toward *Globodera pallida*. *Fundamental and Applied Nematology*, **15**, 449–55.

Sikora, R.A. (1991) The concept of using plant health promoting rhizobacteria for the biological control of plant parasitic nematodes. *Proceedings of the 2nd International Workshop on Plant Growth-Promoting Rhizobacteria, Interlaken, Switzerland*. IOBC Bulletin XIV/8, 3–10.

Sikora, R.A. (1992) Management of the antagonistic potential in agricultural ecosystems for the biological control of plant parasitic nematodes. *Annual Review of Phytopathology*, **30**, 245–70.

Smith, D. and Onions, A.H.S. (1983) A comparison of some preservation techniques for fungi. *Transactions of the British Mycological Society*, **81**, 535–40.

Spiegel, Y., Chet, I. and Cohn, E. (1987) Use of chitin for controlling plant parasitic nematodes. II. Mode of action. *Plant and Soil*, **98**, 337–45.

Stiles, C.M. and Glawe, D.A. (1989) Colonization of soybean roots by fungi isolated from cysts of *Heterodera glycines*. *Mycologia*, **81**, 797–9.

Stirling, G.R. (1979) Effect of temperature on parasitism of *Meloidogyne incognita* eggs by *Dactylella oviparasitica*. *Nematologica*, **25**, 104–10.

Stirling, G.R. (1981) Effect of temperature on infection of *Meloidogyne javanica* by *Bacillus penetrans*. *Nematologica*, **27**, 458–62.

Stirling, G.R. (1984) Biological control of *Meloidogyne javanica* with *Bacillus penetrans*. *Phytopathology*, **74**, 55–60.

Stirling, G.R. (1985) Host specificity of *Pasteuria penetrans* within the genus *Meloidogyne*. *Nematologica*, **31**, 203–9.

Stirling, G.R. (1988) Biological control of plant parasitic nematodes, in *Diseases of Nematodes*, Vol. II, (eds G.O. Poinar and H.-B. Jansson), CRC Press Inc., Boca Raton, pp.93–139.

Stirling, G.R. (1991) *Biological Control of Plant Parasitic Nematodes*. CABI, Wallingford, UK, pp.282.

Stirling, G.R. and Mankau, R. (1978) Parasitism of *Meloidogyne* eggs by a new fungal parasite. *Journal of Nematology*, **10**, 236–40.

Stirling, G.R., McKenry, M.V. and Mankau, R. (1979) Biological control of root-knot nematodes (*Meloidogyne* spp.) on peach. *Phytopathology*, **69**, 806–9.

Stirling, G.R. and Wachtel, M.F. (1980) Mass production of *Bacillus penetrans* for the biological control of root-knot nematodes. *Nematologica*, **26**, 308–12.

Stirling, G.R. and White, A.M. (1982) Distribution of a parasite of root-knot nematodes in South Australian vineyards. *Plant Disease*, **66**, 52–3.

Thomason, I.J. (1987) Challenges facing nematology: environmental risks with nematicides and the need for new approaches, in *Vistas on Nematology*, (eds J.A. Veech and D.W. Dickson), Society of Nematologists, Hyattsville, pp.469 –76.

Tribe, H.T. (1980) Prospects for the biological control of plant parasitic nematodes. *Parasitology*, **81**, 619–39.

Villanueva, L.M. and Davide, R.G. (1984) Evaluation of several isolates of soil fungi for biological control of root-knot nematodes. *The Philippine Agriculturist*, **67**, 361–71.

Walker, G.E. and Wachtel, M.F. (1988) The influence of soil solarization and non-fumigant nematicides on infection of *Meloidogyne javanica* by *Pasteuria penetrans*. *Nematologica*, **34**, 477–83.

Williams, A.B., Stirling, G.R., Hayward, A.C. and Perry, J. (1989) Properties and attempted culture of *Pasteuria penetrans*, a bacterial parasite of root-knot nematode (*Meloidogyne javanica*). *Journal of Applied Bacteriology*, **67**, 145–56.

Zopf, W. (1888) Zur Kenntnis der Infektions-Krankheiten neidener Thiere und Pflanzen. *Nova Acta Acadamiae Caesareoe Leopoldino-Carolinoe Germanicae Naturoe Curiosorum*. Halis Saxonum No. 7, 313–76.

5

Insect viruses as biocontrol agents

Doreen Winstanley and Luciano Rovesti

5.1 INTRODUCTION

Virus diseases of insects and their role in the natural regulation of insect populations have been recognized for many years. When the virulence and insect-specific nature of some viruses was appreciated, research intensified on their potential as biological control agents and numerous field trials were carried out between 1950 and 1960 (Ignoffo, 1973). Viral insecticides, however, failed to become commercially successful at this time probably because of the simultaneous development of numerous synthetic pesticides with broad spectrum, low cost and high insecticidal activity. However, during the late 1970s and throughout the 1980s many unacceptable problems caused by overdependence on chemical insecticides came to light, such as insecticide resistance, environmental pollution by insecticidal residues, emergence of 'new' pest species (i.e. outbreaks of insects which were present but never caused damage in the past) and adverse effects on human health.

At present, over 500 insect pest species are resistant to one or more insecticides (Georghiou and Lagunes, 1988). Even some biological insecticides, namely *Bacillus thuringiensis* toxins, which until recently were thought to be incapable of inducing resistance because of their complex mode of action, are now part of the above mentioned list (McGaughey, 1985; Tabashnik *et al.*, 1990). Some arthropods (e.g. spider mites) reached their present status of pests only after the introduction of broad

Exploitation of Microorganisms Edited by D.G. Jones
Published in 1993 by Chapman & Hall, London ISBN 0 412 45740 7

spectrum insecticides. Pollution by chemicals is a more frequent event. The development of new, alternative, environmentally friendly control agents is urgently needed and some insect viruses, because of their inherent characteristics, appear to be among the promising candidates.

5.2 INSECT VIRUSES

At least 11 groups of viruses, including the Baculoviridae, Reoviridae, Poxviridae, and the Iridoviridae, are known to cause diseases in insects (Entwistle, 1985). Of these, only the Baculoviridae are confined to arthropods; replicating in Lepidoptera (butterflies and moths), Hymenoptera (wasps), Diptera (flies), Coleoptera (beetles), Neuroptera (lacewings), Arachnida (spiders) and Crustacea (prawns and shrimps) (Martignoni and Iwai, 1986). In the other groups of viruses, some members are also pathogenic to vertebrates and/or plants (Gröner, 1986). The safety of insect viruses has been extensively reviewed by Döller (1985) and Gröner (1986). Only one insect-pathogenic virus, Nodamura virus (family: Nodaviridae) is known to affect both a vertebrate and an insect and is transmissible to suckling mice by *Aedes aegypti* (Matthews, 1982). Most of the research and development of viral bioinsecticides has focused on members of the Baculoviridae, because of their specificity and the considerable amount of evidence to suggest that they do not adversely affect vertebrates (Gröner, 1986). This chapter will therefore concentrate on the application of baculoviruses for the control of insect pests.

5.3 BACULOVIRUSES

5.3.1 Characteristics

The family Baculoviridae consists of two sub-families, the *Eubaculovirinae* (the occluded viruses) in which the virions are embedded in a crystalline matrix of protein and the *Nudibaculovirinae* (the non-occluded viruses) (Francki *et al.*, 1991). The occluded viruses can be divided into two genera based on the morphology of their protein occlusion body (OB). In the genus Nuclear Polyhedrosis Virus (NPV), the occlusion bodies are polyhedra-shaped and range in size from 0.5 to 15 μm across (Bilimoria, 1991), and in the genus Granulosis Virus (GV) they are much smaller, ellipsoidal in shape and resemble granules ranging in size from 0.3 to 0.5 μm in length (Crook, 1991). The occlusion body protein, polyhedrin in the case of NPVs and granulin in GVs, has a M_r of approximately 29 000. The occlusion bodies of the NPVs contain many enveloped virions. In the sub-genus multiply-enveloped NPV (MNPV), the virions enclose several nucleocapsids per envelope. However, in the sub-genus singly-enveloped NPV (SNPV), the virions usually contain a single nucleocapsid per envelope. Generally, only one virion containing a single nucleocapsid is embedded in each occluded granulosis virus (GV). The occlusion body (OB) enables the virus to persist in the environment, between host larvae or between generations and may involve an interval of one or more years. In the sub-family *Nudibaculovirinae*, OBs are absent and the virus consists of a single enveloped nucleocapsid.

Baculoviruses have rod-shaped virions, approximately 250 × 50 nm in the case of the *Eubaculovirinae*, containing at least 25 different polypeptides and a double-stranded circular supercoiled DNA genome, packaged with an arginine-rich basic protein into a cylindrical capsid. This is surrounded by an envelope, with an intermediate layer between the capsid and the envelope. Genome sizes of approximately 100 to 180 kilobase pairs (kbp) have been reported for GVs (Crook, 1991) and 90 to 170 kbp for NPVs (Bilimoria, 1991) and probably encode up to 100 genes. The structure and physicochemical properties of baculoviruses have been extensively reviewed by several authors (Kelly, 1985; Federici, 1986; Bilimoria, 1991; Crook, 1991).

5.3.2 Insect hosts

Baculoviruses take their name from the host species from which they were isolated, e.g. *Cydia pomonella* GV (CpGV) was isolated from codling moth larvae.

Of the 600+ baculoviruses reported, the highest proportion have been NPVs, although the number of insect species from which GVs have been identified is currently in the order of 150 (Martignoni and Iwai, 1986; Hull *et al.*, 1989). MNPVs generally have a wider host range than SNPVs and GVs and many can be grown in insect cell culture (Granados and Hashimoto, 1989). For example, *Autographa californica* MNPV (AcMNPV) the most studied baculovirus, has a host range of over 39 insect species in 13 families (Cory and Entwistle, 1990a), and can grow in several insect cell cultures (Wood and Granados, 1991). There is even evidence that AcMNPV can replicate in a mosquito cell line, having crossed the order barrier (Döller, 1985). AcMNPV, however is not regarded as a typical MNPV, having an unusually wide host range.

5.3.3 Diagnosis

The occluded baculoviruses are visible under the light microscope and can be easily classified because of their morphology. It is possible to see if polyhedra are located in the nucleus (NPV) or in the cytoplasm (cytoplasmic polyhedrosis virus and entomopox viruses) by examination of unfixed wet preparations of fresh tissue from virus infected insects under dark field or phase contrast illumination. The GV capsules can be seen as tiny particles in Brownian motion, refringent in dark field and grey with phase contrast. Differential staining and OB morphology are useful tools in diagnosis using the light microscope; however, electron microscopy is needed to differentiate SNPV and MNPV infections as well as to identify possible cytoplasmic polyhedrosis viruses (CPVs) or non-occluded baculoviruses (*Nudibaculovirinae*).

Baculoviruses can be identified more precisely by restriction endonuclease analysis of the viral DNA, and SDS-polyacrylamide gel electrophoresis of virion polypeptides. Serological methods such as radioimmune assay, ELISA, gel diffusion, and fluorescent antibody techniques have also been used both to identify and distinguish between viruses.

5.3.4 Gross pathology

Infection occurs when healthy susceptible larvae feed on baculovirus-infected foliage or other food source. Following ingestion of OBs by a permissive host, there are no apparent signs of disease for a period of time (the eclipse phase). The first signs of infection are colour and behavioural changes. When a high proportion of the fat body cells contain OBs, the insect develops a milky appearance. The infected larvae usually experience a loss of appetite and later cease feeding. The type of behavioural changes observed late in infection include wandering (moving away from their food) or migration to a specific part of the plant, which facilitates the dispersal of the virus. Virus-killed larvae typically hang from the host plant attached by their prolegs, in a straight or an inverted V-shape position. For example, diseased *Pieris rapae* larvae usually attach themselves by their prolegs to the under-surface of a leaf late in infection, and when dead hang in a 'classic' inverted V-shape (Figure 5.1). The insect becomes flaccid and then melanizes; the fragile epidermis easily ruptures, liberating a milky fluid containing up to 10^{11} or 5×10^9 OBs/larva, in the case of GVs and NPVs respectively (Hunter *et al.*, 1984). However, when epidermal cells do not become infected, the integument remains firm and leathery. Although the production of such large quantities of OBs should promote the rapid spread of the disease, epizootics do not usually result from the single application of a viral insecticide and repeated applications are usually required. In some infected insects, particularly where virus replication is slow, there is an inhibition of moulting which prolongs the larval stage, e.g. *Trichoplusia ni* GV (TnGV) (Dougherty *et al.*, 1987), and

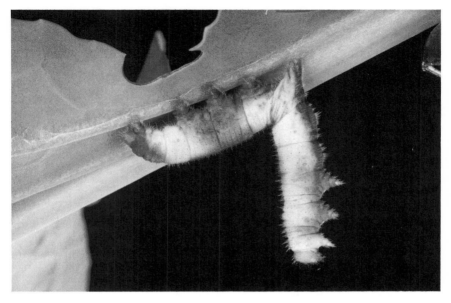

Figure 5.1 Virus-killed *Pieris rapae* larva hanging from the underside of a cabbage leaf. (Courtesy of Norman E. Crook.)

results in an increased yield of OBs. Ecdysteroids, hormones involved in insect moulting, are inactivated by an enzyme ecdysteroid UDP-glucosyltransferase (egt). The *egt* gene has been identified on the genome of AcMNPV (O'Reilly and Miller, 1989), and its expression results in delayed moulting in the infected insect.

5.3.5 Viral replication in the insect

The stages of NPV replication *in vivo* are illustrated in Figure 5.2. The alkaline conditions of the insect midgut result in the breakdown of the OB protein releasing the virions into the midgut, possibly in the area of the microvilli. The envelope of the virion fuses with the membrane of the microvillus resulting in the entry of the nucleocapsids into the midgut epithelial cell. The primary infection of the midgut cells results in the production of many progeny nucleocapsids in the nucleus which pick up a viral envelope when they bud from the distal surface of the midgut cell into the haemolymph (Figure 5.2). The typical result of this primary infection is the amplification of the ingested virus in the form of budded virus, rather than the production of occluded virus. When budded virions reach susceptible cells they are taken up by absorptive endocytosis.

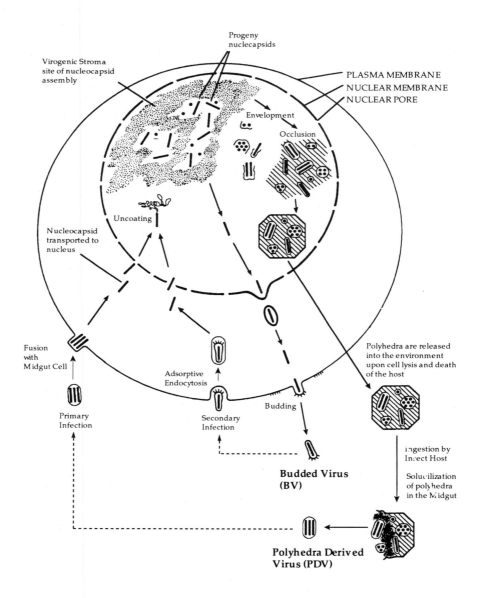

Figure 5.2 Typical cellular infection cycle of the nuclear polyhedrosis virus. (Courtesy of G.W. Blissard and G.F. Rohrmann.)

In NPV infections, **secondary viral replication** occurs within the nucleus. Some of the nucleocapsids either acquire a membrane by *de novo* synthesis in the nucleus and are subsequently occluded (polyhedra derived virus), or migrate to the cell membrane and bud from the cell surface (budded virus). The envelopes acquired in the nucleus and at the cell membrane have a different composition, and the two resulting forms of virus (phenotypes) have different biological properties even though they have the same genotype. The budded virions are far more infectious via the haemolymph than are virions liberated in the gut by alkali digestion of occlusion bodies, and *vice versa*, virions derived from occlusion bodies are more infectious to midgut cells. Only one nucleocapsid is enclosed in each budded virion in both NPVs and GVs. Secondary replication results in both budded virus and occluded virus. Late in NPV infection, insect cells show intact nuclei containing mature polyhedra which are released by lysis of the cell.

Different groups of occluded viruses vary in their tissue tropism; AcMNPV shows wide tissue tropism replicating in a wide range of tissues (Bilimoria, 1991) and other viruses, for example some GVs, show narrow tissue tropism and mainly infect only midgut cells and fat body cells (Crook, 1991). In the case of GVs, there appears to be a correlation between the tissue specificity and the speed of kill. Those GVs which are polyorganotropic involving the tracheal matrix and epidermis generally kill faster than those with limited tissue tropism (Crook, 1991).

5.3.6 Isolates

Genetic variation in baculoviruses is common. The existence of variants in isolates is recognized by the presence of sub-molar fragments in the electrophoretic patterns of DNA restriction endonuclease (REN) digestion products (Gettig and McCarthy, 1982; McIntosh *et al.*, 1987). Ten (wild) isolates of *Heliothis* spp. NPVs (two MNPVs and eight SNPVs) from different geographical locations characterized by REN analysis showed the presence of submolar bands in a number of digests, suggesting that the isolates were mixtures of genotypes (Gettig and McCarthy, 1982). These isolates as well as four other SNPV isolates were bioassayed and showed significant differences in virulence (Hughes *et al.*, 1983), and a relationship between genotype and biological activity was indicated for the *Heliothis* NPVs. There tends to be greater genetic variation between different isolates than within one isolate.

Variants can be cloned by plaque purification in the case of some NPVs, where virus permissive host cells can be grown *in vitro* (Lee and Miller, 1978; Smith and Summers, 1978), and by *in vivo* cloning in the case of GVs (Smith and Crook, 1988), where there is no cell culture system available.

By comparing genotypes with differing insecticidal activity it may be possible to relate DNA sequences to specific biological parameters such as virulence and specificity, as well as other possible changes such as the number and/or size of OBs per cell, host symptoms and host lysis.

5.3.7 Persistence and transmission

Persistence

Insect viruses are widespread, although it is generally acknowledged that soil is their main reservoir where they can persist for long periods being protected from inactivation by UV irradiation and other agents. In one case a baculovirus (*Orgyia pseudotsugata* NPV) was present in the soil 41 years following the last virus epizootic (cyclical outbreak of virus disease) (Thompson *et al.*, 1981). Vast quantities of virus are released from the huge numbers of dead insects accumulating at the peak of a viral epizootic, and could easily result in the production of more than 10^{15} virus OBs ha^{-1} year^{-1}, (Thompson and Scott, 1979). In the USA, autumn cabbage crops are often contaminated with cabbage looper (*T.ni*) NPV, which even survives in sauerkraut (Entwistle, 1983).

On the plant, several factors affect the persistence of the virus such as the adhesion of the OBs to the plant surfaces, inactivation of virus by plant exudates and even acid rain. The principal cause of virus inactivation is UV which results in persistence times normally ranging from 1 hour to a few days.

Transmission

Both biotic and abiotic factors are responsible for the dispersal of virus. An abiotic factor such as rain is responsible for the movement of virus from the soil reservoir onto the plants (through splashing) or from higher to lower positions on the plant. This short range transport probably plays an important role in the development of virus epizootics, which are more frequent in the forest habitat, where pests sporadically reach high populations, and epizootics are usually host-density dependent. Biological agents such as predators and parasitoids also contribute greatly to the dispersal of viruses. Parasitoids which develop in virus infected insects are able subsequently to transmit the disease to healthy hosts. Predators produce virus-contaminated excrement after feeding on infected insects, and it appears that several predators actually prefer virus-infected to healthy prey, for example, *Nabis roseipennis* on *Anticarsia gemmatalis* (Young and Kring, 1991). Among predators, birds are probably more efficient in spreading disease because of the large numbers of insects they consume and the great distances they travel.

5.4 VARIATIONS IN VIRUS SUSCEPTIBILITY IN INSECTS

5.4.1 Viral resistance in insects

The possibility of insects developing resistance to viral insecticides must be considered, since resistance has been one of the major problems arising from the over-use of chemical insecticides. Examples of the development of viral resistance in insects have been reviewed by Briese and Podgaite (1985), Briese (1986) and Fuxa (in press), and indicate that there is considerable variation in the mean dosage of virus required to either infect or kill larvae of different strains of some insect species, with inter-strain differences being observed at all stages of larval development. Antiviral substances have been isolated from gut juice (Funakoshi and Aizawa, 1989a) and haemolymph (Funakoshi and Aizawa, 1989b) of *Bombyx mori* (the silkworm). Although, as yet, there appears to be no evidence of increased viral resistance following the deliberate application of a viral insecticide, there is a suggestion of GV resistance in the cabbage white butterfly *Pieris brassicae* (David and Gardiner, 1960) and in *Eucosma friseana* (Martignoni, in Fuxa (in press)), following a natural epizootic.

5.4.2 Other factors affecting insect susceptibility

Susceptibility is mainly restricted to larvae and is strongly age-related. The neonate larvae may have LD_{50} levels of less than 10 OBs but, in the final instar, the LD_{50} values may be greater than 100 000 OBs. As a result, it is essential to apply the virus at the time of egg hatch to achieve efficient control.

There can be variation in the activity of specific viral isolates dependent on the geographical location of the host. The environment of the insect has also been shown to affect its susceptibility. For example, stress factors such as abnormal temperatures, nutritional deficiencies, or exposure to certain chemicals may increase susceptibility of silkworm to viral infection (Watanabe, 1971). Light may also have an indirect effect on the response to virus since it is required for the synthesis of antiviral protein (Hayashiya *et al.*, 1976).

There is also an indication that differences in virus susceptibility in a particular strain of insect may be affected by differences in the genotype of the plant on which the insect feeds. Silkworms for example, show differences in their susceptibility to *B. mori* NPV when reared on different mulberry genotypes (Sosa-Gomez *et al.*, 1991).

5.5 CURRENT STATUS OF INSECT VIRUSES FOR PEST CONTROL

As previously mentioned, baculoviruses have so far received the most attention for practical use in pest control. The main reason for this is their specificity, which led the WHO to recommend their development as pest control agents, in preference to other insect viruses (WHO, 1973). Table 5.1 lists insect species which have been shown to be susceptible to viral control (≥ 85% effectiveness of treatment) in field conditions. It should be pointed out that, in some cases, the doses used in the reported instances are unlikely to be economically feasible; on the other hand, often the effectiveness rates which can be found in the literature are ≤ 85%, but still judged satisfactory. Table 5.2 provides information on viral products currently in use or close to registration worldwide. It is difficult to give an accurate picture in a field which is changing so rapidly and the products reported here are among those with an official status.

Market share

Viral pesticides have only a tiny share of the huge insecticide market, accounting for 0.2% of total sales (including acaricides and nematicides, data refer to 1985 figures; Jutsum, 1988). They also constitute only a small part of the already limited market of biocontrol agents which, altogether, amounts to *ca*. US$ 34 million (at 1985 values), 90% of which is accounted for by *Bacillus thuringiensis* (Bt) sales. However, this figure does not take into account the use of viruses in the framework of programmes operated by international, non-profit making organizations like FAO (e.g the project for the control of the palm rhinoceros beetle) or the use made locally by many individual farmers or cooperatives in several developing countries. Nevertheless, it is only fair to recognize the fact that viral pesticides have so far had a very poor share of the market, and are of relatively low interest as far as pesticide companies are concerned (Jutsum, 1988).

Examination of Tables 5.1 and 5.2 provides obvious proof that, with very few exceptions, GVs and NPVs are the only viruses so far employed in crop protection. Practical aspects on the use of the more notable baculoviruses have been reviewed by Entwistle and Evans (1985), Huber (1986b) and Cunningham (1988). Other virus groups, notably CPVs and entomopox viruses (EPVs), have attracted some interest for pest control, though very few examples exist of their use in the field. Some details on these groups will be given later.

5.6. STRATEGIES OF USE FOR INSECT VIRUSES

There are two main situations in which viruses are used to control insect pests. In the first, there is a relatively **stable environment**, e.g. pastures and forests, where the crops have a high damage threshold (i.e. the level of damage beyond which it is economically feasible to treat is relatively high), and the virus, either by virtue of its inherent characteristics or the gregarious behaviour of the pest, is able to persist and spread within the insect population. In this instance the virus can be regarded as a classical biocontrol agent effecting longer term control. On the other hand, there is the situation where the environment is **unstable**, the damage threshold is very low and the virus does not persist or spread. This applies to most agricultural environments in which soil is regularly ploughed, crops are harvested annually, crop rotation is practised, etc., and the virus is used as a bioinsecticide. Obviously, these are two extremes and in between is a more or less continuous gradient.

In a stable environment, an equilibrium can be reached between the pest population and the virus, and may even result in cases of long-term or permanent control although this event is not very frequent. The first of such examples is the suppression of populations of *Gylpinia herciniae* in Canada by an NPV accidentally introduced while importing parasites of the European spruce sawfly from Europe. The virus spread rapidly from the points of introduction and, following other voluntary releases to hasten the speed of the epizootic, became established and permanently controlled the pest (Entwistle and Evans, 1985). A second example is the control of the *Oryctes rhinoceros* beetle using its nonoccluded virus. At first, attempts were made to control the beetle by spraying the virus in the breeding sites; this approach was unsuccessful because the virus is very labile in the environment. Eventually, it was realized that the best way to introduce it was through the release of artificially infected adults. Entwistle (personal communication) postulated that viral applications every eight years would be needed to maintain the effective control of *O. rhinoceros*.

Finally, another example of long-term control has been achieved by so-called **environmental manipulation**. In this case the virus is not introduced, but, instead, the conditions are maintained to favour its natural development and to preserve it at a high level, so keeping the pest population under the damage threshold. The best known example for this type of approach is the control of the pasture pests, *Wiseana* spp. After noticing that older leys suffered much less severe attacks by these pests, it was realized that the reason for this was the accumulation of the virus at the top layer of the pastures and that ploughing and hay production caused the removal of most of the virus, with a consequent increase in the pest population. It was therefore concluded that the best

Table 5.1. Insects susceptible to control with viruses in field conditions (control $\geq 85\%$; after Entwistle and Evans, 1985; updated)

Pest	Virus	Crop	Pest	Virus	Crop
Coleoptera					
Oryctes rhinoceros	NOV	Coconut	Malacosoma disstria	NPV	aspen
			M. fragile	NPV	orchards
Lepidoptera			M. neustria	NPV	—
Agrotis ipsilon	GV	okra	Mamestra brassicae	NPV	brassicas
A. segetum	GV	carrots, beets	Orgyia pseudotsugata	NPV	Douglas fir
Adoxophyes orana	NPV	apple	Panolis flammea	NPV	lodgepole pine
	GV	orchards	Papilio xuthus	NPV	Chinese prickly ash
Amathes c-nigrum	GV+NPV	—	Phthorimaea operculella	GV	potatoes
Anomis flava	NPV	cotton	Pieris rapae	GV	brassicas
Anticarsia gemmatalis	NPV	soybean	P. brassicae	GV	cabbages
Choristoneura fumiferana	NPV	spruces	Plodia interpunctella	GV	stored products
C. murinana	GV	spruces	Plusia agnata	NPV	soyabean
Colias philodice eurytheme	NPV	alfalfa	Plutella xylostella	GV	brassicas
Cydia pomonella	GV	apples, pears, walnuts	Spodoptera exempta	NPV	graminaceous crops
			S. exigua	NPV	ornamentals, vegetables

Dendrolimus sibiricus	GV	—
Ectropis obliqua	NPV	tea
Erinnyis ello	GV	cassava
Galleria mellonella	NPV	beehives
Heliothis armigera	NPV	cotton, sorghum, navy beans, tomato
H. punctigera	NPV	cotton
H. virescens	NPV	cotton
H. zea	NPV	cotton, soybean, sorghum
Hyphantria cunea	NPV	broadleaved trees
Kotochalia junodi	NPV	black wattle trees
Lymantria dispar	NPV	broadleaved trees
L. fumida	NPV	Japanese fir
L. monacha	NPV	spruces, etc.
Lobesia botrana	GV	grape
S. litura	NPV	bananas, soybeans, greenhouse strawberries, asparagus
S. littoralis	NPV	cotton
Thaumetopoea pityocampa	NPV+CPV	*Pinus halepensis*
Thymelicus lineola	NPV	Timothy grass
Trichoplusia ni	NPV	brassicas
Wiseana cervinata	NPV	pasture
Wiseana spp.	NPV	pasture
Spilosoma imparilis	NPV	mulberries
Hymenoptera		
Gilpinia hercyniae	NPV	spruces
Neodiprion lecontei	NPV	pines
N. pratti pratti	NPV	pines
N. sertifer	NPV	pines
N. swainei	NPV	jack pine
N. taedae linearis	NPV	loblolby pine

Table 5.2. Viral products currently produced and target pests. (After Deseö and Rovesti, 1992)

Virus	Viral product	Producer	Status	Target pest
Neodiprion sertifer NPV	Monisärmiövirus	Kemira Oy	R	N. sertifer
	Virox	Oxford Virology	R	N. sertifer
	Preserve	Microgenesys	S	N. sertifer
	Neocheck-S	USDA Forest Service	R*	N. sertifer
	Virin-Diprion	—	R	N. sertifer
N. lecontei NPV	Lecontvirus	Canadian Forestry S	R*	N. lecontei
Lymantria dispar NPV	Gypcheck	USDA Forest Service	R*	L. dispar
	Virin-ENSh‡	—	R	L. dispar
Orgyia pseudotsugata NPV	TM Biocontrol-1	USDA Forest Service	R*	O. pseudotsugata
	Virtuss	Canadian Forestry S.	R	O. pseudotsugata
Hyphantria cunea NPV+GV	Virin ABB	—	R	H. cunea
Malacosoma neustria NPV	Virin KSh	—	R	M. neustria
Stilpnotia salicis NPV	Virin LS	—	S	S. salicis
Heliothis armigera NPV	Virin KhS‡	—	R	Heliothis spp.
	78–3	—	R	Heliothis spp.
Mamestra brassicae NPV	Mamestrin	Calliope	S	M. brassicae, Heliothis spp.
	Virin-EKS‡	—	R	Spodoptera spp., Diparopsis vatersei

Virus	Product	Producer	Country	Status	Host range
Spodoptera littoralis NPV	Spodopterin	Calliope	France	S	*Spodoptera littoralis*
S. sunia NPV	VPN 82	Agricola El Sol	Guatemala	R	*S. sunia, S. exigua*
Autographa californica NPV	VPN 80	Agricola El Sol	Guatemala	R	*A. californica, Bucculatrix thuberiella, Estigmene acrea, Plutella xylostella, Pseudoplusia includes, S.exigua, Trichoplusia ni*
Anticarsia gemmatalis NPV	Polygen	Agroggen S/A Biol. Ag.	Brazil	R	*A. gemmatalis*
Cydia pomonella GV	Madex	Andermatt Biocontrol	Switzerland	R	*C. pomonella*
	Granupom	Hoechst	Germany	R	*C. pomonella*
	Carpovirusine	Calliope	France	S	*C. pomonella*
	Decyde	Microgenesys	USA	S	*C. pomonella*
	Virin-Gyap	—	USSR	R	*C. pomonella*
	Virin-OS	—	USSR	R	*A. segetum*
Agrotis segetum GV	Agrovir	—	Denmark	RN	*A. segetum*
Adoxophyes orana GV	Capex	Andermatt Biocontrol	Switzerland	S	*A. orana*
Pieris rapae GV	PrGV (W1–78)	—	China	R	*P. rapae*
Plutella xylostella GV	PxGV	—	China	R	*P. xylostella*

*Products produced and used by the respective government agencies, therefore not available on the market.
R: registered; S: at pre-registration level; RN: registration not requested.
†Presently in course of registration in UK (B.F. Ackers, personal communication).

strategy was the adoption of techniques aimed at the minimal disturbance of soil facilitating the maximum recycling of the virus, e.g. by cultivation of long-term leys, oversowing, etc. (Kalmakoff and Crawford, 1982).

Despite these remarkable examples of long-term control these approaches are not applicable to most agricultural situations, where the viruses are normally applied in a similar fashion to chemical insecticides. Several methods of virus distribution have been investigated (parasite and predator-aided virus dispersal, autodissemination for autocidal control, and others reviewed by Entwistle and Evans, 1985), but spraying is the only method used in practice in agriculture. When it comes to field application, viral insecticides have significant disadvantages, which often make their use more complex an operation: (i) viruses act only after ingestion by the insect pests, while many chemicals have also contact and/or multiple activity; (ii) unlike many recent chemicals, viruses do not have any systemic or locosystemic activity; (iii) viruses are not soluble molecules but particulate entities which may pose peculiar problems of distribution; (iv) viruses have a very limited persistence, but to be effective must remain viable until ingested by the target insect; the timing of application is therefore of paramount importance.

5.7 SPECIFIC EXAMPLES FOR PEST CONTROL

5.7.1 Field crops

Heliothis NPV

The developmental history of this NPV has been comprehensively reviewed by Ignoffo and Couch (1981). Its commercial development began in 1961 and culminated in 1975 with the first registration (at least for Western countries) of a viral insecticide. The commercial development by Sandoz of a viral insecticide based on *Heliothis* NPV, Elcar, has been regarded as a success from a technical standpoint. However, what was miscalculated was the acceptance of such an innovative product by the growers, who in the end preferred the newly marketed synthetic pyrethroids, which are cheaper, easier to use and act faster (even if they are environmentally harmful). The production of Elcar was discontinued in 1982, even though its use continued in many countries at an experimental level. In Europe Elcar was registered in Spain.

The *Heliothis* NPV is still widely used in the former USSR and in China. Of the several cooperative production plants scattered throughout this latter country, the one set up in Hubei Province is currently capable of a monthly production of 10^6 larvae of *H. armigera*. A total output of 23 tons of viral pesticides was obtained from 1988 to 1991, with a protected area of some 40 000 hectares (80% of which was cultivated with cotton) (Zhang Guang-Yu, personal communication).

Anticarsia gemmatalis NPV

The amazing increase in the use of this virus for the protection of soybean crops is an excellent example of how good results can be achieved by a combination of technical ability, efficient extension service and suitable economic conditions. *A. gemmatalis*, the velvetbean caterpillar, is the main pest of this crop, which in Brazil is cultivated over an area of some 12 million hectares. The virus, initially tested in the USA, was used in the field starting from the 1982/83 growing season on about 2000 hectares. In a few years, the area protected with the virus has grown exponentially, with *ca.* 1 000 000 hectares being treated in the 1989/90 season (Moscardi, 1990). Due to a good integration of laboratory and field production methods, the cost for the treatment with *A. gemmatalis* NPV (AgNPV) has worked out at $US 2.00 ha^{-1}, compared with $US 5.00 ha^{-1} for chemical insecticides. With the virus, only one treatment/season is normally necessary, whereas chemical controls require several applications. Together, these factors have resulted in an overall reduction in control costs of up to 70% compared with the use of chemical insecticides (Moscardi, 1990). The production of the virus, initially by the government-funded CNPSo/EMBRAPA, and by farmers' cooperatives, is being passed onto private companies, which have recently registered or are in the process of registering some products.

A strain of AgNPV has been selected for increased activity against the sugarcane borer, *Diatraea saccaralis*. In the course of a 20 passage selection, the infectivity for *D. saccaralis* increased 1000-fold compared with normal type AgNPV (Pavan and Ribeiro, 1989), though still retaining infectivity to the original host. This virus can be produced in *D. saccaralis* which is much easier to rear than *A. gemmatalis*. A commercial product based on this strain is being developed with the trade name Multigen.

Other baculoviruses

No other baculoviruses targeted for field crops have achieved the same usage as *Heliothis* and *Anticarsia* NPVs. Nevertheless, many of them are being used or developed commercially in both developed and developing countries. For example, Mamestrin is in course of registration in France by Calliope (S. Leclant, personal communication), while in China, *P. rapae* GV was being used over an area of more than 5000 hectares annually (in Huber, 1986b) and the NPVs of the tea tussock moth (*Euproctis pseudoconspersa*) and the tung oil tree geometrid (*Buxura suppressaria*) have been used on areas of up to 40 000 and 20 000 hectares respectively (in Li, 1989).

5.7.2 Orchards and plantations

Cydia pomonella *GV*

The development of this virus began as long ago as in the mid-1960s. The virus has been widely studied and tested, initially mostly in the USA and then in Europe in the last 10 years. At present, registration has been granted, or at least filed, in many EEC countries. The registration of this virus has been an important breakthrough, in that it is the first viral pesticide to be registered in Western countries for use on a food crop.

C. *pomonella* GV (CpGV) has been shown to be very effective in many countries. The codling moth is not an easy target, as the neonate larvae enter the fruits generally within 24 hours of hatching. However, because of the very high infectivity of the virus (LD_{50} for first instar (L1) larvae of between 1 to 3 capsules (Crook *et al.*, 1985; Huber, 1986a) it is possible to achieve good control, provided that correct timing and good coverage are implemented. Proper **spray timing** may be difficult, particularly where the codling moth has more then one generation per year or where cool climatic conditions extend the flying period of the moth. For this reason, a new strategy has been successfully tested which involves spraying a reduced amount (1/10) of the virus at weekly intervals. By using this method, the control provided by nine treatments at reduced dose was the same as that obtained with four treatments at the full rate, while the total amount of virus required was reduced by 80% (Dickler and Huber, 1988).

A distinct advantage of the CpGV over the organophosphates, traditionally employed to combat the codling moth, is that the virus does not affect the predators and parasites present in the orchard; consequently, secondary pests like the spider mite, *Panonychus ulmi*, and the woolly aphid, *Eriosoma lanigerum*, are normally kept in control by their natural enemies without the need for specific control measures (Dickler and Huber, 1983; Glen and Payne, 1984). On the other hand, there may be some problems with leafrollers which are normally controlled by wide-spectrum chemical insecticides. In addition, a certain incidence of shallow damage on the fruits is almost unavoidable, and this can be a real drawback in market economies, where the demand is for perfect, totally unblemished produce.

Oryctes *virus*

The control of the rhinoceros palm beetle by use of its NOV is a celebrated example of classical biological control, and has been mentioned previously. The beetle was accidentally introduced, from South-East Asia where it is endemic, to a number of South Pacific countries and into Mauritius (Bedford, 1981). It has been calculated that *O. rhinoceros* cost six South Pacific countries more than US$ 1 million in 1968 alone (in Bedford, 1981), based on direct and indirect costs (loss of copra, control measures, quarantine expenditures, etc.).

The virus has been successfully introduced into numerous South Pacific and Indian Ocean countries, resulting in significant reductions in the beetle population, which, due to its biology, is difficult to control even with chemical insecticides.

5.7.3 Forestry

Forests are good environments for use of viral insecticides, mainly because they are relatively stable ecosystems which permit a high damage threshold. In some cases, where chemicals are banned by law (as in Canada) viruses together with Bt are the only permitted methods of control. Unfortunately, forestry does not provide a good market, accounting for less than 1% of the global insecticide use (Cunningham, 1988). It is not easy to give precise figures on the area treated annually but it is clear that this is rather limited (the overall area in North America and Europe up to 1986 can probably be estimated at 15–20 000 hectares (data extrapolated from Cunningham, 1988)). The unpredictability of the market, due mainly to the cyclic character of infestation by forest pests, has always discouraged industry from entering this field. For this reason, the registration of almost all the products is held by government agencies or government-owned companies. In Eastern Europe, and particularly in the former USSR, the use of viral products in forestry is on a larger scale: Virin ENSh (*Lymantria dispar* NPV), was applied on some 150 000 hectares of forest in 1972–78 (Lipa, personal communication).

In the forest environment are found the most outstanding examples of viral control, particularly against those insects with gregarious behaviour (e.g. sawflies). More details on viral control in forestry can be found in Huber (1986b), Cunningham (1988), and Evans (1990).

5.7.4 Viruses other than baculoviruses

Very little applied work has been done with viruses other than baculoviruses, and most of this has focused on **CPVs** and **EPVs**. CPVs are the only ones to have reached a commercial status, with a product based on the *Dendrolimus spectabilis* CPV (DsCPV) being registered in 1974 in Japan (Katagiri, 1981). At present, this product is not used in Japan; however, this and other CPVs from *D. punctatus* and *D. tabulaeformis* have until recently found large scale application in China (Li, 1989). In addition to the increased mortality caused in the short term by the DsCPV, there is a long-term control effect on the population. In Taiwan, long-term control (up to 10–12 years) of pine caterpillar has apparently been obtained by the combination of the virus with Bt and the fungus, *Isaria farinosa* (Ying, 1986).

Recently, there has been some interest in entomopox viruses (EPVs) for the control of grasshoppers and locusts, which represent a real scourge in vast areas of rangeland in the US (with more than 500 000 hectares infested in 1986; in Agrow, 1989) and in several developing countries. It is difficult to assess the efficacy of these viruses against insects with such great dispersal capability, and currently their use is still at an experimental level. The largest application so far is probably that carried out in collaboration between the USDA's Agricultural Research Service Rangeland Insect Control Research Laboratory and Evans Biocontrol, for which a mixed distribution of an EPV and the microsporidia, *Nosema locustae*, was planned on 10 000 acres in the Western US (Agrow, 1989).

5.8 PRODUCTION AND FORMULATION

5.8.1 Virus production

Aspects of production and processing of baculoviruses have been comprehensively reviewed by Shapiro (1986). Since viruses are obligate pathogens, they can be produced only in living cells. Many hopes lie in the development of an adequate system for production in cell culture; much work has been done in recent years and considerable advances have been made in developing cell lines for the growth of baculoviruses, EPVs and CPVs (Granados *et al.*, 1987). Although the initial costs for constructing industrial plant are high, the actual production process is less labour intensive, and cheaper serum-free culture media are now being developed. However, it is estimated that a 10- to 100-fold increase in the productivity of the present tissue culture systems will be necessary to make this procedure feasible for commercial production of viral insecticides (Shieh, 1989). Consequently, production in whole larvae remains the only practical method for the industry at the present time.

Insect rearing

Laboratory-reared insects are preferred for mass production, as the quality of the final product is often better than that from unchecked field populations. By using standardized rearing procedures, it is easier to optimize the whole process of production, while at the same time minimizing risks of development of unwanted pathogens. However, insect rearing is normally time consuming and expensive, accounting for one- to two-thirds of the total cost of production (Martignoni, 1984). For this reason it may be deemed more economical to infect field-collected larvae or even larvae inoculated in the field and maintained on their host plant until collection. For instance, of the several methods tested in Brazil for the production of AgNPV, the most economical emerged as the treatment of field populations with subsequent collection of dead or dying larvae. Using this method, an average of more than 40 kg/day virus-killed larvae was produced by CNPSo-EMBRAPA, with a total monthly production of 1500 kg, enough for the treatment of 75 000 hectares (Moscardi, 1990). NPVs of sawflies are also produced in the field, as the rearing of these species is difficult. In this case, heavily infested groups of trees are sprayed and, after several days, branches are cut off and the dead or dying larvae collected for further processing.

The insect species used for production does not necessarily have to be that against which the virus is targeted in the field. Attempts have been made to adapt viruses to alternate hosts, which may produce either a higher viral yield and/or be easier to rear, as in the case of AgNPV. By serial passaging of NPVs through alternate hosts, the viruses may tend to become more pathogenic to that host and less pathogenic to the original host. It is therefore of paramount importance constantly to monitor the identity and biological activity of the progeny virus. No such changes have been detected when propagating codling moth GV in false codling moth larvae, *Cryptophlebia leucotreta*, which appears to be a suitable system for the production of this virus (Reiser and Gröner, 1990). However, in most studies it has not been established whether passaging merely selects one or more of the most suitable genotypes from the original isolate, or whether a specific genotype or genotypes are altered as they become adapted. Miller and Dawes (1978) did however passage a plaque purified genotype of AcMNPV through different hosts and found no change in the genotype.

Sometimes, no alternate hosts are available for fastidious insect viruses like *Euproctis chrysorrhoea* or *L.dispar*, whose urticaceous hairs must be eliminated from the final product.

5.8.2 Processing and formulation

At the end of the infection cycle, the virus-containing material must be collected, processed and formulated. This can be done in different ways and the best strategy must be determined on a case by case basis.

Larvae are normally harvested at, or just prior to, death. This may lead to **bacterial contamination** but this is usually within acceptable limits. Virus purification by centrifugation on a gradient is not economically feasible for commercial production; besides, a substantial amount of virus is lost and the resulting purified virus has lower persistence in the field (as it will be explained later). In order to reduce bacterial contamination, it may therefore be convenient to harvest the infected insects before they die. A word of caution, Ignoffo and Shapiro (1978) and Shapiro and Bell (1981) found that the infectivity of polyhedra obtained from living larvae was 6–12 times less than those harvested from dead larvae, and it may be necessary to determine the optimum harvesting time in each situation.

During the process of **formulation**, coformulants such as UV protectants may be added to the virus, to give the final product good stability in storage and optimum properties for tank mixing and field application (Young, 1989). Further reviews of different aspects of formulation are given by Ignoffo and Couch (1981), Entwistle and Evans (1985) and Young and Yearian (1986). Formulation of viruses (and microbials in general) is usually a more delicate process than for chemical insecticides, because of the need to preserve viability of the virus and because of their particulate nature.

The **application** of non-purified virus results in protection of the virus from UV radiation and in rainfastness; properties provided by contaminating insect debris and proteins. In harsh conditions as in the case of cotton fields in Egypt, non-purified *S. littoralis* NPV performed much better than virus purified and formulated with both UV protectants and stickers (Jones, 1988). This fact is obviously of great relevance, as the purification step constitutes a bottleneck in the production process and significantly contributes to the overall cost of the final product.

The last step in production is **quality control**, in which the final product must be standardized for potency and checked for possible contaminating, hazardous agents. Potency of a viral product should best be expressed by both virus particle count and bioassay. The standardization through bioassay is done by comparison to a reference standard, as in the case of Bt. For contamination checks, stringent specifications have been established by regulatory agencies; in particular, the final product must be free from vertebrate pathogenic bacteria (*Salmonella, Shigella, Vibrio*).

5.9 APPLICATION OF VIRAL INSECTICIDES

Viral insecticides have already achieved considerable success in pest control using the same equipment and application criteria employed for chemical insecticides. It may be possible to achieve even better control if certain considerations regarding the biological and physical properties of the viruses are taken into account. For detailed reviews of the factors affecting the application of insect viruses and microbial agents in general, readers are referred to Smith (1971), Lewis (1984), Entwistle and Evans (1985) and Young and Yearian (1986).

5.9.1 Method of application

Although, some work has been done over the last decade to develop specifications for the application of insect viruses, very few workers have compared the efficacy of different equipment under the same operating conditions. Due to the large number of variables involved (biopesticide dose, volume of spray/hectare, viscosity of the tank mix, wind speed, etc.), it is always difficult to compare the results from different trials. As with chemicals, it is generally recognized that the most effective control is achieved by a high density of spray droplet coverage (i.e. by distributing the spray in a high number of small droplets rather than in larger droplets). However, because of the particulate nature of occluded viruses, the generation of small droplets through the use of high pressures and small diameter nozzles might lead to a loss of active material (i.e. OBs) from the droplets, as observed in the case of crystals of Bt (Smith *et al.*, 1977, 1978). Despite the optimization of the volume mean diameter of the droplets, of the density and the volume of spray, the penetration of the treatment within the crop canopy and the deposition of the droplets on the abaxial side of the leaves is often not adequate. Various systems, including electrostatic spray applications, have been devised in an attempt to improve the deposition of the spray to these sites.

5.9.2 Application timing

In order to obtain an effective control of pests, proper **timing** is of paramount importance. This applies to all forms of pest control, but is especially relevant to slow-acting and non-persistent viruses. Treatments must be targeted against the most susceptible stages of the pest, to suppress its population before the development of the less susceptible and more damaging instars. Alam (in Young and Yearian, 1986) calculated that more than 80% of the food consumption by *Pseudoplusia includens* occurred during the last two developmental stages; this figure rises to 95% in case of *Spodoptera exigua* (Smits *et al.*, 1987). As a consequence, much damage will be caused to the crop if the treatment is delayed, even if all the larvae are eventually killed. On the other hand, spraying too early, when only a small proportion of larvae has hatched, is also likely to result in lower effectiveness (Cory and Entwistle, 1990b; Young and Yearian, 1986).

Ideally, treatments should be targeted against **neonate larvae**; however, in some cases a slight delay does not seem to affect the overall efficacy of the treatment (Cory and Entwistle, 1990b), while in others delay is simply not acceptable. *C. pomonella* neonate larvae feed on the leaves or fruit surface for a few hours, then burrow into the fruits and do not re-emerge until pupation; therefore, only the first instar is targetable. On the contrary, *Choristoneura fumiferana* feeds as a needle miner in first and second instars, as bud borer in the third instar and only in the fourth instar are the larvae exposed by flushing of the buds. In this case, best results can be obtained by targetting the fourth instar larvae: damage is high in the year of treatment, but the control can be greater in the following year (Young and Yearian, 1986).

The right timing may also vary with the **host plant**. In case of *Heliothis* spp. on cotton, for example, treatments must be aimed at killing young larvae feeding on the terminal vegetation before they bore into the cotton bolls. On the other hand, on sorghum, larvae feed in the open on panicles and timing is less important.

These few examples illustrate the importance of an adequate know-ledge of the target pest species and of the availability of an adequate monitoring system (young larvae can be very elusive!); however, this applies to all types of control agent.

5.9.3 Tank mixing

A number of tank mixture **adjuvants** have been used to improve the performance of viral sprays. Phagostimulants have been employed in several cases to stimulate feeding insects so that they will ingest more plant material and therefore more virus. This is particularly useful for those pest species whose larvae tend to penetrate rapidly into the plant tissues after hatching. Besides, this may to some extent compensate for a sub-optimal coverage of the virus and for the effect of dilution of the spray deposit (as a result of rapid plant growth) by luring the larvae to the treatment deposit. Molasses and other sugar solutions are among those more commonly used, together with flours and oils from different plants. Besides their effect on feeding they also have sticking, thickening and other properties. However, the effect of tank mixing these and other adjuvants should be carefully evaluated as they may have negative effects on the efficacy of the treatment (through virus inactivation, or inhibition of feeding). The literature is full of information on this topic (reviewed by Entwistle and Evans, 1985, and Young and Yearian, 1986).

Another very important aspect of tank mixing is the **compatibility** of insect viruses with chemical pesticides. Inevitably, viruses are bound (unless used with an autodissemination-type approach) to come in contact with pesticides and other chemicals normally used in agricultu-ral practice. The contact may happen on the crop, if chemical and virus treatments are applied to the same plots, or even in the spray tank as it may be desirable to apply in a single treatment a combination of two, or even more, active ingredients, to control a complex of insect pests, and plant diseases. For detailed reviews on this aspect refer to Benz (1971), Jaques and Morris (1981) and Harper (1986).

5.10 GENETIC IMPROVEMENT OF VIRAL INSECTICIDES

One of the major disadvantages of viral pesticides is the continued damage to the crop by the pest after their application. The virus undergoes several cycles of replication in the insect host during infection, and feeding does not usually cease until late in infection. The LT_{50} can range from as little as four days to several weeks, depending on the virus, dose, insect host and instar. The efficacy of a virus could be improved if it could be genetically modified to produce a decrease in the LT_{50} and/or rapidly inhibit feeding.

Until very recently, improvement of the pathogenicity of baculoviruses by **genetic engineering** was mere speculation; however, there are now several clear examples which demonstrate that this is feasible. Using recombinant DNA technology, it is now possible to insert foreign genes into baculoviruses and achieve expression of those genes in insect cells rapidly and efficiently, (Smith *et al.* 1983b; Pennock *et al.* 1984). Recombinant baculoviruses expressing insecticidal genes could provide more pathogenic viral insecticides.

Inserting foreign genes in place of non-essential genes under the influence of very late promoters, e.g. the polyhedrin and p10 promoters in AcMNPV, results in high expression of the foreign proteins (Smith *et al.*, 1983b; Vlak *et al.*, 1988). The resulting recombinant viruses are either polyhedrin or p10 minus. For bioinsecticidal use, it is desirable to retain polyhedrin, since non-occluded recombinants are so unstable that they cannot be delivered to the field in an active state (Bishop *et al.*, 1989; Wood and Granados, 1991); on the other hand, release of persistent genetically engineered viruses into the environment could pose a risk. This can be partially overcome by producing recombinant viruses which are polyhedrin plus, and p10 minus (Vlak *et al.*, 1988; Weyer and Possee, 1989). It has been demonstrated that occluded p10 minus recombinant AcMNPV has a significantly reduced LD_{50} value compared with wild-type AcMNPV. This is due to unstable polyhedra which lack a polyhedral envelope, and therefore breakdown more efficiently in the gut. The p10 minus polyhedra are also more susceptible to physical damage (Williams *et al.*, 1989) and it is predicted that these recombinants will have lower environmental persistence than wild-type polyhedra (but not as low as polyhedrin minus recombinants), a desirable property for a genetically engineered virus. Another possibility is the co-occlusion of a polyhedrin minus recombinant virus with the wild-type virus by simultaneous infection of insect cells. The proportion of recombinant polyhedrin-minus virus in these occlusion bodies gradually falls after several passages through insects reducing the environmental persistence of the engineered virus (Wood and Granados, 1991).

Recently, AcMNPV recombinants have been produced containing genes encoding insect enzymes, toxins and hormones. The most promising recombinant AcMNPV viruses which have potential to reduce crop damage are occluded and contain insect-active neurotoxin genes, e.g. the toxin gene from the North African scorpion, *Androctonus australis* (Stewart *et al.*, 1991; McCutchen *et al.*, 1991), or the mite toxin gene from the female mite *Pyomotes tritici* (Tomalski and Miller, 1991). Both these recombinant viruses have reduced LD_{50} values and cause paralysis in infected insects as well as reducing food consumption by as much as 50% (in the case of the scorpion toxin) and 40% (in the case of the mite toxin).

The production of recombinant baculoviruses has principally involved AcMNPV and to a lesser extent, *B. mori* NPV. However, the relatively wide host range of AcMNPV, as well as the possibilities of genetic recombination with other baculoviruses in the environment and altered host range which may result from genetically engineering the virus, provide cause for concern. Although the AcMNPV/*Spodoptera frugiperda* cell system is convenient, it will be desirable to engineer viruses with more restricted host ranges. Risk assessment prior to the release of such engineered baculoviruses into the environment is of primary importance.

5.11 CONCLUSIONS

Do insect viruses, and baculoviruses in particular, have the potential to become an effective operative tool in integrated pest management strategies or will they remain an option of governmental agencies and adventurous farmers? Viruses have obvious advantages, particularly of an ecological nature, over chemicals. However, from the point of view of industry, which undoubtedly plays a primary role in the possible development of viruses as pest control agents, their main shortcomings are the lack of contact activity, low speed of action, the reduced host range and the difficulty of production. However, these elements can be more or less serious depending on the economical context in which they are considered. Sometimes, they could even be considered positively:

1 In a developing country where labour is cheap and there is a lack of high technology, the labour intensive *in vivo* production could even be seen as a benefit for the economy, through the creation of new jobs.
2 In conditions where the slow activity of viruses is not so important, because consumers have not (yet) been conditioned to request visually perfect produce.

In addition, the regulations for the use of microbials in many developing countries are less stringent than those of developed countries, and may have contributed to the much greater popularity and success of viral pesticides in developing countries. Baculoviruses are among the most stable entomopathogens and have a long shelf-life as a commercial product when stored under reasonable conditions; the rapid loss of activity occurs when the virus is applied to the plant.

Because chemicals produce a rapid knockdown effect, most farmers expect the same from viruses, but obviously this is not possible. However, in the case of some viruses there can be a bonus of long-term control. Therefore, to assess the performance of viral pesticides properly, checks should be made at longer time intervals than for chemical insecticides and more objective assessment parameters (e.g. marketable production) should be used instead of or together with larval mortality. Also, field trials should be carried out on as large an area as possible if the benefits derived from the use of the selective viruses are to be fully appreciated. Baculoviruses do not affect the complex of natural enemies present in the crop and these may have an important complementary role in regulating pest populations. The resurgence of natural enemies is not possible in a small plot surrounded by a chemically treated area.

Finally, it should be said that although viral products are likely to be more expensive than chemicals, when the so-called environmental costs are considered, the balance may change. Because the market for each viral pest control agent will always be limited to one or at most a few major pests, industry will need help with the research, development and in particular the registration costs, possibly from government agencies. In this way, baculovirus pest control products could become more financially attractive and would help to reduce the dependence on chemical insecticides and their dire consequences. To conclude, baculoviruses do have a role in integrated pest management programmes by offering an effective and environmentally sensitive system of pest management.

Acknowledgements

The authors wish to thank Norman E. Crook, and Phillip F. Entwistle for their discussion and encouragement in preparing this chapter, as well as Carlo M. Ignoffo and many other colleagues who sent useful literature and information.

REFERENCES

Agrow, (1989) *Biological crop protection*. PJB Publications Ltd; Richmond, Surrey (U.K.), pp.109.

Bedford, G.O. (1981) Control of the rhinoceros beetle by baculovirus, in *Microbial Control of Pests and Plant Diseases, 1970–1980*, (ed H.D. Burges), Academic Press, London and New York, pp.409–26.

Benz, G. (1971) Synergism of micro-organisms and chemical insecticides, in *Microbial Control of Insects and Mites*, (eds H.D. Burges and N.W. Hussey), Academic Press, pp.327–55.

Bilimoria, S.L. (1991) The biology of nuclear polyhedrosis viruses, in *Viruses of Invertebrates*, (ed E. Kurstak), Marcel Dekker, Inc, New York, Basel, Hong Kong, pp.1–72.

Bishop, D.H.L., Harris, M.P.G., Hirst, M., *et al.* (1989) The control of insect pests by viruses; opportunities for the future using genetically engineered virus insecticides, in *Progress and Prospects in Insect Control*, (ed N.R. McFarlane), British Crop Protection Council, London, pp.145–55.

Briese, D.T. (1986) Insect resistance to baculoviruses, in *The Biology of Baculoviruses*, (eds R.R. Granados and B.A. Federici), CRC press, Inc. Boca Raton, Florida, vol. II, pp.237–64.

Briese, D.T. and Podgaite, J.D. (1985) Development of viral resistance in insect populations, in *Viral Insecticides for Biological Control*, (eds K. Maramorosch and K.E. Sherman), Academic Press, Inc., Orlando, San Diego, New York, London, Toronto, Montreal, Sydney, Tokyo, pp.361–98.

Cory, J.S. and Entwistle, P.F. (1990a) Assessing the risk of releasing genetically manipulated baculoviruses. *The Exploitation of Micro-organisms in Applied Biology, Aspects of Applied Biology*, **24**, 187–94.

Cory, J.S. and Entwistle, P.F. (1990b) The effect of time of spray on infection of the pine beauty moth, *Panolis flammea* (Den. & Schiff) (Lep., *Noctuidae*), with nuclear polyhedrosis virus. *Journal of Applied Entomology*, **110**, 235–41.

Crook, N.E. (1991) Baculoviridae: Subgroup B. Comparative aspects of granulosis viruses, in *Viruses of Invertebrates*, (ed E. Kurstak), Marcel Dekker, Inc. New York, Basel, Hong Kong, pp.73–109.

Crook, N.E., Spencer, R.A., Payne, C.C. and Leisy, D.J. (1985) Variation in *Cydia pomonella* granulosis virus isolates and physical maps of the DNA from three variants. *Journal of General Virology*, **66**, 2423–30.

Cunningham, J.C. (1988) Baculoviruses: their status compared to *Bacillus thuringiensis* as microbial insecticides. *Outlook on Agriculture*, **17**, 10–17.

David, W.A.L. and Gardiner, B.O.C. (1960) A *Pieris brassicae* (Linnaeus) culture resistant to a granulosis. *Journal of Insect Pathology*, **2**, 106–14.

Deseö Kovács, K.V. and Rovesti, L. (1992) *Lotto Microbiologica Contro 1 Fitofagi Teoria e Practico*, Edagricole, Bologna, pp.156–7.

Dickler, E. and Huber, J. (1983) Microbial control of *Adoxophyes orana* in combination with granulosis virus control of codling moth, in *Programme on Integrated and Biological Control: Progress Report 1979–1981: Commission of the European Communities*; (eds R. Cavalloro and A. Piavaux), pp.14–21.

Dickler, E. and Huber, J. (1988) Modified strategy for use of the codling moth granulosis virus (CpGV), in *Production and Application of Viral Bio-pesticides in Orchards and Vegetables*, (eds H. Audemard and R. Cavalloro), Proceedings of a meeting of the EEC expert group, pp.339–40.

Döller, G. (1985) The safety of insect viruses as biological control agents, in *Viral Insecticides for Biological Control*, (eds K. Maramorosch and K.E. Sherman), Academic Press, Inc., Orlando, San Diego, New York, London, Toronto, Montreal, Sydney, Tokyo, pp.399–439.

Dougherty, E.M., Kelly, T.J., Rochford, R., *et al.* (1987) Effects of infection with a granulosis virus on larval growth, development and ecdysteroid production in the cabbage looper, *Trichoplusia ni*. *Physiological Entomology*, **12**, 23–30.

Entwistle, P.F. (1983) Viruses for insect pest control. *Span*, **26** (2).

Entwistle, P.F. (1985) Viruses – an alternative answer to insect pest control, *NERC News Journal*, September, 1–6.

Entwistle, P.F. and Evans H.F. (1985) Viral control, in *Comprehensive Insect Physiology Biochemistry and Pharmacology*, (eds G.A. Kerkut and L.I. Gilbert), vol. **12**, Insect Control, Pergamon Press, Oxford, pp.347–412.

Evans, H.F. (1990) The use of bacterial and viral control agents in British forestry. *The Exploitation of Micro-organisms in Applied Biology, Aspects of Applied Biology*, **24**, 195–203.

Federici, B.A. (1986) Ultrastructure of baculoviruses, in *The Biology of Baculoviruses*, (eds R.R. Granados and B.A. Federici), CRC press, Inc. Boca Raton, Florida, Vol I, pp.61–88.

Francki, R.I.B., Fauquet, C.M., Knudson, D.L. and Brown, F. (eds) (1991) *Classification and Nomenclature of Viruses*, Fifth report of the international committee on taxonomy of viruses. *Archives of Virology*, Supplementum 2, Springer-Verlag, Wien, New York.

Funakoshi, M. and Aizawa, K. (1989a) Antiviral substance in the silkworm gut juice against a nuclear polyhedrosis virus of the silkworm, *Bombyx mori. Journal of Invertebrate Pathology*, **53**, 135–6.

Funakoshi, M. and Aizawa, K. (1989b) Viral inhibitory factor produced in the haemolymph of the silkworm *Bombyx mori*, infected with nuclear polyhedrosis virus. *Journal of Invertebrate Pathology*, **54**, 151–5.

Fuxa, J.R. (in press). Insect resistance to viruses, in *Parasites and Pathogens of Insects*, (eds N.E. Beckage, S.N. Thompson and B.A. Federici), Academic Press.

Georghiou, G.P. and Lagunes, A. (1988). The occurrence of resistance to pesticides: cases of resistance reported worldwide through 1988. F.A.O., Rome, pp.325.

Gettig, R.R. and McCarthy W.J. (1982) Genotypic variation among wild isolates of *Heliothis* spp. nuclear polyhedrosis virus. *Virology*, **117**, 245–52.

Glen, D.M. and Payne, C.C. (1984) Production and field evaluation of codling moth granulosis virus for control of *Cydia pomonella* in the United Kingdom. *Annals of Applied Biology*, **104**, 87–98.

Granados, R.R., Dwyer, K.G. and Derksen, A.G.C. (1987) Production of viral agents in invertebrate cell culture, in *Biotechnology in Invertebrate Pathology and Cell Culture* (ed. K. Maramorosch), Academic Press, pp.167–81.

Granados, R.R. and Hashimoto, Y. (1989) Infectivity of baculoviruses to cultured cells, in *Invertebrate Cell System Applications*, (ed J. Mitsuhashi), CRC Press, Inc., Boca Raton, Florida, Vol II, pp. 3–14.

Gröner, A. (1986) Specificity and safety of baculoviruses, in *The Biology of Baculoviruses*, (eds R.R. Granados and B.A. Federici), CRC press, Inc., Boca Raton, Florida, Vol I, pp.177–202.

Harper, J.D. (1986) Interactions between baculoviruses and other entomopathogens, chemical pesticides and parasitoids, in *The Biology of Baculoviruses*, (eds R.R. Granados and B.A. Federici). CRC press, Inc., Boca Raton, Florida, Vol II, pp.133–56.

Hayashiya, K., Nishida, J. and Uchida, Y. (1976) The mechanism of formation of red fluorescent protein in the digestive juice of silkworm larvae: the formation of Chlorophyllidae-a. *Japanese Journal of Applied Entomology and Zoology*, **20**, 37–43. (in Japanese)

Huber, J. (1986a) *In vivo* production and standardization. *Proc. IVth International Colloquium on Invertebrate Pathology*, Veldhoven (The Netherlands), pp.87–90.

Huber, J. (1986b) Use of baculoviruses in pest management programs, in *The Biology of Baculoviruses*, (eds R.R. Granados and B.A. Federici), CRC Press, Inc., Boca Raton, Florida, vol. II, pp.181–202.

Hughes, P.R., Gettig, R.R. and McCarthy, W.J. (1983) Comparison of the time-mortality response of *Heliothis zea* to 14 isolates of *Heliothis* nuclear polyhedrosis virus. *Journal of Invertebrate Pathology*, **41**, 256–61.

Hull, R., Brown, F. and Payne, C.C. (eds) (1989) *Directory and Dictionary of Animal, Bacterial and Plant Viruses*, Macmillan, London.

Hunter, F.R., Crook, N.E. and Entwistle, P.F. (1984) Viruses as pathogens for the control of insects, in *Microbiological Methods for Environmental Biotechnology*, (eds J.M. Grainger and J.M. Lynch), Academic Press, Inc., London, pp.323–45.

Ignoffo, C.M. (1973) Development of a viral insecticide: Concept of commercialization. *Experimental Parasitology*, **33**, 386–406.

Ignoffo, C.M. and Couch, T. (1981) The nucleopolyhedrosis virus of *Heliothis* species as a microbial insecticide, in *Microbial Control of Pests and Plant Diseases, 1970–1980*, (ed H.D. Burges), London and York, Academic Press, pp.329–62.

Ignoffo, C.M. and Shapiro, M. (1978) Characteristics of baculovirus preparations processed from living and dead larvae. *Journal of Economic Entomology*, **71**, 186–8.

Jaques, R.P. and Morris, O.N. (1981) Compatibility of pathogens with other methods of pest control and with different crops, in *Microbial Control of Pests and Plant Diseases, 1970–1980*, (ed H.D. Burges), London and York, Academic Press, pp.695–715.

Jones, K.A. (1988) The use of insect viruses for pest control in developing countries. *Environmental Aspects of Applied Biology. Aspects of Applied Biology*, **17**, 425–33.

Jutsum, A.R. (1988) Commercial application of biological control: status and prospects. *Philosophical Transactions of the Royal Society, London* B, **318**, 357–73.

Kalmakoff, J. and Crawford, A.M. (1982) Enzootic virus control of *Wiseana* spp. in the pasture environment, in *Microbial and Viral Pesticides*, (ed E. Kurstak), Marcel Dekker, pp.435–48.

Katagiri, K. (1981) Pest control by Cytoplasmic Polyhedrosis Viruses, in *Microbial Control of Pests and Plant Diseases, 1970–1980*, (ed. H.D. Burges), Academic Press, London and New York, pp.433–40.

Kelly, D.C. (1985) The structure and physical characteristics of Baculoviruses, in *Viral Insecticides for Biological Control*, (eds K. Maramorosch and K.E. Sherman), Academic Press, Inc., Orlando, San Diego, New York, London, Toronto, Montreal, Sydney, Tokyo, pp.469–87.

Lee, H.H. and Miller, L.K. (1978) Isolation of genetic variants of *Autographa californica* nuclear polyhedrosis virus. *Journal of General Virology*, **27**, 754–67.

Lewis, F.B. (1984) Formulation and application of microbial insecticides for forest insect pest management: problems and considerations, in *Pesticide Formulations and Application Systems: Third Symposium*, (eds T.M. Kaneko and N.B. Akesson), American Society for Testing and Materials, Special Technical Publication 828, pp.22–31.

Li, Z. (1989) Successi della lotta a microbiologica contro gli artropodi nocivi in Cina. Atti del convegno: *Applicazioi alternative nella difesa delle piante*, Cesena (Italy), 129–38.

Martignoni, M.E. (1984) Baculovirus: an attractive biological alternative, in *Chemical and Biological Controls in Forestry*, (eds W.Y. Gardner and J. Harvey Jr), American Chemical Society Symposium Series No 238, pp.55–67.

Martignoni, M.E., and Iwai, P.J. (1986) *A catalog of viral diseases of insects, mites, and ticks*, 4th ed., General Technical Report Pacific Northwest 195, U.S. Department of Agriculture, Forest Service, Pacific Northwest Research Station, Portland, OR. pp. 50.

Matthews, R.E.F. (1982) Classification and nomenclature of viruses, in *4th Report of the International Committee on the Taxonomy of Viruses*, S. Karger, Basel.

McCutchen, B.F., Choudary, P.V., Crenshaw, R., *et al.* (1991) Development of a recombinant baculovirus expressing an insect-selective neurotoxin: Potential for pest control. *Bio/Technology*, **9**, 848–52.

McGaughey, W.H. (1985) Insect resistance to the biological insecticide *Bacillus thuringiensis*. *Science*, **229**, 193–5.

McIntosh, A.H., Rice, W.C. and Ignoffo, C.M. (1987) Genotypic variants in wild-type populations of baculoviruses, in *Biotechnology in Invertebrate Pathology and Cell Culture*, (ed K. Maramorosch), Academic Press, Harcourt Brace Jovanovich Publishers, San Diego, New York, Berkeley, Boston, London, Sydney, Tokyo, Toronto, pp. 305–25.

Miller, L.K. and Dawes, K.P. (1978) Restriction endonuclease analysis for the identification of baculovirus pesticides. *Applied Environmental Microbiology*, **35**, 411–21.

Moscardi, F. (1990) Development and use of soybean caterpillar baculovirus in Brazil, in *Proceedings of Vth International Colloquium on Invertebrate Pathology*, Adelaide (Australia), pp.184–7.

O'Reilly, D.R. and Miller, L.K. (1989) A baculovirus blocks insect molting by producing ecdysteroid UDP-glucosyltransferase. *Science*, **245**, 1110–12.

Pavan, O.H.O. and Ribeiro, H.C.T. (1989) Selection of a baculovirus strain with a bivalent insecticidal activity. *Memórias do Instituto Oswaldo Cruz, Rio de Janeiro*, **84** (suppl. III), 63–5.

Pennock, G.D., Shoemaker, C. and Miller, L.K. (1984) Strong regulated expression of *Escherichia coli* beta-galactosidase in insect cells with a baculovirus expression vector. *Molecular and Cellular Biology*, **4**, 399–406.

Reiser, M. and Gröner, A. (1990) Effective production of *Cydia pomonella* granulosis virus in the alternate host *Cryptophlebia leucotreta* (Lep.: *Tortricidae*). *Proceedings of Vth International Colloquium on Invertebrate Pathology*, Adelaide (Australia), p.444.

Shapiro, M. (1986) *In vivo* production of baculoviruses, in *The Biology of Baculoviruses*, (eds R.R. Granados and B.A. Federici), CRC Press, Inc., Boca Raton, Florida, vol.II, pp.31–62.

Shapiro, M. and Bell, R.A. (1981) Biological activity of *Lymantria dispar* nuclear polyhedrosis virus from living and virus killed larvae. *Annals of Entomology Society America*, **74**, 27–8.

Shieh, T.R. (1989) Industrial production of viral pesticides. *Advances in Virus Research*, **36**, 315–43.

Smith, D.B. (1971) Machinery and factors that affect the application of pathogens, in *Microbial Control of Insects and Mites*, (eds H.D. Burges and N.W. Hussey), Academic Press, pp.635–54.

Smith, I.R.L. and Crook, N.E. (1988) *In vivo* isolation of baculovirus genotypes. *Virology*, **166**, 240–4.

Smith, D.B., Hostetter, D.L. and Ignoffo, C.M. (1977) Laboratory performance specifications for a bacterial (*Bacillus thuringiensis*) and a viral (Baculovirus *Heliothis*) insecticide. *Journal of Economic Entomology*, **70**, 437–41.

Smith, D.B., Hostetter, D.L. and Ignoffo, C.M. (1978) Formulation and equipment effects on application of a viral (*Baculovirus Heliothis*) insecticide. *Journal of Economic Entomology*, **71**, 814–17.

Smith, G.E. and Summers, M.D. (1978) Analysis of baculovirus genomes with restriction endonucleases. *Virology*, **89**, 519–27.

Smith, G.E., Summers, M.D. and Fraser, M.J. (1983b) Production of human beta-interferon in insect cells infected with a baculovirus expression vector. *Molecular and Cellular Biology*, **3**, 2156–65.

Smits, P.H., Van der Vrie, M. and Vlak, J.M. (1987) Nuclear polyhedrosis virus for the control of *Spodoptera exigua* larvae on glasshouse crops. *Entomologia Experimentalis et Applicata*, **43**, 73–80.

Sosa-gomez, D.R., Alves, S.B. and Marchini, L.C. (1991) Variation in the susceptibility of *Bombyx mori* to nuclear polyhedrosis virus when reared on different mulberry genotypes. *Journal of Applied Entomology*, **111**(3), 318–20.

Stewart, L.M.D., Hirst, M., Ferber, M.L., *et al.* (1991) Construction of an improved baculovirus insecticide containing an insect-specific toxin gene. *Nature*, **352**, 85–8.

Tabashnik, B.E., Cushi, N.L., Finson, N., *et al.* (1990) Field development of resistance to *Bacillus thuringiensis* in diamond back moth (Lepidoptera:Plutellidae). *Journal of Economic Entomology*, **83**, 1671–6.

Thompson, C.G. and Scott, D.W. (1979) Production and persistence of the nuclear polyhedrosis virus of the Douglas-fir tussock moth, *Orgyia pseudotsugata* (Lepidoptera:*Lymantriidae*), in the forest ecosystem. *Journal of Invertebrate Pathology*, **33**, 57–65.

Thompson, C.G., Scott, D.W. and Wickman, B.E. (1981) Long term persistence of nuclear polyhedrosis virus of the Douglas-fir tussock moth, *Orgyia pseudotsugata* (Lepidoptera: Lymantriidae) in forest soil. *Environmemtal Entomology*, **10**, 254–5.

Tomalski, M.D. and Miller, L.K. (1991) Insect paralysis by baculovirus-mediated expression of a mite neurotoxin gene. *Nature*, **352**, 82–5.

Vlak, J.M., Klinkenberg, F.A., Zall, K.J.M., *et al.* (1988) Functional studies on the p10 gene of *Autographa californica* nuclear polyhedrosis virus using a recombinant expressing a p10-β-gal fusion gene. *Journal of General Virology*, **69**, 765–6.

Watanabe, H. (1971) Resistance of the silkworm to the cytoplasmic polyhedrosis virus, in *The Cytoplasmic Polyhedrosis Virus of the silkworm*. (eds H. Aruga and Y. Tanada), University of Tokyo Press, Tokyo, pp.169–84.

Weyer, U. and Possee, R.D. (1989) Analysis of the promoter of the *Autographa californica* nuclear polyhedrosis virus p10 gene. *Journal of General Virology*, **70**, 203–8.

Weyer, U., Knight, S. and Possee, R.D. (1990) Analysis of very late gene expression by *Autographa californica* nuclear polyhedrosis virus and the further developments of multiple expression vectors. *Journal of General Virology*, **71**, 1525–34.

WHO (1973) The use of viruses for the control insect pests and disease vectors. Report of a joint FAO/WHO meeting on insect viruses. World Health Organization; technical report series No 531; pp.48.

Williams, G.V., Rohel, D.Z., Kuzio, J., *et al.* (1989) A cytopathological investigation of AcMNPV p10 gene function using insertion/deletion mutants. *Journal of General Virology*, **70**, 187–202.

Wood, H.A. and Granados, R.R. (1991) Genetically engineered baculoviruses as agents for pest control. *Annual Review of Microbiology*, **45**, 69–87.

Ying, S.L. (1986) A decade of successful control of pine caterpillar, *Dendrolimus punctatus* Walker (Lepidoptera: Lasiocampidae), by microbial agents. *Forest Ecology and Management*, **15**, 69–74.

Young S.Y. (1989) Problems associated with the production and use of viral pesticides. *Memórias do Instituto Oswaldo Cruz, Rio de Janeiro*, **84** (suppl. III), 67–73.

Young, S.Y. and Kring, T.J. (1991) Selection of healthy and NPV infected *Anticarsia gemmatalis* (Lep: *Noctuidae*) as prey by nymphal *Nabis roseipennis* (Hymenoptera:*Nabidae*) in laboratory and in soybean. *Entomophaga*, **36**, 265–73.

Young S.Y. and Yearian, W.C. (1986) Formulation and application of baculoviruses, in *The Biology of Baculoviruses*, (eds R.R. Granados and B.A. Federici), CRC Press, Inc., Boca Raton, Florida, vol. II, pp.157–79.

6

Cyanobacteria and *Azolla*

B.A. Whitton

6.1 INTRODUCTION

Cyanobacteria (blue-green algae) contribute to the fertility of many agricultural ecosystems, either as free-living organisms or in symbiotic association with the water-fern *Azolla* (Fay, 1983). Several species are also taken directly from nature as a food (Jassby, 1988) and, although these natural populations are of only very local importance, strains of the genus *Spirulina* are cultivated in large open-air ponds in several countries (Richmond, 1988). Cyanobacteria are important primary producers in many freshwater fish-ponds and are overall probably the most important primary producers in the oceans (Carr and Wyman, 1986).

Much of their agricultural importance has been attributed to the ability of many strains to fix atmospheric nitrogen, but a number of other types of interaction have been described between cyanobacteria, soils and crops (Roger and Kulasooriya, 1980). Perhaps the most widely reported aspect of cyanobacteria in agricultural systems is the addition of inocula to rice-fields to enhance grain yields; this received considerable publicity in the late 1970s, both in scientific journals (Agarwal, 1979) and in the popular press. Re-evaluation of the literature has, however, made some recent authors (e.g. Roger, 1991) much more cautious about accepting earlier conclusions. Nevertheless cyanobacteria are so widespread in soils and water and associated with such a wide range of symbiotic associations, that it is easy to speculate about ways in which their contribution to crop fertility might be enhanced.

The aim of the present chapter is to review the ways in which cyanobacteria are known to influence crops or might perhaps do so in

Exploitation of Microorganisms Edited by D.G. Jones
Published in 1993 by Chapman & Hall, London ISBN 0 412 45740 7

the future. A brief introduction to the general features of the group is included.

6.2 FEATURES OF CYANOBACTERIA

2.1 General biology

Cyanobacteria are **prokaryotes** possessing the ability to synthesize chlorophyll *a* and at least one phycobilin accessory pigment; typically water is the hydrogen donor during photosynthesis. The text by Fay (1983) provides an easy, though rather dated, introduction to the group. The most recent general reviews are by Castenholz and Waterbury (1989) and Whitton (1992). Detailed accounts of particular topics are given in three edited volumes: Carr and Whitton (1982); Fay and Van Baalen (1987); Mann and Carr (1992). Only those aspects especially relevant to crops are discussed here.

The ecological success of cyanobacteria is favoured by a number of features widespread in the group. The **temperature optimum** is often higher by at least several degrees than that for most eukaryotic algae (Castenholz and Waterbury, 1989), though this cannot fully explain why cyanobacteria tend to be more important in tropical than the equivalent temperate ecosystems, because cyanobacteria are also important in many arctic and antarctic ecosystems. Many cyanobacteria tolerate high levels of **ultra-violet irradiation** (Garcia-Pichel and Castenholz, 1991), whereas others photosynthesize at low photon flux densities (van Liere and Walsby, 1982). Tolerance of desiccation and water stress is widespread (Whitton, 1987) and cyanobacteria are among the most successful organisms in highly saline environments (Borowitzka, 1986), including soils (Singh, 1950, 1961). The ability of many species to fix (di)nitrogen gives a competitive advantage where combined nitrogen levels are low (Howarth *et al.*, 1988).

With the exception of planktonic cyanobacteria, most species tolerate intermittent drying, and soil strains are often highly tolerant of **desiccation** (Whitton, 1987). Those species, which have to withstand long periods of physiological inactivity at the soil surface, usually have cells or filaments surrounded by a sheath containing a brown pigment, scytonemin, absorbing in the near ultra-violet region of the spectrum (Garcia-Pichel and Castenholz, 1991). The evidence strongly suggests that scytonemin production is an adaptive strategy for photoprotection against short-wavelength solar irradiaton. Many filamentous forms survive drying without other obvious external morphological features, although extensive molecular, biochemical and ultrastructural changes are known to occur (Scherer *et al.*, 1984; Potts and Bowman, 1985). Other filamentous forms, all of which possess the ability to form heterocysts and fix nitrogen, form large **akinetes** (spores), which include a high content of stored nitrogen reserve polymer (Whitton, 1987), but apparently not phosphate (as quoted by Roger, 1991). The akinetes develop as a response to nutrient (usually phosphate) or other limitation (Whitton, 1987). It is unclear whether the ordinary filaments of akinete-forming strains are as tolerant of drying as the akinetes, but there is no evidence that akinete-forming strains as a whole are more tolerant of desiccation than those which lack the ability to form akinetes. There are several records of prolonged survival of dried material of the common heterocystous soil species, *Nostoc commune*, which does not form akinetes. One such study, which was conducted with particular care, was that by Trainor (1985), who showed *N. commune* to be viable after 25 years; it seems probable that many common soil species can survive for at least this period, provided reasonable care is taken during storage.

All heterocystous cyanobacteria (excluding laboratory mutants) can apparently fix nitrogen, but a number of non-heterocystous species can also do so (Gallon and Chaplin, 1988; Gallon *et al.*, 1991). Most of the latter are filamentous, but there is at least one colonial unicellular nitrogen-fixer (*Aphanothece*) common in some rice-fields (Roger, 1991). The possession of a heterocyst is one mechanism by which cyanobacteria can protect the dinitrogenase enzyme from damage by oxygen and heterocystous nitrogen-fixers are always associated with well aerated environments. Nitrogen fixation in most non-heterocystous filamentous nitrogen-fixers is enhanced at low ambient oxygen concentrations, so it is not surprising that non-heterocystous soil nitrogen-fixers have mostly been reported from the surface of waterlogged soils (Roger, 1991).

Symbiotic associations

Examples of symbiotic associations are known between cyanobacteria and most major groups of plants and animals (Whitton, 1992). In most, but not all cases which have been investigated, the cyanobacterium fixes nitrogen and part of this fixed nitrogen passes to the partner (Smith and Douglas, 1987). In addition to the *Anabaena–Azolla* symbiosis, other vascular plants with a highly developed symbiotic cyanobacterial association are cycads (*Nostoc*) and *Gunnera* (*Nostoc*). The *Anabaena–Azolla* association differs markedly from the *Gunnera* (and probably most other) association(s), however, in that the *Anabaena* appears highly specialized to this relationship (Zimmerman *et al.*, 1989), whereas re-infection experiments with *Gunnera* indicate that there is considerable, but not absolute promiscuity of the isolated strains (Bonnett and Silvester, 1981; Quispel, 1991). They differ also in that *Anabaena azollae* grows in a well-illuminated environment, whereas the symbiotic *Nostoc* in cycads receives only a very low light flux and that inside *Gunnera* receives no light, although both continue to synthesize chlorophyll; the cyano-bacterium obtains fixed carbon from the host plant. No explanation has been provided for the fact that these successful symbiotic associations have evolved with a photosynthetic nitrogen-fixer rather than the non-photosynthetic nitrogen-fixers found in other vascular plants.

The ability of many cyanobacteria forming dense **water-blooms** to produce extremely toxic substances has been known for many years (Francis, 1878), although, for a variety of reasons, the topic has received particular publicity recently. It is therefore surprising that there are few studies concerning the possibility that other cyanobacteria closely asso-ciated with human activities are toxic. Many of the species which are important in fish-ponds are the same as those forming water-blooms, and aquaculture practice in warmer climates often encourages the growth of cyanobacteria. Cyanobacteria were implicated in Haff's dis-ease, which was common in the 1920s and 1930s along part of the Baltic coast that had a substantial component of fish in the diet (Utkilen, 1992), so presumably toxins are not necessarily destroyed by fish which toler-ate them. Cyanobacteria are conspicuous in many rice-growing areas (refer to section 6.3.1), where farmers often come into contact with dense growths, dried material is distributed by the wind and the water supply may include obvious growths. If toxic strains are widespread, these may contribute to the overall health problems found in regions such as low-lying parts of south-east Asia.

In view of the proposals to develop new symbiotic associations involving cyanobacteria and crop plants (refer to section 6.5), it is worth speculating whether any of the partners in symbiotic cyanobacterial associations may gain resistance to grazing as a result of toxins released by the cyanobacterium. **Cycad sap** is used in Madagascar to produce an alcoholic beverage, which can be dangerously toxic (Murphy, 1985), so studies are needed to establish whether or not cyanobacterial toxins pass to the xylem along with amino acids (Pate *et al.*, 1988). If attempts to establish new symbiotic associations between nitrogen-fixing cyano-bacteria and crops are eventually successful, the possibility that toxins might be introduced into the host plant will need to be considered.

6.2.2 Soil

The general features of cyanobacteria, including some of the key experi-mental methods used in their study and comments on the particular taxonomic diffulties posed by the group, have been summarized by Whitton (1992). A detailed review of soil algae by Metting (1981) includes many other references to cyanobacteria, including a list of 37 genera with species recorded from soils.

Most cyanobacterial **biomass** usually occurs at the surface, with some cells or filaments penetrating several millimetres into the soil (Schwabe, 1963). However, there are many reports of cyanobacteria at greater depths in agricultural soils (Metting, 1981), but it is not clear to what extent populations are able to persist or even increase in the absence of light. Some soil cyanobacteria can grow photoheterotrophically (Reynaud and Franche, 1986) and some can also grow heterotrophically (Khoja and Whitton, 1975). Filamentous forms are probably all capable of movement under some conditions, giving them the potential to reach the surface due to their own physiological activity. **Phototaxis** has been demonstrated in many cyanobacteria (Castenholz, 1982), including strains isolated from soil. However, species differ markedly in the extent to which their filaments exhibit motility in nature. Many heterocystous forms show motile stages for only a short period after nutrient enrichment (Whitton, 1992).

According to Granhall (1975), the major factors (in addition to light) influencing the occurrence of cyanobacteria in soil are moisture, pH, mineral nutrients and combined nitrogen. **Soil moisture** has been shown to be especially important in a number of studies (Zimmerman *et al.*, 1980), with wet depressions containing much higher populations than surrounding dry areas.

Cyanobacteria have never been reported from **pH** values below 4.0 and, in temperate regions, are infrequent in soils at pH values up to about 6.0 (Lund, 1947; Granhall and Henriksson, 1969) and perhaps sometimes entirely absent; otherwise they are widespread in soils, though tending to become increasingly important the higher the pH value. However, some caution is needed when considering the records from soils with low pH values, because unsuitable media may sometimes have been used in attempts to quantify the organisms. This may explain the absence of cyanobacteria reported for many soils in Japan by Watanabe (1959), a result which encouraged that author to initiate studies on the addition of cyanobacterial inocula to soils. Among the **mineral nutrients,** calcium and phosphate are especially effective at enhancing cyanobacterial populations (Roger and Kulasooriya, 1980; Roger, 1991). Although the greater abundance of cyanobacteria in calcareous environments has long been known, there is still no general explanation for it. The importance of added phosphate may be explained in part by the high requirement of nitrogen fixation for phosphate.

Inorganic phosphate mobilization

Several reasons have been suggested for the importance of cyanobacteria in soils in addition to the ability of many to fix nitrogen. Because of the frequent presence of a mucilaginous sheath, they can be highly effective at binding soil particles (Bailey *et al.*, 1973). Another reason is the ability to mobilize insoluble forms of inorganic phosphate. Enough reports have been presented to regard this as well established, but they are brief and need to be followed by more critical studies. All but one of 18 strains tested by Bose *et al.* (1971) solubilized tricalcium phosphate; other materials utilized as P sources by cyanobacteria include Mussorie rock phosphate (Roychoudhury and Kaushik, 1989) and hydroxyapatite (Cameron and Julian, 1988). It has been suggested by several authors (e.g. Natesan and Shanmugasundaram, 1989) that the mechanism by which fixed soil phosphates are solubilized is by means of extracellular phosphatases, but much more critical studies are needed. Another suggestion (Arora, 1969) is that excretion of organic acids by cyanobacteria may play a role in enhancing the availability of phosphate to rice.

Most cyanobacteria can mobilize organic phosphates by means of cell-bound ('surface') phosphatase and often also extracellular phosphatase; this is usually an inducible activity (Healey, 1982). All 50 strains, including many from rice-fields, assayed for their ability to grow with organic phosphates as their sole P source used phosphate monoesters, almost all used diesters, but only some used phytic acid (Whitton *et al.*, 1991). There is some evidence (Singh, 1961; Kaushik and Subhashini, 1985) that cyanobacteria may be particularly effective at increasing phosphorus availability in saline soils, but it is unclear which cyanobacterial features are responsible for this property. The availability of iron and manganese may also be influenced by cyanobacteria. Das *et al.* (1991) showed that cyanobacterial growth in submerged rice soils caused a decrease in ammonium acetate-extractable forms of these elements and increases in other forms. These changes were explained as due to release of O_2, addition of organic matter and liberation of extracellular organic compounds. The decomposition of the cyanobacterial biomass led to further changes in the various fractions, which were ascribed to the development of strong reducing conditions and the formation of organic acids. It was suggested that a decreased content of the readily available form of Fe in soil might help to minimize Zn deficiency in rice.

There has been considerable discussion in the literature on sports turf that cyanobacteria may be involved in the serious nuisance termed 'black layer' (Hodges, 1987), which occurs on dry turf overlying sandy soils subject to frequent irrigation and unsatisfactory drainage. The black layer consists of a subsurface anoxic zone and sometimes a surface dark layer. It seems likely that the cyanobacteria (narrow Oscillatoriaceae) are only associated with the surface layer, but that this growth may enhance the poor drainage and thus reinforce the conditions favouring the black subsurface layer (Baldwin and Whitton, 1992).

6.2.3 Morphological associations with crop plants

Where cyanobacteria are abundant in rice-fields (refer to section 6.3.1), they are often **epiphytic** on submerged parts of the plants themselves. This occurs on the lower parts of rice stems in rice paddies and on the aquatic roots and older parts of leaf sheaths of deepwater rice (Whitton *et al.*, 1988a). The fact that they are not more abundant on aquatic roots is probably due largely to a combination of low light flux and grazing. However, the occurrence of dense masses of one species of cyanobacterium, *Gloeotrichia pisum*, is so widespread on the aquatic roots of the floating grass *Hygroryza aristata* (a popular species for cattle feed) in Bangladesh, but not the roots of other floating plants, as to suggest that the host plant may play a role in encouraging this relationship.

The possibility that cyanobacteria may live **endophytically** in deepwater rice plants was investigated by Kulasooriya *et al.* (1980) and Whitton *et al.* (1988a). Both reported endophytic *Nostoc* and *Calothrix* inside senescent leaf sheaths, but not elsewhere. Even if all the nitrogen fixed by these organisms passed to the rice plant, it would probably supply less than 0.05% of the plant's N requirements (Whitton and Roger, 1989).

In contrast to observations on field grown plants, a number of experimental laboratory studies have demonstrated that cyanobacteria can occur inside higher plant tissues, including the roots of deepwater rice (Kozyrovskaya, 1990). Filaments of a laboratory isolate of *Anabaena thermalis* invaded root hairs, parenchyma, intercellular spaces and xylem. During studies on the regeneration of tobacco plants from mixed cultures of tobacco callus and *A. variabilis*, cyanobacterial filaments were found in intercellular spaces of the primary cortex (Pivovarova *et al.*, 1986). Other studies on cyanobacteria and callus tissues are described in Section 6.5.

Recent laboratory studies have shown that cyanobacteria may also occur in **wheat roots** (Gantar *et al.*, 1991a, b). In addition to loose associations of *Anabaena* with root hairs, much tighter associations occurred with two of the *Nostoc* isolates tested; these isolates were capable of penetrating both the root epidermis and cortex. The restriction of the ability to form tight associations to only a few of the strains tested suggests that specific cyanobacterial and root exudate molecules may be involved in recognition (Gantar *et al.*, 1991a). *Nostoc* 2S9B was found inside root cells, but apparently only cells empty due to the presence of large vacuoles or to degeneration (Gantar *et al.*, 1991b). Co-cultivation of wheat with these cyanobacteria led to a considerable increase in root length, but not root dry weight (Obrecht *et al.*, 1993). However, the nitrogen contents of both roots and shoots increased following co-cultivation. The extent of nitrogen increase apparently depends on both the wheat cultivar and cyanobacterial isolate.

These various reports suggest that careful studies should be made to determine whether or not intracellular cyanobacteria are of quantitative importance in the roots of crop plants at field sites where cyanobacteria are conspicuous at the soil surface. Even if intracellular cyanobacteria do not reach a sufficient density to make a significant contribution to nitrogen fixation, they might have an influence on other micro-organisms inside the roots or provide an inoculum to re-invade soil or root surface populations depleted by grazing.

6.2.4 Other interactions between cyanobacteria and crop plants

In addition to increasing the availability of combined N to rice plants (Section 6.3.2), it has been reported that cyanobacteria can influence the **physiology** of rice and other crops in a variety of other ways. Much of the early documentation from field experiments was based on indirect evidence and should be treated with caution (Roger and Kulasooriya, 1980). However, there are many reports of the presence of cyanobacteria or cyanobacterial extracts influencing **seed germination**. Pre-soaking rice seeds in both cultures and extracts enhanced germination and seedling growth at a site where sulphate reduction was important (Jacq and Roger, 1977). Pre-soaking rice seeds with extracts of *Phormidium* promoted root and shoot growth (Gupta and Shukla, 1969) and a variety of other responses (refer to bibliography in Roger and Kulasooriya, 1980). Based almost entirely on circumstantial evidence, several agents have been suggested as responsible for these effects: gibberellin (Gupta and Shukla, 1969); vitamin B_{12} and amino acids (Venkataraman and Neelakantan, 1967). In a comparison of the effect of two *Anabaena variabilis* strains on rice growing in a greenhouse, Albrecht *et al.* (1991) ascribed the much higher root, shoot and grain dry weights found with one of the strains to its high rate of ammonia release to the environment.

In spite of numerous reports of **growth regulation** by cyanobacteria, there has been no definitive study in which a cyanobacterial regulator has been isolated and characterized (Metting and Pyne, 1986). Further, there must be a suspicion that only positive results have been published, following the screening by Pedurand and Reynaud (1987) of 135 non-axenic cyanobacterial strains in logarithmic growth phase for their effects on rice germination and growth; 30% of strains had no effect on germination and 30% caused inhibition. In contrast, growth of the rice was stimulated by 21% of the isolates and inhibited by 12%. Among eight strains of *Anabaena* stimulating growth, only one remained effective after the cultures had been made axenic. The authors suggested that pre-soaking rice seeds in a cyanobacterial culture should be done only with caution, or avoided altogether.

The use of cyanobacterial extracts to enhance **growth processes** has also been extended to somatic embryos and callus tissues. The frequency of germination of carrot somatic embryos was markedly enhanced when the calcium alginate in which they were embedded was enriched with a hot-water extract of a marine *Synechococcus* (Wake *et al.*, 1991, 1992). The activity was associated largely with the non-dialysable fraction of the extract. The enhancement of shoot and root growth was much greater than that with other methods which have been tested for artificial seed systems. A 'plant restoration accelerator' from *Synechococcus*, *Nostoc* and several other phototrophic microorganisms, which can acclerate growth and shoot formation from callus tissue, has been patented (Japanese patent 3,227,905, 8 October 1991).

6.3 FREE-LIVING CYANOBACTERIA IN RICE-FIELDS

6.3.1 Occurrence and ecology

The importance of nitrogen fixation by cyanobacteria in maintaining the natural fertility of rice-fields was first recognized in Bengal by De (1939). Subsequently, numerous studies have confirmed the importance of these and other nitrogen-fixing organisms in maintaining a moderate, but constant, rice production in fields receiving no nitrogen fertilizer (Roger, 1991). According to Roger and Watanabe (1986), spontaneous biological nitrogen fixation, without fertilizer addition, permits moderate grain yields of about 2 t ha^{-1}. Accounts of cyanobacteria in rice-fields come from many countries (Table 6.1), but much of the earlier information came from India. Singh (1961) not only summarized the early studies, but also pointed out ways in which cyanobacterial growth can be encouraged and thus improve soil fertility. He applied this to saline soils potentially suitable for growing wheat or rice (Singh, 1950) and to more typical rice-fields. More recently, most of the practical interest has focused on rice-fields. Much of the literature to 1980 on cyanobacteria and rice was summarized in the review by Roger and Kulasooriya (1980).

The paddy-field ecosystem provides an especially suitable environment for cyanobacteria, at least during the earlier stages of the crop, since it combines water, light, high temperature and often conditions of reduced Eh and relatively high phosphate. There are reports from many countries of abundant cyanobacteria in rice-paddies, though green algae may largely replace them where high levels of nitrogen

Table 6.1. Literature on cyanobacteria in rice-fields of various regions. (Where many studies have been reported, only the most recent or important are included.)

Africa
Egypt	El-Nawawy, 1973
Mali	Traore *et al.*, 1975
Morocco	Renaut *et al.*, 1975
Nigeria	Moore, 1963

Asia
Bangladesh	Catling *et al.*, 1981; Whitton *et al.*, 1988a,b
India	Gupta, 1966; Tiwari, 1972, 1975; Saha and Mandal, 1979; Mahajan and Patel, 1984; Patel and Mahajan, 1984; Jha *et al.*, 1986; Anand, 1990
Iraq	Al-Kaisi, 1976; Al-Mousawi and Whitton, 1983
Japan	Watanabe, 1965
Pakistan	Ali *et al.*, 1978
Philippines	Roger *et al.*, 1986a
Sri Lanka	Fritsch, 1907; Thurukkanasan *et al.*, 1977; Kulasooriya and Silva, 1981
Thailand	Heckman, 1979
Australia	Bunt, 1961

Europe
Greece	Anagnostidis *et al.*, 1981
Italy	Ciferri, 1963a,b; Bisiach, 1970
Spain	Batalla, 1975

North America
USA	Chapman *et al.*, 1972

fertilizer are used, such as in many fields in Australia, southern Europe and the USA (refer to Table 6.1). Isolates from many countries are included in a culture collection (together with *Azolla* strains) maintained at the International Rice Research Institute in the Philippines (Roger and Ardales, 1991; Watanabe *et al.*, 1992). Cyanobacteria probably contribute the highest proportion of the total algal biomass in saline and/or calcareous soils, such as occur in south-east Iraq (Al-Kaisi, 1976; Maulood *et al.*, 1981). Examples of some more detailed floristic and ecological accounts are given in Table 6.1.

Much of the literature on cyanobacteria in paddy-fields is descriptive, but a number of quantitative studies have been made since plating techniques and other methods were developed (Roger and Reynaud, 1976, 1977; Reynaud and Roger, 1977). Values have been recorded as wet or dry weight, chlorophyll content per unit area and cell counts per unit area or mass of soil. Biomass measurements in rice paddies frequently show values up to several tonnes ha^{-1} wet weight (Roger and Kulasooriya, 1980), though rising as high as 24 tonnes ha^{-1} with a dense *Gloeotrichia* cover in the Philippines (Watanabe *et al.*, 1977). Measurements of chlorophyll from soils with a visually obvious cyanobacterial cover gave values of 2–68 μg chlorophyll *a* cm^{-2} (Rother and Whitton, 1989). Counts (based on colony-forming units, i.e. cells or filaments) range from a few to 10^7 g^{-1} dry soil (Roger and Kulasooriya, 1980).

Marked changes in **biomass** have been reported during the growth season, but the stage at which the peak is reached depends on availability of water, available nutrients and the light flux reaching the soil or water. Mass growths of cyanobacteria occurred in wet fields in Bangladesh before the rice was planted (Rother and Whitton, 1989), especially where the fields were bunded to retain standing water; however, the highest biomass often occurs several weeks after the rice has been planted, especially if fertilizer is added at the same time, as was reported for fields in Japan (Ichimura, 1954). An experimental mixture of *Aulosira*, *Aphanothece* and *Gloeotrichia* was used to measure seasonal changes in biomass and N yield at Cuttack, Orissa (Bisoyi and Singh, 1988b). Biomass production ranged from 3.3 to 366 kg dry weight ha^{-1} month^{-1}, with the highest rates occurring from March to May; analysis of the relationship between seasonal changes in environmental variables and rate of increase in biomass suggested that **solar radiation** was the most important factor.

6.3.2 Nitrogen fixation and nitrogen cycling

Many of the common cyanobacteria in rice-fields are nitrogen fixers, e.g. *Anabaena*, *Aulosira*, *Calothrix*, *Gloeotrichia*, *Scytonema*, *Tolypothrix*. The extent to which cyanobacteria contribute to the nitrogen requirement of the crop is probably determined by the cyanobacterial biomass at different stages of the season, rate of nitrogen fixation, turnover of the fixed nitrogen and the extent to which cyanobacterial nitrogen becomes available to the plant (Roger, 1991). Standing crops of nitrogen-fixing cyanobacteria range from a few kg to 0.5 tonnes ha^{-1} dry weight (Roger *et al.*, 1987). However, values expressed as fresh or dry weight give little information on the agronomic significance of the biomass, because of the wide range of dry matter and ash contents of field-grown cyanobacteria (Roger *et al.*, 1986b). Nitrogen in one tonne fresh weight averages 1.2 kg, but can range from 0.1–4 kg. (The N content of laboratory cultures is usually in the range 5–7% dry weight: Islam and Whitton, 1992.) A visible growth usually corresponds to less than 10 kg N ha^{-1}, while a dense 'bloom' may correspond to 10–20 kg N ha^{-1} (Roger et al., 1986b).

Rates of nitrogen fixation

The rates of cyanobacterial nitrogen fixation have been reported for a number of sites, though mostly in the Philippines by staff of the International Rice Research Institute. The majority of studies have been conducted using the acetylene reduction assay technique without any cross-calibration to values obtained using $^{15}N_2$, so they can be regarded as only approximate. Watanabe et al. (1977) concluded that values of about 30 kg N ha^{-1} (per rice crop) could be expected when environmental conditions are favourable for cyanobacterial growth, and this value has since been taken by most researchers as the amount which ought to be obtained in well-managed paddy-fields, when addition of N fertilizer is kept to a minimum. Values for deepwater rice-fields in Bangladesh were, however, well below this value (Rother et al., 1988). Two fields studied in detail gave values below 10 kg N ha^{-1} during the deepwater rice season; much of this fixation was associated with epiphytic cyanobacteria (mainly *Gloeotrichia*) near the edges of the fields, the cyanobacterial population further into the field being much lower, presumably due to reduced light flux. It is suggested that the dense cyanobacterial epiphyte cover reported (Kulasooriya et al., 1980) for experimental rice plants was due to a much higher light flux than typically encountered in the field.

In spite of the low values for cyanobacterial nitrogen fixation inside deepwater rice-fields and the usual lack of fertilizer, the fields maintain a moderate fertility (Whitton et al., 1988b). This is almost certainly due in part to nitrogen in silt and water contributed by the flood river water. However, cyanobacterial nitrogen fixation probably makes an important contribution, even that occurring inside the fields is minor. This could occur in two ways. Cyanobacterial nitrogen fixation on wet or shallow flooded soils prior to rice planting exceeded that during the flood season (10.2 and 3.8 kg ha^{-1} at the two sites: Whitton et al., 1988b). Fallow areas within the overall deepwater rice-growing area may also make a significant contribution, as submerged weeds such as *Utricularia* often have a very extensive cover of N_2-fixing cyanobacteria. Any contribution of this fixed nitrogen to the rice-fields would depend not only on the rate at which this nitrogen is cycled and released as soluble combined nitrogen, but also on the extent of lateral water movement. Water levels rise and fall several times during a flood season, so lateral nutrient movement seems likely to be important.

Only a few studies have been reported on how the nitrogen fixed by cyanobacteria reaches the rice plant **inside** paddies. Release of extracellular nitrogen may be one source, but the impact of grazers and pathogens, followed by subsequent recycling as soluble combined nitrogen, is probably much more important. It has been suggested that the contribution of nitrogen by cyanobacteria might be enhanced by controlling the grazers (Section 6.3.4). This would be most important at the beginning of the rice season, since nitrogen released later may have no influence on grain yield, though it might be important for the subsequent agricultural crop. Uptake of ^{15}N by rice from that fixed by cyanobacteria and the *Anabaena – Azolla* symbiosis (Section 6.4) has been the subject of several studies. Field and pot trials on the rice cultivar IR32 by Tirol *et al.* (1982) showed that 23–28% of ^{15}N fixed by *Nostoc* sp. reached the first crop. A comparison of the amount of ^{15}N from *Nostoc* and labelled ammonium sulphate remaining in the soil after two crops showed values of 57% and 30–40%, respectively. Mian and Stewart (1984) found that IR8 rice plants received 26–32% of the applied ^{15}N of *Azolla caroliniana* ^{15}N in 60 days. During another 15-day study with IR8 (Mian and Stewart, 1985) with *Azolla*, free-living *Anabaena* and *Nostoc*, 26, 49 and 53%, respectively, of the N applied at the start was released. Of the total N assimilated by the rice plants, 48, 61 and 62% was supplied by *Azolla*, *Anabaena* and *Nostoc*, respectively. ^{15}N from cyanobacteria in a deepwater rice plot has also been shown to reach the rice plant (Watanabe and Ventura, 1982).

6.3.3 Inoculation

The use of cyanobacterial inocula to enhance nitrogen fixation appears to have been first tested systematically in Japan (Watanabe, 1962), because a survey had shown very low or no cyanobacterial populations in some soils (Watanabe, 1959). As discussed in Section 6.2.2, this may have been due to unsuitable culture media being used to isolate organisms from some soils. Watanabe focused his early investigations on *Tolypothrix tenuis*, but subsequently included other species such as *Aulosira fertilissima* (Watanabe, 1973).

Algalization

Venkataraman (1961) first introduced the term, algalization, for the practice of adding cyanobacteria to soils and this is now used widely. According to Metting (1988), algalization is applied to some two million hectares in India, using mixtures of *Anabaena*, *Aulosira*, *Nostoc*, *Plectonema*, *Scytonema* and *Tolypothrix*, and to some 25 000 hectares in China, most often using *Anabaena azotica* and *Nostoc sphaeroides*. The lack of success with preliminary algalization experiments in Japan (Watanabe, 1973) and Taiwan (Huang, 1978) has discouraged further studies there.

Algalization was initiated and promoted in India by the All India Coordinated Project on Algae (Venkataraman, 1972). The methodology to prepare inocula was described by Venkataraman (1981) and publicized widely elsewhere. Cyanobacteria are grown from March to May in open-air, shallow tanks, into which farm soil, superphosphate, starter inocula and sometimes also insecticides are added. In view of the observations of Grant *et al.* (1983: see Section 6.3.4), the last may be especially important if the insecticide activity persists sufficiently long to control grazing after the inoculum has been added to the field. If necessary, lime is added to adjust the soil pH to 7.0–7.5. A thick cyanobacterial mat develops within 15–20 days and the contents of the trays are then allowed to dry, producing flakes which are harvested for distribution. A tray with a surface area of $1.6\,m^2$ produces sufficient material each season to inoculate one hectare. Typically the inocula are added about one week after the rice plants have been transplanted (Sharma and Gupta, 1983).

Most algalization trials have been made using inocula developed from mixes of laboratory isolates, but it is uncertain to what extent these still predominate at the end of the growth period in the open trays. On the assumption that the fastest-growing organisms can double in number every day, it seems likely that marked changes in species composition can occur by the end of 20 days. This is especially likely when farmers prepare inocula on their farms, since the added soil is from local fields. Algalization trials with locally-produced inocula may therefore be much more suitable to local conditions than inocula prepared in another region. Another problem is that inocula have often been supplied without quality control, so the viable cyanobacterial count may be low. An investigation of the ratio of indigenous heterocystous cyanobacteria in 102 soils to heterocystous cyanobacteria in 22 soil-based inocula showed that in 90% of the cases, the indigenous cyanobacteria were more abundant than the inoculum (Roger *et al.*, 1987).

The data also appear to be lacking to support the claim by Agarwal (1979) that the cyanobacteria introduced as a result of algalization can establish themselves almost permanently if inoculation is performed repeatedly for 3–4 cropping seasons. The one quantitative study of the fate of inoculated strains is that of Reddy and Roger (1988). The fate of five laboratory-grown heterocystous strains representing 75% of the inoculum was studied for one month in $1\,m^2$ plots of five different soils. During the month following inoculation, the inoculated strains multiplied to some extent in all soils, but rarely dominated the indigenous cyanobacteria and did so only when the growth of the indigenous nitrogen-fixing species was poor or after the population of indigenous species declined. In a comparison (Bisoyi and Singh, 1988a) of the effects of phosphate fertilizer on production of cyanobacterial biomass in test plots, use of an inoculum from a distant site led to much lower values than obtained with local inocula. There are, however, several situations where the introduction of special inocula may have an advantage. One is where there is marked environmental change brought about by a new farming practice and the native cyanobacteria might be expected to be slow to adapt to the change. This could occur if there is a sudden increase in the rate of fertilizer application or pesticides are introduced into an area where they have not been used previously.

Several authors have suggested that strains with particular features likely to be valuable in the field should be incorporated in the inoculum. Attempts to obtain these have been made by screening a range of strains from enrichment cultures and to obtain mutants in strains already in culture. The former approach has provided fast-growing laboratory strains (Antarikanonda and Lorenzen, 1982), strains releasing ammonia to the environment (Albrecht *et al.*, 1991; Boussiba, 1991; Rai and Prakasham, 1991; Tiwari *et al.*, 1991) and a nitrogen-fixing *Gloeocapsa* highly resistant to the herbicides Machete and Basalin (Singh *et al.*, 1986). In spite of many comments in the literature about the value of strains showing high ammonia release as inocula, there is no evidence that these strains can compete under field conditions. However, the introduction of herbicide-resistant strains may prove valuable, because some herbicides are markedly inhibitory to cyanobacterial growth (Padhy, 1985b; Leganés and Fernandez-Valiente, 1992) and perhaps nitrogen fixation in particular (Kashyap and Pandey, 1982). A number of isolates resistant to herbicides or other pesticides have been reported (Padhy, 1985b; Stratton, 1987), most being the result of laboratory selection during culture with increasing levels of the pesticide or the use of induced mutants. The rapid spread of such strains could be encouraged by inclusion in algalization inocula. Padhy (1985a) warned that care should be taken not to introduce pesticide-resistant strains of species forming nuisance water-blooms. However, this seems unrealistic, both because it must be assumed there is a some risk of *in situ* transfer of pesticide-resistance between species and because resistance is in any case likely to evolve naturally in fields subject to frequent pesticide application.

Another situation where algalization may be useful is where fields undergo especially marked environmental changes during the year and where farming practice during the non-rice season is likely to lead to a marked decrease in suitable inocula for the next wet season. Deepwater rice fields might provide a favourable situation, since some, though not all, species, found at the time of peak flood differ from those found earlier in the season on wet soils or in shallow water (Whitton *et al.*, 1988a).

Success of algalization

Is there evidence that algalization works? A summary of the results from all studies to 1980 (Roger and Kulasooriya, 1980) showed a mean increase in grain yield due to algalization of 28% in pot experiments and 15% in field experiments. The authors pointed out that most experiments had been designed simply to obtain data about grain yield and that little information was available on the N_2-fixing flora, nitrogen-fixing activity or the nitrogen balance in an inoculated paddy-field. Some reviews have accepted the success of algalization as well-established (Agarwal, 1979) and this has encouraged further uncritical studies in other countries. Nevertheless, since the monograph by Roger and Kulasooriya (1980) there have been further apparently well-designed experiments reporting increased grain yield and/or nitrogen content and straw nitrogen content (Venkataraman, 1981; Sharma and Gupta, 1983; Ahmed and Ahmedunnisa, 1984; Singh and Singh, 1985).

In addition to the use of dried inocula, floating gelatinous colonies of *Nostoc* are added to some fields in China in much the same was as *Azolla* (Section 6.4), with populations allowed to develop in ponds and then released into paddy fields when the rice is planted (T.A. Lumpkin, personal communication, quoted in Whitton and Roger, 1989). The system appears to be only local. Experimental studies with similar *Nostoc* spp. in the Philippines led to an increase in grain yield up to 22% (Pantastico and Gonzales, 1976). The extent to which the cyanobacterium increased was influenced markedly by grazers, but this problem was overcome in pot trials with *Nostoc* and rice by including the fish *Tilapia* in the system (Martinez, 1978). It is suggested that this approach has considerable potential in areas where an extensive use of ponds is maintained throughout the year, especially if these are used for aquaculture.

The latest statistical analysis of the general data on algalization is that of Roger (1991). He concluded that (i) the response to inoculation varies, (ii) the response is small, and (iii) the experimental error is larger than the response. The most common design of cyanobacterial inoculation experiments has been 4 × 4 m plots with four replicates, which usually gives a coefficient of variation higher than 10% and a minimum detectable difference of 14.5%, larger than the average yield increase reported after algalization. In addition, Roger (1991) points out that farmers have probably often not bothered to report unsuccessful results, such as in the case described by Pillai (1980). Roger concluded that algalization can increase rice yield, but its effects often seem to be erratic and limited, which may explain the limited adoption of the technique by farmers.

6.3.4 Other methods of increasing cyanobacterial biomass

In spite of the author's scepticism about many results quoted in the literature, there seems little doubt that increasing the cyanobacterial biomass in rice-fields would often enhance nitrogen supply for the rice-plants and perhaps also the subsequent crop. There are a variety of ways this might be achieved in addition to the use of cyanobacterial inocula.

Any extension of the period when the field is **wet** or **flooded to the fallow season** is likely to increase cyanobacterial growths (Singh, 1961; Whitton and Roger, 1989), though it may also encourage weed growths and the need for extra ploughing. Addition of phosphate fertilizer and in some cases also liming (Saha and Mandal, 1980a; Watanabe and Cholitkul, 1982) is likely to enhance nitrogen fixation, but the addition of nitrogen fertilizer has been shown in several studies to **decrease** cyanobacterial growth (Roger and Kulasooriya, 1980) or to influence species composition, though the results are somewhat erratic. Possible explanations for this are the rapid mobilization of N fertilizers in the soil and the frequent simultaneous addition of phosphate and other fertilizers.

The presence of **combined nitrogen** reduces or inhibits cyanobacterial nitrogen fixation, but laboratory strains are known to differ markedly in their response to the sensitivity or the particular N source (Saha and Mandal, 1980b; Prosperi *et al.*, 1992). Inhibition can, however, be largely overcome in the field by deep placement of the N fertilizer. For instance, a field experiment by Roger *et al.* (1980) showed that surface application of urea completely inhibited nitrogenase activity, whereas deep placement of urea supergranules permitted 70% of the nitrogenase of the control. Assessment of the influence of phosphate may be complicated by the fact that species can respond differently. Highly significant differences were found in the production of cyanobacterial biomass in test plots with *Aulosira* spp., *Aphanothece* spp. or *Gloeotrichia* spp. to which phosphate fertilizer had been added (Bisoyi and Singh, 1988a). In an experiment with $17.4 \, kg \, ha^{-1}$ P, the yields for the three cyanobacterial groups were, respectively, 2.5, 3.5 and 5 times higher than the control.

Another means of enhancing the contribution of cyanobacteria to the rice-field ecosystem may be to control the **level of grazing**, but not to reduce it to such a low level that nitrogen cycling is hindered. Such use of controlled grazing was tested by Grant *et al.* (1983) in experimental rice plots by means of commercial pesticides and neem (*Azadirachta indica*) seeds. Suppression of ostracod grazing by Perthane or neem tripled the cyanobacterial biomass and increased N_2 fixation rates 10-fold. In the absence of ostracods, cyanobacteria multiplied rapidly early in the rice cultivation cycle to be succeeded by green algae. Suppression of molluscan grazing had little effect. Total rice grain N increased by up to 37% when grazing was arrested. The authors suggested that early consumption of inocula could explain many reported failures in algalization. They concluded that use of commercial pesticides to control grazers would be uneconomic, but use of cheap preparations of neem to enhance N_2 fixation may be realistic.

6.4 *ANABAENA–AZOLLA* SYMBIOSIS

Cultivation of *Azolla*, with its associated N_2-fixing *Anabaena*, provides another means of adding combined nitrogen to ecosystems in the subtropics. *Azolla* is mostly intercropped with rice in flooded fields, contributing combined nitrogen, green manure and sometimes acting as food for fish in ditches within the rice-fields (Lumpkin, 1986). The surface floating mat of *Azolla* plants can bring about a marked reduction in weed density, including some serious rice-field nuisances (Fosberg, 1942; Satapathy and Singh, 1985). *Azolla* may also be grown with other crops such as *Sesbania* and *Colocasia* (Kannaiyan, 1987). Harvested *Azolla* is used as a green manure and also as a feed for poultry and other animals (Lumpkin and Plucknett, 1982). In contrast to free-living cyanobacteria, it is usually essential to inoculate *Azolla* in order to achieve a significant biomass at the required time. The very extensive literature on *Azolla* and its use in agricultural systems has been reviewed by Lumpkin and Plucknett (1982) and by authors in papers from two symposia (International Rice Research Institute, 1987; Dutta and Sloger, 1991).

There is a long history of *Azolla* cultivation in Vietnam (Moore, 1969) and China (Lumpkin and Plucknett, 1980) and one report (Gopal, 1967) of its use in India, but extensive cultivation elsewhere commenced only in the early 1970s. The recent introductions to other regions have mostly started with trials at research institutes, followed by help from agricultural extension workers, but mixed *Azolla*-rice cultivation is now well established in parts of India (Kannaiyan, 1987) and the Philippines (Mabbayad, 1987). However, *Azolla* shows a number of requirements, which markedly restrict the area suited to its culture. These include the need for permanent ponds to maintain inocula, the requirement for free-standing water for growth in the field and the frequent intolerance of high temperatures under field conditions (see below). On the other hand, *Azolla* is more tolerant of lower pH values than free-living cyanobacteria (Roger *et al.*, 1986c). Lumpkin and Plucknett (1982) estimated that mixed *Azolla*-rice cultivation in China occurred in 2% of the total 34 million ha used for rice. The proportion of the present-day global 150 million ha under rice production is probably rather less than this

(Whitton and Roger, 1989). Nevertheless, *Azolla* often has a very important role in those agricultural systems where it is used.

Many studies have been conducted to find strains especially suited to particular regions and to improve cultivation techniques to enhance grain yield. Under optimum laboratory conditions the doubling time for species widely used in agriculture is about 1.9–2.0 days (Peters *et al.*, 1980; Watanabe and Berja, 1983) and growth rates approaching this probably often occur in the field. However, sustained growth of dense populations in the field is frequently poor at high temperatures, even though the laboratory optima for strains of three widely used species was about 30°C (Peters *et al.*, 1980) or 33°C for *A. mexicana* (Watanabe and Berja, 1983). In the subtropics (Gopal, 1967; Watanabe, 1982) or parts of the tropics with marked seasonal shifts in temperature (Watanabe *et al.*, 1980), field *Azolla* has often been reported to show poor growth during seasons when the temperature is highest. The most likely explanation for this is not a direct response to temperature, but an indirect one due to increased grazing and disease under conditions of high temperature and humidity (Lumpkin, 1987) and increased competition with free-living cyanobacteria. Fiore and Gutbrod (1987) also noted that insect attack occurred mainly in the warm season in Brazil. If grazing and disease are key factors, it should be possible to select or breed for tolerant strains, thus increasing the geographical regions where *Azolla* cultivation is practicable. Because high temperatures do not appear to be directly limiting, *Azolla* has an excellent potential for successful cultivation in irrigated deserts where humidity is relatively low and alternate host plants for insects are limited (Lumpkin, 1987).

As with the free-living nitrogen-fixing cyanobacteria, phosphorus is usually the major limiting element in the field (Roger and Watanabe, 1986); growth is often enhanced markedly by added phosphate fertilizer, nitrogen fixation showing an especially rapid response (Watanabe *et al.*, 1980; Tung and Watanabe, 1983). Marked differences in the phosphate requirement for healthy growth were found between species and, in the case of *A. pinnata*, between strains (Subudhi and Watanabe, 1981). As the plant can accumulate about 10 times the amount of phosphorus required to support its normal nitrogen requirement, the use of phosphorus fertilizer in ponds where the inocula are grown permits the plant to increase in biomass when transferred to paddy-fields, even if the concentration of phosphorus in the field is low (Watanabe *et al.*, 1988). Several studies have also shown a positive response to applications of molybdenum and iron (Singh, 1977; Singh *et al.*, 1991). The *Anabaena–Azolla* association has an advantage over most freeliving cyanobacteria in being less sensitive to inhibition by combined nitrogen, so less caution is needed if a nitrogen fertilizer is added. In one study, the presence of 2.5 mM ammonium led to only a

30% reduction in N_2 fixation (Ito and Peters, quoted by Watanabe, 1982).

6.5 ATTEMPTS TO DEVELOP NEW SYMBIOTIC ASSOCIATIONS

The most direct way to enhance nitrogen fixation in crops would be either to develop new symbiotic associations or to introduce a functional nitrogenase system into cells of the crop. Cyanobacteria would appear to be especially well suited for these purposes, because of the great variety of symbiotic associations that occur between them and other organisms and because most of the cyanobacteria involved are nitrogen-fixers. The well-documented cases of symbiotic nitrogen-fixers all possess heterocysts, which are presumably the sites where the fixation occurs. However, several possible non-heterocystous symbiotic nitrogen-fixers have been reported and a relatively wide range of free-living non-heterocystous cyanobacteria have been studied (Gallon and Chaplin, 1988), including a few records among unicellular forms. These last would seem especially suitable for incorporation into eukaryotic cells.

There have been several successes at incorporating unicellular cyano-bacteria into higher plant protoplasts (Burgoon and Bottino, 1976), but apparently not yet a nitrogen-fixing strain and not yet in a long-term stable relationship (Gamborg and Bottino, 1981). An attempt (Meeks *et al.*, 1978) to produce tobacco protoplasts incorporating the filamentous nitrogen-fixer, *Anabaena variabilis*, was relatively unsuccessful. The pro-toplasts disintegrated in five days, although some cyanobacterial fila-ments remained intact. The most successful systems for mixed culture of a cyanobacterium and higher plant callus tissue have been developed by a research group at Moscow University (Gusev *et al.*, 1984). Mixed cultures of tobacco callus and *A. variabilis* are capable of nitrogenase activity (Gusev *et al.*, 1986). It is not essential for the cyanobacterium to act as an autotroph in such relationships, since the nitrogen-fixing *Chlorogloeopsis fritschii* can grow in mixed culture in the dark (Gusev *et al.*, 1980).

6.6 DISCUSSION AND CONCLUSIONS

It is clear that free-living cyanobacteria play an important role in main-taining soil fertility in many rice-growing areas and that the addition of combined nitrogen to the system is the most important reason for this. Many rice-field ecosystems can be managed such that the nitrogen input from cyanobacteria is maximized, but it is still difficult to assess the extent to which algalization is a useful way of achieving this, because so many investigations have been conducted uncritically. Nevertheless, some researchers with considerable local experience of the topic seem firmly convinced of the success of the approach (Tiwari *et al.*, 1991; L. Venkataraman, personal communication). As discussed in Section 6.3.3, inocula are likely to be of most use when there are sharp fluctuations in the natural cyanobacterial populations or when marked changes in environmental conditions occur, especially if the conditions are those to which the natural population has not previously been subjected. It is suggested that the most useful way for algalization studies to progress is to conduct experiments under conditions where success is most likely. This should, of course, be borne in mind when the overall value of the method is considered.

Whether or not algalization in rice-fields proves to be of widespread importance, there is considerable scope for increasing cyanobacterial biomass by other means, judging by the studies on phosphate addition and grazing (Section 6.3.4). Unfortunately, such work is hindered by the lack of straightforward ecological studies combining reliable taxonomic and quantitative data on cyanobacteria-related processes. However, even where it is possible to enhance the availability of N to the rice plant by algalization or other means, it is unrealistic to believe that modern high-yielding cultivars can be grown entirely without fertilizer (Kulasooriya and Hirimburegama, 1991). Nevertheless, rice cultivars less demanding of nitrogen are frequently grown successfully with *Azolla* alone (Roger *et al.*, 1986c).

The importance of cyanobacteria in other agricultural ecosystems is also well established (Section 6.2.2), but few detailed studies have been reported since that by Singh (1961). The situations where cyanobacteria are most likely to be important are where the soils are moist or flooded for a period before or during the early growth of a crop such as wheat. A combination which might be expected to favour cyanobacterial growth would be waterlogged soil and slight salinity at a season with rising temperatures and light flux. The role of cyanobacteria should be evaluated carefully in regions where such conditions occur, such as parts of south-east Europe and south-east and central Asia. Surveys are needed to establish to what extent the laboratory phenomenon of cyanobacteria inside wheat tissues (Section 6.2.3) also occurs in the field. If climatic changes should lead to warm, moist winters in some temperate regions, this might also be expected to enhance growth of cyanobacteria during the fallow season.

Although the spread of *Azolla* cultivation resulting from recent research has been quite slow, it will probably eventually occur throughout the subtropics, except for countries like Australia and the USA where labour costs are too high. Such expansion will be favoured by breeding new strains and conducting training programmes adapted to local environmental conditions. Acceptance of *Azolla* for purposes additional to its use as a green manure, such as an animal feed, will also be important. The area incorporating *Azolla* cultivation is unlikely to exceed much more than 2% of the world's total rice area, unless the costs of nitrogen fertilizers increase markedly, or strains can be obtained which thrive under fully tropical conditions (Whitton and Roger, 1989).

REFERENCES

Agarwal, A. (1979) Blue-green algae to fertilize Indian rice paddies. *Nature*, **279**, 181.

Ahmed, S.I. and Ahmedunnisa (1984) Utilization of blue-green algae as biofertilizer for paddy cultivation. *Pakistan Journal of Science*, **27**, 355–8.

Albrecht, S.L., Latorre, C., Baker, J.T. and Shanmugam, K.T. (1991) Genetically altered cyanobacteria as nitrogen fertilizer supplier for growth of rice, in, *Biological Nitrogen Fixation Associated with Rice Production* (eds S.K. Dutta and C. Sloger) Oxford and IBH Publishing Co., New Delhi., pp.163–75.

Ali, S., Rajoka, M.I. and Sandhu, G.R. (1978) Blue-green algae of different rice growing soil series of the Punjab. *Pakistan Journal of Botany*, **10**, 197–207.

Al-Kaisi, K.A. (1976) Contributions to the algal flora of the rice fields of Southeastern Iraq. *Nova Hedwigia*, **27**, 813–27.

Al-Mousawi, A.H.A. and Whitton, B.A. (1983) Influence of environmental factors on algae in rice-field soil from the Iraqi marshes. *Arab Gulf Journal of Scientific Research*, **1**, 237–53.

Anagnostidis, K., Economou-Amilli, A. and Tsangridis, A. (1981) Taxonomic and floristic studies on algae from rice-fields of Kalochorion-Thessaloniki, Greece. *Nova Hedwigia*, **34**, 1–189.

Anand, N. (1990) *Handbook of Blue-green Algae (of Rice Fields of South India)*, Bishen Singh Mahendra Pal Singh, 23-A Connaught Place, Dehra Dun, India, pp.80.

Antarikanonda, P. and Lorenzen, H. (1982) N_2-fixing blue-green algae (Cyanobacteria) of high efficiency from paddy soils of Bangkok, Thailand. Characterization of species and N_2-fixing capacity in the laboratory. *Archiv für Hydrobiologie, Supplemente*, **63**, 653–70.

Arora, S.K. (1969) The role of algae on the availability of phosphorus in paddy fields. *Riso*, **18**, 135–8.

Bailey, D., Mazurak, A.P. and Rosowski, J.R. (1973) Aggregation of soil particles by algae. *Journal of Phycology*, **9**, 99–101.

Baldwin, N.A. and Whitton, B.A. (1992) Cyanobacteria and eukaryotic algae occurring in sports turf and amenity grasslands: a review. *Journal of Applied Phycology*, **4**, 39–47.

Batalla, J.A. (1975) *Las Algas de los Arrozales y el Empleo de los Alguicidas*. Valenica, Federacion Sindical de Agricultores de Espana, pp.57.

Bisiach, M. (1970) Primi dati sulle infestazioni algali da Cianoficee nelle risiae italiane. *Riso*, **19**, 129–34.

Bisoyi, R.N. and Singh, P.K. (1988a) Effect of phosphorus fertilization on blue-green algal inoculum production and nitrogen yield under field conditions. *Biology and Fertility of Soils*, **5**, 338–43.

Bisoyi, R.N. and Singh, P.K. (1988b) Effect of seasonal changes on cyanobacterial production and nitrogen-yield. *Microbial Ecology*, **16**, 149–54.

Bonnett, H.T. and Silvester, W.B. (1981) Specificity in the *Gunnera – Nostoc* endosymbiosis. *New Phytologist*, **89**, 121–8.

Borowitzka, L.J. (1986) Osmoregulation in blue-green algae. *Progress in Phycological Research*, **4**, 243–56.

Bose, P., Nagpal, U.S., Venkataraman, G.S. and Goyal, S.K. (1971) Solubilization of tricalcium phosphate by blue-green algae. *Current Science*, **7**, 165–6.

Boussiba, S. (1991) Nitrogen fixing cyanobacteria potential uses. *Plant and Soil*, **137**, 177–80.

Bunt, J.S. (1961) Nitrogen-fixing blue-green algae in Australian rice soils. *Nature*, **192**, 479–80.

Burgoon, A.C. and Bottino, P.J. (1976) Uptake of the nitrogen fixing blue-green algae *Gloeocapsa* into protoplasts of tobacco and maize. *Journal of Heredity*, **67**, 223–6.

Cameron, H.J. and Julian, G.R. (1988) Utilization of hydroxyapatite by cyanobacteria as their sole source of phosphate and calcium. *Plant and Soil*, **109**, 123–4.

Carr, N.G. and Wyman, M. (1986) Cyanobacteria: Their biology in relation to oceanic picoplankton. *Canadian Bulletin of Fisheries and Aquatic Science*, **214**, 59–204.

Carr, N.G. and Whitton, B.A. (1982) *The Biology of Cyanobacteria*, Blackwell, Oxford, and University of California Press, Berkeley, pp.688.

Castenholz, R.W. (1982) Motility and taxes, in *The Biology of Cyanobacteria*, (eds N.G. Carr and B.A. Whitton), Blackwell, Oxford, and University of California Press, Berkeley, pp.413–39.

Castenholz, R.W. and Waterbury, J.B. (1989) Cyanobacteria, in *Bergey's Manual of Systematic Bacteriology* Volume 3, (eds J.T. Staley, M.P. Bryant, N. Pfennig and J.G. Holt), Williams and Wilkins, Baltimore, pp.1710–27.

Catling, H.D., Martinez, M.R. and Islam, Z. (1981) Survey of algae associated with deepwater rice in Bangladesh. *Cryptogamie, Algologie*, **II(2)**, 109–21.

Chapman, R.L., Bayer, D.E. and Lang, N.J. (1972) Observations on the dominant algae in experimental California rice fields. *Journal of Phycology, Supplement*, **8**, 17.

Ciferri, T. (1963a, b) Le alghe delle acque dolci pavesi e vercellesi e le loro associazoni nelle risaie 1. *Riso*, **12(2)**, 30–63; *Riso*, **12(3)**, 31–5.

Das, S.C., Mandal, B. and Mandal, L.N. (1991) Effect of growth and subsequent decomposition of blue-green algae on the transformation of iron and manganese in submerged soils. *Plant and Soil*, **138**, 75–84.

De, P.K. (1939) The role of blue-green algae in nitrogen fixation in rice fields. *Proceedings of the Royal Society of London, Series B*, **127**, 121–39.

Dutta, S.K. and Sloger, C. (1991) *Biological Nitrogen Fixation Associated with Rice Production*, Oxford and IBH Publishing Co., New Delhi, pp.63–75.

El-Nawawy, A.S. (1973) Nitrogen fixing algae from Egyptian soils, in *Global Impacts of Applied Microbiology*, 4th International Conference, São Paulo, Brazil, July 23–28, New York, Unipublications.

Fay, P. (1983) *The Blue-greens, (Studies in Biology No. 160)*, Edward Arnold, London, pp.88.

Fay, P. and Van Baalen, C. (1987) *The Cyanobacteria – A Comprehensive Review*, Elsevier, Amsterdam, pp.543.

Fiore, M.F. and Gutbrod, K.G. (1987) Use of *Azolla* in Brazil, in *Azolla Utilization, Proceedings of the Workshop on Azolla Use, Fujian, China, 31 March-5 April 1985*, International Rice Research Institute, PO Box 933, Manila, Philippines, pp.123–30.

Fosberg, F.R. (1942) Uses of Hawaian ferns. *American Fern Journal*, **32**, 15–23.

Francis, G. (1878) Poisonous Australian lakes. *Nature*, **18**, 11–12.

Fritsch, F.E. (1907) A general consideration of aerial and fresh water algal flora of Ceylon. *Proceedings of the Royal Society of London, Series B,* **11**, 79–197.

Gallon, J.R. and Chaplin, A.E. (1988) Recent studies on N_2 fixation by non-heterocystous cyanobacteria, in *Proceedings of the 7th International Congress on Nitrogen Fixation,* (eds H. Bothe, F.J. de Bruijn and W.E. Newton), Fischer, Stuttgart., pp.183–8.

Gallon, J.R., Hashem, M.A. and Chaplin, A.E. (1991) Nitrogen fixation by *Oscillatoria* spp. under autotrophic and photoheterotrophic conditions. *Journal of General Microbiology,* **137**, 31–9.

Gamborg, O.I. and Bottino, P.J. (1981) Protoplasts in genetic modifications of plants, in *Advances in Biochemical Engineering 19,* (ed. A. Fiechter) Springer-Verlag, Berlin, pp.239–63.

Gantar, M., Kerby, N.W., Rowell, P. and Obrecht, Z. (1991a) Colonization of wheat (*Triticum vulgare* L.) by N_2-fixing cyanobacteria: I. A survey of soil cyanobacterial isolates forming associations with roots. *New Phytologist,* **118**, 477–83.

Gantar, M., Kerby, N.W. and Rowell, P. (1991b) Colonization of wheat (*Triticum vulgare* L.) by N_2-fixing cyanobacteria: II. An ultrastructural study. *New Phytologist,* **118**, 485–92.

Garcia-Pichel, F. and Castenholz, R.W. (1991) Characterization and biological implications of scytonemin, a cyanobacterial sheath pigment. *Journal of Phycology,* **27**, 395–409.

Gopal, B. (1967) Contribution of *Azolla pinnata* R. Br. to the productivity of temporary ponds at Varanasi. *Tropical Ecology,* **8**, 126–9.

Granhall, U. (1975) Nitrogen fixaton by blue-green algae in temperate soils, in *Nitrogen Fixation by Free-living Microrganisms,* (ed. W.D.P. Stewart), Cambridge University Press, Cambridge, UK, pp.189–98.

Granhall, U. and Henriksson, E. (1969) Nitrogen-fixing blue-green algae in Swedish soils. *Oikos,* **20**, 175–8.

Grant, I.F., Tirol, A.C., Aziz, T. and Watanabe, I. (1983) Regulation of invertebrate grazers as a means to enhance biomass and nitrogen fixation of Cyanophyceae in wetland rice fields. *Soil Science Society of America Journal,* **47**, 660–75.

Gupta, A.B. (1966) Algal flora and its importance in the economy of rice fields. *Hydrobiologia,* **28**, 213–22.

Gupta, A.B. and Shukla, A.C. (1969) Effect of algal extracts of *Phormidium* species on growth and development of rice seedlings. *Hydrobiologia,* **34**, 77–84.

Gusev, M.V., Butenko, R.G. and Korzhenevskya, T.G. (1984) Cyanobacteria in association with cultivated cells of higher plants. *Soviet Scientific Reviews, Section D,* 1–40.

Gusev, M.V., Butenko, R.G., Korzhenevskya, T.G., Lobakova, E.D. and Baulina, O.I. (1980) Intercellular symbiosis of suspension ginseng cultures and cyanobacteria. *European Journal of Cell Biology,* **22**, 503.

Gusev, M.V., Korzhenevskaya, T.G., Pyvovarova, L.V., Baulina O.I. and Butenko, R.G. (1986) Introduction of a nitrogen-fixing cyanobacterium into tobacco shoot regenerates. *Planta,* **167**, 1–8.

Healey, F.P. (1982) Phosphate, in *The Biology of Cyanobacteria,* (ed. N.G. Carr and B.A. Whitton), Blackwell, Oxford, and University of California Press, Berkeley, pp.105–24.

Heckman, C.W. (1979) *Rice Field Ecology in Northeastern Thailand. The Effect of Wet and Dry Seasons on a Cultivated Aquatic Ecosystem,* Dr W. Junk, The Hague, pp.228.

Hodges, C.F. (1987) Blue-green algae and black layer. *Landscape Management,* **26**(10), 42–4; **26**(11), 30–1.

Howarth, R.W., Marino, R., Lane, J. and Cole, J.J. (1988) Nitrogen fixation in freshwater, estuarine, and marine ecosystems. 1. Rates and importance. *Limnology and Oceanography*, **33**, 669–87.

Huang, Chi-Ying (1978) Effects of nitrogen fixing activity of blue-green algae on the yield of rice plants. *Botanical Bulletin of Academia Sinica*, **19**, 41–52.

Ichimura, S. (1954) Ecological studies on the plankton in paddy fields. I. Seasonal fluctuations in the standing crop and productivity of plankton. *Japanese Journal of Botany*, **14**, 269–79.

International Rice Research Institute (1987) *Azolla Utilization. Proceedings of the Workshop on Azolla Use, Fuzhou, Fujian, China, 31 March – 5 April 1985*, International Rice Research Institute, PO Box 933, Manila, Philippines, pp.296.

Islam, M.R. and Whitton, B.A. (1992) Cell composition and nitrogen fixation by the deepwater rice-field cyanobacterium (blue-green alga) *Calothrix* D764. *Microbios*, **69**, 77–88.

Jacq, V. and Roger, P.A. (1977) Diminution des fontes de semis dues à la sulfatoréduction, par un prétraitement des graines de riz avec des cyanophycées. *Cahiers ORSTOM Ser. Biol.* **12 (2)**, 1–11.

Jassby, A. (1988) *Spirulina*: a model for microalgae as human food, in *Algae and Human Affairs*, (eds C.A. Lembi and J.R. Waaland), Cambridge University Press, Cambridge, UK, pp.149–79.

Jha, M.N., Jha, U.N., Ahmad, N. and Mallil, M.K. (1986) Cyanobacterial flora of rice field soils of Pusa and its adjoining areas. *Phykos*, **25**, 97–101.

Kannaiyan, S. (1987) Use of Azolla in India, in *Azolla Utilization. Proceedings of the Workshop on Azolla Use, Fuzhou, Fujian, China, 31 March – 5 April 1985*, International Rice Research Institute, PO Box 933, Manila, Philippines, pp.109–118.

Kashyap, A.K. and Pandey, K.D. (1982) Inhibitory effects of rice-field herbicide Machete on *Anabaena doliolum* Bharadwaja and protection by nitrogen sources. *Zeitschrift für Pflanzenphysiologie*, **107**, 339–45.

Kaushik, B.D., and Subhashini, D. (1985) Amelioration of salt-affected soils with blue-green algae. II. Improvement in soil properties. *Proceedings of the Indian National Academy B*, **51**, 386–9.

Khoja, T.M. and Whitton, B.A. (1975) Heterotrophic growth of filamentous blue-green algae. *British Phycological Journal*, **10**, 139–41.

Kozyrovskaya, N. (1990) *Cyanonews*, **6**(3), 5. (Newsletter distributed by MSUDOE Plant Research Laboratory, Michigan State University, East Lansing.)

Kulasooriya, S.A. and Hirimburegama, W.K. (1991) Phototrophic nitrogen fixation in wetland rice fields, in *Biological Nitrogen Fixation Associated with Rice Production*, (eds S.K. Dutta and C. Sloger), Oxford and IBH Publishing Co., New Delhi., pp.191–210.

Kulasooriya, S.A., Roger, P.A., Barraquio, W.L. and Watanabe, I. (1980) Biological nitrogen fixation by epiphytic microrganisms in rice fields. *IRRI Research Papers*, **47**, 1–10. (International Rice Research Institute, PO Box 933, Manila, Philippines.)

Kulasooriya, S.A. and Silva, R.S.Y. (1981) Multivariate interpretation of the distribution of nitrogen-fixing blue-green algae in rice soils in central Sri Lanka. *Annals of Botany, New Series*, **47**, 31–52.

Leganés, F. and Fernandez-Valiente, E. (1992) Effects of phenoxyacetic herbicides on growth, photosynthesis, and nitrogenase activity in cyanobacteria from rice fields. *Archives for Environmental Contamination and Toxicology*, **22**, 130–4.

Lumpkin, T.A. (1986) Current Chinese *Azolla* research. *Azolla Newsletter*, **2(1)**,

1. (Published by International Rice Research Institute, PO Box 933, Manila, Philippines.)

Lumpkin, T.A. (1987) Environmental requirements for successful *Azolla* growth, in *Azolla Utilization, Proceedings of the Workshop on Azolla Use, Fujian, China, 31 March–5 April 1985*, International Rice Research Institute, PO Box 933, Manila, Philippines, pp.89–97.

Lumpkin, T.A. and Plucknett, D.L. (1980) *Azolla*: botany, physiology, and use as a green manure. *Economic Botany*, **34**, 111–53.

Lumpkin, T.A. and Plucknett, D.L. (1982) *Azolla as a Green Manure: Use and Management in Crop Production*. Westview Tropical Agricultural Series, Westview Press, Boulder, CO, USA, pp.230.

Lund, J.W.G. (1947) Observations on soil algae. 2. Notes on groups other than diatoms. *New Phytologist*, **46**, 35–60.

Mabbayad, B.B. (1987) The *Azolla* program in the Philippines, in *Azolla Utilization. Proceedings of the Workshop on Azolla Use, Fuzhou, Fujian, China, 31 March – 5 April 1985*, International Rice Research Institute, PO Box 933, Manila, Philippines, pp.101–8.

Mahajan, A.D. and Patel, R.J. (1984) Some Rivulariaceae RABENH. From paddy fields of Kaira District, Gujarat, India. *Feddes Repertorium*, **93**, 533–9.

Mann, N.H. and Carr, N.G. (eds) (1992) *Photosynthetic Prokaryotes*. Plenum Press, London, pp.275.

Martinez, M.R., Evangelista, C.L. and Pantastico, J.B. (1978) *Nostoc commune* Vauch. as a potential fertilizer in rice-fish culture: a preliminary study. *The Philippine Journal of Crop Science*, **2**, 252–6.

Maulood B.K., Hinton G.C.F., Whitton B.A. and Al-Saadi H.A. (1981) On the algal ecology of the lowland Iraqi marshes. *Hydrobiologia*, **80**, 269–76.

Meeks, J.C., Malmberg, R.L. and Wolk C.P. (1978) Uptake of auxotrophic cells of a heterocyst-forming cyanobacterium by tobacco protoplasts and the fate of their association. *Planta*, **133**, 55–60.

Metting, B. (1981) The systematics and ecology of soil algae. *The Botanical Review*, **47**, 195–312.

Metting, B. (1988) Micro-algae in agriculture, in *Micro-Algal Biotechnology*, (eds M.A. Borowitzka and L.J. Borowitzka), Cambridge University Press, Cambridge. UK, pp.288–304.

Metting, B. and Pyne, J.W. (1986) Biologically active compounds from micro-algae. *Enzyme and Microbial Technology*, **8**, 385–448.

Mian, M.H. and Stewart, W.D.P. (1984) A study on the availability of biologically fixed atmospheric dinitrogen by *Azolla–Anabaena* complex to the flooded rice crops, in *Proceedings of First International Workshop on Practical Application of Azolla for Rice Production*, (eds W.S. Silver and E.C. Schroder), Martinus Nijhoff/Dr W. Junk, The Netherlands, pp.168–75.

Mian, M.H. and Stewart, W.D.P. (1985) Fate of nitrogen applied as *Azolla* and blue-green algae (Cyanobacteria) in waterlogged rice soils – a ^{15}N tracer study. *Plant and Soil*, **83**, 363–70.

Moore, A.W. (1963) Occurrence of non-symbiotic nitrogen-fixing microorganisms in Nigerian soils. *Plant and Soil*, **19**, 385–95.

Moore, A.W. (1969) *Azolla*: biology and agronomic significance. *The Botanical Review*, **35**, 35–57.

Murphy, D. (1985) *Muddling Through in Madagascar*, John Murray, pp.274.

Natesan, R. and Shanmugasundaram, S. (1989) Extracellular phosphate solubilization by the cyanobacterium *Anabaena* ARM310. *Journal of Bioscience*, **14**, 203–8.

Obrecht, Z., Kerby, N.W., Gantar, M. and Rowell, P. (1993) Effects of root associated N$_2$-fixing cyanobacteria on the growth and nitrogen content of wheat (*Triticum vulgare* L.) seedlings. *Biology and Fertility of Soils*, **15**, 68–72.

Padhy, R.N. (1985a) Cyanobacteria employed as fertilizers and waste disposers. *Nature*, **317**, 475–6.

Padhy, R.N. (1985b) Cyanobacteria and pesticides. *Residue Reviews*, **95**, 1–44.

Pantastico, J.B. and Gonzales, J.L. (1976) Culture and use of *Nostoc commune* as biofertilizer. *Kalikasan, Philippines Journal of Biology*, **5**, 221–34.

Pate, J.S., Lindblad, P. and Atkins, C.A. (1988) Pathways of assimilation and transfer of fixed nitrogen in coralloid roots of cycad-*Nostoc* symbioses. *Planta*, **176**, 461–71.

Patel, R.J. and Mahajan A.D. (1984) Some interesting *Anabaena* Bory from paddy fields of Kaira District, Gujarat, India. *Pakistan Journal of Botany*, **16**, 87–93.

Pedurand, P. and Reynaud, P.A. (1987) Do cyanobacteria enhance germination and growth of rice? *Plant and Soil*, **101**, 235–40.

Peters, G.A., Toia, Jr, R.E., Evans, W.R., Christ, D.K., Mayne B.C. and Poole, R.E. (1980) Characterization and comparisons of five N$_2$-fixing *Azolla–Anabaena* associations. I. Optimization on growth conditions for biomass increase and N content in a controlled environment. *Plant Cell Environment*, **3**, 261–9.

Pillai, K.G. (1980) *Biofertilizers in Rice Culture. Problems and Perspectives for Large Scale Adoption*, ACRIP Publication No. 196. (Quoted by Roger, 1991)

Pivovarova, L.V., Korzhenevskaya, T.G., Butenko, R.G. and Gusev, M.V. (1986) Localization of cyanobacteria growing in association with callus tissue and with regenerated plants of tobacco. *Akademia Nauk USSR*, **33**, 74–81. (in Russian)

Potts, M. and Bowman, M.A. (1985) Sensitivity of *Nostoc commune* UTEX 584 (Cyanobacteria) to water stress. *Archives for Microbiology*, **141**, 51–6.

Prosperi C., Boluda, L., Luna, C. and Fernandez-Valiente, E. (1992) Environmental factors affecting nitrogenase activity of cyanobacteria isolated from rice fields. *Journal of Applied Phycology*, **4**, 197–204.

Quispel, A. (1991) A critical evaluation of the prospects for nitrogen fixation with non-legumes. *Plant and Soil*, **137**, 1–11.

Rai, A.N. and Prakasham, R. (1991) Transport of inorganic nitrogen in cyanobacteria and its relevance in use of cyanobacteria as biofertilizers, in *Biological Nitrogen Fixation Associated with Rice Production*, (eds S.K. Dutta and C. Sloger), Oxford and IBH Publishing Co., New Delhi., pp.183–9.

Reddy, P.M. and Roger, P.A. (1988) Dynamics of algal populations and acetylene-reducing activity in five rice soils inoculated with blue-green algae. *Biology and Fertility of Soils*, **6**, 14–21.

Renaut, J., Sasson, A., Pearson, H.W. and Stewart, W.D.P. (1975) Nitrogen-fixing algae in Morocco, in *Nitrogen Fixation by Free-living Microorganisms*, (ed. W.D.P. Stewart), Cambridge University Press, Cambridge, pp.229–46.

Reynaud, P.A. and Franche, C. (1986) Isolation and characterization of non-heterocystous tropical cyanobacteria growing on nitrogen-free medium. *MIRCEN Journal*, **2**, 427–43.

Reynaud, P.A. and Roger, P.A. (1977) Milieux sélectifs pour la numération des algues eucaryotes, procaryotes et fixatrices d'azote. *Revue d'Ecologie et Biologie du Sol*, **14**, 421–8.

Richmond, A. (1988) *Spirulina*, in *Micro-algal Biotechnology*, (eds M.A. Borowitzka and L.J. Borowitzka) Cambridge University Press, Cambridge, 477 pp., pp.85–121.

Roger, P.A. (1991) Reconsidering the utilization of blue-green algae in wetland rice cultivation, in *Biological Nitrogen Fixation Associated with Rice Production*, (eds S.K. Dutta and C. Sloger), Oxford and IBH Publishing Co., New Delhi., pp.119–41.

Roger, P.A. and Ardales, S. (1991) The IRRI blue-green algae collection and computerized information on the strains available for distribution. *Journal of Applied Phycology*, **3**, 375–6.

Roger, P.A. and Kulasooriya, S.A. (1980) *Blue-green Algae and Rice*. International Rice Research Institute, PO Box 933, Manila, Philippines.

Roger, P.A., Kulasooriya, S.A., Tirol, A.C. and Craswell, E.T. (1980) Deep placement: a method of nitrogen fertilizer application compatible with algal nitrogen fixation in wetland rice soils. *Plant and Soil*, **57**, 137–42.

Roger, P.A., and Reynaud, P.A. (1976) Dynamique de la population algale au cours d'un cycle de culture dans une rizière Sahelienne. *Revue d'Ecologie et Biologie du Sol*, **13**, 545–60.

Roger, P.A. and Reynaud, P.A. (1977) La biomasse algale dans les rizières de Sénégal: importance relative des cyanophycées fixatrice de N_2. *Revue d'Ecologie et Biologie du Sol*, **14**, 519–30.

Roger, P.A., Santiago-Ardales, S., Reddy, P.M. and I. Watanabe (1987) The abundance of heterocystous blue-green algae in rice soils and inocula used for application in rice fields. *Biology and Fertility of Soils*, **5**, 98–105.

Roger, P.A., Santiago-Ardales, S. and Watanabe, I. (1986a) Nitrogen fixing blue-green algae in rice soils of Northern Luzon (Philippines). *The Philippine Agriculturalist*, **69**, No 4B, 589–98. (Special Blue-green Algae Issue of the Journal)

Roger, P.A., Tirol, A., Ardales, S. and Watanabe, I. (1986b) Chemical composition of cultures and natural samples of N_2-fixing blue-green algae from rice fields. *Biology and Fertility of Soils*, **2**, 131–46.

Roger, P.A., Voggesberger, M. and Margraf, J. (1986c) Nitrogen-fixing phototrophs in the Ifugao rice terraces (Philippines). *The Philippine Agriculturalist*, **69**, No 4B, 599–609. (Special blue-green Algae Issue of the Journal.)

Roger, P.A. and Watanabe, I. (1986) Technologies for utilizing biological nitrogen fixation in wetland rice: potentialities, current usage, and limiting factors. *Fertilizer Research*, **9**, 39–77.

Rother J.A., Aziz, A., Hye Karim, N. and Whitton, B.A. (1988) Ecology of deepwater rice-fields in Bangladesh. 4. Nitrogen fixation by blue-green algal communities. *Hydrobiologia*, **169**, 43–56.

Rother, J.A. and Whitton, B.A. (1989) Nitrogenase activity of blue-green algae on seasonally flooded soils in Bangladesh. *Plant and Soil*, **113**, 47–52.

Roychoudhury P. and Kaushik, B.D. (1989) Solubilization of Mussorie rock phosphate by cyanobacteria. *Current Science*, **58**, 569–70.

Saha, K.C. and Mandal, L.N. (1979) Distribution of nitrogen fixing blue-green algae in some rice soils of West Bengal. *Journal of Indian Society for Soil Science*, **27**, 470–7.

Saha, K.C. and Mandal, L.N. (1980a) Influence of urea on the nitrogen accretion due to blue-green algae. *Indian Journal of Agricultural Science*, **50**, 431–3.

Saha, K.C. and Mandal, L.N. (1980b) Fixation of nitrogen by blue-green algae in acidic and lateritic rice soils of West Bengal. *Journal of Indian Society for Soil Science*, **28**, 98–102.

Satapathy, K.B. and Singh, P.K. (1985) Control of weeds by *Azolla* in rice. *Journal of Aquatic Plant Management*, **13**, 40–2.

Scherer, S., Ernst, A., Chen, T-W. and Böger, P. (1984) Rewetting of drought-resistant blue-green algae: time course of water uptake and reappearance of respiration, photosynthesis, and nitrogen fixation. *Oecologia (Berlin)*, **62**, 418–23.

Schwabe, G.H. (1963) Blaualgen der photographen Grenzschict (Blaualgen und Lebensraum VII) *Pedobiologia* **2**, 132–52.

Sharma, B.M. and Gupta, R.S. (1983) Effect of algal application on rice yield in Jammu Division. *Phykos*, **22**, 176–9.

Singh, A.L. and Singh, P.K. (1985) *Proceedings of the Indian Academcy of Science (Plant Science)*, **96**, 147–52.

Singh, L.J., Tiwari, D.N. and Singh, H.N. (1986) Evidence for genetic control of herbicide resistance in a rice-field isolate of *Gloeocapsa* sp. capable of aerobic diazotrophy under photoautotrophic conditions. *Journal of General and Applied Microbiology*, **81**, 81–8.

Singh, P.K. (1977) Multiplication and utilization of fern 'Azolla' containing nitrogen-fixing algal symbiont; a green manure in rice cultivation. *Riso*, **26**, 125–37.

Singh, P.K., Singh, D.P., Manna, A.B., Singh, R.P. and Bisoyi, R.N. (1991) Performance of *Azolla caroliniana* in Indian rice fields, in *Biological Nitrogen Fixation Associated with Rice Production*, (eds S.K. Dutta and C. Sloger), Oxford and IBH Publishing Co., New Delhi, pp.95–107.

Singh, R.N. (1950) Reclamation of usar lands in India through blue-green algae. *Nature*, **165**, 325–6.

Singh, R.N. (1961) *Role of Blue-green Algae in Nitrogen Economy of India Agriculture.* Indian Council of Agricultural Research, New Delhi.

Smith, D.C. and Douglas, A.E. (1987) *The Biology of Symbiosis*, Edward Arnold, London, pp.302.

Stratton, G.W. (1987) The effects of pesticides and heavy metals towards phototrophic microorganisms. *Reviews of Environmental Toxicology*, **3**, 71–147.

Subudhi B.P.R. and Watanabe, I. (1981) Differential phosphorus requirements of *Azolla* species and strains in phosphorus-limited continuous culture. *Soil Science and Plant Nutrition*, **27**, 237–47.

Thurukkanasan, A., Kulasooriya, S.A. and Theivendirarajah, K. (1977) A periodic survey of the blue greens found abundantly during the monsoon period (from Sept. to January) in a paddy field near Jaffa Campus at Vaddukoddai and a general survey of blue greens in the Jaffna peninsula during the same period, in, *Proceedings of 33rd Annual Session of the Sri Lanka Association for Advancement of Science*, pp.50–1.

Tirol, A.C., Roger, P.A. and Watanabe, I. (1982) Fate of nitrogen from a blue-green alga in a flooded rice soil. *Soil Science and Plant Nutrition*, **28**, 559–69.

Tiwari, G.L. (1972) A study of the blue-green algae from paddy field soils of India. *Hydrobiologia*, **29**, 335–50.

Tiwari, G.L. (1975) A study of the blue-green algae from paddy field soils of India II. Taxonomic considerations of non-heterocystous blue-green algae. *Nova Hedwigia*, **25**, 765–98.

Tiwari, D.N., Kumar, A. and Mishra, A.K. (1991) Use of cyanobacterial diazotrophic technology in rice agriculture. Scientific note. *Applied Biochemistry and Biotechnology*, **28/29**, 387–96.

Trainor, F.R. (1985) Survival of algae in a desiccated soil: a 25 year study. *Phycologia*, **24**, 79–82.

Traore, K.A., Sasson, A. and Renaut, J. (1975) Contribution à l'étude floristique des cyanophytes du Mali. *Revue d'Ecologie et Biologie du Sol*, **12**, 567–78.

Tung, H.F. and Watanabe I. (1983) Differential response of *Azolla–Anabaena* associations to high temperature and minus phosphorus treatments. *New Phytologist*, **93**, 423–31.

Utkilen, H. (1992) Cyanobacterial toxins, in *Photosynthetic Prokaryotes*, (eds N.H. Mann and N.G. Carr), Plenum Press, New York, pp.211–31.

van Liere, L. and Walsby, A.E. (1982) Interactions of cyanobacteria with light, in *The Biology of Cyanobacteria*, (eds N.G. Carr and B.A. Whitton), Blackwell, Oxford, and University of California Press, Berkeley, pp.9–45.

Venkataraman, G.S. (1961) The role of blue-green algae in agriculture. *Science and Culture*, **27**, 9–13.

Venkataraman, G.S. (1972) *Algal Biofertilizers and Rice Cultivation*, Today and Tomorrow's Printers and Publishers, New Delhi, pp.75.

Venkataraman, G.S. (1981) *Blue-green Algae for Rice Production. A Manual for its Promotion, FAO Soil Bulletin No. 46*, FAO, Rome.

Venkataraman, G.S. and Neelakantan, S. (1967) Effect of the cellular constituents of the nitrogen fixing blue-green alga *Cylindrospermum muscicola* on the root growth of rice plants. *Journal of General and Applied Microbiology*, **13**, 53–61.

Wake, H., Umetsu, H., Ozeki, Y., Shimomura, K. and Matsunaga, T. (1991) Extracts of marine cyanobacteria stimulated somatic embryogenesis of *Daucus carota* L. *Plant Cell Reports*, **9**, 655–8.

Wake, H., Akasaka, A., Umetsu, H., Ozeki, Y., Shimomura, K. and Matsunaga, T. (1992) Enhanced seed germination of artificial seeds by marine cyanobacterial aextract. *Applied Microbiology and Biotechnology*, **36**, 684–8.

Watanabe, A. (1959) Distribution of nitrogen-fixing blue-green algae in various areas of south and east Asia. *Journal of General and Applied Microbiology*, **5**, 21–9.

Watanabe, A. (1962) Effect of nitrogen-fixing blue-green alga *Tolypothrix tenuis* on the nitrogenous fertility of paddy soil and on the crop yield of rice plant. *Journal of General and Applied Microbiology*, **8**, 85–91.

Watanabe, A. (1965) Studies on the blue-green algae as green manure in Japan. *Proceedings of the National Academy of Sciences of India*, **35**, 361–9.

Watanabe, A. (1973) On the inoculation of paddy fields in the Pacific area with nitrogen-fixing blue-green algae. *Soil Biology and Biochemistry*, **5**, 161–2.

Watanabe, I. (1982) *Azolla–Anabaena* symbiosis – its physiology and use in tropical agriculture, in *Microbiology of Tropical Soils and Plant Productivity*, (eds Y.R. Dommergues and H.G. Diem), Martinus Nijhoff/Dr W. Junk Publishers, The Hague, pp.169–85.

Watanabe, I., Berja, N.S. and del Rosario, D.C. (1980) Growth of *Azolla* in paddy field as affected by phosphorus fertilizer. *Soil Science and Plant Nutrition*, **26**, 301–7.

Watanabe, I. and Cholitkul, W. (1982) Nitrogen fixation in acid sulfate soils. *Tropical Agricultural Research Series* No. 15, 219–26.

Watanabe, I. and Berja, N.S. (1983) The growth of four species of *Azolla* as affected by temperature. *Aquatic Botany*, **15**, 175–85.

Watanabe, I., Lee, K.K., Alimagno, B.V., Sato, M., Del Rosario, D.C. and De Guzman, M.R. (1977) Biological N_2 fixation in paddy field studied by *in situ* acetylene-reduction assays. *IRRI Research Papers Series*, **3**, 1–16.

Watanabe, I., Lupis, H.T., Oliveros, R. and Ventura, W. (1988) Improvement of phosphate fertilizer application to *Azolla*. *Soil Science and Plant Nutrition*, **34**, 557–69.

Watanabe, I., Roger, P.A., Ladha, J.K. and van Hove, C. (1992) *Biofertilizer Germplasm Collections at IRRI*, International Rice Research Institute, PO Box 933, Manila, Philippines, pp.66.

Watanabe, I. and Ventura, W. (1982) Nitrogen fixaton by blue-green algae asociated with deepwater rice. *Current Science*, **51**, 462–5.

Whitton, B.A. (1987) Survival and dormancy of blue-green algae, in *Survival and Dormancy of Micro-organisms*, (ed Y. Henis), Wiley, New York, pp.209–66.

Whitton, B.A. (1992) Diversity, ecology and taxonomy of the cyanobacteria, in

Photosynthetic Prokaryotes, (eds N.H. Mann and N.G. Carr), Plenum Press, New York, pp.1–51.

Whitton, B.A., Aziz, A., Kawecka, B. and Rother, J.A. (1988a) Ecology of deepwater rice-fields in Bangladesh. 3. Associated algae and macrophytes. *Hydrobiologia*, **169**, 31–42.

Whitton B.A., Grainger, S.L.J. and Simon, J.W. (1991) Cell-bound and extracellular phosphatase activities of cyanobacterial isolates. *Microbial Ecology*, **21**, 85–98.

Whitton B.A., Hye Karim, N. and Rother J.A. (1988b) Nutrients and nitrogen fixation in Bangladesh deepwater rice-fields, in *Proceedings of the 1987 Workshop on Deepwater Rice, Bangkok*, International Rice Research Institute, PO Box 933, Manila, Philippines, pp.67–74.

Whitton B.A. and Roger, P.A. (1989) Use of blue-green algae and *Azolla* in rice culture, in *Microbial Inoculation of Crop Plants. Special Publication of the Society for General Microbiology 25*, (ed. R. Campbell and R.M. Macdonald), IRL Press at Oxford University Press, Oxford, New York, Tokyo, pp.89–100.

Zimmerman, W., Metting, B. and Rayburn, W. (1980) The occurrence of blue-green algae in silt loams of Whitman County, Washington. *Soil Science*, **130**, 11–18.

Zimmerman, W.J., Rosen, B.H. and Lumpkin T.A. (1989) Enzymatic, lectin and morphological characterization and classification of presumptive cyanobionts from *Azolla* Lam. *New Phytologist*, **113**, 497–503.

7

Mycorrhizal associations

Derek T. Mitchell

7.1 INTRODUCTION

Root systems of plants are known to have a profound effect on soil microorganisms, particularly fungi. Some fungi develop an intimate symbiotic association producing what is termed a mycorrhiza ('fungus root'). Mycorrhizal fungi are largely dependent on the host plant for their energy and carbon requirements and generally have poor ability to exist saprotrophically in the soil. A further distinction between mycorrhizal and other (saprotrophic and pathogenic) fungi is that they are supposed to aid the establishment, growth and development of plants. They also benefit the host plant by improving mineral nutrition, increasing drought tolerance, supplying growth-regulating substances and vitamins and protecting against pollutants and soil-borne pathogens. The role of mycorrhizal fungi in the improvement of growth and mineral nutrition of plants has been studied extensively (Harley, 1969; Harley and Smith, 1983). Mycorrhizas help the plant to utilize soil nutrients, especially if they are growing in low nutrient soils. They have been shown to utilize recalcitrant forms of nutrients such as organic nitrogen and phosphorus, which would normally be unavailable to the root. Mycorrhizal fungi can cleave organic nitrogen and phosphorus compounds by secreting enzymes such as proteases and phosphatases. Either hyphae (the vegetative form of fungi) or mycelial-cords (an aggregation of hyphae – Cairney, 1991) grow across nutrient depletion zones in the soil around the root, exploring the soil and conveying nutrients back to the root. Cords have also been implicated in the transport of water to the root, thus allowing the plant to survive drought

Exploitation of Microorganisms Edited by D.G. Jones
Published in 1993 by Chapman & Hall, London ISBN 0 412 45740 7

(Duddridge *et al.*, 1980). The production of vitamins by mycorrhizal fungi has been shown to increase root size and longevity (Ek *et al.* 1983). Examples of pollutants, from which the mycorrhiza may protect the plant, are acid rain, aluminium toxicity, sulphur dioxide and nitrous oxide (Jeffries, 1987). The amelioration of heavy metal toxicity by mycorrhizas may be associated with the protective fungal mantle, binding of metals to roots or fungi, or interactions between heavy metals and anions (Dixon and Buschena, 1988). The mycorrhiza may protect the root against pathogens by either acting as a mechanical barrier against infection or secreting antibiotics (Marx, 1969; Marx and Krupa, 1978). The defence against pathogen attack by the mycorrhiza can be indirect by restricting release of exudates or altering the microbial populations in the rhizosphere (Marx, 1973). There is evidence that some mycorrhizal fungi are involved in the elicitation of phytoalexins and associated isoflavonoids in the root (Morandi and Gianinazzi–Pearson, 1986; Gianinazzi *et al.*, 1990).

7.1.1 Historical aspects and classification

Mycorrhizal associations have existed since the Devonian period and probably evolved to allow both partners to survive extreme environmental conditions (Pirozynski and Malloch, 1975). Some primitive plant orders, e.g. Magnoliales, are obligately dependent on mycorrhizas (Baylis, 1975). Most flowering plants, Conifers and Ferns are mycorrhizal (Harley and Harley, 1987) but some flowering plant families contain non-mycorrhizal members. These include the Cruciferae (Cabbage family), Chenopodiaceae (Goosefoot family including a number of weeds as well as Sugar beet), Resedaceae (Mignonette family), Cyperaceae (Sedge family) and two families found in the southern hemisphere (Proteaceae and Restionaceae). In members of the Cruciferae, the host roots lack a diffusible growth stimulus for mycorrhizal infection to take place (Glenn *et al.*, 1988). In other families, there are special structures such as proteoid roots in the Proteaceae and dauciform roots in the Cyperaceae, which allow them to perform similar functions to mycorrhizas (Lamont, 1983).

Mycorrhizas are divided into two broad groups known as **ectotrophic** and **endotrophic**. There is a small group of ecto/endotrophic mycorrhizas, which are termed arbutoid if they are associated with *Arbutus* and *Arctostaphylos* (Ericaceae) or monotropoid if associated with members of the Monotropaceae (Harley and Harley, 1987). Endotrophic mycorrhizas are the most widespread and are further divided into vesicular-arbuscular (VA), ericoid, and orchidaceous mycorrhizas. The vesicular-arbuscular mycorrhizas (VAM) are the most common and have been found in roots of over 1000 genera of 200 families of vascular plants and are very important in that they are associated with roots of economically-important crop plants, shrubs and trees. Ectotrophic mycorrhizas are only found on about 10% of the world flora. These are mainly on trees of temperate regions such as Pinaceae (Pine, Fir, Larch), Fagaceae (Oak, Chestnut, Beech), Betulaceae (Alder, Birch) and Salicaceae (Willow) and a southern Hemisphere Family (Myrtaceae, Gum family) (Meyer, 1973; Harley and Harley, 1987).

7.2 TYPES OF MYCORRHIZAL ASSOCIATION

7.2.1 Ectomycorrhizas

Ectomycorrhizas are mainly short roots, which become modified by the fungus into sheathing mycorrhizas. There is a characteristic anatomy in which a sheath (mantle) and Hartig net are formed in the root (Figure 7.1). The sheath surrounds the root, varies in thickness from one to many layers and can be up to 100 μm thick. The Hartig net is a network of hyphae, which penetrates between the cortical cells, the extent of penetration of the cortex varying but never going beyond the endodermis. The Hartig net of Conifer mycorrhizas may penetrate as far as the endodermis but in others such as Beech, the penetration may only be

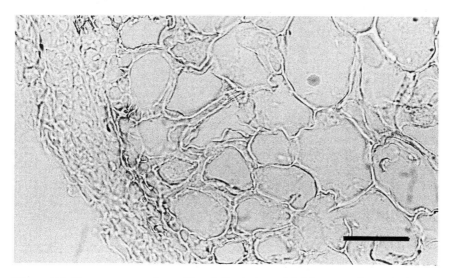

Figure 7.1 Transverse section of beech/*Lactarius subdulcis* mycorrhiza. Scale = 25 μm.

as far as the outer two layers of the cortex. Some mycorrhizal fungi such as the E-strain fungus (*Complexipes moniliformis*), form sheathing mycorrhizas on nursery-grown Spruce and ectendomycorrhizas with nursery-grown Pine (Thomas and Jackson, 1979, 1982). The host root may also be modified, the outer cortical cells becoming brown due to secretion of polyphenols.

The fungi mainly belong to the Agaricales, Hymenomycetes, Basidiomycotina and form conspicuous fruit bodies such as mushrooms, toadstools, puffballs and earthballs on the forest floor during autumn and spring. Some are edible, such as the Cep (*Boletus edulis*), Chantarelle (*Cantharellus cibarius*) and Pine ring (*Lactarius deliciosus*) and these could form a secondary crop in forests, but further research must first be undertaken (Jeffries, 1987). A few, such as *Humaria hemispherica* and *Tricharina gilva* are found in the Ascomycotina; another well-known ectomycorrhizal Ascomycete is *Tuber melanosporum*, which produces truffles and is in commercial production in France, Italy and New Zealand. Truffle plantations (truffières) are usually established with mixed plantings of mycorrhizal Oak and Hazel seedlings (Chevalier and Frochot, 1980). If these truffières are properly maintained, fruit bodies can be cropped in the third to fifth year and production can continue up to 35–50 years (Anon, 1989).

Morphology

The morphology of ectomycorrhizas ranges from being unbranched to pinnate through to varying degrees of branching. Some have striking colours (e.g. *Cenococcum geophilum* are jet black, *Paxillus involutus* have a silvery appearance and *Laccaria amethystina* have violet tips). The mycorrhizal surface may be either smooth or woolly with dense wefts of hyphae. Hyphal aggregates may emanate from the surface as either cords (*Thelephora terrestris*), sclerotia (almost rounded bodies of hyphae – *C. geophilum*) or both (*P. involutus*). The presence of cords indicates that the mycorrhizal associations are actively involved in the transport of nutrients and water (Duddridge *et al.*, 1980; Finlay and Read, 1986), whereas the formation of sclerotia is an indication that the fungus can survive extreme environmental conditions (Marx and Cordell, 1989). In inoculation programmes, one of the initial criteria for selection of ectomycorrhizal fungi has been the ability of the fungus to produce cords and sclerotia (Marx and Cordell, 1989).

The fungi associated with ectomycorrhizas can now be identified according to specific features of their morphology and anatomy (Agerer, 1987–1990; Ingleby *et al.*, 1990). In the future, the detection of diagnostic fungal isozymes by means of polyacrylamide gel electrophoresis (PAGE) will also help in their identification (Sen, 1990).

7.2.2 Vesicular-arbuscular mycorrhizas (VAM)

These have been described as the most ubiquitous mycorrhizas (Amijee, 1989) and are associated with the two largest flowering plant families (Asteraceae and Poaceae). Some examples of the Bryophyta (Liverworts such as *Pellia* and *Marchantia* and Mosses such as *Funaria*), Pteridophyta (Ferns such as *Pteridium*, *Polysticum* and *Osmunda*) and Gymnosperms (Conifers such as *Taxus*, *Sequoia* and *Gingko*) are also VA mycorrhizal. Deciduous trees, which are VA mycorrhizal, include *Acer* (Maples and Sycamore), *Fraxinus* (Ash), *Juglans* (Walnut) and *Sorbus* (Rowan). Most indigenous trees in the tropical rain forests are VA mycorrhizal.

Morphology

The overall morphology of the root is not affected by the VAM association in that root hairs are still present, although maize and onion VAM roots may be a bright yellow colour. Their characteristic features are the production of arbuscules (Figure 7.2) and vesicles (Figure 7.3) within the cortex of the root. The arbuscules are the active part of the association in that they are analogous to haustoria of rust and mildew pathogens in being involved in nutrient exchange. The vesicles function as storage structures for nutrients. The fungi forming VAM cannot be identified on the basis of their root infection, but they have been distinguished as either coarse or fine endophytes (McConigle and Fitter, 1990). It is only

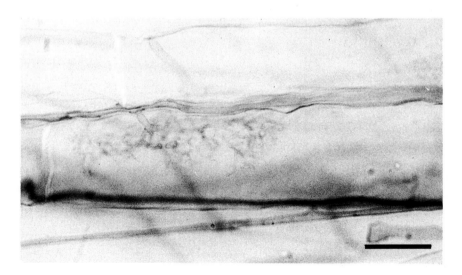

Figure 7.2 Cortical cells of onion root showing intercellular hyphae and an arbuscule of VAM fungus. Scale = 25 μm.

Mycorrhizal associations

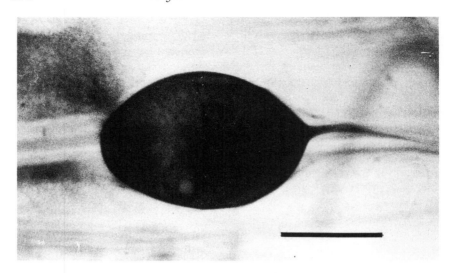

Figure 7.3 Cortical cell of acacia root showing vesicle of VAM fungus. Scale = 25 μm.

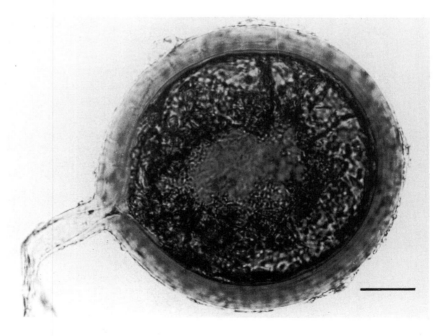

Figure 7.4 Chlamydospore of *Glomus* sp., a VAM fungus. Scale = 25 μm.

structures outside the root such as spores, e.g. azygospores and chlamy-dospores (Figure 7.4) and extramatricle vesicles that can be used for identification purposes. VAM fungi belong to the Endogonales, Zygomycetes and the genera are *Glomus, Sclerocystis, Gigaspora, Scutello-spora, Acaulospora, Glaziella* and *Entrophospora*. They are identified by means of their azygospores and chlamydospores on the basis of their size, shape, pigmentation, wall ornamentation, germination characteris-tics, hyphal connection, method of spore formation and mode of occlu-sion of spore contents. The wall structure can be graphically represented (muragraph) and is a means of identification (Walker, 1983). Both sexual and asexual spores have been observed in *Gigaspora decipiens* (Tom-merup and Sivasithamparam, 1990). Since identification based on mor-phological characters has been difficult, other methods are being devel-oped such as detection of protein isozymes by PAGE (Sen and Hepper, 1986) and use of serological methods with fluorescent antibodies (Wright *et al.*, 1987).

7.2.3 Ericoid endomycorrhizas

These are confined to members of the Ericales such as *Calluna* (Ling), *Erica* (Heather), *Rhododendron* and *Vaccinium* (Cranberry, Blueberry and Bilberry) and have a world-wide distribution (Read, 1983). These plants have a dense root system ending in very fine, branched rootlets termed **hair roots**. The hair roots have a narrow central stele surrounded by

Figure 7.5 Root squash of cranberry showing ericoid endomycorrhizal infection. Scale = 25 μm.

1–3 layers of cortical cells, which contain the endomycorrhizal infection (Figure 7.5). One of the features of these roots is that they lack root hairs and an epidermis and rely entirely on the external hyphae of their mycorrhizal association to obtain nutrients from the soil. One fungus, which was isolated from an ericoid endomycorrhiza, has produced fruit bodies (apothecia) (Read, 1974) and is now identified as *Hymenoscyphus ericae* belonging to the Discomycetes, Ascomycotina (Kernan and Finocchio, 1983). *Oidiodendron maius* has also been identified as an ericoid endomycorrhizal fungus of *Rhododendron* (Douglas *et al.*, 1989).

7.2.4 Orchidaceous mycorrhizas

These are the mycorrhizas of orchids and have characteristic coils of septate hyphae or less regular hyphal aggregates within the cortical cells of the roots, called pelotons (Figure 7.6). The pelotons persist for a time and then degenerate. Orchids have small seeds, which contain little food reserves and on germination, the majority have to be colonized by a fungus before they develop further. The orchid may rely on its fungal partner for its organic carbon and energy. All the fungi known to be mycorrhizal on orchids are saprotrophic in the soil but some act as root pathogens of crops and trees. The fungi are mainly species of *Rhizoctonia* belonging to the Deuteromycotina or *Armillaria, Ceratobasidium, Fomes, Sebacina, Thanatephorus* and *Tullasnella* of the Basidiomycotina (Harley and Harley, 1987).

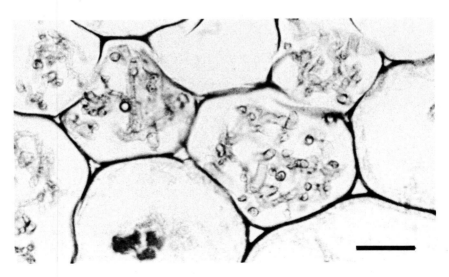

Figure 7.6 Cortical cells of an orchid root showing orchidaceous endomycorrhizal infection. Scale = 25 μm.

7.3 ROLE OF MYCORRHIZAS IN AGRICULTURE, FORESTRY AND HORTICULTURE

Apart from a few families in the flowering plants, mycorrhizal infection of most plants is indispensible for their establishment and growth and it may be essential to inoculate seedlings with the requisite symbiont under particular circumstances. One common horticultural and forestry practice is **transplantation** and there is good evidence that mycorrhizal plants survive transplantation better than non-mycorrhizal ones (Parke *et al.*, 1983). Seedlings of most naturally-occurring Orchids are entirely dependent on the presence of their mycorrhizal partner for their development (Harley and Harley, 1987). However, research on the nutrient requirements for germination has been undertaken (Knudson, 1922) and efficient culture media are now routinely used instead of mycorrhizal inoculation in commercial horticulture. Further work has to be pursued to determine whether mycorrhizal inoculation of adult orchid plants is necessary (Gianinazzi *et al.*, 1990). In the case of other mycorrhizal associations, the need to inoculate will depend on the presence or absence of indigenous mycorrhizal populations in the soil and also on their ability to form mycorrhizas on the roots. Exotic trees may require inoculation before they are introduced to new sites, which have either not been afforested or in the case of the Tropics, lack the essential mycorrhizal fungi (Marx, 1980a). Another example is the introduction of ericaceous fruit crops such as Blueberry to pastures, which lack the ericoid endomycorrhizal fungus (Powell, 1981). The inoculation of Blueberry has been advocated (Powell, 1982) and inoculation at outplanting can increase fruit yields (Powell and Bates, 1981).

7.3.1 Soil disinfection

Most modern commercial nurseries either disinfect soils before planting or use artificial potting mixtures and pursue micropropagation techniques. In France, soil **fumigation** in shrub and herbaceous plant nurseries is standard practice to eradicate soil-borne pathogens, but indigenous VAM fungi may also be eliminated (Gianinazzi *et al.*, 1989). In such situations, very patchy growth is known to occur and it may be essential to inoculate the plants with VAM fungi. In France, some success appears to have been achieved with VAM inoculation in the propagation of woody ornamentals such as *Ampelopsis quinquefolia*, *Liquidambar styraciflua* and lilac (*Syringa vulgaris*) (Gianinazzi *et al.*, 1989). Soil fumigation is also practised in the tree bare-root nurseries of Ireland and the United Kingdom. There is evidence that populations of nursery-type ectomycorrhizal fungi alter as a result of Dazomet treatment in a brown earth soil (Grogan and Mitchell, 1990). In another inoculation experiment in a light sandy soil, the same fumigation appears to eliminate all the nursery-type ectomycorrhizal fungi (Wilson *et al.*, 1990).

7.3.2 Nutrient uptake

Many demonstrations have shown VAM to be more efficient than non-mycorrhizal roots at taking up nutrients, particularly **phosphorus**, but most of these studies have been undertaken in sterilized soils (Powell, 1984). However, the application of phosphate fertilizer may negate the effect of VAM infection on plant growth such as barley (Clarke and Mosse, 1981) and the number of VAM infection entry points may be reduced under high phosphate conditions (Amijee, 1989). In well-developed agricultural systems, the most direct effect that VAM fungi may have on the physiology of crop plants is in the uptake of phosphorus (Stribley *et al.*, 1980). In such situations, mycorrhizal inoculation is unlikely to have much impact on crop production, especially when the cost of superphosphate fertilizer is not high (Stribley, 1990). In undeveloped soils, VAM inoculation may also be unnecessary, as crop husbandry could be modified to make a more efficient use of the indigenous VAM fungi (Stribley, 1990). Either crop rotation, which includes the use of a readily VA-infected plant, or the application of selective fungicides, may be ways of stimulating indigenous VAM fungi in the soil, but these require further investigation. It will only be in those soils where there is a complete lack of VAM fungi that inoculation will be necessary. The determination of the VAM inoculum potential of the soil is thus a prerequisite before any attempt to inoculate is considered (Powell, 1980).

In forests, particularly on organic soils such as Conifer stands in the Tundra, there is growing evidence of ectomycorrhizal fungi influencing the uptake of **nitrogen** as well as phosphorus in trees. Some ectomycorrhizal fungi can utilize protein-nitrogen (Abuzinadah and Read, 1986) and they might provide direct access to simple organic nitrogen reserves in the soil (Harley, 1969; Laiho, 1970; Lundeberg, 1970; Bowen and Smith, 1981; Alexander, 1983). Specific ectomycorrhizal associations of trees may lead to a more effective competition with the decomposer populations and to an overall tightening of the nitrogen cycle (Abuzinadah *et al.*, 1986). The distinction between different ectomycorrhizal fungi in their ability to utilize various forms of nitrogen and other nutrients may be important in their selection for inoculation of tree seedlings prior to transplanting onto either mineral or organic soils.

7.3.3 Forest and soil characteristics

Forests are known to have characteristic populations of ectomycorrhizas, which may vary between tree species and differ with age of individual trees (Zak and Marx, 1964). One species may form numerous biotypes or clones in a very limited area of a pure stand (Fries, 1987; Dahlberg and Stenlid 1990). Most ectomycorrhizal fungi have a broad host range (e.g. *P. involutus*) but some are restricted to few species (e.g. *Suillus grevillei* on Larch) (Duddridge, 1986). Forests with a mixture of tree species have a more diverse ectomycorrhizal population than mono-cultural stands and the majority of the ectomycorrhizal fungi are shared between tree species (Heslin *et al.*, 1992). Mycorrhizal populations may also be influenced by the seasonal changes in rainfall and temperature (Wilkins and Harris, 1946). The development of mycorrhizal coloniza-tion of Sitka spruce seedlings on Irish peaty and podsolic soils may vary according to the number of forest rotations each soil has been subjected to (Grogan and Mitchell, 1990). Sitka spruce seedlings growing in soil, which has had two forest rotations, showed a better shoot dry mass-to-mycorrhizal relationship than seedlings growing in soil which was either not previously afforested or had only experienced one forest rotation (Figure 7.7). In south east Oregon and northern California, mycorrhizal formation was poor in those clearcut sites where burning is a standard practice (Parke *et al.*, 1984).

In the United Kingdom, there is a distinct ectomycorrhizal succession of early-stage and late-stage fungi in birch woodlands and Sitka spruce and Lodgepole pine plantations (Dighton and Mason, 1985). Young stands are dominated by *Laccaria*, *Hebeloma* and *Inocybe*, whereas on maturation of the forest, these are replaced by *Cortinarius*, *Russula* and *Amanita* (Mason *et al.*, 1982). Under sterile conditions, both early-stage and late-stage fungi readily form ectomycorrhizas in tree seedlings but in non-sterile soils, only early-stage fungi will colonize the roots (Dighton and Mason, 1985). Early-stage fungi may not disappear totally from mature stands but instead, may be suppressed because of changes in canopy and root soil characteristics (Dighton *et al.*, 1986). In view of these findings, early-stage fungi have been selected in the ectomycorrhizal inoculation of Sitka spruce seedlings for planting on first rotation sites (Wilson *et al.*, 1987). It remains to be seen whether similar successions occur in forests in other parts of the world.

Virgin sites such as **coal spoils** or **waste tips** usually lack the necessary soil microflora for normal plant growth. Such situations are ideal for the introduction of plants which have previously been inoculated with a mycorrhizal fungus tolerant of the soil conditions. In North America, *Pisolithus tinctorius*, an ectomycorrhizal fungus, is known to increase the survival of pines planted on a range of industrial waste sites (Marx, 1980b; Rhodes, 1980), while the VAM fungus, *Gigaspora gigantea*, improves the growth of red maple, when grown on anthracite waste (Jackson and Mason, 1984). In the United Kingdom, *Scleroderma citrinum* and *P. involutus* are the most common ectomycorrhizal fungi used to inoculate trees for transplanting on coal tips (Jackson and Mason, 1984).

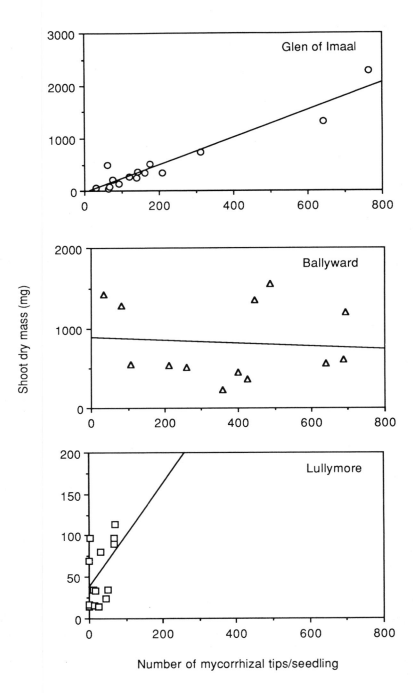

Number of mycorrhizal tips/seedling

Ericaceous endomycorrhizal fungi have been shown to tolerate a range of **heavy metals** (Bradley *et al.*, 1981). *Calluna vulgaris* dominates acid soils caused by copper and zinc mining activities in Europe and *Vaccinium* and *Gaultheria* are prominent on soils contaminated by nickel/copper smelting in Canada. It has been suggested that these plants are able to survive these conditions because of their ericoid endomycorrhizal association (Read, 1983).

7.4 SELECTION AND PRODUCTION OF ECTOMYCORRHIZAL FUNGI FOR TREE INOCULATION

If the inoculation of trees is to be commercially successful, there must be a careful selection of the most appropriate fungus or fungi with respect to tree species and planting sites. Several isolates from different tree hosts and different sites should be used in order to select the best symbiont for the tree or trees in question. The most suitable inoculant fungi are most likely to be those early-stage ectomycorrhizal ones in the forest, which either produce spores or can be cultured and readily form mycorrhizas with the desired host tree(s). Although host specificity does occur, one tree may form mycorrhizas with several species of fungi (Trappe, 1977).

7.4.1 Propagation systems

There are two propagation systems for rearing tree seedlings in nurseries and these have been termed **bare-root** and **container** systems. The bare-root system is the propagation of seedlings in soil and initially, various fumigation procedures are used to eradicate weeds and soil-borne pathogens before sowing. The ectomycorrhizal inoculum is applied to the soil surface at 0.5–1.5 litres/m^2 (Le Tacon *et al.*, 1987). The container system is more reliable as the potting medium, which is usually peat moss with vermiculite or perlite, can be steam-sterilized, although this may not be necessary as mycorrhizal formation can occur successfully under non-sterile conditions (Shaw *et al.*, 1982). Inoculum is mixed with the potting medium and then transferred to 'Ray Leach' containers (65 ml capacity, Ray Leach Con-tainer Nursery, Oregon, USA). Less inoculum is necessary than with the bare-root propagation

---

Figure 7.7 The relationships between shoot dry mass and number of mycorrhizal tips per seedling in seven-month-old Sitka spruce seedlings grown in soil from Glen of Imaal (two forest rotations), Ballyward (one forest rotation) and Lullymore (not previously afforested), Counties Wicklow and Kildare, Ireland. (Grogan and Mitchell, unpublished data.)

method and studies using *Laccaria laccata* S238A have indicated similarities in mycorrhizal formation and performance of Douglas fir seedlings at inoculation rates of 1:4 to 1:128 (Hung and Molina, 1986).

One problem is that the inoculant fungi may not be sufficiently aggressive to compete with the naturally-occurring fungi, particularly in bare-root nurseries. In two inoculation programmes in Ireland, only 3 out of 11 fungi formed mycorrhizas on Sitka spruce seedlings (Table 7.1); All 11 inoculant fungi have been successful in other inoculation trials (Le Tacon and Bouchard, 1986). Tree seedlings are normally reared in nurseries before they are transplanted onto the forest site. The mycorrhizas produced from the inoculant fungus must be suited to the soil conditions and compete with the mycorrhizal fungi of both nursery and

Table 7.1. The ability of ectomycorrhizal fungal isolates to form mycorrhizas in Sitka spruce seedlings in trials at a bare-root and container nursery in Ireland. (+) successful; (*) some success (Grogan and Mitchell, unpublished)

Isolate	Origin	Bare-root nursery	Container nursery
Hebeloma crustuliniforme	SIV, France		
H. crustuliniforme	S166, USA	+	+
Laccaria laccata	S238A, USA	+	+
Laccaria bicolor	KROP, Canada	+	+
L. bicolor	MUEL, France		
L. proxima	DIGH, Britain		
L. tortilis	7PDA, France		
Lactarius rufus	GRAM2, France		*
Paxillus involutus	LITT, USA		
P. involutus	QBC, Canada		
P. involutus	PEP, France		

new forest site. One example is *C. geophilum*, which is tolerant of drought conditions and a wide range of soil pH conditions (Trappe, 1977). It is difficult for *C. geophilum* to establish on pine seedlings in the irrigated nurseries, where there is normally the more competitive naturally-occurring *T. terrestris* (Marx *et al.*, 1978). In other studies, *S. bovinus* (Levisohn, 1956) and *P. involutus* (Laiho, 1970) form abundant mycorrhizas in the nurseries with soils of a high organic matter content, but on transplanting onto low organic matter soils, they are replaced by other mycorrhizal fungi. The upper and lower **temperature** limits for the survival and growth of the fungi should be determined. Mountain ecotypes such as *S. variegatus* have lower temperature optima than lowland ecotypes (Moser, 1958a).

7.4.2 Inoculation

Tree seedlings will naturally form mycorrhizas in nurseries without the necessity of inoculation. Examples of nursery-type mycorrhizal fungi in the United Kingdom are *Amphinema byssoides*, *T. terrestris*, *H. hemisphaerica*, and *T. gilva* (Ingleby *et al.*, 1990) but they may not necessarily be the ideal fungi for transplanting onto new forest sites. Large numbers of spores occur in the atmosphere and may be carried long distances. Even though spore dissemination occurs mainly during autumn, sowing at most bare-root nurseries takes place in spring at a time when mycorrhizal colonization may be erratic. One mycorrhizal fungus, which sporulates during spring, is *T. terrestris*. It is well adapted to nursery conditions and is usually the first to form mycorrhizas (Le Tacon *et al.*, 1987).

Many methods of inoculation have been used and the crudest has been the transfer of soil or humus collected from an established forest (Mikola, 1973; Marx, 1980a). One major drawback to this method is that not all ectomycorrhizal species are beneficial and soil-borne pathogens such as *Pythium*, *Fusarium* and *Phytophthora cinnamomi* may be present. Soil from a mature forest may contain late-stage mycorrhizal fungi, which will not readily form mycorrhizas on the seedlings in the nursery. Table 7.2 presents results of an experiment carried out in a new field of an Irish nursery showing that soil from the established nursery may contain more beneficial mycorrhizal fungi than soil from a forest. Excised mycorrhizas or mycorrhizal seedlings have also been used instead of soil as sources of inoculum in new nurseries (Mikola, 1970).

Another method has been the use of **spores** of various fungi, especially members of the Gasteromycetes (e.g. *Rhizopogon luteolus*). One advantage is that spores can readily be obtained from fruit bodies. However, fruit bodies usually occur seasonally during autumn and

Table 7.2. Shoot dry mass (mean ± standard error of mean) and percentage mycorrhizal infection of four-month-old seedlings of Sitka spruce growing in the Irish Forestry Board (Coillte) bare-root nursery and an adjacent new field at Ballintemple, Co Carlow, Ireland (Grogan and Mitchell, unpublished)

	Shoot dry mass (mg)	Mycorrhizas (%)
Old nursery	73.4 ± 10.4	100
New field		
Soil from old nursery added	53.1 ± 11.3	36
Soil from local forest added	37.1 ± 7.4	0
Untreated	36.8 ± 2.6	0

spores may not survive storage before inoculation of the spring-sown nursery beds. There is generally a lack of information on the viability and genetic diversity of spores and the ideal conditions necessary to store them (Marx and Cordell, 1989). A further problem is that the formation of mycorrhizas from basidiospores takes three to four weeks longer than vegetative inoculum of the same fungus (Theodorou and Bowen, 1970; Marx *et al.*, 1976). This delay may be important in reducing the effectiveness of the introduced inoculum and may be sufficient time to allow invasion of roots by pathogens. Improvements can be made by using spore-encapsulated seed (Theodorou and Benson, 1983).

The use of vegetative inoculum in the form of **cultured mycelium** has been recommended as the most suitable material (Marx and Cordell, 1989). Firstly, the fungi are isolated from either fruit bodies or surface-sterilized, excised mycorrhizas and then grown initially on agar media containing antibiotics (e.g. Aureomycin). The isolation from fruit bodies is ideal, since a precise identification can be made. Isolation from excised mycorrhizas is more tedious, requiring surface sterilization and considerable time to allow the fungus to grow out onto the agar medium. Fungal hyphae only emanate from a small proportion of excised mycorrhizas and identification is more difficult (Heslin *et al.*, 1992). Moser (1958 a, b, c) in Austria was one of the first to produce inoculum, which was mycelium of *Suillus plorans*. Unfortunately, not all ectomycorrhizal fungi can be easily cultured and many grow very slowly on artificial media. Examples, which are difficult to culture, are members of *Inocybe, Cortinarius* and *Russula*. Specific growth media containing simple sugars and growth substances such as thiamine have been used and one (modified Melin–Norkrans medium) is used routinely (Marx and Kenney, 1982). It is easy to produce 30–40 litres of fungus for small inoculation studies, but difficulties are encountered in culturing the considerable quantities required for large scale commercial programmes (Marx and Cordell, 1989). It has been calculated that tens of thousands of litres of inoculum are required to inoculate over 1.5 billion pine seedlings per annum in one nursery in southern USA (Marx and Cordell, 1989). Various methods of culturing have been recommended and submerged liquid culture methods appear to be better than solid substrate systems (Harvey *et al.*, 1989). Inoculum entrapped in alginate polymeric gels enables the ectomycorrhizal fungus, *Hebeloma cylindrosporum*, to maintain a high metabolic activity with improved mycorrhizal formation, even in the presence of *T. terrestris* (Le Tacon *et al.*, 1985).

7.4.3 Pisolithus tinctorius

One ectomycorrhizal fungus, which has been commercially produced, is *Pisolithus tinctorius*, and has been marketed under the name

MycoRhiz. This ectomycorrhizal fungus was chosen because it has a wide host and geographic range and tolerates various habitats, including waste sites. *P. tinctorius* can grow at temperatures as high as 42°C (Hung and Chien, 1978) and has a hyphal thermal death point of 45°C (Lamb and Richards, 1971). Its mycorrhizas may have to tolerate soil temperatures in excess of 60°C on exposed coal spoils, but also withstand low temperature conditions, including frozen soil (Marx and Bryan, 1975). Yellow-gold **cords** are produced and aid water and nutrient absorption. Cords have been observed at least four metres away from young seedling roots in coal spoils (Marx and Cordell, 1989). **Sclerotia** are also produced and will allow the fungus to survive under harsh conditions (Marx *et al.*, 1982). However, it has rarely been used in inoculation programmes in Europe, although fruit bodies have been recorded in either sandy or well-drained, gravelly soil in fields or roadsides in Britain (Phillips, 1981). Vermiculite and peat moss moistened with modified Melin–Norkrans medium is an ideal substrate for production of inoculum (Marx, 1969). Leaching the inoculum to remove any remaining soluble carbohydrates improves its effectiveness (Marx, 1980a). Its bulk density is reduced to 320–390 g l^{-1} with a final pH of 4.4–5.2, which mixes more uniformly in soil than the undried product. The dried inoculum can be stored at 5°C for nine weeks and at room temperature for five weeks (Marx, 1980a). It was initially produced by Abbott Laboratories, North Chicago and then by Sylvan Spawn Laboratory, Worthington, Pennsylvania but Mycorr Tech Inc., University of Pittsburg Applied Research Centre, Pittsburg is the current producer (Marx and Cordell, 1989). Only one isolate was used in the Abbott and Sylvan Spawn inoculum development programme and was originally obtained from a fruit body under pine in northeast Georgia over 25 years ago (Marx and Cordell, 1989). After repeated inoculations and re-isolations from either ectomycorrhizas or fruit bodies, the isolate has become more aggressive in forming mycorrhizas (Marx and Daniel, 1976; Marx 1981).

7.5 SELECTION AND PRODUCTION OF VESICULAR-ARBUSCULAR MYCORRHIZAL (VAM) INOCULUM

There have been few attempts to rationalize the selection of VAM fungi for inoculation of plants. It has already been discussed that suitable VAM fungi may already be present in the soil and modifications to husbandry practices may be all that is necessary to stimulate VAM production. It is therefore essential to determine whether it is necessary to inoculate, and also which VAM fungi to use.

7.5.1 Selection

There are various approaches to assess the VAM potential of a soil and the effectiveness of the indigenous VAM populations in stimulating plant growth (Gianinazzi et al., 1989). Populations have been assessed as number of spores present (Hayman et al., 1975; Black and Tinker, 1977), extent of mycorrhizal infection in host roots (Powell, 1976) and total number of infective mycorrhizal propagules (spores, hyphae and VA-infected root fragments) in the soil (Powell 1980). The first two methods have been criticized as spore numbers were not related to infectivity of a soil containing inoculum and high root infection levels can develop with time from few mycorrhizal propagules (Moorman and Reeves, 1979). Using the third method, Powell (1980) was able to show that disturbed situations such as eroded banks and slip faces had lower mycorrhizal potential than established pastures. There appears to be some correlation between mycorrhizal infectivity and spore numbers in soil of undisturbed shrub vegetation (Berliner et al., 1989). The measurement of VA mycorrhizal potential is performed by a dilution method and although the calculated infectivity levels are probably underestimates, the method is both suitable and useful (Powell, 1980). It will thus indicate the necessity to inoculate with more efficient VAM fungi.

7.5.2 Production

One major problem in the production of inoculum is that VAM fungi cannot be cultured (Hepper, 1984a). The addition of inorganic sulphur-containing compounds, in particular potassium sulphite and potassium metabisulphite, to the agar medium may increase hyphal growth from germinating spores of *Glomus caledonicum*, but the VAM fungus cannot be grown further (Hepper, 1984b). This has led to the belief that VAM fungi survive by relying entirely on the host roots for their existence. Thus, inoculum has to be produced by growing-up plants in media containing spores, hyphae or VAM-infected roots (pot culture) (Powell, 1984).

Spores can be extracted from the soil by one of the following methods: wet-sieving method (Gerdemann, 1955; Gerdemann and Nicolson, 1963), sucrose density centrifugation (Daniels and Skipper, 1982) and flotation-adhesion technique (Sutton and Barron, 1972). They can then be separated, microscopically examined and finally surface sterilized before they are used in axenic plant culture to produce inoculum (Hepper, 1984a). The final form of inoculum may be either roots or spores and hyphae trapped in soil or artificial/inorganic carriers (Menge, 1984). In France, the established method is the soil-based type in which disinfected soil is inoculated, clover is grown for three to five months and then the soil and roots are used as inoculum (Gianinazzi *et al.*, 1989). Peat-based inoculum can be produced by growing VAM plants in peat blocks standing in a shallow nutrient-flow culture system (Warner *et al.*, 1985). In Germany, expanded clay is the substratum, in which spores and hyphae are lodged (Dehne and Backhaus, 1986). Other carriers have been used such as Nutri-link, which is developed by Native Plant Inc. for the American market (Wood, 1987). Further developments have been the use of aeroponic culture media, in which the nutrient solution is atomized onto the root system suspended in a chamber (Hung and Sylvia, 1988), the use of aseptically-grown root organ cultures (Mosse and Hepper, 1986) and entrapment of vesicles and mycelial fragments in alginate beads (Strullu and Plenchette, 1991).

The introduction of VAM inocula into potting compost and field soil on a commercial scale has limitations (Linderman, 1988). In general, inoculation trials have tended to use excessive amounts of inoculum to guarantee rapid infection (Powell, 1984; Hall, 1988). More research needs to be undertaken to standardize the mycorrhizal potential of the inoculum and to determine the edaphic requirements of each VAM fungus (Gianinazzi *et al.*, 1989). Endemic VAM fungi may be more affected by soil disturbance than introduced ones, but VAM fungi isolated from one soil may have limited effect on plant growth in another (Linderman, 1988). Plants with coarse roots appear to respond to inoculation better than plants with fine roots (Linderman, 1988). The ability of VAM fungi to form mycorrhizas may be dependent on the genotype of the host (Bertheau *et al.*, 1980; Krishna *et al.*, 1985; Lackie *et al.*, 1988). The identification and manipulation of host genes regulating the formation of mycorrhizas has begun as a result of studies with *myc* plant mutants (Duc *et al.*, 1989). Future research should investigate the genetic control of mycorrhizal formation.

7.6 SELECTION AND PRODUCTION OF ERICOID ENDOMYCORRHIZAL INOCULUM

Inoculation of members of the Ericales appears to be limited to one species (i.e., the development and growth of Blueberry crops on pasture soils) (Powell, 1981). Heathers, Azaleas and Rhododendrons are not inoculated on a commercial scale. The selection of inoculum has been made easy by the fact that few fungal species (mainly *H. ericae*) are involved but is hampered because fruit bodies have rarely been seen in the field. The fungus has been isolated directly from the roots by means of a root washing and maceration technique (Pearson and Read, 1973). The fungus can be easily grown on malt agar or general nutrient medium (Mitchell and Read, 1981). Various methods have been used to synthesize mycorrhizas on roots of ericaceous seedlings under axenic conditions (Moore-Parkhurst and Englander, 1981) but there have been few attempts to do this on a large commercial scale.

7.7 CONCLUSIONS AND FUTURE DEVELOPMENTS

Considerable progress has been made during the past three decades to improve the growth of plants by exploiting the mycorrhizal symbionts. There have been numerous field inoculation studies and these have been reported in previous reviews for ectomycorrhizal (Marx and Cordell, 1989), VAM (Powell and Bagyaraj, 1984) and ericoid endomycorrhizal (Powell, 1982) associations. The potential uses of mycorrhizal inoculum are considerable, although the best results from inoculation programmes will be on those sites where the requisite mycorrhizal fungi are absent or some toxic chemical (biocide) has eradicated the original symbionts. In general, inoculum has been produced in various forms and mycorrhizal seedlings have been reared in nurseries prior to transplantation onto sites, or the inoculum has been directly applied (e.g. inoculation of pastures). Major problems have resulted from this approach, such as the inability of the inoculum to survive in competition with naturally-occurring mycorrhizal fungi and rhizosphere microorganisms. Thus, the inoculum fungus or 'cocktail' must be sufficiently aggressive enough to overcome this competition and be adaptable to different soil conditions such as nutrients, pH, temperature, moisture and aeration (Slankis, 1974). The indigenous VA mycorrhizas can be reduced by leaving the soil fallow, growing a non-host plant or sterilizing the soil. The crop may then readily respond to inoculation (Daft, 1991). Inoculation programmes have been further hampered by the limitations of producing sufficient quantities of inoculum for large scale commercial ventures. The technology has been developed but more research is required to increase the scale of operation. More superior strains or species of mycorrhizal fungi must be selected and improvements may be made by genetic engineering (Beringer *et al.*, 1987). The first demonstration of genetic transformation within an ectomycorrhizal fungus (*Laccaria laccata*) was reported by Lemke (1990). The next step will be the identification of specific genes in the fungus, which will be beneficial to the symbiosis. In the case of VA mycorrhizas, the identification and manipulation of host genes involved in the symbiosis has already commenced as a result of work on *myc* plant mutants (Duc *et al.*, 1989). Since VAM fungi cannot be cultured, host plants will have to be selected to improve the quality of the inoculum, which on introduction to the field soil, will compete with populations of non-effective indigenous VA fungi (Gianinazzi *et al.*, 1989).

Mycorrhizal exploitation need only entail the manipulation of effective indigenous populations in the soil rather than application of inoculum. Mycorrhizal fungi may be stimulated by altering crop rotation regimes, using crop mixtures and applying selective fungicides. The growth of Sitka spruce on low nutrient sites is enhanced by the presence of a 'nurse' tree such as Lodgepole pine, Japanese larch and Scots pine and this may be as a result of stimulating a more diverse ectomycorrhizal flora than occurs in pure stands of Sitka spruce (Heslin *et al.*, 1992). The fungicide, Benomyl, has been shown to improve disease tolerance and ectomycorrhizal formation in White pine (*P. strobus*) (Bastide and Kendrick, 1990). These manipulations may provide a more simple approach and may be less expensive than the use of inoculation practices.

Acknowledgements

The author thanks Helen Grogan, Eimear Hughes, Angela Kennedy, Laura Jackson, Rosemary Mitchell, Elizabeth Murphy and John O'Neill for critically reading the manuscript.

REFERENCES

Abuzinadah, R.A. and Read, D.J. (1986) The role of proteins in the nitrogen nutrition of ectomycorrhizal plants. I. Utilization of peptides and proteins by ectomycorrhizal fungi. *The New Phytologist*, **103**, 481–93.

Abuzinadah, R.A., Finlay, R.D. and Read, D.J. (1986) The role of proteins in the nitrogen nutrition of ectomycorrhizal plants. II. Utilization of protein by mycorrhizal plants of *Pinus contorta*. *The New Phytologist*, **103**, 495–506.

Agerer, R. (1987–1990) *Colour Atlas of Ectomycorrhizae*, 1–4 delivery, Einhorn-Verlag Eduard Dietenberger, Schabisch Gmund.

Alexander, I.J. (1983) Ectomycorrhizas in the nitrogen cycle, in *Nitrogen as an Ecological Factor*, (eds J.A. Lee, S. McNeill and I.H Rorison), Blackwell, Oxford, pp.69–93.

Amijee, F. (1989) Vesicular-arbuscular mycorrhizas: an ubiquitous symbiosis between fungi and roots of vascular plants. *The Mycologist*, **3**, 176–80.

Anon (1989) Production of truffles at Ivermy Agricultural Centre, New Zealand. *Report*.

Bastide, P.Y. and Kendrick, B. (1990) The *in vitro* effects of benomyl on disease tolerance, ectomycorrhiza formation and growth of white pine (*Pinus strobus*) seedlings. *Canadian Journal of Botany*, **68**, 444–8.

Baylis, G.T.S. (1975) The magnolioid mycorrhiza and mycotrophy in root systems derived from it, in *Endomycorrhiza*, (eds F.E. Sanders, B. Mosse and P.B. Tinker), Academic Press, London, pp.373–89.

Beringer, J.E., Burggraaf, A.J.P., Reddell, P. and Turner, G. (1987) The role of mycorrhizas in crop growth and prospects for producing modified strains of mycorrhizal fungi, in *Genetics and Plant Pathogenesis*, (eds P.R. Day and G.J. Jellis), Blackwell Scientific Publications, Oxford, pp.91–100.

Berliner, R., Mitchell, D.T. and Allsopp, N. (1989) The vesicular-arbuscular mycorrhizal infectivity of sandy soils in the south-western Cape, South Africa. *South African Journal of Botany*, **55**, 310–13.

Bertheau, Y., Gianinazzi-Pearson, V. and Gianinazzi, S. (1980) Dévelopement et expression de l'association endomycorhizienne chez le blé. I. Mise en évidence d'un effet variétal. *Annales d'Amélioration des Plantes*, **30**, 67–78.

Black, R.L.B. and Tinker, P.B. (1977) Interactions between effects of vesicular-arbuscular mycorrhiza and fertilizer phosphorus on yield of potatoes in the field. *Nature*, **267**, 510–11.

Bowen, G.D. and Smith, S.E. (1981) The effects of mycorrhizas on nitrogen uptake by plants, in *Terrestrial Nitrogen Cycles*, (eds F.E. Clarke and T. Rosswall), *Ecological Bulletin*, **33**, 237–47.

Bradley, R.A., Burt, A.J. and Read, D.J. (1981) Mycorrhizal infection and resistance to heavy metal toxicity in *Calluna vulgaris*. *Nature*, **292**, 335–7.

Cairney, J.W.G. (1991) Rhizomorphs: organs of exploration or exploitation. *The Mycologist*, **5**, 5–10.

Chevalier, G. and Frochot, H. (1980) Truffle production from artificially mycorrhizal plants: first results. *Abstract of the 5th North American Conference on Mycorrhizas*, **1**, 35.

Clarke, C. and Mosse, B. (1981) Plant growth responses to vesicular-arbuscular mycorrhiza. XII. Field inoculation responses of barley to two soil P levels. *The New Phytologist*, **87**, 695–703.

Daft, M.J. (1991) Use of VA mycorrhizas in agriculture: problems and prospects. *Abstracts of the Third European Symposium on Mycorrhizas*.

Dahlberg, A. and Stenlid, J. (1990) Population structure and dynamics in *Suillus bovinus* as indicated by spatial distribution of fungal clones. *The New Phytologist*, **115**, 487–93.

Daniels, B.A. and Skipper, H.D. (1982) Methods for the recovery and quantitative estimation of propagules from soil, in *Methods and Principles of Mycorrhizal Research*, (ed N.C. Schenck), The American Phytopathological Society, St Paul, pp.29–35.

Dehne, H.W. and Backhaus, G.F. (1986) The use of vesicular-arbuscular mycorrhizal fungi in plant production. I. Inoculum production. *Zeitscrift fur Pfanzenkrankheiten und Pfanzenschutz*, **93**, 415–24.

Dighton, J. and Mason, P.A. (1985) Mycorrhizal dynamics during forest tree development, in *Developmental Biology of Higher Fungi*, (eds D. Moore, L.A. Casselton, D.A. Wood and J.C. Frankland), Cambridge University Press, Cambridge, pp.117–39.

Dighton, J., Poskitt, J.M. and Howard, D.M. (1986) Changes in occurrence of basidiomycete fruit bodies during forest stand development: with specific reference to mycorrhizal species. *Transactions of the British Mycological Society*, **87**, 163–71.

Dixon, R.K. and Buschena, C.A. (1988) Response of ectomycorrhizal *Pinus banksiana* and *Picea glauca* to heavy metals in soil. *Plant and Soil*, **105**, 265–71.

Douglas, G.C., Heslin, M.C. and Reid, C. (1989) Isolation of *Oidiodendron maius* from *Rhododendron* and ultrastructural characterization of synthesized mycorrhizas. *Canadian Journal of Botany*, **67**, 2206–12.

Duc, G., Trouvelot, A., Gianinazzi-Pearson, V. and Gianinazzi, S. (1989) First report of non-mycorrhizal plant mutants (*Myc-*) obtained in pea (*Pisum sativum* L.) and faba bean (*Vicia faba* L.). *Plant Science*, **60**, 215–22.

Duddridge, J.A. (1986) Specificity and recognition in mycorrhizal associations, in *Physiological and Genetical Aspects of Mycorrhizae*, (eds V. Gianinazzi-Pearson and S. Gianinazzi), Institut National de la Recherche Agronomique, Paris, pp.45–58.

Duddridge, J.A., Malibari, A. and Read, D.J. (1980) Structure and function of mycorrhizal rhizomorphs with special reference to their role in water transport. *Nature*, **287**, 834–36.

Ek, M., Ljungquist, P.O. and Stenstrom, E. (1983) Indole-3-acetic acid production by mycorrhizal fungi determined by gas chromatography mass spectrometry. *The New Phytologist*, **94**, 401–7.

Finlay, R.D. and Read, D.J. (1986) The structure and function of the vegetative mycelium of ectomycorrhizal plants. I. Translocation of C-14 labelled carbon between plants interconnected by a common mycelium. *The New Phytologist*, **103**, 143–56.

Fries, N. (1987) Somatic incompatibility and field distribution of the ectomycorrhizal fungus *Suillus luteus* (Boletaceae). *The New Phytologist*, **107**, 735–9.

Gerdemann, J.W. (1955) Relation of a large soil-borne spore to phycomycetous mycorrhizal infections. *Mycologia*, **47**, 619–32.

Gerdemann, J.W. and Nicolson, T.H. (1963) Spores of mycorrhizal *Endogone* species extracted from soil by wet-sieving and decanting. *Transactions of the British Mycological Society*, **46**, 235–44.

Gianinazzi, S., Gianinazzi-Pearson, V. and Trouvelot, A. (1989) Potentialities and procedures for the use of endomycorrhizas with special emphasis on high value crops, in *Biotechnology of Fungi for Improving Plant Growth*, (eds J.M. Whipps and R.D. Lumsden), Cambridge University Press, Cambridge, pp.41–54.

Gianinazzi, S., Trouvelot, A. and Gianinazzi-Pearson, V. (1989) Conceptual approaches for the rational use of VA endomycorrhizae in agriculture: possibilities and limitations. *Agriculture, Ecosystems and Environments*, **29**, 153–61.

Gianinazzi, S., Trouvelot, A. and Gianinazzi-Pearson, V. (1990) Role and use of mycorrhizas in horticultural crop production. *XXIII International Horticultural Congress, Plenary Lectures*, Firenze, Italy, pp.25–30.

Glenn, M.G., Chew, F.S. and Williams, P.H. (1988) Influence of glucosinolate content of *Brassica* (Cruciferae) roots on growth of vesicular-arbuscular mycorrhizal fungi. *The New Phytologist*, **110**, 217–25.

Grogan, H. and Mitchell, D.T. (1990) The mycorrhizal status of some forest sites and the propagation of ectomycorrhizal Sitka spruce seedlings in Ireland. *Aspects of Applied Biology*, **24**, 123–30.

Hall, I.R. (1988) Potential for exploiting vesicular-arbuscular mycorrhizas in agriculture. *Advances in Biotechnological Processes*, **9**, 141–74.

Harley, J.L. (1969) *The Biology of Mycorrhizas*, Leonard Hill, London.

Harley, J.L. and Harley, E.L. (1987) A check-list of mycorrhiza in the British Flora. *The New Phytologist*, **105** (2), 1–102.

Harley, J.L. and Smith, S.E. (1983) *Mycorrhizal Symbiosis*, Academic Press, London.

Harvey, L.M., Smith, J.E., Kristiansen, B., Neill, J. and Senior, E. (1989) The cultivation of ectomycorrhizal fungi, in *Biotechnology of Fungi for Improving Plant Growth*, (eds J.M. Whipps and R.D. Lumsden), Cambridge University Press, Cambridge, pp.27–39.

Hayman, D.S., Johnson, A.M. and Ruddlesdin, I. (1975) The influence of phosphate and crop species on *Endogone* spores and vesicular-arbuscular mycorrhiza under field conditions. *Plant and Soil*, **43**, 489–95.

Hepper, C.M. (1984a) Isolation and culture of VA mycorrhizal (VAM) fungi, in *VA mycorrhiza* (eds C.L. Powell and D.J. Bagyaraj), CRC Press, Boca Raton, Florida, pp.95–112.

Hepper, C.M. (1984b) Inorganic sulphur nutrition of the vesicular-arbuscular mycorrhizal fungus *Glomus caledonicum*. *Soil Biology and Biochemistry*, **16**, 669–72.

Heslin, M.C., Blasius, D., McElhinney, C. and Mitchell, D.T. (1992) Mycorrhizal and associated fungi of Sitka spruce in Irish forest mixed stands. *European Journal of Forest Pathology*, **22**, 46–57.

Hung, L.L. and Chien, C.Y. (1978) Physiological studies on two ectomycorrhizal fungi, *Pisolithus tinctorius* and *Suillus bovinus*. *Transactions of the Mycological Society of Japan*, **19**, 121–7.

Hung, L.L. and Molina, R. (1986) Use of the ectomycorrhizal fungus *Laccaria laccata* in forestry. III. Effects of commercially produced inoculum on container-grown Douglas-fir and Ponderosa pine seedlings. *Canadian Journal of Forest Research*, **16**, 802–6.

Hung, L.L. and Sylvia, D.M. (1988) Production of vesicular-arbuscular mycorrhizal fungus inoculum in aeroponic culture. *Applied Environmental Microbiology*, **54**, 353–7.

Ingelby, K., Mason, P.A., Last, F.T. and Fleming, L.V. (1990) *Identification of ectomycorrhizas*, ITE Research Publication No. 5, Her Majesty's Stationery Office, London.

Jackson, R.M. and Mason, P.A. (1984) *Mycorrhiza*, Edward Arnold, London.

Jeffries, P. (1987) Use of mycorrhizae in agriculture. *CRC Critical Reviews in Biotechnology*, **5**, 319–57.

Kernan, M.J. and Finocchio, A.F. (1983) A new discomycete associated with roots of *Monotropa uniflora* (Ericaceae). *Mycologia*, **75**, 916–20.

Knudson, L. (1922) Non-symbiotic germination of orchid seeds. *Botanical Gazette*, **73**, 1–25.

Krishna, K.R., Shetty, K.G., Dart, P.J. and Andrews, D.J. (1985) Genotype dependent variation in mycorrhizal colonization and response to inoculation in pearl millet. *Plant and Soil*, **86**, 113–25.

Lackie, S.M., Bowley, S.R. and Peterson, R.L. (1988) Comparison of colonization among half-sib families of *Medicago sativa* L. by *Glomus versiforme* (Daniels & Trappe) Berch. *The New Phytologist*, **108**, 477–82.

Laiho, O. (1970) *Paxillus involutus* as a mycorrhizal symbiont of forest trees. *Acta Forestalia Fennica*, **106**, 1–73.

Lamb, R.L. and Richards, B.N. (1971) Effect of mycorrhizal fungi on the growth and nutrient status of slash and radiata pine seedlings. *Australia Forestry*, **35**, 1–7.

Lamont, B.B. (1983) Strategies for maximizing nutrient uptake in two mediterranean ecosystems of low nutrient status, in *Mediterranean-type Ecosystems: The Role of Nutrients*, (eds F.J. Kruger, D.T. Mitchell and J.U.M. Jarvis), Springer-Verlag, Berlin, pp.246–73.

Lemke, P.A. (1990) Transgenic manipulation of a mycorrhizal fungus, in *Abstracts of the Fourth International Mycological Congress*, (eds A. Reisinger and A. Bresinsky), Regensburg, p.198.

Le Tacon, F. and Bouchard, D. (1986) Effects of different ectomycorrhizal fungi on growth of larch, Douglas fir, Scots pine and Norway spruce seedlings in fumigated nursery soil. *Acta Oecologica*, **7**, 389–402.

Le Tacon, F., Garbaye J. and Carr, G. (1987) The use of mycorrhizas in temperate and tropical forests. *Symbiosis*, **3**, 179–206.

Le Tacon, F., Jung, G., Mugnier, J., *et al.* (1985) Efficiency in a forest nursery of an ectomycorrhizal fungus inoculum produced in a fermenter and entrapped in polymeric gels. *Canadian Journal of Botany*, **63**, 1664–8.

Levisohn, I. (1956) Growth stimulation of forest tree seedlings by the activity of free-living mycorrhizal mycelia. *Forestry*, **29**, 53–9.

Linderman, R.G. (1988) VA (Vesicular-arbuscular) mycorrhizal symbiosis. *ISI Atlas of Science: Animal and Plant Sciences*, pp.183–8.

Lundeberg, G. (1970) Utilization of various nitrogen sources, in particular bound soil nitrogen, by mycorrhizal fungi. *Studia Forestalia Suecica*, **79**, 5–95.

McConigle, T.P. and Fitter, A.H. (1990) Ecological specificity of vesicular-arbuscular mycorrhizal associations. *Mycological Research*, **94**, 120–2.

Marx, D.H. (1969) The influence of ectotrophic mycorrhizal fungi on the resistance of pine roots to pathogenic infections. I. Antagonism of mycorrhizal fungi to root pathogenic fungi and bacteria. *Phytopathology*, **59**, 153–63.

Marx, D.H. (1973) Mycorrhizae and feeder root diseases, in *Ectomycorrhizae: Their Ecology and Physiology*, (eds G.C. Marks and T.T. Kozlowski), Academic Press, New York, pp.351–82.

Marx, D.H. (1980a) Ectomycorrhizal fungus inoculations: a tool for improving forestation practices, in *Tropical Mycorrhizal Research*, (ed P. Mikola), Oxford University Press, London, pp.13–71.

Marx, D.H. (1980b) Role of mycorrhizae in forestation of surface mines, in *Proceedings of a Symposium on Trees for Reclamation in the Eastern United States, Forest Service General Technical Report, NE-61*, US Department of Agriculture, Lexington, pp.109–16.

Marx, D.H. and Bryan, W.C. (1975) Growth and ectomycorrhizal development of loblolly pine seedlings in fumigated soil infested with fungal symbiont *Pisolithus tinctorius*. *Forest Science*, **21**, 245–54.

Marx, D.H., Bryan, W.C. and Cordell, C.E. (1976) Growth and ectomycorrhizal development of pine seedlings in nursery soils infested with the fungal symbiont *Pisolithus tinctorius*. *Forest Science*, **22**, 91–100.

Marx, D.H. and Cordell, C.E. (1989) The use of specific ectomycorrhizas to improve artificial forestation practices, in *Biotechnology of Fungi for Improving Plant Growth*, (eds J.M. Whipps and R.D. Lumsden), Cambridge University Press, Cambridge, pp.1–25.

Marx, D.H. and Daniel, W.J. (1976) Maintaining cultures of ectomycorrhizal and plant pathogenic fungi in sterile water cold storage. *Canadian Journal of Microbiology*, **22**, 338–41.

Marx, D.H. and Kenney, D.S. (1982) Production of ectomycorrhizal fungus inoculum, in *Methods and Principles of Mycorrhizal Research*, (ed N.C. Schenck), The American Phytopathological Society, St Paul, pp.131–46.

Marx, D.H. and Krupa, S.V. (1978) Mycorrhizae A. Ectomycorrhizae, in *Interactions Between Nonpathogenic Soil Microorganisms and Plants*, (eds Y.R. Domergues and S.V. Krupa), Elsevier Scientific Publishing Company, Amsterdam, pp.373–400.

Marx, D.H., Morris, W.G. and Mexel, J.G. (1978) Growth and ectomycorrhizal development of loblolly pine seedlings in fumigated and nonfumigated soil infested with different fungal symbionts. *Forest Science*, **24**, 193–203.

Marx, D.H., Ruehle, J.L. Kenney, D.S. *et al.* (1982) Commercial vegetative inoculum of *Pisolithus tinctorius* and inoculation techniques for development of ectomycorrhizae on container-grown tree seedlings. *Forest Science*, **28**, 373–400.

Mason, P.A., Last, F.T., Pelham, J. and Ingleby, K. (1982) Ecology of some fungi associated with an ageing stand of birches. *Forest Ecology and Management*, **4**, 19–40.

Menge, J.A. (1984) Inoculum production, in *VA mycorrhiza* (eds C.L. Powell and D.J. Bagyaraj), CRC Press, Florida, pp.187–203.

Meyer, F.H. (1973) Distribution of ectomycorrhizae in native and man-made forests, in *Ectomycorrhizae: Their Ecology and Physiology*, (eds G.C. Marks and T.T. Kozlowski) Academic Press, New York, pp.79–105.

Mikola, P. (1970) Mycorrhizal inoculation in afforestation. *International Review of Forest Research*, **3**, 123–96.

Mikola, P. (1973) Application of mycorrhizal symbiosis in forestry practice, in *Ectomycorrhizae: Their Ecology and Physiology*, (eds G.C. Marks and T.T. Kozlowski), Academic Press, New York, pp.383–411.

Mitchell, D.T. and Read, D.J. (1981) Utilization of inorganic and organic phosphates by the mycorrhizal endophytes of *Vaccinium macrocarpon* and *Rhododendron ponticum*. *Transactions of the British Mycological Society*, **76**, 255–60.

Moore-Parkhurst, S. and Englander, L. (1981) A method for the synthesis of a mycorrhizal association between *Pezizella ericae* and *Rhododendron maximum* seedlings growing in a defined medium. *Mycologia*, **73**, 994–7.

Moorman, T. and Reeves, F.B. (1979) The role of endomycorrhizae in the revegetation practices in the semi-arid West. II. A bioassay to determine the effect of land disturbance on endomycorrhizal populations. *American Journal of Botany*, **66**, 14–18.

Morandi, D. and Gianinazzi-Pearson, V. (1986) Influence of mycorrhizal infection and phosphate nutrition on secondary metabolite contents of soybean roots, in *Physiological and Genetical Aspects of Mycorrhizae*, (eds V. Gianinazzi-Pearson and S. Gianinazzi), Institut National de la Recherche Agronomique, Paris, pp.787–91.

Moser, M. (1958a) Die kunstliche Mykorrhizaimpfung an Forstpflanzen. I. Erfahrungen bei der Reinkultur von Mykorrhizapilzen. *Forstwissenschaftliches Centralblatt*, **77**, 32–40.

Moser, M. (1958b) Die kunstliche Mykorrhizaimpfung und Forstpflanzen. II. Torfstreukultur von Mykorrhizapilzen. *Forstwissenschaftliches Centralblatt*, **77**, 273–8.

Moser, M. (1958c) Die Einfluss tiefer Temperaturen auf das Washstum und die Lebenstatigkeit horerer Pilze mit spezieller Berucksichtigung von Mykorrhizapilzen. *Sydowia*, **12**, 386–99.

Mosse, B. and Hepper, C.M. (1986) Vesicular-arbuscular mycorrhizal infections in root organ cultures. *Physiological Plant Pathology*, **5**, 215–23.

Parke, J.L., Linderman, R. and Black, C.H. (1983) Role of ectomycorrhizas in drought tolerance of Douglas fir seedlings. *The New Phytologist*, **95**, 83–95.

Parke, J.L., Linderman, R.G. and Trappe, J.M. (1984) Inoculum potential of ectomycorrhizal fungi in forest soils of Southwest Oregon and Northern California. *Forest Science*, **2**, 300–4.

Pearson, V. and Read, D.J. (1973) The biology of mycorrhiza in the Ericaceae. I. The isolation of the endophyte and synthesis of mycorrhizas in aseptic culture. *The New Phytologist*, **72**, 371–9.

Phillips, R. (1981) *Mushrooms and Other Fungi of Great Britain and Europe*, Pan Books Ltd., London.

Pirozynski, K.A. and Malloch, D.W. (1975) The origin of land plants: a matter of mycotrophism. *Biosystems*, **6**, 153–64.

Powell, C.L. (1976) Mycorrhizal fungi stimulate clover growth in New Zealand hill country soils. *Nature*, **264**, 436–8.

Powell, C.L. (1980) Mycorrhizal infectivity of eroded soils. *Soil Biology and Biochemistry*, **12**, 247–50.

Powell, C.L. (1981) Mycorrhizal fungi and blueberries – how to introduce them. *New Zealand Journal of Agriculture*, **143**, 33–5.

Powell, C.L. (1982) The effect of ericoid mycorrhizal fungus, *Pezizella ericae* (Read) on the growth and nutrition of seedlings of blueberry (*Vaccinium corymbosum*). *Journal of American Society of Horticultural Science*, **107**, 1012–15.

Powell, C.L. (1984) Field inoculation with VA mycorrhizal fungi, in *VA mycorrhiza* (eds C.L. Powell and D.J. Bagyaraj), CRC Press, Boca Raton, FA, pp.205–22.

Powell, C.L. and Bagyaraj, D.J. (1984) *VA Mycorrhiza*, CRC Press, Boca Raton, FA.

Powell, C.L. and Bates, P.M. (1981) Ericoid mycorrhizas stimulate fruit yield of blueberry. *Hortscience*, **16**, 655–6.

Read, D.J. (1974) *Pezizella ericae* sp. nov., the perfect state of a typical mycorrhizal endophyte of Ericaceae. *Transactions of the British Mycological Society*, **63**, 381–419.

Read, D.J. (1983) The biology of mycorrhiza in the Ericales. *Canadian Journal of Botany*, **61**, 985–1004.

Rhodes, L.H. (1980) The use of mycorrhizae in crop production systems. *Outlook on Agriculture*, **10**, 275–81.

Sen, R. (1990) Isozymic identification of individual ectomycorrhizas synthesized between Scots pine (*Pinus sylvestris* L.) and isolates of two species of *Suillus*. *The New Phytologist*, **114**, 617–26.

Sen, R. and Hepper, C.M. (1986) Characterization of vesicular-arbuscular mycorrhizal fungi (*Glomus* spp.) by selective enzyme staining following polyacrylamide gel electrophoresis. *Soil Biology and Biochemistry*, **18**, 29–34.

Shaw III, C.G., Molina, R. and Walden, J. (1982) Development of ectomycorrhizae following inoculation of containerized Sitka and White spruce seedlings. *Canadian Journal of Forest Research*, **12**, 191–5.

Slankis, V. (1974) Soil factors influencing formation of mycorrhizae. *Annual Review of Phytopathology*, **12**, 437–57.

Stribley, D.P. (1990) Do vesicular-arbuscular mycorrhizal fungi have a role in plant husbandry? *Aspects of Applied Biology*, **24**, 117–21.

Stribley, D.P., Tinker, P.B. and Rayner, J.H. (1980) Relation of internal phosphorus concentration and plant weight in plants infected by vesicular-arbuscular mycorrhizas. *The New Phytologist*, **86**, 261–6.

Strullu, D.G. and Plenchette, C. (1991) The entrapment of *Glomus* sp. in alginate beads and their use as root inoculum. *Mycological Research*, **95**, 1194–6.

Sutton, J.C. and Barron, G.L. (1972) Population dynamics of *Endogone* spores in soil. *Canadian Journal of Botany*, **50**, 1909–14.

Theodorou, C. and Benson, A.D. (1983) Operational mycorrhizal inoculation of nursery beds with seed-borne fungal spores. *Australian Forestry*, **46**, 43–7.

Theodorou, C. and Bowen, G.D. (1970) Mycorrhizal responses of radiata pine in experiments with different fungi. *Australian Forestry*, **34**, 183–91.

Thomas, G.W. and Jackson, R.M. (1979) Sheathing mycorrhizas of nursery grown *Picea sitchensis*. *Transactions of the British Mycological Society*, **73**, 117–25.

Thomas, G.W. and Jackson, R.M. (1982) *Complexipes moniliformis* – Ascomycete or Zygomycete? *Transactions of the British Mycological Society*, **79**, 149–86.

Tommerup, I.C. and Sivasithamparam, K. (1990) Zygospores and asexual spores of *Gigaspora decipiens*, an arbuscular mycorrhizal fungus. *Mycological Research*, **94**, 897–900.

Trappe, J.M. (1977) Selection of fungi for ectomycorrhizal inoculation in nurseries. *Annual Review of Phytopathology*, **15**, 203–22.

Walker, C. (1983) Taxonomic concepts in the Endogonaceae: spore wall characteristics in species descriptions. *Mycotaxon*, **18**, 443.

Warner, A., Mosse, B. and Dingermann, L. (1985) The nutrient film technique for inoculum production, in *Proceedings of the 6th North American Conference on Mycorrhizae*, (ed R. Molina), Forest Research Laboratory, Corvallis, p.85.

Wilkins, W.H. and Harris, G.C.M. (1946) The ecology of larger fungi. V. An investigation into the influence of rainfall and temperature on the seasonal production of fungi in a beechwood and a pinewood. *Annals of Applied Biology*, **33**, 179–88.

Wilson, J., Mason, P.A., Last, F.T., Ingelby, K. and Munro, R.C. (1987) Ectomycorrhiza formation and growth of Sitka spruce seedlings on first-rotation forest sites in northern Britain. *Canadian Journal of Forest Research*, **17**, 957–63.

Wilson, J., Ingleby, K. and Mason P.A. (1990) Ectomycorrrhizal inoculation of Sitka spruce; survival of vegetative mycelium in nursery soil. *Aspects of Applied Biology*, **24**, 109–15.

Wood, T. (1987) Commercial production of VA mycorrhiza inoculum: axenic versus non axenic techniques, in *Mycorrhizae in the Next Decade: Practical Applications and Research Priorities*, (eds D.M. Sylvia, L.L. Hung and J.H. Graham), University of Florida, Gainsville, p.274.

Wright, S.F., Morton, J.B. and Sworobuk, J.E. (1987) Identification of a vesicular-arbuscular mycorrhizal fungus by using monoclonal antibodies in an enzyme-linked immunosorbant assay. *Applied Environmental Microbiology*, **53**, 2222–5.

Zak, B. and Marx, D.H. (1964) Isolation of mycorrhizal fungi from roots of individual slash pines. *Forest Science*, **10**, 214–22.

8

Rhizobium inoculation of crop plants

D. Gareth Jones and D. Mark Lewis

8.1 INTRODUCTION AND HISTORICAL BACKGROUND

In these increasingly environmentally aware times, re-cycling of industrial products has become very fashionable and, indeed desirable. However, man is only now attempting what living organisms have been practising over millions of years. The most familiar re-cycling is that of the element carbon and the pathway from atmospheric carbon dioxide via photosynthesis into plant material is one of the first taught aspects of biology at secondary education level. The role of microorganisms in the decomposition of plant material is also an area which is reasonably well understood but it has not been subjected to the intense research activity as has the contribution of microorganisms to the nitrogen cycle. The nitrogen cycle has been described as being second only in importance to the carbon cycle. The truth must be, however, that both cycles and many others are essential for the continuation of life on this planet.

8.1.1 Nitrogen fixation and symbiosis

The element nitrogen is an essential component of all living things, being the basis of proteins and the genetically important nucleic acids. It is an unfortunate quirk of nature, therefore, that very few living organisms have the ability to utilize the enormous resevoir of nitrogen,

Exploitation of Microorganisms Edited by D.G. Jones
Published in 1993 by Chapman & Hall, London ISBN 0 412 45740 7

nearly 80%, in the earth's atmosphere. The nitrogen cycle is obviously of considerable importance in the biosphere and the nitrogen fixation component demonstrates the tremendous potential for replacing artificial nitrogen fertilizers by biologically-fixed nitrogen, especially from the Rhizobium–legume symbiosis.

Indeed, it is the environmental implication of the too liberal use of fertilizer nitrogen, coupled with the need to reduce crop input costs, that has given the study of biological nitrogen fixation an impetus in recent years. The economic and environmental pressures also coincided with exciting developments in the field of molecular biology and, in particular, our understanding of the genetic control of nitrogen fixation in *Rhizobium*. The discovery of plasmids and the elucidation of the *nif* and *nod* genes in nodule-forming bacteria, opened up the opportunity for scientists to genetically manipulate these essential functions with the possibility that improved strains of rhizobia may be 'engineered' with enhanced levels of fixation, increased competitive ability, adaptation to specific environments or unique compatibility with a particular legume host genotype. The ultimate could be a new strain which combined all these attributes. The objectives of this approach are highly desirable, the science underlying the strategy impeccable, the results however, have been most disappointing but an evaluation of the present position and a view of the possible future will be presented later in this chapter.

Given the potential rewards from an efficient symbiotic association, it is imperative that both components of the symbiosis are considered and the limiting factors to the host legume and the bacteria fully understood. The vast literature on the practice of legume inoculation with strains of rhizobia provides a good insight into the implications of these interactions. Reference to the most recent publication on legume inoculant production and quality control (Anon, 1991) will reveal global coverage of these subjects and also provide a very comprehensive bibliography with the paper by J.A. Thompson (1991) posing and answering many questions on production, standards and quality control.

Inoculation

Inoculation is the technology used to increase the number of desirable rhizobia in the legume rhizosphere where they will be in a position to invade the root hairs and stimulate early nodule formation. The review by Newbould (1983), cites much of the early literature on the subject of inoculation although, as a concluding chapter in a book based on a symposium on temperate legumes, it is not surprising that it is the inoculation of the latter that takes priority in the text.

As for host plants, the Leguminosae is a very large family, adapted to many and extremely contrasting soil types and climate (Allen and

Allen, 1981). There is evidence of their cultivation in the early Chinese and Roman civilizations and their role in enhancing soil fertility has long been appreciated. However, it was the ability to isolate and mass culture the symbiotic micro-symbiont *Rhizobium* by Beijerinck (1888) that resulted in their deliberate use in sown pastures. Nowadays, clovers and lucerne are most valuable sources of protein for animals in temperate zones as are beans, peas and soyabeans for animals and man in the hotter climates. The Leguminosae has been divided into three major sub-families, the *Papilionaceae* (comprising most of the agriculturally important legumes), the *Mimosaceae* and the *Caesalpinaceae*. Between 80–90% of the species in the *Papilionaceae* form nodules but only about a quarter of the *Mimosaceae* and relatively few of the *Caesalpinaceae*. With well over 12 000 species of Leguminosae, most of which are known to fix nitrogen, it is surprising that so few, perhaps fifty or so, have been exploited for agricultural purposes (Postgate, 1978).

During the course of evolution, as legume species developed and became distinct from one another, their symbiotic partners also evolved and became more and more adapted to particular host species. Such specificity, or compatibility, has allowed *Rhizobium* to be classified into so-called 'cross-inoculation groups', divisions depending upon the host species that each group can successfully nodulate. The original grouping was expedient but could be faulted on the basis of the level of 'promiscuity' between groups and the fact that the classification placed both 'fast-growing' rhizobia, e.g. *Rhizobium trifolii*, and the 'slow-growing' rhizobia, e.g. *R. japonicum*, in the same genus.

A later classification (Table 8.1) described by Jordan (1984) partially corrected these deficiencies and also re-classified some of the original

Table 8.1. *Rhizobium*–plant assocations. (After Jordan, 1984.)

Rhizobium	*Plant*
Rhizobium meliloti	Lucerne (or Alfalfa)
Rhizobium leguminosarum	
biovar *viceae*	Pea, Vetch
biovar *trifolii*	Clover
biovar *phaseoli*	Bean
Rhizobium fredii	Soybean
Bradyrhizobium japonicum	Soybean
Rhizobium loti	Lotus
Bradyrhizobium spp.	Parasponia (a non-legume)
Azorhizobium caulinodans	Sesbania (forms stem nodules)

Fast-growing species are now placed in the genus *Rhizobium*, slow-growers in the genus *Bradyrhizobium*.

species into *R. leguminosarum* where they are more accurately described as biovars, e.g. *R. leguminosarum* biovar (bv.) *phaseoli*. With the even more recent developments in *Rhizobium* genetics, this classification certainly has more merit, especially when it is considered that the genes for host specificity are now known and can relatively easily be transferred so that one biovar becomes transformed into another in respect of its host range.

With this degree of specificity, it is not surprising that any attempt to introduce a new legume species into an area often failed due to the absence of the appropriate, matching rhizobia. Similarly, there have been many surveys of the distribution of rhizobia in soils which showed that a particular location was lacking a rhizobial population of sufficient numbers or of an adequately high level of nitrogen fixing efficiency (Holding and King, 1963; Jones, *et al.*, 1964; Jones, 1966). Nutman and Ross (1970) also carried out rhizobial counts in the soils of the experimental farms at Rothamsted and Woburn in the UK. They found a wide distribution of clover, pea, bean, lucerne and lupin rhizobia with the two former being far more abundant than the rest. As might be expected, numbers were much lower in soils where the matching host had not been cultivated for a few years although in Holland, van Schreven (1972) found that the specific rhizobia for sainfoin (*Onobrychis vicii folia*) were present in 30 out of 51 Dutch soils of varying pH, in many of which sainfoin had not been grown previously.

On a global basis, it is very apparent that there would be many areas where, for a number of reasons, the rhizobial population is very low, poorly effective in fixation or inappropriate to the legume to be sown. It is in these areas that the greatest response to inoculation might be expected, although the use of inoculants has also been shown to produce a significant improvement in plant establishment and early seedling growth (Jones and Thomas, 1966; Hely *et al.*, 1980).

Historical aspects

Rhizobium inoculants now have a very respectable history as they approach the centenary of their introduction. In 1895, the German scientists, Nobbe and Hiltner, filed patents for the process of inoculating seed with pure cultures of rhizobia. The first commercial inoculants were gelatin-based and would be suspended in water at the time of application. They were marketed under the name *Nitragin* but many other carriers have been used since (Fred *et al.*, 1932; Newbould, 1983) including sterile soil, peat and a variety of relatively inert materials plus the addition of some form of adhesive (Nethery, 1991; Somasegaran, 1991; Wadoux, 1991).

The early history of inoculant production has been briefly described above with the reports of Fred *et al.* (1932) providing a good summary of the scope and potential of this practice. Before World War II, inoculation of legumes was of limited distribution throughout the world and utilized a variety of methods which produced very variable results. Postwar, interest grew in this area and Bjalfve (1949) reported on 873 clover inoculation trials in Sweden in the five years 1944–8, in which no negative results were obtained. In the Netherlands, inoculation was always recommended for the newly reclaimed polder areas, for any soils where the legume had not been grown for at least four years and also, for any markedly acid soils (van Schreven *et al.*, 1954) In Australia and New Zealand, inoculation rapidly became the normal custom with several legumes (Paton, 1957; Cass Smith and Goss, 1958; Toms, 1960; Roughley, 1961; Hastings and Drake, 1962).

The technology of the early inoculation trials at this time was none too sophisticated with the work of Poulter (1933) a classical example. She described the deficiency of the white clover nodule organism in some Welsh soils and produced good results simply by applying inoculum to the area in the form of soil taken from a productive clover pasture from the nearby town of Aberystwyth. However, inoculation often resulted in mixed success as, for example, on a mineral soil adjacent to Miss Poulter's first experiment which was on acid peat under bent/Fescue and Nardus/Molinia, she failed to produce any inoculation response.

8.2 THE NATURE OF RHIZOBIAL INOCULANTS

In a recent review on the use of *Rhizobium* inoculants, Eaglesham (1989) reported the rapid increase in interest in the legume-*Rhizobium* symbiosis after the 1973–4 energy crisis. The increased costs of nitrogenous fertilizers, coupled with a growing appreciation of the environmental implications of their too liberal use, stimulated a great deal of research into the more efficient utilization of legumes and, in tandem, the production and use of highly effective inoculants.

Much has been written regarding the constituents of the ideal inoculant but, in reality, there are many successful combinations with availability of local ingredients often the most important factor in their choice. The production and quality control of legume inoculants has been the subject of many reviews with Thompson (1980) covering topics such as (a) the type of inoculant produced, (b) production techniques, (c) quality control, and (d) quality control of pre-inoculated seed. The preparation, use and quality control of inoculants with special reference to developing countries has been covered in some detail by Burton (1981) and for Australia by Vincent (1981). Roughley and Pulsford (1982) updated the latter publication but the recent publication from FAO (Anon, 1991) gives a much more detailed and globally inclusive account of the subject with discussion of selection of strains, packaging, labelling and use of commercial inoculants being very well described by Hardarson and Bowen (1991).

However, with such a long and important history, it is a sad comment on the 'state of the art' that the same basic techniques are still in use today. In the majority of instances, and in many countries, rhizobial strains are still selected only on the basis of a high level of nitrogen fixing ability with the host species in general and the inoculum applied in a finely ground peat as a slurry plus an adhesive. The **selection** of strains of rhizobia for use as inoculants and their characterization and maintenance requires a skilful and increasingly technological approach. Date (1982) has described these processes in some detail. Their initial isolation from the nodules of suitable legumes, their evaluation for efficiency of nitrogen fixation, competitive ability, saprophytic competence and even compatibility with the particular genotype of legume to be inoculated, are all aspects of typical screening protocols.

8.2.1 Carriers

In early experimental work and, to some extent on the farm, the rhizobia were applied to seed in a suspension in skimmed milk. This not only acted in an adhesive capacity but also as a carrier providing some protection and sustenance. The situation has now changed, with peat being used as the dominant carrier throughout the world. Many alternatives have been tried with varying success. Odevemi and Okoronkwo (1985) investigated the suitability of several local materials as rhizobial carriers in inoculants in Nigeria. While peat was very good, they also found that a mixture of lignite and calcium carbonate was also successful and even sub-bituminous coal and cowdung could be used. The survival of *R. phaseoli* in coal-based inoculants was also studied by Crawford and Berryhill (1983). They reported that peat was undoubtedly the best carrier but all eight coal-based inoculants provided more than 10^4 viable rhizobia per seed after four weeks. A similar exercise was carried out by Dube *et al.* (1980) in which they compared coal, lignite and a coal and lignite mixture with a commercial inoculant, but only the mixture increased the soyabean seed yields more than the purchased product. Anyango *et al.* (1985) introduced another variation using locally found filter mud-based inoculants which compared favourably with those more conventionally based on a peat carrier.

In France and Senegal, Jung *et al.* (1982) compared polyacrylamide, alginate or a mixture of xanthan and carob gum for their suitabilities as inoculant carriers. Whether dried or semi-dried, they were all successful and gave a good 'shelf-life' of up to 90 days. Soyabean seed inoculated up to 48 days with the xanthan mixture resulted in plants comparable in nodulation and growth to plants given peat inoculation at the time of sowing. However, a word of caution must be introduced for, in Vietnam, Duong *et al.* (1984) conducted several inoculation trials in acid soils ranging from pH4.5 to 5.1. With soyabean as the host crop, these authors found that locally produced inoculants were no better than the uninoculated controls, whereas a commercial granular inoculant from the Nitragin Company gave a significant improvement in nodulation and growth.

Kremer and Peterson (1982) compared peat-based inoculants with those based on oil derivatives. In particular, they quantified the recovery of *R. phaseoli* from seed inoculated with these inoculants. They recovered 3×10^6 cells/seed from the oil-based product but only 8×10^2 from the peat inoculant. They also found that the oil-based treatment gave higher numbers of nodules and weights and a higher total nitrogen content. However, these results are somewhat misleading as the overall results from the use of peat inoculants in many countries clearly indicates a very satisfactory protocol. The use of mineral oils as carriers has also been discussed by Chao and Alexander (1984).

The feasiblity of developing carrier-based inoculants from a two-year-old freeze-dried cell concentrate of a cowpea *Rhizobium* was examined by Jauhri (1989). Maximum survival of lyophilized rhizobia was obtained when their suspensions in yeast-extract-mannitol broth were aerated for 24 hours prior to mixing in the carrier.

Survival

The survival of rhizobia in the inoculant and, eventually, in the soil is of the utmost importance. Somasegaran *et al.* (1984) studied the influence of high temperatures on the growth and survival of *Rhizobium* spp. in peat inoculants during preparation, storage and distribution. High temperatures and consequent desiccation can be lethal to rhizobia and their use in tropical areas can be severely limited by these factors. The inoculants were sent to 13 tropical countries with an initial cell count of 1×10^8 cells/g. Laboratory data showed that 28°C was optimal for multiplication to maximal numbers, more than 1×10^{10} cells/g peat. Six inoculants showed excellent viability after storage at at 28°C for 24 weeks.

Wolf and Hoflich (1986) found that the growth rate of rhizobia in peat was increased by adding molasses, but they emphasized that storage of the inoculant at cool temperatures was very important if viability of the bacteria was to be maintained for long periods. Paczkowski and Berryhill (1979) measured the survival of *R. phaseoli* strains in coal-based inoculants. They found that most strains sustained a population of 10^6

cells g^{-1} for at least seven months and some even had 10^6 cells/g after one year. Salema *et al.* (1982) compared mixtures of sucrose or maltose and sodium glutamate for their effect on the survival of rhizobia on mung bean (*Vigna radiata*) and found that the mixtures were superior to the single solutions in enhancing viability of rhizobial cells.

Catroux (1991) described the quality control standards for inoculants in France. On the basis that a legume will produce a number of nodules positively correlated with the log number of rhizobia on the seed (with *B. japonicum* in France, r = 0.71), the standard imposed demands 10^6 viable cells per seed. For practical purposes, the standard is expressed per hectare. Assuming 400 000 soybean seeds are sown per hectare, a soybean inoculant must contain at least 4×10^{11} viable *B. japonicum* per hectare packaging. In the same way, the standard for alfalfa inoculant was established at 10^3 viable *R. meliloti* per seed, which corresponds to 0.5×10^{11} bacteria per hectare packaging. In the United Kingdom, there is now only one commercial company producing inoculants (Day, 1991). This company, New Plant Products Limited which is a subsidiary of Agricultural Genetics Company, has no quality standards imposed by legislation but insist on rigorous standards (at least 2×10^9 cells/g inoculant to be maintained for at least 18 months) and so successfully competes for sales in other countries.

The survival of inoculants is dependent upon many factors, both on the seed and in the soil. The application of crop protection chemicals as seed treatments is an obvious candidate for study for possible adverse effects on inoculant viability. In fact, the opposite was found when Jauhri and Agarwel (1984) evaluated various fungicide seed treatments such as thiram, captan and mancozeb; they found that mancozeb alone or with thiram gave the best yield and seed emergence in soyabeans growing on *Macrophomina phaseolina*-infested soil. There were no adverse effects of the fungicides on the inoculant. Similar work was carried out by Welty *et al.* (1988) who also found that the combined inoculation and fungicide treatments mostly increased seed yield, nodule weight and plant establishment, but there was variation between years.

'Shelf-life'

Much work has been carried out on the 'shelf-life' of inoculants and it has been shown that the survival of rhizobia in peat cultures is dependent on the type of peat, its pH, its sterilization and the storage temperature of the inoculated seed (Date, 1970). Roughley (1981) found that where legumes were to be established in *Rhizobium*-free soil under good conditions, 100 rhizobia per seed was normally satisfactory but, if soil conditions were adverse for rhizobial survival or there was a large indigenous population of ineffective rhizobia, numbers in excess of 10^6

per seed would be required. Obviously, the actual numbers needed depend upon the size of the seeds and the external factors, but there is no doubting that the higher the numbers the greater the chance of inoculation success. Roughley and Thompson (1978) compared the number of rhizobia in commercial broths containing more than $5 \times 10^8/$ ml rhizobia with the number in commercial unsterilized and sterilized peat cultures prepared from the broths and showed that there was no correlation between the numbers in the broths and the corresponding peat cultures. A comprehensive study of the survival of the inoculant with different carrier treatments was carried out by Sparrow (1981). **Peat** was consistently the best with charcoal and vermiculite also satisfactory but the **temperature of storage** was a significant factor in the ranking of treatments.

In the USA, Weaver and Frederick (1974) suggested that numbers of soybean nodule bacteria, 1000 times greater than the indigenous population, would be required for successful nodulation by the introduced strain. The success of inoculation should not be difficult to evaluate, with many trials producing outstanding, highly visible increases in crop growth. The criteria for a successful inoculant are also not difficult to detail although the actual composition and method of application might differ in individual cases.

8.2.2 Adhesives

Much work has also been carried out to evaluate adhesives in inoculant production. Elegba and Rennie (1984) compared many candidate adhesives including gum arabic (40% w/v) and carboxymethyl cellulose (4% w/v). They both performed well, binding over 800 mg of inoculant/seed. Readily available wallpaper glue (10% w/v) bound almost 900 mg inoculant/seed, as did two commercial adhesives, *Nutrigum* and *Nitracoat*, but all the test materials resulted in more than 10^5 viable rhizobia/ seed and thus satisfied the requirements of the Fertilizers Act of Canada. Adhesives are of obvious importance in ensuring the initial retention of the inoculant and carrier, but they must also contribute to the persistence of the rhizobia. In other words they probably determine the 'shelf life' of the inoculant. Weaver *et al.* (1985) studied the survival of rhizobia on inoculated Arrowleaf clover (*Trifolium vesiculosum*). The use of the adhesive gum arabic gave better survival than sucrose or water but even then one strain tested declined from 10 000 cells/seed to less than 10 cells/seed in six days. The better of the two strains gave 200 cells/seed after six days. Gum arabic also enhanced survival under lime pellets, but survival was still poor.

Inevitably, there is a debit side to the use of adhesives and Jauhri (1988) showed that, as the concentration of adhesives increased, there was a corresponding loss of the inoculant. However, this loss was reduced with increasing soil moisture.

8.2.3 Seed pelleting

Rhizobia are very sensitive to acidity, heat, drought and light and it is, therefore, essential that the process of inoculation is undertaken as close to the time of sowing as possible. In commercial practice, it is this time factor that most probably causes the majority of inoculation failures. Methods to prolong the viability of rhizobia after seed inoculation include protective coats and various forms of pellets.

Seed pelleting is a relatively recent practice and has found much popularity in Australasia. Various pelleting materials have been tested with **lime** being the most used as it served the dual purpose of inoculant protection and acid soil amelioration. Anderson and Spencer (1948), Cass Smith (1959) and Hastings and Drake (1960) all demonstrated that legumes could be established easily in pastures with the minimum of cultivation and lime by using seed pelleted with lime. Jones *et al.* (1967) compared the nodulation and growth of white clover when inoculated and pelleted with either lime or lime and phosphate and sown on to areas with varying applications of lime. Both high lime dressing and lime pellets gave greater clover content in the sward and the total increase in clover due to inoculation was also highly significant.

Hastings and Drake (1963) reported beneficial effects of inoculation, lime pelleting and pelleting with Gafsa phosphate/dolomite on the less fertile soils in their trials. After the treated seed had been stored for four months, the Gafsa/dolomite pellet was superior to the other treatments in all the soils tested. Similarly, Norris (1971) reported a very variable response from eight legumes pelleted with lime and rock phosphate, the best results being achieved where the conditions had been contrived to force a bias in favour of lime pelleting.

The authors have tested variously commercially-prepared samples of inoculated and pelleted seed with very variable results, the best only being satisfactory for a few weeks but, even with the best, the results often did not match up to freshly inoculated seed. In another study with *R. phaseoli*, Barkdoll *et al.* (1983) found that the numbers of rhizobia in a granular inoculant were always higher than on seed of *P.vulgaris* treated with a pellet inoculant. The problem of the length of the storage period by the farmer after purchase is the most critical factor, with too long a period reducing rhizobial viability to below a minimum threshold level (Brockwell *et al.*, 1975).

The upsurge of interest in biologically-fixed nitrogen over the past 20 years stimulated a more technological approach to the problem of rhizobial protection on the seed. Much of the work in this area is covered by patent or, at least, the details of the processes have not been published by the commercial firms responsible for their development. It is, therefore, difficult to make precise evaluations but there are several such products currently on the market some of which incorporate additional compounds with plant nutrient value and/or fungicidal properties.

8.2.4 Other practices

Inoculation is normally carried out to the seed before sowing but there are other alternatives. Hely *et al.* (1980) found that spraying the inoculant in a vertical band into the soil below the seed produced better establishment of *Trifolium subterraneum* than did the more conventional drill sowing with lime-pelleted seed. They also confirmed that banded spray inoculation plus fungicidal seed dressing increased the number and size of nodules and enhanced young plant growth. Another variation was the development of 'fluid drilling' where both the seeds and the inoculant were suspended in sloppy gel and then extruded into the furrow (Hardaker and Hardwick, 1978). Jensen (1987) also reported on the successful placement of the rhizobia directly into the furrow for the inoculation of the pea crop. Another interesting variation is provided by the work of Rogers *et al.* (1982), who used a spray inoculant technique as a post-emergence remedial treatment when nodulation and establishment was deficient.

Dual inoculation of legumes with rhizobia and vesicular-arbuscular mycorrhiza has also attracted attention in the past 15 years or so. Azcon-Aguilar and Barea (1981) inoculated lucerne (*Medicago sativa*) with *R. meliloti* and the endomycorrhizal fungus *Glomus mosseae* and produced a doubling of the yield. Similarly, Manjunath and Bagyaraj (1984) inoculated pigeon peas and cowpeas with both *Rhizobium* spp. and *Glomus fasciculatum* in phosphate-deficient soils and significantly increased nodulation and shoot dry weight. With Phaseolus beans, Kucey and Bonetti (1988) increased nitrogen fixation by 50–82 mg N/ plant after seed inoculation with rhizobia, mycorrhiza and a seed dressing of captan.

An early decision would have to be made as to whether a **single strain** or a **multi-strain** inoculant should be used. The insurance nature of the 'cocktail' type has its attractions although there is little evidence that there is a consistent and significant benefit above that from the use of single strains. Rennie and Dubetz (1984) compared both types of inoculant on soyabean. Eight single strains and two multi-strain inoculants were evaluated over a two year period on soils containing no indigenous rhizobia. Two single strains were consistently superior in nitrogen fixation with all the cultivars tested and one of the multi-strain inoculants gave poor results due to inter-strain competition but the other gave good results.

Inoculant rates

There have been many experiments to determine the effect of inoculant rates on nodulation with most results indicating a related beneficial effect as the rate was increased. As an example, Smith *et al.* (1981) conducted a soyabean inoculation trial with inoculum rates varying from *ca.* 10^3 to 10^9 cells/cm row. The results demonstrated a linear relationship between total numbers of nodules, weight of nodules per plant and inoculum rate. Martensson (1990) did similar work with *R. legumino-sarum* bv. *trifolii* with the modification that repeated inoculation was also included. In the field where native ineffective red clover bacteria were present, yield from plants inoculated with strain 7612 increased with increasing inoculum concentration, as did the number of early nodules occupied by this strain. In the glasshouse, where strain 285 was the exclusive nodule occupant in competition with strain 7612, the latter increased its occupancy after repeated inoculation. However, Rys and Bonish (1984) inoculated white clover in soils where the native popula-tion of *R. trifolii* was *ca.* 10^6/g soil. Here, the inoculant did not cause nodulation even at 15 times the recommended rate, but this was most probably due to nematode infestation in the area.

Inter-strain competition

Competition between strains is also a critical factor in the success of a strain in forming nodules. Early work with biovar *trifolii* (Nicol and Thornton, 1942) showed significant differences between strains in com-petition and it was suggested that this might be due to differential growth rates of the strains. Gibson *et al.* (1976) made an interesting comparison of competitiveness and persistence amongst five strains of *R. trifolii*. With subterranean clover as the host, they found that, in the first year, all inoculant strains were present at high frequency in the sampled nodule population from the field plots. Where an inoculant containing a mixture of equal parts of the five strains was used, one strain predominated and this was found at two separate trial sites. After three years at one site, only three of the five strains were persistent. In contrast, Marques Pinto *et al.* (1974), Labandera and Vincent (1975) and Russell and Jones (1975) have demonstrated that some strains may be very successful in nodulation even when they are very much in the minority in the soil population.

The competitive ability of a strain is put to its most severe test when the strain is used in the field. The extent of its success is undoubtedly influenced by soil type and soil nutrient status, the host legume and, of course, its own innate colonizing ability. Brockwell *et al.* (1982) compared inoculant strains by measuring their success in forming nodules in the first year (competitiveness) and in the second year (persistence). They found that inoculated strains formed more nodules in the first year and that they persisted better at sites where there were few indigenous rhizobia. There can be no doubt that the most important result from all the published work on competitiveness of inoculants is that there is very marked variation and this character should always be checked when screening inoculants.

'Pre-inoculation'

The concept and practice of 'pre-inoculation' has its strength in the reluctance of farmers to carry out the inoculation procedures on the farm and their preference for purchasing 'ready to use' seed. The review by Eaglesham (1989) surveys the use of rhizobial inoculants in 70 countries and the response to his questionnaire indicated a huge range of practices and levels of success. Some countries, such as Australia, where peat-based inoculants were introduced in the 1940s, have a rigorous quality control system which not only persuaded farmers of their quality but also generated much excellent research into the technology of inoculant production. In contrast, in South Vietnam, peat-based inoculants were initially introduced but their use has declined almost to the point of disappearance due to the decline in world soyabean prices and the pressures that inflicted on the farming industry. In many areas, inoculum is now produced by the village headman using a portable fermenter with all the possible associated risks of serious contamination (Lie, personal communication). The limitations as exemplified in the latter example and the production of pre-inoculated seed of dubious quality, compounded by bad farmer practice, has resulted in a high level of mistrust of seed inoculation.

8.3 THE EXTENT AND SUCCESS OF *RHIZOBIUM* INOCULATION

Despite the vast amount of successful research carried out on the legume/*Rhizobium* symbiosis, inoculation has not increased in the way one might have expected. We now know much about many aspects of the ecology and, in recent years, the molecular biology of the interaction but still this has not resulted in the wider use of inoculation.

There are a few simple criteria that allow us to predict where we may expect inoculation of legumes to be successful. Due to the co-evolution of host plant species and compatible strains or species, legumes grown away from their centre of origin can expect to be in soil lacking the appropriate *Rhizobium* species. The same can be true of plants grown in virgin, marginal or reclaimed land. Certain soils, for example acid types, are unsuitable for the saprophytic survival of rhizobia, leading to the lack of nodulating strains. Obviously, there is a complex interaction of environment, microbe and plant. Of the environmental factors, it is the **availability of nitrogen** in the soil that is one of the most important influences on nodulation and nitrogen fixation.

The literature has many examples of inoculation of virtually every legume of economic importance and, in many .cases, the treatment resulted an increase in yield. The only paper which actually shows the practical applications of the technology on a global scale is that of Eaglesham (1989). In this work, he sent questionnaires to research workers throughout the world in order to ascertain the use of inoculants. What his survey found was that the cultural and environmental factors suggested above are strongly influencing the practical application. There is no doubt that the most widely used inoculant is that of *Bradyrhizobium japonicum* on the soybean crop. This is particularly the case in many parts of the Americas. Soybean is exotic to these areas and therefore no indigenous rhizobia can effectively nodulate the crop. Soybean was first grown in north America in the eighteenth century but was not really successful until the introduction of soil from Japan containing the compatible rhizobia (Allen and Allen, 1981). This early success has continued, the *B. japonicum* inoculant market in the US alone being worth $18 million dollars in 1987 (Lethbridge, 1989). With the widespread use of *B. japonicum* inoculants, it may be expected that it would become established in the soil and further inoculation would not be required. In many areas, however, soybean is still being grown for the first time. On land which soybean cultivation is established, inoculation is not as common but can still be justified as an 'insurance' against poor nodulation. It should be remembered that soybean is the single largest source of both edible oil and protein meal, providing 30% of the former and 50% of the latter for animals. The United States is the biggest market for *B. japonicum* inoculants, although Brazil and Argentina also use large quantities.

In many respects, Australasia is a similar case to America in that all agriculturally important legumes are exotic. Inoculation has been a considerable success, particularly with forage crops which have become an integrated part of farming in many places. The use of these exotic forage legumes and their inoculation has resulted, in many places, in the rhizobia becoming naturalized with inoculation no longer being required.

In Europe, the majority of the agriculturally useful legumes are indigenous and therefore inoculation is not often required. Lucerne is the most important inoculated legume but, as with many other parts of the world, soybean is increasing in production. The regions that can grow it are limited and, of those that can, Italy's output is growing rapidly with a corresponding increase in the requirement for inoculant.

Eaglesham (1989) concluded from his survey that inoculation with rhizobia had made a large contribution to agriculture on a worldwide scale. Most significant results have been achieved with soybean/*B. japonicum*, but there have also been successes with a number of other legumes. There are still, however, many areas of the world where inoculation has made little or no impact.

8.4 FUTURE PROSPECTS

There are two quite different ways in which *Rhizobium/Bradyrhizobium* inoculation can be developed in the future. Firstly, there is the expansion of its use in developing countries where it is not presently carried out and, secondly, the development of superior rhizobia/host interactions in terms of increased nitrogen fixation, better nodulation competitive ability of rhizobial strains and other useful characteristics.

8.4.1 Inoculation in developing countries

Most biological nitrogen fixation technology transfer is directed at the large-scale farmer who, relatively speaking, needs it least. In countries where farmers do not have the finances to take advantage of fertilizers, the cost and sustainability of legumes along with inoculation, where necessary, may be particularly attractive (Bohlool, 1988). It must not be assumed, however, that inoculation is the only improvement needed. In many cases, nitrogen fixation is just one of a number of limiting factors to successful cultivation. However, it could be an extremely important part of a package of measures capable of increasing yields. Criticism has been made of the fact that inoculation research is carried out at research institutions rather than in farmer's fields and it is, therefore, difficult to truly assess the results (Somasegaran *et al.*, 1984; Wynne, *et al.*, 1987).

In an attempt to identify what gains in production could be made, a large collaborative research project, INLIT (International Network of Legume Inoculation Trials), was set up (Bohlool, 1988). From 1980–1986, 200 standardized experiments were set up in 28 developing tropical countries. A selection of 13 tropical legumes, including chickpea, lentil, soybean and urd bean, were assessed and responses to additions of lime and fertilizers were also recorded. Overall, there was a significant benefit gained from inoculation, but responses ranged from insignificant to highly significant depending on the species and environment, reflecting the problems in predicting responses due to the interactions mentioned earlier.

A similar project based on forage legumes was carried out by CIAT (International Centre for Tropical Agriculture) in South America and by ICARDA (International Centre for Agricultural Research in Dry Areas) in North Africa. In both situations, inoculation has been used to great benefit (Sylvester-Bradley, 1984; Cocks, 1988). The work of these groups clearly indicates that inoculation of both grain and forage legumes could play a very important role in increasing crop yields in developing countries. The challenge is to transfer the appropriate technology to these areas. A number of organizations are addressing themselves to these problems. Methods of inoculant production, quality control and packaging are well established and simple inexpensive fermenters are available. It is education and training, outreach and information services that are important in facilitating the transfer of the appropriate technologies to tropical developing countries (Hubbell, 1988; Mytton, 1988).

8.4.2 Improvement of the symbiosis

This section will discuss firstly, the improvements in the attributes of the bacteria that will give increased legume productivity and, secondly the manipulation of competitive ability to allow the better strains to achieve nodulation, even in the presence of indigenous bacteria.

It has long been known that strains of rhizobia vary in their ability to fix nitrogen. Selection of such strains can and is carried out for the production of inoculants (refer to Section 8.1.2). The basis of such variation in effective strains will be due, in part, to the complex interaction between the host and the micro-symbiont (Mytton, 1984). There are, however, **specific genes** that may be involved in a number of cases. One such gene is the hydrogenase gene (*hup*) present in some strains of *Bradyrhizobium*, but less common in *Rhizobium* strains. Hydrogen is a product of the nitrogen fixation process during the reduction of nitrogen by nitrogenase. The recapture of the hydrogen generates ATP (Evans *et al.*, 1987). This process should make nitrogen fixation more efficient. The gene has been cloned and experiments to verify its potential carried out (Lambert *et al.*, 1985). Isogenic strains of *B. japonicum*, either hup^+ or hup^- were compared for dry weight production in symbiosis with soybean (Evans *et al.*, 1985). The results showed significant increases in the levels of nitrogen in the seed (8.6%), leaf (27%) and total plant (11%). These results, however, have been difficult to reproduce consistently (Evans *et al.*, 1987).

There are **other genes** involved in the symbiosis which show some promise as targets for manipulation for increased nitrogen fixation, these include *nifA*, *dctA* and *nodD*. Alterations to the expression of *nifA*, which is involved in nitrogen fixation regulation in the nodule, can affect plant growth both positively and negatively (Cannon *et al.*, 1988). C_4-dicarboxylic transport (*dct*) is of fundamental importance in the functioning of the nodule and it would appear that manipulation of one of the genes in this system (*dctA*) can lead to greater legume productivity (Birkenhead *et al.*, 1988; Cannon *et al.*, 1988). *NodD* is the gene required for the induction of the other nodulation genes and interacts with plant-produced flavonoids to do so (Spaink *et al.*, 1989a).

There are undoubtedly other ways in which the symbiosis can be improved. Evidence for this comes from reports on **mutagenesis treatment** of rhizobia which resulted in increases in host productivity. Most of this work has been carried out on *B. japonicum*. Mutants isolated all had increased nitrogenase activities when compared with the wild-type strain (Maier and Brill, 1978; Scott *et al.*, 1979; Paau, 1989). Such mutants have been shown to work in the field, although they did not have to compete with an indigenous population.

Genetic manipulation

Improvements to a strain's nitrogen fixation efficiency can be made by replacing its resident symbiotic (Sym) plasmid with one which is more effective. This is particularly attractive if the strain has useful characteristics other than those encoded by its Sym plasmid. Chen *et al.* (1991) improved the symbiotic properties of an acid-tolerant strain of *Rhizobium leguminosarum* bv. *trifolii* ANU 1173. They demonstrated that the acid tolerance was not due to the Sym plasmid which was then cured and replaced with a Sym plasmid pBRIAN, a plasmid known to be more efficient at nitrogen fixation. The resulting strain gave a 17% increase in nitrogen fixation under acid conditions compared with the original strain. Many of the original strain's characteristics have, in this way, been combined with a higher nitrogen fixation capacity. What was very encouraging about this work was that the conjugated plasmid was maintained stably in the recipient strain. This suggests that the transfer of megaplasmids offers a potentially rewarding method of manipulating strains in improvement programmes.

Increasing nitrogen fixation in order to increase yields in this way is only one approach, there are other possible methods. Some legumes suffer from **pest damage**, for example from the *Sitona* weevil to nodules. Clover, pea and bean can be attacked in this way. Control of such pests has been shown to be possible through the use of the *Bacillus thuringiensis* subsp. *tenebriosis* (Bt) toxin. The Bt toxin gene has been transferred to *Rhizobium* species and it has been shown that it may be possible to control these pests in this manner (Skot *et al.*, 1990), providing an interesting alternative method to maintain nodule productivity.

Such genetic manipulations have their merit but the main problem encountered with rhizobia is achieving nodulation with an introduced strain, particularly when indigenous strains are also present in a soil. Improvements in the ability to fix nitrogen are of little consequence if that strain fails to form nodules when inoculated in the field. The ability to compete successfully with indigenous populations, in itself, can be of great benefit as, in many situations, indigenous strains are often not very effective (Jones, 1963). In such circumstances, a strain possessing a high competitive ability in forming nodules is required. In soil, an introduced rhizobial strain has to survive variations in temperature, moisture and pH. It also has to compete for nutrients with other microorganisms and other rhizobia. These are just a few of the stresses imposed on an inoculant strain before it can begin to infect the plant. The problem of achieving nodulation with the required strain has been approached in a number of ways but they fall into two broad categories. First, is the utilization of the selectivity of the host and, second, is the manipulation of a component of competitiveness in the rhizobia.

The theory behind the first method is that the particular host legume cannot be nodulated by the indigenous population but only by the selected strain which is used as the inoculant. The utilization of such specificity has been proposed by Forbert *et al.* (1991). *Rhizobium leguminosarum* bv. *viciae* strains can nodulate peas, vetches, lentils and *Lathyrus* spp. (Vincent, 1982). Within the biovar *viciae*, most natural isolates do not nodulate the primitive cultivar called Afghanistan (Lie, 1978; Ma and Iyer, 1990). This apparent resistance to nodulation has been found to be due to the presence of the recessive gene *sym* 2 which can be crossed into conventional pea cultivars (Holl, 1975; Le Gal *et al.*, 1989). Rare isolates of *Rhizobium* have been found which can nodulate Afghanistan pea as well as the common cultivars of pea. These strains all have a gene called *nodX* as an extension of the *nod* operon present in all biovar *viciae* isolates (Davis *et al.*, 1988). Because *nodX* is rare in natural populations of *R. leguminosarum*, it was thought that the manipulation of the *sym* 2 gene into commercial cultivars of pea along with the addition of *nodX* to a selected strain of *R. leguminosarum* would mean that nodulation with the selected strain could be guaranteed. There is, however, a major problem with this technique. It would appear that strains that do not contain the *nodX* gene can block nodulation by the strains containing *nodX*. Genes involved in this competitive nodulation blocking phenotype (Cnb+) have been identified by mutagenesis of *R. leguminosarum* bv. *viciae* strain PF2 (Dowling *et al.*, 1987, 1989). The *nodX* gene is no guarantee of nodulation in the competitive soil environment. The system may assist the specificity of nodulation, but the strain involved will have to be competitive in other ways.

Flavonoids, isoflavones and **chalcones** are all known to interact with the *nodD* gene product of *Rhizobium* and *Bradyrhizobium* to induce the expression of the nodulation genes which lead to the formation of nodules (Firmin *et al.*, 1986; Peters *et al.*, 1986; Redmond *et al.*, 1986; Djordjevic *et al.*, 1987; Kosslak *et al.*, 1987; Peters and Long, 1988; Zaat *et al.*, 1987; Maxwell *et al.*, 1989). Work on *nodD* suggests that it is open to manipulation, with mutants of the gene becoming flavonoid independent or losing specificity to particular inducer molecules (Burn *et al.*, 1987, 1989; Johansen *et al.*, 1988; Spaink *et al.*, 1989a, b).

It has also been demonstrated that certain substances can inhibit the flavonoid/*nodD* interaction (Cunningham *et al.*, 1991). These inhibitors were shown to inhibit nodulation of soybean by *B. japonicum*. The plant could be nodulated by flavonoid independent strains while in the presence of indigenous rhizobia and the appropriate inhibitor of nodulation gene induction. There are, however, a number of practical problems. One of these is that to prevent induction of *nod* genes in all strains of *Bradyrhizobium japonicum* present, a number of inhibitors would be necessary (Cunningham *et al.*, 1991). These disadvantages would certainly reduce the commercial potential of this approach.

Manipulation of the bacteria alone would be easier than altering both partners in the symbiosis. The most obvious method to improve nodulation competitive ability would be to increase selected strain numbers in the rhizosphere or to decrease the numbers of 'opposition' rhizobia. Some strains of rhizobia have been found to produce bacteriocins. Hodgson *et al.* (1985) worked on a bacteriocin-producing strain of *Rhizobium leguminosarum* bv. *trifolii* CB782. This strain increased the nodule occupancy of a bacteriocin-resistant strain when co-inoculated with a bacteriocin-sensitive strain. These experiments were not carried out in the presence of an indigenous population or with isogenic strains. Another *R. leguminosarum* bv. *trifolii* strain, T24, was found to produce a bacteriocin referred to as trifolitoxin (Schwinghamer and Belkengren, 1968; Triplett and Barta, 1987). This strain was found to be a good competitor when co-inoculated with a bacteriocin-sensitive *R. leguminosarum*, bv. *trifolii* (Schwinghamer and Belkengren, 1968). This was very encouraging as far as competitive ability was concerned but, unfortunately, T24 is ineffective in N-fixation. Confirmation of the importance of the trifolitoxin was obtained with the production of Tn5 mutants lacking toxin production (Triplett and Barta, 1987). These strains lost advantage in competition with toxin sensitive strains. Using Tn5 mutagenesis, genes for trifolitoxin production and sensitivity were mapped and cloned (Triplett, 1988; Triplett *et al.*, 1989). The *tfx* region was found to be 4.4 kb and was transferred to an effective, trifolitoxin-sensitive strain of *R. leguminosarum* bv. *trifolii* TA1 (Triplett, 1990). This region was stably transferred by marker exchange. The strain expressed the introduced genes in a stable manner and showed increased competitiveness when co-inoculated with a trifolitoxin sensitive strain (Triplett, 1990). All these experiments were carried out in sterile conditions. It is not known whether these results can be repeated in non-sterile soil.

Many *mutants* have been found which are associated with altered competitivity ability, unfortunately the majority reduce this attribute. These include motility (Ames and Bergman, 1981; Mellor *et al.*, 1987), cell surface characteristics such as non-mucoid strains (Araujo and Handelsman, 1989) and extracellular polysaccharide (EPS) mutants (Zdor and Pueppke, 1989; Bhagwat *et al.*, 1989).

Rhizopines

The discovery of substances referred to as rhizopines in *R. legumino-sarum* bv. *viciae* (Skot and Egsgaard, 1984) and *R. meliloti* (Murphy *et al.*, 1987) inhabited nodules could be of great interest to scientists looking for characteristics that may lead to increased competitive ability. Production of these amino acid derivatives is encoded by the *Rhizobium* only when it has differentiated into a bacteroid within the nodule. The rhizobia also contain genes encoding for catabolism of the rhizopines, but these are expressed in the free-living bacteria (Murphy *et al.*, 1987, 1988). It would appear that the bacteroid produces substrates for the free-living bacteria that formed the nodule. As to whether a competitive advantage is given to the free-living rhizobia has not yet been demonstrated. This interaction does offer opportunities for manipulation by the molecular geneticist in the improvement of competitive ability of strains.

The development of rhizobia consistently high in competitive ability for nodulation of a given host has not been achieved. The main reason for this is due to the highly complex soil environment in which introduced rhizobia have to compete. In this respect, the symbiotic improvement of intrinsically competitive strains (Chen *et al.*, 1991) may be an important step towards achieving this aim.

REFERENCES

Allen, O.N. and Allen, E.K. (1981) *The Leguminosae. A Source Book of Characteristics, Uses and Nodulation*. Macmillan, London.

Ames, P. and Bergman, K. (1981) Competitive advantage provided by bacterial motility in the formation of nodules by *Rhizobium meliloti. Journal of Bacteriology*, **148**, 728–9.

Anderson, A. J. and Spencer, D. (1948) Lime in relation to clover nodulation at sites on the Southern Tablelands of New South Wales. *Journal of the Australian Institute of Agricultural Science*, **14**, 39.

Anon. (1991) *Expert consultation on legume inoculant production and quality control*. Report, Food and Agriculture Organisation of the United Nations, Rome.

Anyango. B., Keya, S.O. and Balasundarum, V.R. (1985) Assessment of filter-mud as a carrier for legume seed inoculants: physico-chemical properties and *Rhizobium phaseoli* survival, in *Biological Nitrogen Fixation in Africa* (eds H. Ssali and S.O. Keya), MIRCEN, Nairobi, Kenya, pp.180–96.

Araujo, R.S. and Handelsman, J. (1989) Nodulation competitiveness of *Rhizobium leguminosarum* bv. *phaseoli* mutants with altered cell surface characteristics, in *Proceedings of the 12th North American Symbiotic Nitrogen Fixation Conference*, p.72.

Azcon – Aguilar, C. and Barea, J.M. (1981) Field inoculation of *Medicago* with V-A mycorrhiza and *Rhizobium* in phosphate-fixing agricultural soil. *Soil Biology and Biochemistry*, **13**, (1), 48–52.

Barkdoll A.W., Sartain, J.B. and Hubbell, D.H. (1983) Effect of soil implanted granular pellet *Rhizobium* inoculant on *Phaseolus vulgaris* L. in Honduras. *Proceedings of the Soil and Crop Science Society of Florida*, **42**, 184–9.

Beijerinck, M.W. (1888) Die Bacterien der Papilionaceeknollchen, *Botanische Zeitung*, **46**, 725–804.

Bhagwat, A.A., Sloger, C. and Kister, D.L. (1989) Isolation of extracellular polysaccharide defective mutants of *B. japonicum* 61A89 and USDA. 438, in *Proceedings of the 12th North American Symbiotic Nitrogen Fixation Conference*, p.97.

Birkenhead, K., Manian, S.S. and O'Gara, F. (1988) Dicarboxylic acid transport in *Bradyrhizobium japonicum*: use of *R. meliloti dct* gene(s) to enchance nitrogen fixation. *Journal of Bacteriology*, **170**, 184–9.

Bjalfve, G. (1949) Inoculation trials of leguminous plants 1914–1948. Lucerne and clover trials. *Annals Royal Agricultural College*, Sweden, **16**, 603–17.

Bohlool, B.B. (1988) *Rhizobium* technology: Applications for international agricultural development in the tropics, in *Nitrogen Fixation – One Hundred Years After*, (eds H. Bothe, F.J. de Bruijin and W.E. Newton), Gustav Fischer, pp.759–64.

Brockwell, J., Herridge, D.F., Roughley, R.J., *et al.* (1975) Studies on seed pelleting as an aid to legume inoculation. 4. Examination of pre-inoculated seed. *Australian Journal of Experimental Agriculture and Animal Husbandry*, **15**, 780–7.

Brockwell, J., Diatloff, A., Roughley, R.J., *et al.* (1982) Selection of rhizobia for inoculants, in *Nitrogen Fixation in Legumes*, (ed J.M.Vincent), Academic Press, Sydney, Australia, pp.173–91.

Burn, J.E., Rossen, L. and Johnston, A.W.B. (1987) Four classes of mutations in the *nodD* gene of *Rhizobium leguminosarum* biovar *viciae* that affect its ability to autoregulate and/or activate other *nod* genes in the presence of flavonoid inducers. *Genes and Development*, **1**, 456–64.

Burn, J.E., Hamilton, W.D., Wooton, J.C. and Johnston, A.W.B. (1989) Single and multiple mutations affecting properties of the regulatory gene *nodD* of *Rhizobium*. *Molecular Microbiology*, **3**, 1567–77.

Burton, J.C. (1981) *Rhizobium* inoculants for developing countries. *Tropical Agriculture*, **58**, (4), 291–5.

Cannon, F.C., Benyon, J., Hankinson, T., *et al.* (1988) Increasing biological nitrogen fixation by genetic manipulation, in *Nitrogen Fixation – One Hundred Years After*, (eds H. Bothe, F.J. de Bruijin and W.E. Newton), Gustav Fischer, pp.735–40.

Cass Smith, W.P. (1959) Topical notes on seed pelleting. *Journal of the Department of Agriculture, Western Australia*, **8**, 447.

Cass Smith, W.P. and Goss, D.M. (1958) A method of inoculating and lime-pelleting leguminous seeds. *Journal of Agriculture, Western Australia*, **7**, 119–21.

Catroux, G. (1991) Inoculant quality standards and controls in France, in *Expert consultation on legume inoculant production and quality control*. Report Food and Agriculture Organization of the United Nations, Rome, pp.113–20.

Chao, W.L. and Alexander, M. (1984) Mineral soils as carriers for *Rhizobium* inoculants. *Applied and Environmental Microbiology*, **47**, (1), 94–7.

Chen, H., Richardson, A.E., Gartner, E., *et al.* (1991) Construction of an acid-tolerant *Rhizobium leguminosarum* bv. *trifolii* strain with enhanced capacity for nitrogen fixation. *Applied and Environmental Microbiology*, **57**, 2005–11.

Cocks, P.S. (1988) The role of forage legumes in livestock based farming systems, in *Nitrogen Fixation by Legumes in Mediterranean Agriculture*, (eds D.P. Beck and L.A. Materon), Martinus Nijhoff, Lancaster; pp.3–10.

Crawford, S.L. and Berryhill, D.L. (1983) Survival of *Rhizobium phaseoli* in coal-based legume inoculants applied to seeds. *Applied and Environmental Microbiology*, **45**, (2), 703–5.

Cunninghan, S., Kollmeyer, W.D. and Stacey, G. (1991) Chemical control of interstrain competition for soybean nodulation by *Bradyrhizobium japonicum*. *Applied and Environmental Microbiology*, **57**, 1886–92.

Date, R.A. (1970) Microbiological problems in the inoculation and nodulation of legumes. *Plant and Soil*, **32**, 703–25.

Date, R.A. (1982) Collection, isolation, characterization and conservation of *Rhizobium*, in *Nitrogen Fixation in Legumes*, (ed J.M.Vincent), Academic Press, Sydney, pp.95–109.

Davis, E.O., Evans, I.J. and Johnston, A.W.B. (1988) Identification of *nodX*, a gene that allows *Rhizobium leguminosarum* bv. *viciae* strain TOM to nodulate Afghanistan peas. *Molecular and General Genetics*, **212**, 531–5.

Day, J.M. (1991) Inoculant production in the UK, in *Expert consultation on legume inoculant production and quality control*. Report, Food and Agriculture Organization of the United Nations, Rome, pp.75–85.

Djordjevic, M.A., Innes, R.W. Wijffelman, C.A., *et al.* (1987) Clovers secrete specific phenolic compounds which either stimulate or repress *nod* gene expression in *Rhizobium trifolii*. *EMBO Journal*, **6**, 1173–9.

Dowling, D.N., Samrey, U., Stanley, J., *et al.* (1987) Cloning of *Rhizobium leguminosarum* genes for competitive nodulation blocking on peas. *Journal of Bacteriology*, **169**, 1345–8.

Dowling, D.N., Stanley, J. and Broughton, W.J. (1989) Competitive nodulation blocking of Afghanistan pea is determined by *nodDABC* and *nodFE* alleles in *Rhizobium leguminosarum*. *Molecular and General Genetics*, **216**, 170–4.

Dube, J.N., Mahere, D.P. and Rawat, A.K. (1980) Development of coal as a carrier for rhizobial inoculants. *Science and Culture*, **46**, (8), 304.

Duong, T.P., Diep, C.N., Khiem, N.T., *et al.* (1984) *Rhizobium* inoculant for soybean (*Glycine max* (L) Merrill) in the Mekong Delta. II. Response of soybean to chemical nitrogen fertilizer and *Rhizobium* inoculation. *Plant and Soil*, **79**, (2), 241–7.

Eaglesham A.R.J. (1989) Global importance of *Rhizobium* as an inoculant, in *Microbial Inoculation of Crop Plants*, (eds R. Campbell and R.M. Macdonald), I.R.L. Press, London, pp.29–48.

Elegba, M.S. and Rennie, R.J. (1984) Effect of different inoculant adhesive agents on rhizobial survival, nodulation and nitrogenase (acetylene-reducing) activity of soybeans (*Glycine max* L. Merrill). *Canadian Journal of Soil Science*, **64**, (4), 631–6.

Evans, H.J., Hanus, F.J., Haugland, R.A., *et al.* (1985) Hydrogen recycling in nodules affects nitrogen fixation and growth of soybeans, in *World Soybean Research Conference III, Proceedings*. Westview Press, Boulder, Colorado, pp.935–42.

Evans, H.J., Harker, A.R., Papen, H., *et al.* (1987) Physiology, biochemistry and genetics of the uptake hydrogenase in rhizobia. *Annual Review of Microbiology*, **41**, 335–61.

Firmin, J.L., Wilson, K.E., Rossen, L., *et al.* (1986) Flavonoid activation of nodulation genes in *Rhizobium* reversed by other compounds present in plants. *Nature*, **324**, 90–2.

Forbert, P.R., Roy, N., Nash, J.H.E., *et al.* (1991) Procedure for obtaining efficient root nodulation of a pea cultivar by a desired *Rhizobium* strain and preempting nodulation by other strains. *Applied and Environmental Microbiology*, **57**, 1590–4.

Fred, E.B., Baldwin, I.L. and McCoy, E.M. (1932) *Root Nodule Bacteria and Leguminous Plants*. University of Wisconsin, Madison, USA.

Gibson, A.H., Date, R.A., Ireland, J.A. *et al.* (1976) A comparison of competitiveness and persistence amongst five strains of *Rhizobium trifolii*. *Soil Biology and Biochemistry*, **8**, 395–401.

Hardaker, J.M. and Hardwick, R.C. (1978) A note on *Rhizobium* inoculation of beans (*Phaseolus vulgaris*) using the fluid drill technique. *Experimental Agriculture*, **14**, (1), 17–21.

Hardarson, G. and Bowen, G. (1991) Rhizobial inoculants for leguminous crops. Package labelling and quality control, in *Expert consultation on legume inoculant production and quality control*. Report, Food and Agriculture Organization of the United Nations, Rome, pp.131–42.

Hastings, A. and Drake, A.D. (1960) Inoculation and pelleting of clover seed. *New Zealand Journal of Agriculture*, **101**, 619.

Hastings, A. and Drake, A.D. (1962) Pelleting inoculated clover seed. *New Zealand Journal of Agriculture*, **104**, 330.

Hastings, A. and Drake, A.D. (1963) Pelleted inoculated clover seed aids pasture establishment on problem soils. *New Zealand Journal of Agriculture*, **106**, (6) 463–8.

Hely, F.W., Hutchings, R.J. and Zorin, M. (1980) Methods of rhizobial inoculation and sowing techniques for *Trifolium subterraneum* L. establishment in a harsh winter environment. *Australian Journal of Agricultural Research*, **31**, (4), 703–12.

Hodgson, A.L.M., Roberts, W.P. and Waid, J.S. (1985) Regulated nodulation of *Trifolium subterraneum* inoculated with bacteriocin producing strains of *Rhizobium trifolii*. *Soil Biology and Biochemistry*, **17**, 475–8.

Holding, A.J. and King, J. (1963) The effectiveness of indigenous populations of *Rhizobium trifolii* in relation to soil factors. *Soil Biology and Biochemistry*, **1**, (1), 57–61.

Holl, F.B. (1975) Host plant control of the inheritance of dinitrogen fixation in the *Pisum–Rhizobium* symbiosis. *Euphytica*, **24**, 767–70.

Hubbell, D.H. (1988) Extension/transfer of BNF technology, in *Nitrogen Fixation by Legumes in Mediterranean Agriculture*, (eds D.P. Beck and L.A. Materon), Martinus Nijhoff, Lancaster, pp.3–10.

Jauhri, K.S. (1988) Seed bed inoculants to overcome the loss of rhizobia from germinating seed. *Zentralblatt fur Mikrobiologie*, **143**, (4), 285–92.

Jauhri, K.S. (1989) Feasibility of utilising freeze-dried cell concentrates of *Rhizobium* in commercial production of inoculants. *Zentralblatt fur Mikrobiologie*, **144**, (3), 187–9.

Jauhri, K.S. and Agarwel, D.K. (1984) Effect of seed dressing fungicides on soybean inoculated with *Rhizobium japonicum* in charcoal rot infested soil. *Pesticides*, **18**, (8), 47–8.

Jensen, E.S. (1987) Inoculation of pea by application of *Rhizobium* in the planting furrow. *Plant and Soil*, **97**, (1), 63–70.

Johansen, E., Stencel, L. and Appelbaum, E. (1988) Isolation of *Bradyrhizobium japonicum* mutants with hyperinducible common *nod* genes, in *Nitrogen Fixation: One Hundred Years After*, (eds H. Bothe, F.J. de Bruijin and W.E. Newton), Gustav Fischer, Stuttgart, p.464.

Jones, D. Gareth (1963) Symbiotic variation of *Rhizobium trifolii* with S100 Nomark white clover (*Trifolium repens* L.). *Journal of the Science of Food and Agriculture*, **14**, 740–3.

Jones, D. Gareth (1966) The contribution of white clover to a mixed upland sward. II. Factors affecting the density and effectiveness of *Rhizobium trifolii*. *Plant and Soil*, **24**, 250–60.

Jones, D. Gareth and Thomas, S.B. (1966) The use of inoculation and pelleting in the establishment of white clover under mountain conditions. *Journal of Applied Bacteriology*, **29**, (2), 430–7.

Jones, D. Gareth, Druce, R.G. and Thomas, S.B. (1967) Comparative trials of seed pelleting, inoculation and the use of high lime dressings in upland reclamation. *Journal of Applied Bacteriology*, **30**, 511–17.

Jones, D. Gareth, Munro, J.M.M., Hughes, R., *et al.* (1964) The contribution of white clover to a mixed upland sward. I. The effect of *Rhizobium* inoculation on the early development of white clover. *Plant and Soil*, **21**, (1), 63–9.

Jordan D.C. (1984) *Rhizobeaceae*, in *Bergey's Manual of Systematic Bacteriology*, Volume 1, (eds N.R. Krieg and J.G. Holt), Williams and Williams, Baltimore, USA, pp.234.

Jung, G., Mugnier, J., Diem, H.G., *et al.* (1982) Polymer-entrapped *Rhizobium* as an inoculant for legumes. *Plant and Soil*, **65**, (2), 219–31.

Kosslak, R.M., Schroth, R., Paaren, H.E., *et al.* (1987) Induction of *Bradyrhizobium japonicum* common *nod* genes by isoflavones isolated from *Glycine max*. *Proceedings of the National Academy of Science (USA)*, **84**, 7428–32.

Kremer, R.J. and Peterson, H.L. (1982) Field evaluation of selected *Rhizobium* in an improved legume inoculant. *Agronomy Journal*, **75**, (1), 139–43.

Kucey, R.M.N. and Bonetti, R. (1988) Effect of vesicular-arbuscular mycorrhizal fungi and captan on growth and N-fixation by *Rhizobium*-inoculated field beans. *Canadian Journal of Soil Science*, **68**, (1), 143–9.

Labandera, C.A. and Vincent, J.M. (1975) Competition between an introduced strain and native Uruguayan strains of *Rhizobium trifolii*. *Plant and Soil*, **42**, 327–47.

Lambert, G.R., Cantrell, M.A., Hanus, F.J. *et al.* (1985) Intra- and interspecies transfer and expression of *Rhizobium japonicum* hydrogen uptake genes and autotrophic growth capability. *Proceedings of the National Academy of Science (USA)*, **82**, 3232–6.

Le Gal, M.F., Hobbs, S.L.A. and Delong, C.M.O. (1989) Gene expression during the infection process in nodulating and non-nodulating pea genotypes. *Canadian Journal of Botany*, **67**, 2535–8.

Lethbridge, G. (1989) An industrial view of microbial inoculants for crop plants, in *Microbial Inoculation of Crop Plants*, (eds R. Campbell and R.M. Macdonald), IRL Press, Oxford, pp.11–28.

Lie, T.A. (1978) Symbiotic specialization in pea plants: the requirement of specific *Rhizobium* strains for peas from Afghanistan. *Annals of Applied Biology*, **88**, 462–5.

Ma, S.W. and Iyer, V.N. (1990) New field isolates of *Rhizobium leguminosarum* biovar *viceae* that nodulate the primitive pea cultivar Afghanistan in addition to modern cultivars. *Applied and Environmental Microbiology*, **56**, 2206–12.

Maier, R. and Brill, W.J. (1978) Mutant strains of *Rhizobium japonicum* with increased ability to fix nitrogen for soybean. *Science*, **201**, 448–50

Manjunath, A. and Bagyaraj, D.J. (1984) Response of pigeon pea and cowpea to phosphate and dual inoculation with vesicular-arbuscular mycorrhiza and *Rhizobium*. *Tropical Agriculture*, **61**, (1), 48–52.

Marques Pinto, C., Yao, P.Y. and Vincent, J.M. (1974) Nodulating competitiveness amongst strains of *Rhizobium meliloti* and *Rhizobium trifolii*. *Australian Journal of Agricultural Research*, **25**, 317–29.

Martensson, A.M. (1990) Competition of inoculated strains of *Rhizobium leguminosarum* bv. *trifolii* in red clover using repeated inoculation and increased inoculum levels. *Canadian Journal of Microbiology*, **36**, (2), 136–9.

Maxwell, C.A., Hartwig, U.A, Joseph, C.M., *et al.* (1989) A chalcone and two related flavonoids released from alfalfa roots induce *nod* genes of *Rhizobium meliloti*. *Plant Physiology*, **91**, 842–7.

Mellor, H.Y., Glenn, A.R., Arwas, R., *et al.* (1987) Symbiotic and competitive properties of motility mutants of *Rhizobium trifolii* TA1. *Archives of Microbiology*, **148**, 34–9.

Murphy, P.J., Heycke, N. Banfalvi, Z., *et al.* (1987) Genes for the catabolism and synthesis of an opine-like compound in *Rhizobium meliloti* are closely linked and on the Sym plasmid. *Proceedings of the National Academy of Science (USA)*, **84**, 493–7.

Murphy, P.J., Heycke, N., Trenz, S.P., *et al.* (1988) Synthesis of an opine-like compound, a rhizopine, in alfalfa nodules is symbiotically regulated. *Proceedings of the National Academy of Science (USA)*, **85**, 9133–7.

Mytton, L.R. (1984) Developing a breeding strategy to exploit quantitative variation in symbiotic nitrogen fixation. *Plant and Soil*, **82**, 329–35.

Mytton, L.R. (1988) Workshop synthesis and recommendations developed from group discussions, in *Nitrogen Fixation by Legumes in Mediterranean Agriculture* (eds D.P. Beck and L.A. Materon), Martinus Nijhoff, Lancaster, pp.373–9.

Nethery, A.A. (1991) Inoculant production with non-sterile carriers in, *Expert consultation on legume production and quality control*. Report, Food and Agriculture Organization of the United Nations, Rome, pp.43–50.

Newbould, P. (1983) The application of inoculation in agriculture, in *Temperate Legumes: Physiology, Genetics and Nodulation*, (eds D.G. Jones and D.R. Davies), Pitmans, London, pp.417–42.

Nicol, H. and Thornton, H.G. (1942) Competition between related strains of nodule bacteria and its influence on infection of the legume host. *Proceedings of the Royal Society*, Series B, **130**, 32–59.

Norris, D.O. (1971) Seed pelleting to improve nodulation of tropical and subtropical legumes. *Australian Journal of Experimental Agriculture and Animal Husbandry*, **11**, 194–201.

Nutman, P.S. and Ross, G.J.S. (1970) *Rhizobium* in the soils of the Rothamsted and Woburn farms. *Rothamsted Experimental Station Report for 1969*, Part 2, 141–7.

Odevemi, O. and Okoronkwo, N. (1985) The suitability of local materials as carriers for rhizobia in legume inoculant production in Nigeria, in *Biological Nitrogen Fixation in Africa*, (eds H.Ssali and S.O.Keya), MIRCEN, University of Nairobi, Kenya, pp.135–50.

Paau, A.S. (1989) Improvement of *Rhizobium* inoculants. *Applied and Environmental Microbiology*, **55**, 862–5.

Paczkowski, M.W. and Berryhill, D.L. (1979) Survival of *Rhizobium phaseoli* in coal-based legume inoculants. *Applied and Environmental Microbiology*, **38**, (4), 612–15.

Paton, D.F. (1957) Responses to seed inoculation of pasture legumes in Tasmania. *Tasmanian Journal of Agriculture*, **28**, 389–98.

Peters, N.K., Frost, J.W. and Long, S.R. (1986) A plant flavone, luteolin, induces expression of *Rhizobium meliloti* nodulation genes. *Science*, **233**, 977–80.

Peters, N.K. and Long, S.R. (1988) *Rhizobium meliloti* nodulation gene inducers and inhibitors. *Plant Physiology*, **88**, 396–400.

Postgate, J. (1978) *Nitrogen Fixation*. Studies in Biology No.92. Edward Arnold, Ltd., London.

Poulter, A.A. (1933) Deficiency of the nodule organism on some Welsh soils. *Welsh Journal of Agriculture*, **9**, 145–59.

Redmond, J.W., Batley, M.A., Djordjevic, R.W., *et al.* (1986) Flavones induce expression of nodulation genes in *Rhizobium. Nature*, **323**, 632–4.

Rennie, R.J. and Dubetz, S. (1984) Multi-strain vs single strain *Rhizobium japonicum* inoculants for early maturing soybean cultivars: N fixation quantified by ^{15}N isotope dilution. *Agronomy Journal*, **76**, (3), 498–502.

Rogers, D.D., Warren, R.D. Jr. and Chamblee, D.S. (1982) Remedial post-emergence legume inoculation with *Rhizobium. Agronomy Journal*, **74**, (4), 613–19.

Roughley, R.J. (1961) Inoculation of legumes with nitrogen fixing bacteria. *Agriculture Gazette, New South Wales*, **72**, 40–4.

Roughley, R.J. (1981) Quality control of legume inoculants, in *Current Perspectives in Nitrogen Fixation*, (eds A.H. Gibson and W.E. Newton), Australian Academy of Sciences, Canberra, pp.340–1.

Roughley, R.J. and Pulsford. D.J. (1982) Production and control of legume inoculants, in *Nitrogen Fixation in Legumes*, (ed J.M. Vincent), Academic Press, Sydney, Australia, pp.39–209.

Roughley, R.J. and Thompson, J.A. (1978) The relationship between the numbers of rhizobia in broth and the quality of peat-based legume inoculants. *Journal of Applied Bacteriology*, **44**, 317–9.

Russell, P.E. and Jones, D. Gareth (1975) Immunofluorescence studies of selection of strains of *R. trifolii* by S184 white clover (*T. repens* L.). *Plant and Soil*, **42**, 119–29.

Rys, G.J. and Bonish, P M. (1984) Influence of inoculation and additional nutrients on establishment of the white clover symbiosis in a cultivated soil. *New Zealand Journal of Experimental Agriculture*, **12**, (4), 295–301.

Salema, M.P., Parker, C.A., Kidby, D.K., *et al.* (1982) Death of rhizobia on inoculated seed. *Soil Biology and Biochemistry*, **14**, (1), 13–14.

Schwinghamer, E.A. and Belkengren, R.P. (1968) Inhibition of rhizobia by a strain of *Rhizobium trifolii*: some properties of the antibiotic and of the strain. *Archiv fur Mikrobiologie*, **64**, 130–45.

Scott, D.B., Hennecke, H. and Lim, S.T. (1979) The biosynthesis of nitrogenase MoFe protein polypeptides in free-living culture of *Rhizobium japonicum. Biochimica et Biophysica Acta*, **565**, 365–78.

Skot, L. and Egsgaard, H. (1984) Identification of ononitol O-methyl-*scyllo*-inositol in pea root nodules. *Planta*, **161**, 32–6.

Skot, L., Harrison, S.P., Nath, A., *et al.* (1990) Expression of insecticidal activity in *Rhizobium* containing the endotoxin gene cloned from *Bacillus thuringiensis* subsp. *tenebrionis. Plant and Soil*, **127**, 285–95.

Smith, R.S., Ellis, M.A. and Smith, R.E. (1981) Effect of *Rhizobium japonicum* inoculant rates on soybean nodulation in a tropical soil. *Agronomy Journal*, **73**, (3), 505–8.

Somasegaran, P. (1991) Inoculant production with emphasis on choice of carriers, methods of production and reliability testing/quality assurance guidelines, in *Expert consultation on legume inoculant production and quality control*. Report, Food and Agricultural Organization, Rome, pp.87–106.

Somasegaran, P., Reyes, V.G. and Hoben, H.J. (1984) The influence of high temperature on the growth and survival of *Rhizobium* spp. in peat inoculants during preparation, storage and distribution. *Canadian Journal of Microbiology,* **30**, (19), 23–30.

Spaink, H.P., Okker, R.J.H., Wijffelman, C.A. *et al.* (1989a) Symbiotic properties of rhizobia containing a flavonoid-independant hydrid *nodD* product. *Journal of Bacteriology,* **171**, 4045–53.

Spaink, H.P., Wijffelman, C.A., Okker, R.J.H., *et al.* (1989b) Localization of functional regions of the *nodD* product using hybrid *nodD* genes. *Plant Molecular Biology,* **12**, 59–73.

Sparrow, S.D.Jr. (1981) Survival of *Rhizobium phaseoli* in inoculants and inoculation of field grown navy beans. *Dissertation Abstracts B,* **42**, (1), 26.

Sylvester-Bradley, R. (1984) *Rhizobium* inoculation trials designed to support a tropical forage selection programme. *Plant and Soil,* **82**, 377–86.

Thompson, J.A. (1980) Production and quality control of legume inoculants, in *Methods for Evaluating Nitrogen Fixation,* (ed F.J. Bergersen), John Wiley & Sons, London, pp.489–533.

Thompson, J.A. (1991) Legume inoculant production and quality control, in *Expert consultation on legume inoculant production and quality control.* Report, Food and Agriculture Organization of the United Nationa, Rome, pp.15–32.

Toms, W.J. (1960) The use of seed inoculum for subterranean clover. *Journal of Agriculture, Western Australia,* **1**, 925–6.

Triplett, E.W. (1988) Isolation of genes involved in nodulation competitiveness from *Rhizobium leguminosarum* bv. *trifolii* T24. *Proceedings of the National Academy Science (USA),* **85**, 3810–14.

Triplett, E.W. (1990) Construction of a symbiotically effective strain of *Rhizobium leguminosarum* bv. *trifolii* with increased nodulation competitiveness. *Applied and Environmental Microbiology,* **56**, 98–103.

Triplett, E.W. and Barta, T.M. (1987) Trifolitoxin production and nodulation are necessary for the expression of superior nodulation competitiveness of *Rhizobium leguminosarum* bv. *trifolii* strain T24 on clover. *Plant Physiology,* **85**, 335–42.

Triplett, E.W., Schink, M.J. and Noelder, K.L. (1989) Mapping and subcloning of the trifolitoxin production and resistance genes from *Rhizobium leguminosarum* bv. *trifolii* T24. *Molecular Plant–Microbe Interactions,* **2**, 202–8.

van Schreven, D.A. (1972) Note on the specificity of the rhizobia of crown vetch and sainfoin. *Plant and Soil,* **36**, 325–30.

van Schreven, D.A., Otzen, D. and Lindbergh, D.J. (1954) On the production of legume inoculants in a mixture of peat and soil. *Antonie van Leeuwenhoek,* **20**, 33.

Vincent, J.M. (1981) Modern concepts in legume inoculation, in *Biological Nitrogen Fixation Technology for Tropical Agriculture,* (eds P.H. Graham and S.C. Harris), Centro Internacional de Agricultura Tropical, Cali, Columbia, pp.105–14.

Vincent, J.M. (1982) *Nitrogen Fixation in Legumes.* Academic Press, Sydney, Australia.

Wadoux, P. (1991) Inoculant production in industry using sterile carriers, in *Expert consultation on legume production and quality control.* Report, Food and Agriculture Organization of the United Nations, Rome, pp.33–42.

Weaver, R.W. and Frederick, L.R. (1974) Effect of inoculum rate on competitive nodulation of *Glycine max* L. Merrill. II Field studies. *Agronomy Journal,* **66**, 233–6.

Weaver, R.W., Materon, L.A., Krautmann, M.E., *et al.* (1985) Survival of *Rhizobium trifolii* in soil following inoculation of arrowleaf clover. MIRCEN, *Journal of Applied Microbiology and Biotechnology,* **1**, (4), 311–18.

Welty, L.E., Prestbye, L.S., Hall, J.A., *et al.* (1988) Effect of fungicide seed treatment and *Rhizobium* inoculation on chickpea production. *Applied Agricultural Research*, **3**, (1), 17–20.

Wolf, H.J. and Holfich G. (1986) International knowledge of the technology of production and storage of *Rhizobium* inoculants with peat as a carrier. *Zentrablatt fur Mikrobiologie*, **14**, (3), 169–76.

Wynne, J.C., Bliss, F.A. and Rosas, J.C. (1987) Principles and practice of field designs to evaluate symbiotic nitrogen fixation, in *Symbiotic Nitrogen Fixation*, (ed G.H. Elkan), Marcel Dekker, New York, pp.371–89.

Zaat, S.A.J., Wijffelman, C.A., Spaink, H.P. *et al.* (1987) Induction of the *nodA* promoter of *Rhizobium leguminosarum* Sym plasmid pRL1J1 by plant flavanones and flavones. *Journal of Bacteriology*, **169**, 198–204.

Zdor, R.E. and Pueppke, S.G. (1989) Nodulation competitiveness of *Tn5*-induced mutants of *Rhizobium fredii* 208 that are altered in motility and extracelluar polysaccharide production, in *Proceedings of the 12th North American Symbiotic Nitrogen Fixation Conference*, p.33.

9

The use of microorganisms in plant breeding

Gareth M. Evans

9.1 INTRODUCTION

The manipulation of genes between individuals within the same species and even between individuals from different species and genera has been the foundation of plant breeding for the last 100 years. The whole process relies on natural recombination of genes during meiosis in hybrid genotypes resulting from sexual hybridization of suitable parents. Selected recombinants having the desired gene combination are isolated over a period of one or more generations and exploited as pure lines in inbreeding crops and as hybrids or heterozygous open pollinated populations (synthetics) in outbreeding crops.

Remarkable advances have been made and will undoubtedly continue to be made using these well tried and proven breeding techniques. Yield increases of about 1% per year in both wheat and barley have been consistently achieved through conventional plant breeding over the last 80 years (Austin *et al.*, 1980; Riggs *et al.*, 1981). New cultivars are derived either by the wholesale reassortment of a multitude of polygenes or by the manipulation of one or two genes of major effect, e.g. the introduction of the Norin 10 dwarfing genes in wheat.

Plant breeding programmes are lengthy and expensive; it takes 10 to 15 years from the first cross to the release of a cultivar. Firstly, there is the need to assemble or generate a range of segregant individuals which

Exploitation of Microorganisms Edited by D.G. Jones
Published in 1993 by Chapman & Hall, London ISBN 0 412 45740 7

the breeder thinks will contain one or more plants which satisfy the breeding objectives. In the majority of programmes, this is achieved by controlled hybridization of chosen parents. Secondly, the breeder has to identify and separate out the 'superior' individuals from the mass of others which are not quite so good or even poor. This naturally involves a period of assessment and selection over several generations. Finally, the selected genotypes or their progeny have to be assembled as a cultivar which then has to be further assessed through extensive performance trials ultimately leading up to the statutory National List and Recommended List Trials.

The success or failure of a breeding programme depends initially on having suitable germplasm with which to work. In the main, this originates from within the cultivated gene pool of the crop. Introgression of desired genes from non-cultivated forms of the same species or from related species results in considerable disruption of the genetic balance in the recipient genotypes and this can be particularly frustrating when only a few genes are required from a donor genotype. Many generations of backcrossing are then needed to reassemble the desired gene combinations.

Systems of **direct transformation** of elite genotypes by non-sexual means have always been attractive to the breeders as a way of avoiding the problems indicated above. Progress in recombinant DNA technology offers some hope that this can now be achieved on a regular basis. It has been demonstrated that a gene sequence can be extracted from one plant, cloned in a bacterial system and then introduced into new host cells. Moreover, it has been shown that these sequences function normally in their new environment and are also transmitted to the progeny. Single genes of known effect are prime candidates for such technology and it is argued that the problems associated with conventional breeding would be avoided by using this type of technology. Problems such as linked inheritance of undesirable genes and disruption of genetic balance would not arise.

Several methods are now available for the introduction of DNA into host cells although not all of these result in the total integration of the transferred DNA into chromosomes of the host cells. Although methods such as direct uptake of DNA by plant protoplasts, micro-injection of DNA into host cells or the use of micro-projectiles to convey the DNA to its target (particle gun), are all useful methods of creating transgenic plants, this chapter will concentrate on the use of delivery methods based on microorganisms. Discussion of *Agrobacterium*-mediated gene transfer will form the major part of the work with some reference being made also to the possibility of using vectors based on viral genomes. The transformation systems themselves will be described, together with a discussion of methods of recovering transformed cells and of identifying target genes. Finally, possible applications to crop improvement will be examined and compared with more conventional systems of breeding. The title of the chapter is interpreted in its narrow sense. No reference will be made to screening for resistance to viral, bacterial and fungal pathogens in conventional breeding programmes.

9.2 *AGROBACTERIUM*-MEDIATED GENE TRANSFER

The soil bacteria *Agrobacterium tumefaciens* and *Agrobacterium rhizogenes* cause crown gall tumours and hairy roots respectively on dicotyledonous plants. Of great significance, however, is that genes from the bacterial plasmids are transferred to and become integrated into the genome of the host plant cells (Chilton *et al.*, 1977, 1982). Derivatives of these pathogens, in particular of *A. tumefaciens*, have been used extensively as efficient vectors for the production of transgenic plants. The process of tumour formation in *A. tumefaciens* has been studied in some detail and has been reviewed by Kahl and Schell (1982). Nevertheless, it is relevant here to give a very brief description of the process as it forms the background to the development of *Agrobacterium*-mediated gene transfer.

9.2.1 Tumour formation by *A. tumefaciens*

A virulent strain of *A. tumefaciens* contains a large plasmid called the Ti (tumour inducing) plasmid. This is a circular, double stranded DNA structure of some 200–250 kb (kilobases). Ti plasmids found in different virulent strains of *A. tumefaciens* have four regions of homology of which two are involved in the process of gene transfer and tumour formation (Walden, 1988). These are the T-DNA (Transferred) region which, as the name implies, is transferred to the host and the *vir* region which is not transferred. Compounds released from wounded plant cells attract the bacterium and also result in the transcription of genes from the *vir* region of the *Agrobacterium* plasmid. *Vir* gene activation also results in the appearance of single stranded T-DNA, known as the T strand. This is a linear copy of the T-DNA region of the Ti plasmid and is synthesized following single stranded nicks within the 25 bp repeat sequences which borders the T-DNA region. Evidence suggests that an endonuclease, the product of *vir*D gene, is responsible for the initial nicks (Yanofski *et al.*, 1986). The T-strand, together with part of its flanking 25 bp repeats, is transferred to the plant cell by a process analogous to bacterial conjugational transfer and becomes stably inserted in the host genome (Stachel and Zambryski, 1986). The precise nature of the integration process is still not known but it is clear that one or more copies of the T-DNA can be present in the host genome.

Although it has been generally accepted that the site of T-DNA integration into the plant cell was random, there is now some evidence that this might not be so (Koncz *et al.*, 1989). Within the plant cell, the T-DNA is fully transcribed, resulting in the tumorous growth crown gall of *A. tumefaciens* or hairy roots of *A. rhizogenes*. Tumour tissues contain large amounts of chemicals known as **opines**, the type of opine synthesized being a function of the type of *Agrobacterium* strain responsible for tumour formation. Two of the most common opines are octopine and nopaline, and it is convention to designate the *Agrobacterium* strains as either **octopine** or **nopaline** types. In nopaline strains of *A. tumefaciens*, the size of the T-DNA insert which produces crown gall is in the region of 2 kb in length, while in crown galls caused by octopine strains there are two non-contiguous segments of integrated T-DNA of 14 and 7 kb (Walden, 1988). The T-DNA region is also known to contain three genes which code for phytohormones which, when transferred to plant cells, produce the oncogenic phenotype, i.e. the characteristic tumour growth.

9.2.2 *Agrobacterium* plasmids as transformation vectors in higher plants

The ability of *Agrobacterium* to transfer part of its plasmid DNA into plant chromosomes made it an ideal candidate for development as a transformation vector. Several series of observations during the early 1980s confirmed that modified Ti plasmids could be used as vectors for the introduction of heterologous DNA into plant genomes. In the first place, it was shown that the T-DNA oncogenes responsible for tumour growth were neither essential for transfer of T-DNA into plant cells, nor for its insertion into plant chromosomes. This resulted from the construction of Ti plasmids which lacked the *onc* genes and which subsequently did not produce tumorous growth even though it was evident that the remaining T-DNA construct had been incorporated in the host genome (Zambryski *et al.*, 1983). Secondly, it was shown that heterologous DNA engineered into the T-DNA region of the Ti plasmid was transferred to the host cell chromosomes together with the remaining T-DNA region (Bevan *et al.*, 1983). Moreover, it was shown that such DNA was inherited in the normal Mendelian manner from plants regenerated from transformed tissues. It was also shown that the *vir* genes need not be present on the same plasmid as the T-DNA for efficient transformation to be achieved. The *vir* genes can, therefore, act in *trans* (de Framond *et al.*, 1983).

A wide variety of non-oncogenic vectors based on the Ti plasmid are now available for use with *Agrobacterium*. They are divided into two types, the *cointegrative* and the *binary* vectors (for review, refer to Draper *et al.*, 1988 and Walden, 1988). They both have the common property that any DNA located between the 25 bp T-DNA border repeats is transferred from the Ti plasmid into the plant genome. The basic difference between them lies in the position of the *vir* genes. In cointegrative vectors, the *vir* region of the Ti plasmid is unmodified and is located on the same plasmid as the inserted foreign DNA. Consequently, such vectors are sometimes called *cis* vectors (Walden, 1988). The DNA region to be transferred is initially sub-cloned in a small intermediate vector in *E. coli* and the plasmid transferred into *Agrobacterium* by conjugation and the use of a third type of plasmid, a conjugation-proficient helper plasmid. Recombination between the resident disarmed Ti plasmid and the intermediate vector within *Agrobacterium* results in the required DNA sequences being inserted in the appropriate site on the vector Ti plasmid, thus producing a cointegrate. A suitable selectable marker such as antibiotic resistance is included to reclaim the recombinant plasmids. Neither the intermediate nor the helper plasmids are capable of replication within *Agrobacterium*.

As the name implies, a binary vector system involves two separate plasmids. The modified T-DNA region including any inserted foreign DNA is constructed in one plasmid, while the *vir* genes are on a separate plasmid and function in *trans* (de Framond *et al.*, 1983). As such, the modified T-DNA plasmid is smaller than the cointegrative one described previously; it is capable of replication within *E. coli* and can be transferred into *Agrobacterium* by conjugation through tri-parental mating involving a second *E. coli* strain carrying a suitable helper plasmid.

9.2.3 Genetic markers

The recognition and the recovery of transformed cells, tissues or plants requires specialized assay systems. Unique DNA sequences detected by Southern analysis can be proof of transformation, although this is of limited use for selection purposes. It is often the case that transformed cells will be mixed with normal non-transformed cells or protoplasts and it is necessary to separate the transformed from the normal tissue. An increasing number of dominant selectable markers which confer resistance to antibiotics or some other chemical are now becoming available for this purpose. Many, although not all, are of bacterial origin. Other markers allow for easy assay of transformed cells or tissue through a specific enzymatic reaction which can be easily recognized; an extensive list of both types is given by Walden and Schell (1990). One of the most common of the selectable markers is the NPT II (*neo*) gene that codes for the enzyme phosphotranferase II which in turn confers resistance to the aminoglycoside antibiotics kanamycin, neomycin and G418 (Herrera-Estrella *et al.*, 1983). A gene conferring resistance to the herbi-

cide glyphosate has also been used for such a purpose (Shah *et al.*, 1986). This will be discussed in more detail at a later stage. Such genes are constructed as part of chimaeric gene complexes within the T-DNA borders of the disarmed Ti plasmid.

It is of advantage, although not essential, that markers or 'reporter' genes which are used to demonstrate gene activity in transformed tissue do not necessitate the destruction of the tissue during the assay. In choosing the correct reporter gene for use with a particular tissue, it has to be established that the enzyme encoded by the gene is not already present in the tissue and, moreover, that the marker gene is 'switched on' in the target cells. Promoter sequences are normally included in the T-DNA construct of the disarmed Ti plasmid. These commonly are either the opine synthase promoters *ocs* or *nos* from the T-DNA itself or the 35S RNA promoter from Cauliflower Mosaic Virus (CAMV). It is also of advantage if the assay system is simple and easy to perform. Several such systems are now available, the most useful ones allowing histochemical detection of the enzyme in the transformed tissue. Examples are β-glucuronidase (GUS) which cleaves a variety of commercially available substrates giving fluorometric or colour reactions and luciferase genes from both firefly and luminescent bacteria (*Vibrio harveyi* and *V. fischeri*) which catalyse bioluminescent reactions (Ow *et al.*, 1986). β-galactosidase, another useful enzyme for assay purposes, is produced by the *lacZ* gene and can be used in a variety of cells to produce a blue colour reaction with IPTG (isopropyl-thiogalactoside) and X-gal (5-bromo-4-chloro-3-indolyl-β-galactopyranoside (Teeri *et al.*, 1989).

The essential features of typical disarmed cointegrative and binary Ti plasmid vectors are illustrated in Figure 9.1. They consist of:

1. Two resistance genes, one for selection in bacteria (e.g. Kan^R) and the other encoding antibiotic resistance in the target tissue (e.g. *neo*);
2. *vir* genes necessary for infection of host plant tissue and transfer of T-DNA to the plant genome;
3. End border sequences;
4. Multi-cloning site for insertion of the plant DNA, and
5. Promoter sequence such as the P*nos* (nopaline synthase promoter) or the 35S RNA promoter from CAMV.

The *lac* α peptide gene with inserted polylinker region is often included to detect insertion of foreign DNA using the blue/white system.

9.2.4 Methods of transforming plant cells and tissues using *Agrobacterium*

The variety of techniques used to introduce foreign genes into plant cells using vectors derived from the Ti plasmid can be summarized under three headings.

(a) Cointegrative Ti plasmid

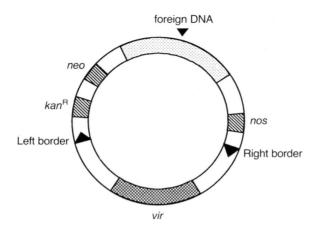

(b) Binary Ti plasmids

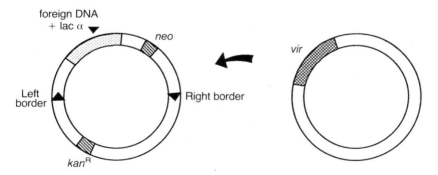

Figure 9.1 Generalized schematic diagram of disarmed (a) cointegrative and (b) binary vectors of *Agrobacterium tumefaciens*. *vir* = virulence genes; *neo* = NPT II gene, selectable marker for use in plant tissue; *Kan*^{R} = kanamycin resistance for selection of recombinant plasmid; *nos* = nopaline synthase gene, supplies the promoter system for marker genes.

Inoculation of explants

This is the most common and probably the easiest to perform. Explants of cotyledons, leaves, stems, roots or tubers are inoculated for a short period in a liquid culture of *Agrobacterium tumefaciens* containing the appropriate plasmid vector with its plant gene insert (Horsch *et al.*, 1985) before being transferred to a solid agar based medium to complete the process of 'infection'. Further transfer to a regeneration medium results in callus formation at the periphery of the explants and this can be removed and placed on a fresh medium. Selection for transformed tissue can be carried out at any stage of culture if a selectable marker was originally included in the plasmid construct. For example, if the NPT II gene (*neo*) is incorporated, kanamycin can be included in the regeneration medium to kill off the untransformed cells. Carbenicillin or a suitable substitute is also included to remove the *Agrobacterium*. The process is summarized in Figure 9.2. It should be noted that the induction of callus on the explants is not due to ·the activity of the Ti plasmid; it is in fact due to the culture medium used. Leaf or cotyledon

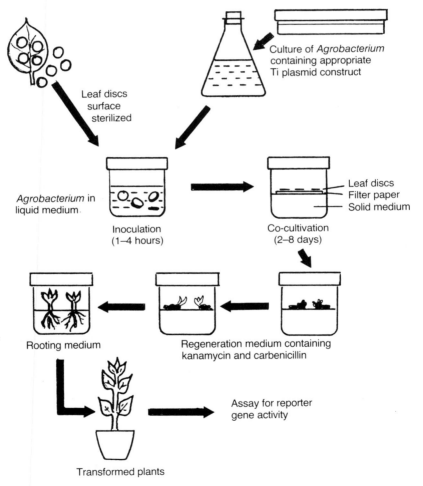

Figure 9.2 Schematic diagram of a transformation protocol using *Agrobacterium* as a vector.

discs from actively growing plants or young seedlings are a convenient source of explant tissue. Finally, after root induction on a suitable growth medium the appropriate assay for the foreign gene product can be carried out.

Co-cultivation of protoplasts

This technique involves inoculation of isolated protoplasts with a culture of *Agrobacterium* containing the appropriate vector and its DNA insert (De Block *et al.*, 1984). Callus induction is followed by regeneration of shoots and roots on appropriate agar media. If the DNA contains a selectable marker, the transformed tissue can be selected for either at the callus phase or during plant regeneration.

Embryo inoculation

Transformation of imbibed embryos of *Arabidopsis thalina* inoculated with a culture of *A. tumefaciens* containing a plasmid vector carrying a kanamycin resistance gene was reported by Feldmann and Marks (1987). This raises the possibility that successful transformation can be achieved, in some cases at least, without resorting to tissue culture and a callus phase with the danger of unwanted somaclonal variation. However, the true potential of this technique remains to be established.

Agrobacterium constitutes an excellent vector for introducing genes into target cells or tissue and explant inoculation appears to be the simplest way of achieving the appropriate transfer. However, there are limitations to the use of this system for the production of transgenic plants. Although it has wide applicability to dicotyledonous plants, problems of regeneration of plants from explants and protoplasts has restricted progress in some crops. Of greater significance is that *Agrobacterium* is not normally a pathogen of monocotyledonous species such as those from the Gramineae. Although there are some claims that *Agrobacterium*-mediated gene transfer has been successfully achieved in cereals (Raineri *et al.*, 1990) there are very few confirmed successes. In addition to the difficulty of introducing *Agrobacterium* into cells of graminaceous plants, there is also considerable difficulty in regenerating plants from explants or from protoplasts of this group and, where this has been achieved, it is often genotype dependent. Examples of regeneration from cultured protoplasts are given by Simmonds (1991).

9.3 VIRAL GENOMES AS TRANSFORMATION VECTORS

Interest in the possible use of plant viruses as a means of introducing genetic material into plant cells has arisen because of the limitations of the *Agrobacterium*-based system when applied to monocotyledons. It is well known that when viruses infect plant cells they express their own genome and can achieve high copy number. It is argued that if a suitable viral genome could be engineered so as to contain foreign DNA instead of, or in addition to, part of its own genome without affecting its infectivity, this construct would replicate inside the plant cell resulting in multiple copies of the introduced gene. The foreign gene is not, of course, integrated into the host genome and is, therefore, not normally inherited. The ability of many viruses to spread within the plant guarantees that once a gene has been introduced on board such a vector it will become systemic within the plant.

Most interest has been placed in the potential of the Caulimoviruses as vectors and in particular in Cauliflower Mosaic Virus (CAMV). This virus has the attraction that it can be introduced through wounded tissue and then become systemic within the plant. Moreover, the genome of CAMV is itself a double stranded DNA circle. However, there are severe limitations to its use as a vector. The genome is relatively small being only 8 kb and most of this is required for infectivity (Delseny and Hull, 1983). Because of this and packaging limitations, only small sections of foreign DNA can be inserted in the viral genome (Gronenborn *et al.*, 1981). Only two small regions of the genome can be modified without affecting infectivity. Nevertheless, there have been some established reports of transfer of small genes into plants and, moreover, that these were expressed at a high level within the host (Brisson *et al.*, 1985; Lefebvre *et al.*, 1987).

Considerable interest has also been shown in the Gemini viruses. Again, their genomes consist of DNA but unlike the caulimoviruses it is single stranded. The genome is even smaller than that of CAMV, consisting of one or two units of about 2.0 to 2.5 kb in length. Interest in this group stems from the fact that they infect a wide range of crop plants, both dicotyledons and monocotyledons. Examples are tomato golden mosaic virus (TGMV), bean golden mosaic virus (BGMG), beet curly top virus (BCTG), maize streak virus (MSV) and wheat dwarf virus (WDV). They are transmitted by whitefly (BGMV and TGMV) or by leaf hoppers (BCTV, MSV and WDV). The genomes of those gemini viruses transmitted by whitefly comprise two components, while those transmitted by leaf hoppers consist of a single component only (Walden, 1988).

An interesting recent innovation is that of Agroinfection (Grimsley *et al.*, 1988) or, as it is sometimes called, Agroinoculation. In this context, it refers to the introduction of viruses or viroids into plant cells through the use of *Agrobacterium*. One or more copies of the viral genome is introduced into the Ti plasmid of the bacterium and thus into the host plant. This provides an alternative route for the introduction of viruses independent of insect vectors and could be useful in the introduction of cloned DNA of geminiviruses such as MSV, WDV and digitaria streak virus (DSV) into their respective hosts for the purpose of analysing host pathogen interactions (Grimsley, 1990). However, it is difficult to envisage any direct use of this technique in plant breeding. It is more likely that it will be useful in the development of systems which will allow detailed analysis of viral biology. It is also possible that modified viral constructs could be used to induce resistance in plants to certain viral diseases. This resistance would not be heritable unless the viral constructs are integrated into the host chromosomes.

9.4 APPLICATION TO PLANT BREEDING

9.4.1 General considerations

Plant breeding has been concerned traditionally with the transfer of genes between individuals and, moreover, that this has been achieved by sexual hybridization followed by recombination. This has limited the size of the effective gene pool of a particular crop to genotypes of the same species together with those from closely related species with which they are sexually compatible. Although techniques such as embryo rescue have increased the scope for interspecific and intergeneric hybridization, the accessible gene pool is still limited and particularly so in diploids. In theory, the new techniques of gene transfer involving recombinant DNA and vectors such as *Agrobacterium* removes these interspecific barriers to gene introgression. However, in practice there are many difficulties. Problems associated with the technology of gene transfer are paralleled by problems with the identification and extraction of suitable target genes. It is only natural that, in the first instance, the characteristics being targeted for modification will be those already being addressed by conventional plant breeding but in which progress towards achieving the objectives has been limited. It is also unfortunate that precise details of the genetic control of many of the so called 'desirable' attributes are not accurately described. This is particularly so where characteristics are controlled by what has been traditionally called **polygenes** but which are now known as quantitative trait loci (QTL).

Even where characteristics are controlled by single genes of major effect, recombinant DNA technologies demand that these can be identified at the molecular level, the DNA sequence extracted, cloned into vectors for multiplication and then for insertion into the host genome. Nevertheless, characterized single genes are the likely targets for manipulation using vectors such as *Agrobacterium* plasmids. Moreover, most of these genes will originate from within the same species or from closely related species or genera. At the same time, it has to be borne in mind that it is possible that genes from bacteria and viruses and even from animals could be transferred into plant cells and be transcribed to produce unique properties or useful products. For example, there is considerable interest in the possibility of transforming potato with an animal gene such as that for insulin. It is argued that with the appropriate promoters there is no reason why insulin should not be synthesized in the plant cells and stored in the tubers from where it could be extracted.

Despite the immediate problems, non-sexual gene transfer is potentially a very powerful technique which could be of considerable benefit to plant breeding. Attention is being focused on many aspects of research and development. A substantial proportion of present research is merely aimed at developing efficient transfer systems with the only genes transferred being the reporter genes themselves, genes such as NPT II, GUS and dihydrofolate reductase (DHFR). Recent publications such as those of Mullins *et al.* (1990), Dong *et al.* (1991), Lu *et al.* (1991) and Mante *et al.* (1991) illustrate this well. In contrast, the production of insect-resistant or herbicide-tolerant genotypes by *Agrobacterium*-mediated gene transfer has the immediate aim of producing new cultivars.

The paucity of genes which can be readily identified, isolated and cloned demonstrates clearly the the need for more information on metabolic and developmental pathways in our major crop plants and how these are genetically controlled. An important strategy for doing this is the development of comprehensive genetic maps using molecular markers such as isoenzymes, RFLPs (Restriction Fragment Length Polymorphism) and RAPD (Random Amplification of Polymorphic DNA). In RFLP mapping, specific genomic or c-DNA probes are ligated into modified plasmids and used to transform an appropriate strain of *E. coli* for multiplication in culture. Following extraction and labelling of the probe DNA, it is used to detect homologous regions of the genome in individuals from a segregating family of plants using Southern hybrization (Helentjaris *et al.*, 1985). A comprehensive map of these markers can then be used to analyse complex characters, those controlled by polygenes (QTL) through linkage to the marker loci (refer to Tanksley *et al.*, 1989 for review). In this way, it is hoped that the genetic control of many of the important morpho-physiological characters can be allocated to specific regions of the chromosomes. Further refined techniques such as 'chromosome walking' would allow the precise coding region associated with a particular gene to be identified and cloned.

To date, however, the technology for routinely isolating genes of interest is not available. Some success has been achieved by constructing c-DNA libraries and then screening the clones with an appropriate probe derived from purified RNA obtained from tissue where there is a high level of expression of the gene. This type of protocol is unlikely to be successful where the gene product is very rare. Other novel approaches to the isolation of plant genes which are now being considered are transposon tagging, insertion mutagenesis and gene rescue by DNA complementation of characterized mutants (Walden and Schell, 1990).

Despite this constraint on gene manipulation, some significant successes have been achieved using *Agrobacterium* as the vector. Particular progress has been made in engineering plants tolerant to certain broad spectrum herbicides and to the production of insect and disease resistant plants. Engineering male sterile plants for use in hybrid seed production is another objective.

9.4.2 Herbicide tolerance

Traditionally, the control of weeds in crops has been based upon selective herbicides, with broad spectrum herbicides being used on fallow land or to remove complete swards before cultivation. This strategy has undoubtedly been successful but it is particularly difficult to obtain a single selective herbicide which will remove both broad leaved and grassy weeds. Repeated applications using different herbicides is becoming increasingly unacceptable on environmental as well as economic grounds. A completely new strategy is that of engineering crops which are resistant to some of the broad spectrum herbicides, ones which are incidentally of low toxicity, rapidly degraded in the soil and of high activity. In this way, a growing crop can be sprayed with one of these herbicides thus removing all plants except the resistant crop.

Two approaches have been taken to creating herbicide resistant crops (Gasser and Fraley, 1989). Firstly, **tolerance** can be created by altering the level and sensitivity of the target enzyme and, secondly, by incorporating genes that will **detoxify** the herbicide. Glyphosate, the well-known broad spectrum herbicide, acts by inhibiting the enzyme 5-enol-pyruvylshikimate-3-phosphate synthase (EPSP). A gene coding for a glyphosate-tolerant variant of this enzyme was introduced into tobacco plants thus producing glyphosate-resistant plants (Comai *et al.*, 1985) while tolerance was produced in *Petunia* by introducing a gene construct which resulted in overproduction of ESPS enzyme (Shah *et al.*, 1986). On the other hand, resistance to gluphosinate and bialaphos was induced in tobacco, tomato and potato by the introduction of a gene which produced a detoxifying enzyme (De Block *et al.*, 1987). The resistance gene *bar* was isolated by Murakami *et al.* (1986) from *Streptomyces hygroscopicus* and encodes for the enzyme phosphinothricin acetyl transferase (PAT) which converts phosphinothricin (PPT), the active herbicidal ingredient in both gluphosinate and bialaphos, into a non-herbicidal acetylated form. Excellent resistance to gluphosinate was expressed by transgenic tobacco and potato lines in extensive field trials (De Greef *et al.*, 1989). Similarly constructed herbicide-resistant lines of cabbage, poplar and sugar beet have also been engineered (Dekeyser *et al.*, 1990). Release of these transgenic herbicide resistant plants is only a matter of time. Much will depend on regulatory approval for the introduction of recombinant organisms into the environment and on the field performance of these lines. In the end it is likely that such genes will be introduced into a range of high yielding superior cultivars, an excellent example of genetic engineering supplementing traditional breeding.

9.4.3 Insect resistance

The high toxicity of many insecticides used in agriculture and the consequent effects on non-harmful insects and other organisms within the crop canopy gives cause for concern. The development of biocontrol systems is, therefore, attractive. Of particular interest in the context of breeding is the construction of systems based on Bt toxins from *Bacillus thuringiensis*. Although this was covered in some detail in Chapter 2 it is relevant to refer to it briefly here again. The Bt toxin is active against most lepidopteran larvae (moth and butterfly) but is not harmful to other insects and higher animals. Nevertheless, the insecticidal spectrum of Bt toxins from different *B. thuringiensis* strains shows considerable variation and use could be made of this in the control of specific larval pests. The Bt genes have been extracted from the bacterium, cloned in disarmed Ti plasmids and tranferred to several crop plants (Gasser and Fraley, 1989; Dekeyser *et al.*, 1990). Transgenic tomato, tobacco, cotton and potato plants resistant to caterpillar pests have been tested in the laboratory and in greenhouse and field tests. Commercial exploitation of this innovation will again be dependent on regulatory approval as well as evidence of stability of the constructs over many seed generations.

9.4.4 Virus resistance

Progress in engineering plants resistant or tolerant to viruses through the use of 'coat mediated protection' has been rapid. Again, this topic was covered in Chapter 2. Suffice to comment here that gene constructs encoding for viral coat protein have been introduced into host plants thus inducing resistance to the viruses from which the genes were taken. Resistance to TMV infection in tobacco plants was achieved by introducing only the coat protein gene of TMV (Powell-Abel *et al.*, 1986).

Transgenic plants of potato resistant to potato virus X and Y (review by Beachy, 1990) and of alfalfa resistant to alfalfa mosaic virus (Hill *et al.*, 1991) have also been produced through transfer of coat protein genes. This approach to constructing virus tolerant crop plants should make a significant contribution to breeding for virus resistance since conventional resistance genes are often difficult to locate.

9.4.5 Male sterility

Hybrid varieties are extensively used in outbreeding crops and are also becoming increasingly attractive as a method of maximizing performance in self-pollinating crops. It is the proven method of exploiting specific combinations of genes, combinations which result in hybrid vigour (heterosis) and which cannot be assembled in homozygous lines. In their simplest form, they are F1 hybrids between two parental inbred (homozygous) lines. Production strategy requires the separate multiplication of the parental lines followed by crossing on a large scale, often involving large acreage. One of the parental lines (the A line) must not be allowed to produce pollen so that the seed formed on this line will be totally hybrid. The simplest way of achieving this is by mechanical emasculation, e.g. detasselling in maize. This is difficult, if not impossible, in most other crops and pollination control is then achieved by genetically-controlled male sterility.

Two basic types of genetic control are recognized, that under the control of **cytoplasmic genes** (*cms*) and that under the control of **nuclear genes** (*ms*). Cytoplasmic male sterility is maternally inherited, but normal pollen development can be restored by nuclear restorer genes (*Rf*). These are dominant and are inherited in the normal Mendelian manner. Hybrid seed production involves using a cytoplasmic male sterile A line and a pollinator R (restorer) line which carries the *Rf* genes. Good sources of male sterile cytoplasms are rare and often involve introducing the nuclear genotype of a cultivated species into the cytoplasm from a non-cultivated related species. This type of alien cytoplasm, although conferring male sterility, can also carry other genes which have a detrimental effect on the phenotype. One possible future development is that of cloning these *cms* genes into a suitable vector and introducing them together with their appropriate promoter sequence into a normal cytoplasm. The constructs and vectors will have to be such that they remain in the cytoplasm. Viral vectors would appear to be the more suitable for this purpose but, since viruses are not inherited through seed, the *cms* construct will have to be removed from the vector within the plant.

Of greater interest as a possible target character is the second type of male sterility, that controlled by nuclear genes. Such genes are quite common in many species and are usually recessive (Kaul, 1988). Since many of these mutations are associated with a defective tapetum the suggestion is that this is essential for normal pollen development and that male sterility is brought about through abnormal tapetum development (Mariani *et al.*, 1991). However, exploitation of this system is particularly difficult since multiplication of the male sterile stocks can only be achieved by crossing the homozygous male sterile line (*ms/ms*) with an isogenic heterozygous (+/*ms*) male fertile counterpart. The difficulty arises in separating out the *ms/ms* from the +/*ms* segregants before flowering. The idea of selective markers linked to the male sterile gene in order to facilitate separation is not new (Wiebe, 1960) but complete linkage is difficult to achieve.

A recent innovative development which could have far reaching implications is that of engineering male sterile genotypes through interference with tapetal development (Mariani *et al.*, 1990). Initially, a tapetal specific gene was extracted from tobacco (*Nicotiana tabacum*) and the promoter fused to a ribonuclease gene from either *Aspergillus oryza* or from *Bacillus amyloliquefaciens*. These two chimaeric genes were individually introduced into the T-DNA region of modified Ti plasmid of *Agrobacterium* and separately introduced into tobacco (Mariani *et al.*, 1990). The ribonuclease gene was specifically expressed in the anther tissue resulting in degeneration of the tapetum with consequent male sterility. The same constructs were also introduced with the same result into oil seed rape (*Brassica napus*). These constructs act as dominant male sterile genes, in contrast with the recessive nature of most natural mutants. Further restructuring was needed to allow separation of the male sterile segregants from their male fertile counterparts. Mariani *et al.* (1991) described the construction of an even more complex chimaeric gene involving the tapetal specific promoter and the ribonuclease genes described previously together with the bialaphos resistant gene construct described in Section 9.4.2. These were transferred to both tobacco and oil seed rape through *Agrobacterium*-mediated transformation and, since the male sterile (*Ms*) and the herbicide-resistant gene (*Hs*) were in the same construct, they behave as one dominant 'supergene' confering both male sterility and herbicide resistance when present. Moreover, they are inherited as one gene. An outline plan for their use in hybrid seed production is given in Figure 9.3. Significantly, only half the hybrid progeny would be male fertile. Although this is of no consequence in crops cultivated for their vegetative parts and is probably not a limiting factor in cross-pollinating crops, it would not be acceptable in self-pollinating grain crops. The R line will have to be engineered to make it a truly restorer line so that all the hybrid progeny are male fertile. One possibility would be to introduce a gene construct which would

Stage I. Construction of male sterile
herbicide-tolerant 'A Line'

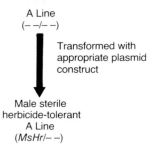

A Line
(– –/– –)

Transformed with
appropriate plasmid
construct

Male sterile
herbicide-tolerant
A Line
(*MsHr*/– –)

Stage II. Multiplication of 'A Line'

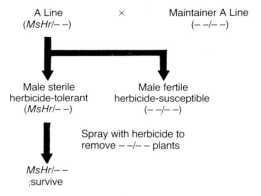

A Line   ×   Maintainer A Line
(*MsHr*/– –)   (– –/– –)

Male sterile   Male fertile
herbicide-tolerant   herbicide-susceptible
(*MsHr*/– –)   (– –/– –)

Spray with herbicide to
remove – –/– – plants

MsHr/– –
;survive

Stage III. Hybrid seed production

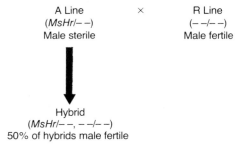

A Line   ×   R Line
(*MsHr*/– –)   (– –/– –)
Male sterile   Male fertile

Hybrid
(*MsHr*/– –, – –/– –)
50% of hybrids male fertile

Figure 9.3 Overall plan for production of hybrids using male sterile herbicide-resistant transformed plants. (After Mariani *et al.*, 1991.)

interfere with the expression of the ribonuclease gene, perhaps an antisense gene.

There is no doubt that engineered male sterile genes will be introduced to many dicotyledonous plants in the immediate future through the use of *Agrobacterium*. Progress in modifying monocotyledons such as cereals will depend on other transformation systems, probably the ballistic method.

9.4.6 Antisense genes

Smith *et al.* (1988) showed that antisense RNA produced in stably transformed tomato plants inhibited the expression of an endogenous gene. A 730 bp fragment of the developmentally regulated gene for polygalacturonase (PG), an enzyme which causes fruit softening during ripening, was fused in inverted orientation to the CAMV 35S RNA promoter and to the nopaline synthase gene (*nos*). This chimaeric gene was introduced into a binary vector system of *Agrobacterium tumefaciens* and the modified bacterium then used to transform tomato stem segments. Three regenerated plants were shown to produce antisense PG RNA which led to a reduction in PG activity in ripe fruits. The aim of this manipulation was to produce transgenic tomato plants with fruits which do not deteriorate quickly during storage. This type of approach could also be important in analysing the effect of gene expression as well as in plant breeding.

9.5 FUTURE PROSPECTS

The introduction of genetically engineered crops into the market place will be influenced as much by non-technical factors as by the success of the techniques themselves and the results achieved. Public perception of recombinant DNA technologies, statutory regulations and degree of patent rights on the engineered products will have a profound influence on the speed of future development and, indeed, on the nature of the development.

9.5.1 Public perception

Although breeders have been manipulating genes by means of hybridization for the last 100 years or so, the public are still nervous of the prospect of genes being extracted and then reintroduced into quite different plants. To some extent this is understandable. Not only are the genotypes themselves being restructured but the genes themselves are being modified and in some cases they are extracted from prokaryotes and used for an entirely different function in eukaryotes. Examples have been quoted in this chapter of bacterial genes being used for producing male sterility in higher plants. There is no inherent danger in such manipulations but the public need to be convinced.

In addition to public reservations on the grounds of safety and ethics there is also concern over where such research is likely to lead to in terms of overall production of many crops. At a time of overproduction in Europe and North America, there are questions being asked about the need for such research. The same questions could be asked of the need for conventional plant breeding and agricultural research in general. However, it can be argued that it is important under any system to produce crops which can utilize the inputs as efficiently as possible. It is highly likely in future that more attention will have to be given to the nature and the amount of inputs. Lowering inputs while retaining production levels will pose a new challenge to the breeders and biotechnologists.

Considerable debate has taken place regarding the field testing of engineered plants and the general consensus is that it should be allowed provided it is under strict control. The ultimate test will come when products from engineered plants are actually marketed. Will they be labelled as such and if so, how will they be accepted by the public?

9.5.2 Regulations

Strict regulations governing testing of genetically engineered crops are in place in most countries, although their stringencies vary considerably. They have been formulated to take account of public concern over the risks, however slight, associated with recombinant DNA technology. It is perhaps ironic that if a disease-resistant gene was transferred from one species to another using sexual hybridization and embryo rescue through culture on a suitable medium, it would not be subject to any regulations. If the same gene could be extracted and introduced into a suitable vector such as *Agrobacterium* and then introduced into another species the products would be subject to statutory regulations in force in that country. However, there is a move in the USA to include the exploitation of both non-sexual and sexual recombinants under the same regulations (Gavaghan, 1992).

One facet of safety which has to be addressed is that referring to expression of the selectable markers in commercial crops. As we saw earlier, these are essential for the separation of transformed from non-transformed tissue during the gene transfer phase. Unfortunately, the most efficient and consequently the most widely used of these selectable marker genes is NPT II (*neo*) which confers resistance to the antibiotics neomycin and kanamycin. Not only is it expressed in the transformed tissue in culture, it is also expressed in the regenerated plants and in descendants of such plants. The questions being asked concern the fate of the enzyme neomycin phosphotransferase II when present in fresh plant tissue consumed directly by humans. Flavell *et al.* (1992) conclude that the oral use of antibiotics could in no way be compromised by the NPT II enzyme in the diet.

At present, even the most advanced trials are at the small plot stage of assessment. Development of gene transfer technology will be influenced by the nature of the statutory trials demanded and their cost in relation to the returns to be expected on the total investment.

9.5.3 Plant Breeder's Rights

Considerable debate exists within the seed industry regarding the level of protection that should be afforded genetically engineered crops. Any cultivar that is granted Plant Breeders Rights based on the appropriate trials would qualify for protection under that act. However, this only relates to the multiplication and marketing of that variety. To date, it does not necessarily protect it from use in further breeding by conventional means. In the USA, attention has been focused on 'patent rights' relating to genes, processes of isolating genes and the creation of the genetically modified plants themselves (Gasser and Fraley, 1989). It appears that in the USA at least, patent rights affords greater protection than that afforded by Plant Varieties Rights. The large multinational companies are pressing for the same protection in Europe, but at present plant varieties are not patentable although this is being reviewed by the European Patent Office. Further investment by large companies will certainly be influenced by the amount of protection afforded. If, as it appears in the USA, considerable protection can be obtained under Patent Rights, plant breeding of this kind is likely to become more and more the domain of the large companies who will have the necessary finance for research and development.

9.5.4 Genetic engineering and conventional plant breeding

The remarkable progress being made in *Agrobacterium*-mediated gene transfer in dicotyledonous plants has focused attention on its role in relation to existing techniques in plant breeding. Although there is no doubt that it is a very powerful new technology it is not likely to completely replace existing schemes based on sexual hybridization. Indeed it is most unhelpful to consider them as alternatives; they should be linked into an overall strategy of crop improvement. The exciting developments reported earlier in this chapter have mostly involved interspecific and intergeneric gene transfer. Such transfer is not new; it has been practised by breeders for a considerable period of time. The major contribution of recombinant DNA technology is to widen the scope for this breeding strategy. With conventional breeding techniques, the limit is set by sexual incompatibilty barriers which usually means that only genotypes from related species can be hybridized. In theory, there is no limit to the scope of gene transfer using the new molecular techniques. We have already seen bacterial genes being introduced into higher plants and viral promoters (CAMV 35S) being used in chimaeric gene constructs.

Even within existing crop plant gene pools there will be candidate genes which could be transferred more effectively by recombinant DNA technologies and the number of these target genes will increase with time. However, the manipulation of complex characters will still rely on conventional breeding techniques of hybridization and selection. The linking of the two technologies is typified by the example of engineered male sterility for use in the production of hybrid seed. The parental inbreds would be obtained by conventional means and their combining ability would have been assessed by the appropriate progeny tests, although RFLP analysis could aid the grouping of inbreds. The chosen parental A line would then be engineered to make it male sterile and herbicide-resistant to facilitate the production of hybrid seed.

9.6 CONCLUSIONS

Non-sexual gene transfer offers exciting opportunities in crop improvement. Whether it is used for direct substitution of existing alleles in very much the same way as in conventional plant breeding or, whether it is used for introducing 'alien' genes from a completely unrelated species, it will undoubtedly become a powerful new tool in plant breeding. Apart from this direct application to the goal of producing the 'ideal' ideotype, there is the added prospect of producing single gene substitutions for determining the genetic control of complex metabolic pathways which ultimately determine the many morpho-physiological characters which concern the plant breeder. Although vectors based on the *Agrobacterium* Ti plasmid are by far the most efficient non-sexual method of achieving complete integration of genes into dicotyledonous plants, other systems involving the introduction of naked DNA by electroporation or by the particle gun will probably be used for gene transfer in monocotyledons. The paucity of target genes demands that considerable effort in both time and money is needed to unravel the complex genetic control of the characters relevant to improvement of our crop plants.

REFERENCES

Austin, R.B., Bingham, J., Blackwell, R.D., *et al.* (1980) Genetic improvement in winter wheat yields since 1900 and associated physiological changes. *Journal of Agricultural Science, Cambridge*, **94**, 675–9.

Beachy, R.N. (1990) Plant transformation to confer resistance against virus infection, in *Gene Manipulation in Plant Breeding II*, (ed J.P. Gustafson), Plenum Press, New York, pp.305–12.

Bevan, M., Flavell, R.B. and Chilton, M.D. (1983) A chimaeric antibiotic resistance gene as a selectable marker for plant cell transformation. *Nature*, **304**, 184–7.

Brisson, N., Paszkowski, J., Penswick, J., *et al.* (1985) Expression of a bacterial gene in plants by using a viral vector. *Nature*, **310**, 511–14.

Chilton, M.-D., Drummond, M.H., Merlo, D.J., *et al.* (1977) Stable incorporation of plasmid DNA into higher plant cells: molecular basis of crown gall tumorigenesis. *Cell*, **11**, 263–71.

Chilton, M.-D., Tepfer, D.A., Petit, A., *et al.* (1982) *A. rhizogenes* inserts T-DNA into genomes of host plant root cells. *Nature*, **295**, 432–4.

Comai, L., Facciotti, D., Hiatt, W.R., *et al.* (1985) Expression in plants of a mutant *aro*A gene from *Salmonella typhimurium* confers tolerance to glyphosate. *Nature*, **317**, 741–4.

De Block, M., Botterman, J., Vandewiele, M., *et al.* (1987) Engineering herbicide resistance in plants by expression of a detoxifying enzyme. *EMBO Journal*, **6**, 2513–18.

De Block, M., Herrera-Esrella, L., Van Montagu, M., *et al.* (1984) Expression of foreign genes in regenerated plants and their progeny. *EMBO Journal*, **3**, 1681–9.

De Framond, A.J., Barton, K.A., and Chilton, M.-D. (1983) Mini-Ti: a new vector strategy for plant genetic engineering. *Bio/Technology*, **1**, 262–9.

De Greef, W., Delon, R., De Block, M., *et al.* (1989) Evaluation of herbicide resistance in transgenic crops under field conditions. *Bio/Technology*, **7**, 61–4.

Dekeyser, R., Inze, D. and van Montagu, M. (1990) Transgenic plants, in *Gene Manipulation and Crop Improvement II*, (ed J.P. Gustafson), Plenum Press, New York, pp.237–50.

Delseny, M. and Hull, R. (1983) Inoculation and characterization of faithful clones of the genomes of cauliflower mosaic virus isolation Cabb B-J1, CM-184 and Bari 1. *Plasmid*, **9**, 31–41.

Dong, Jin-Zhuo, Yang, Mir-Zhu, Jia, Shi-Rong, *et al.* (1991) Transformation of melon (*Cucumus melo* L.) and expression of the cauliflower mosaic virus 35S promoter in transgenic melon plants. *Bio/Technology*, **9**, 864–8.

Draper, J., Scott, R., Armitage, P. and Walden, R. (1988) *Plant Genetic Transformation and Gene Expression. A laboratory manual*, Blackwell Scientific Publications, Oxford.

Feldmann, K.A. and Marks, M.D. (1987) *Agrobacterium*-mediated transformation of germinating seeds of *Arabidopsis thalina*: A non tissue culture approach. *Molecular General Genetics*, **208**, 1–9.

Flavell, R.B., Dart, E., Fuchs, R.L., *et al.* (1992) Selectable marker genes: safe for plants? *Bio/Technology*, **10**, 141–4.

Gasser, C.S. and Fraley, R.T. (1989) Genetically engineered plants for crop improvement. *Science*, **244**, 1293–9.

Gavaghan, H. (1992) Washington takes a stand on Biotechnology. *New Scientist*, **133**, No. 1811, 10.

Grimsley, N.H. (1990) Agroinfection. *Physiologia Plantarum*, **79**, 147–53.

Grimsley, N.H., Hohn B., Hohn, T., *et al.* (1988) Agroinfection, an alternative route for viral infection of plants by using the Ti-plasmid. *Proceedings of the National Academy of Sciences, USA*, **83**, 3283–6.

Gronenborn, B., Gardner, R.C., Schaefer, S., *et al.* (1981) Propagation of foreign DNA in plants using cauliflower mosaic virus as a vector. *Nature*, **294**, 773–6.

Helentjaris, T., King, G., Slocum, M., *et al.* (1985) Restriction fragment polymorphism as probes for plant diversity and their development as tools for plant breeding. *Plant Molecular Biology*, **5**, 109–18.

Herrera-Estrella, L., De Block, M., Messens, E., *et al.* (1983) Chimaeric genes as dominant selectable markers in plant cells. *EMBO Journal*, **2**, 987–95.

Hill, K.K., Jarvis-Egan, N., Halk, E.L., *et al.* (1991) The development of virus resistant alfalfa, *Medicago sativa* L. *Bio/Technology*, **9**, 373–7.

Horsch, R.B., Fry, J.E., Hoffman, N.L., *et al.* (1985) A simple and general method for transforming genes into plants. *Science*, **227**, 1229–31.

Kahl, G. and Schell, J. (1982) *Molecular Biology of Plant Tumours*, Academic Press, London.

Kaul, M.L.H. (1988) *Male Sterility in Higher Plants*, Monograph on Theoretical and Applied Genetics, Vol. 10, Springer, Berlin.

Koncz, C., Martini, N., Mayerhofer, R., *et al.* (1989) High frequency T-DNA-mediated gene tagging in plants. *Proceedings of the National Academy of Sciences, USA*, **86**, 8467–71.

Lefebvre, D.D., Miki, B.L. and Laliberte, J.F. (1987) Mammalian metallothionein functions in plants. *Bio/Technology*, **5**, 1053–6.

Lu, Ching-Yi, Nugent, G., Wardley-Richardson, T., *et al.* (1991) *Agrobacterium*-mediated transformation of carnation (*Dianthus caryophyllus*). *Bio/Technology*, **9**, 864–8.

Mante, S., Morgens, P.H., Scorza, R., *et al.* (1991) *Agrobacterium*-mediated transformation of plum (*Prunus domestica* L.) hypocotyl slices and regeneration of transgenic plants. *Bio/Technology*, **9**, 853–7.

Mariani, C., Beuckeleer, M. de., Truettner, J., *et al.* (1990) Induction of male sterility in plants by a chimaeric ribonuclease gene. *Nature*, **347**, 737–41.

Mariani, C., Goldberg, R.B. and Leemans, J. (1991) Engineered male sterility in plants, in *Molecular Biology of Plant Development, Symposium XLV, Society of Experimental Biology*, (eds G.I. Jenkins and W. Schuch), Portland Press, Colchester, England, pp.271–9.

Mullins, M.G., Tang, F.C.A. and Facciotti, D. (1990) *Agrobacterium*-mediated transformation of grapevines: Transgenetic plants of *Vitis rupestris* Scheele and buds of *Vitis vinifera* L. *Bio/Technology*, **8**, 1041–5.

Murakami, T., Anzai, H., Imai, S., *et al.* (1986) Bialaphos biosynthetic genes of *Streptomyces hygroscopicus*: molecular cloning and characterization of the gene cluster. *Molecular General Genetics*, **205**, 42–50.

Ow, D., Wood, K.V., De Luca, L., *et al.* (1986) Transient and stable expression of fierfly luciferase gene in plant cells and transgenic plants. *Science*, **234**, 856–9.

Powell-Abel, P.A., Nelson, R.S., De, B., *et al.* (1986) Delay of disease development in transgenic plants that express the tobacco mosaic virus protein gene. *Science*, **232**, 738–43.

Raineri, D-M., Bottino, P., Gordon, M.P., *et al.* (1990) *Agrobacterium*-mediated transformation of rice (*Oryza sativa* L.). *Bio/Technology*, **8**, 33–8.

Riggs, T.J., Hanson, P.R., Start, N.D., *et al.* (1981) Comparison of spring barley varieties grown in England and Wales between 1880 and 1980. *Journal of Agricultural Science, Cambridge*, **97**, 599–610.

Shah, D.M., Horsch, R.B., Klee, H.J., *et al.* (1986) Engineering herbicide tolerance in transgenic plants. *Science*, **233**, 478–81.

Simmonds, J. (1991) Gene transfer in the Gramineae. *Plant Breeding Abstracts*, **61**, (12), 1369–76.

Smith, C.J.S., Watson, C.F., Ray, J., *et al.* (1988) Antisense RNA inhibition of polygalacturonase gene expression in transgenic tomatoes. *Nature*, **334**, 724–6.

Stachel, S.E. and Zambryski, P.G. (1986) *Agrobacterium tumefaciens* and the susceptible plant cell: A novel adaptation of extracellular recognition and DNA conjugation. *Cell*, **47**, 155–7.

Tanksley, S.D., Young, N.D., Paterson, A.H., *et al.* (1989) RFLP mapping in plant breeding: New tools for an old science. *Bio/Technology*, **7**, 257–64.

Teeri, T.H., Lehraslaiho, H., Franck, M., *et al.* (1989) Gene fusion to *lacZ* reveals expression of chimaeric genes in transgenic plants. *EMBO Journal*, **8**, 343–50.

Walden, R. (1988) *Genetic Transformation in Plants*, Open University Press, Milton Keynes.

Walden, R. and Schell, J. (1990) Techniques in plant molecular biology – progress and problems. *European Journal of Biochemistry*, **192**, 563–76.

Wiebe, G.A. (1960) A proposal for hybrid barley. *Agronomy Journal*, **52**, 181–2.

Yanofski, M.F., Porter, S.G. and Young, C. (1986) The *virD* operon of *Agrobacterium tumefaciens* encodes a site specific endonuclease. *Cell*, **47**, 471–7.

Zambryski, P., Joos, H., Genetello, C., *et al.* (1983) Ti plasmid vector for the introduction of DNA into plant cells without alteration of their normal regeneration capacity. *EMBO Journal*, **2**, 2143–50.

10

The mushroom industry

J.F. Smith

'When we say that the Mushroom ranks amongst the most esteemed esculents: that it is one of the most delicious and at the same time is highly nutritious – that it, in fact approaches nearer the animal food than any other vegetable does: that the supply generally is quite inadequate of the demand, and that mushrooms can be grown in nearly every village and in the suburbs of almost all cities and town of this country, we must at once concede that their increased cultivation is highly desirable'.

J.Wright FRHS (1894) Mushrooms for the Million, Fleet St. London.

10.1 HISTORY OF MUSHROOM CULTURE

Mushrooms have been esteemed as a food delicacy for centuries and are frequently referred to in ancient Greek and Roman literature. The first detailed records on cultivation of the white button mushroom (*Agaricus bisporus*), occurred during the reign of Louis XIV when Tounefort (1707), at the Royal Academy of Science, France, described a successful method of growing mushrooms on stable manure. Knowledge of compost pre-treatment was extremely vague at this time and emphasis was directed at the design and preparation of ridge beds (Figure 10.1) made from stable manure rather than at the composting process itself. Mushroom colonization of such substrates was allowed to occur naturally, as it was an accepted fact that 'mushroom seeds' were abundant in horse-manure droppings. While there was much truth in this belief, as horses inevitably would consume plant material contain-

Exploitation of Microorganisms Edited by D.G. Jones
Published in 1993 by Chapman & Hall, London ISBN 0 412 45740 7

Figure 10.1 Out-door mushroom ridge bed.

ing mushroom spores during the course of summer and autumn grazing, the concentration of spores surviving equine digestion was likely to be extremely variable. Although it has been an accepted fact for over 200 years that horse manure is an ideal medium for the germination of mushroom spores, to be reliant on naturally occurring spores as the sole source of inoculum understandably meant unpredictable mushroom production in these very early days. If the colonization of the composted manure was successful, mushroom fruit-bodies were produced (initiation) by covering the ridge beds with a layer of mature dry leaf mould.

By the end of the 18th century, the composting of manure substrates was regarded as an essential pre-requisite before its inoculation with mushroom mycelium (spawn). Abercrombie (1779) described a method of composting stable manure in stacks with a cross-sectional dimension of 1.5 metres wide and 1.5 metres high, a method similar to that used today. He also emphasized the importance of wetting and frequent mixing of the stable manure compost during a three to four week period in order 'that it may meliorate, by discharging the rank obnoxious steam' and 'give an additional vent to the fierce ferment'. Such actions were deemed necessary to produce a uniform substrate on which the mushroom would survive and flourish. Mushroom growers of the time soon realized that inoculation of composted substrates with the 'stringy fibres' (mycelium/spawn) taken from dung heaps, or spent cucumber or melon beds, previously colonized by mushrooms, improved their chances of producing good mushroom crops even further. Generally, mushroom spawn obtained from such locations gave better results than that obtained from pasture land, but it was many years later before the taxonomy of mushrooms was attempted and failures of the mushroom crop could be attributed to selecting the wrong spawn. In order to assert a degree of predictability into crop production it soon became common practice to save the mushroom mycelium from exhausted but successful mushroom beds to re-inoculate fresh beds, but successive isolations and re-inoculations resulted in a gradual decline in the productivity and quality of the crops.

10.1.1 Spawn preparation

More consistent crops were obtained in the late 19th century, when a method of preparing manure spawn was perfected. Mushroom mycelium was inoculated on to pre-formed bricks made from moist soil and horse manure which had previously been partially dried. Once completely colonized, the bricks could be kept cool and dry in well aerated sheds for many years before use. While this was a major step forward in the consistency of mushroom production, from time to time failure of apparently well prepared spawn was attributed to 'inherent exhaustion consequent of unintermittent propagation' (Wright, 1894).

The preparation of 'virgin spawn' can possibly be attributed to a UK spawn company, Hamlin & Co., who claimed to have prepared brick spawn from germinated mushroom spores as early as 1886 (Cayley, 1938). French spawn companies soon followed this approach (Constatin and Matruchot, 1894 cited in Atkins, 1972), but it was Ferguson (1902) who first gave a detailed description of how mushroom spores could be germinated in the laboratory. Within three years, Duggar (1905) described a method of making spawn from mushroom tissue. This method was quickly exploited by the mushroom industry as it made possible for the first time selection and maintenance of a particular strain.

The use of more reliable spawns led to a surge in mushroom production in the latter part of the 19th century, such that mushrooms became a popular food in many of the European countries. As demand for regular supplies increased, emphasis was directed at all-the-year-round growing. Mushroom growers in France had been exploiting the uniform environmental conditions found in underground quarries of Paris since the beginning of the 19th century to such an extent that by the 1880s, 25 tonnes of mushrooms were being produced daily around Paris (Wright, 1894). Many of the major mushroom growers in the UK persisted with ridge bed production until the turn of the century, producing the bulk of their crops mainly out-of-doors, between March and September, but the demand for good quality mushrooms at all seasons soon encouraged tomato growers in England to exploit their empty glasshouses throughout the winter months (Rettew *et al.*, 1941). While it was French mushroom growers who pioneered protected cropping, the idea of designing purpose-built mushroom houses in which the environment could, to some degree, be controlled, can largely be attributed to an English man, Edward Callow. As head gardener to Lord Glastonbury during the early part of the 19th century he described a growing house that accommodated both floor and shelf beds, a facility that could produce mushrooms throughout the colder months of the year (Callow, 1831). This was achieved by insulating the growing house roof with a 'thick layer of thatch' which resisted the cold of the winter as well as the heat of the summer. Filling a floor trench with 'hot dung' during very cold weather also created a more favourable environment in which to grow mushrooms.

Although in the years that followed there were many variations on this theme, it was not until the 1920s that the **shelf bed** concept was fully exploited in the USA. Mushroom growers in Pennsylvania designed and developed wooden growing houses accommodating a double row of fixed shelves, in tiers five to six high. This standard growing facility was quickly adopted in the UK and other parts of Europe, although there was a tendency to build smaller houses accommodating only three to four shelves per tier (Atkins, 1966).

10.1.2 Substrate preparation

For over 200 years, the composting of stable manure was regarded as more of an art rather than a science. Improvements to the method of substrate preparation had slowly evolved from grower experience and empirical trials by mushroom researchers. As compost stacks were generally prepared in a single-phase operation out-of-doors, the period of composting was influenced to a great extent by environmental factors, but the main contributory factor determining compost duration was compost formulation itself. Formulations rich in animal manures had higher nitrogen contents and ammonia generated as a consequence of microbial activity generally took longer to clear. Substrates were turned periodically until microbial activity subsided and this was generally judged when the compost had lost its self-heating capacity, was free of ammonia and became sweet smelling. The single-phase system of composting remained common practice until about 1915, when growers in the United States, discovered by accident that allowing compost to complete its thermogenesis in shelved compost beds was beneficial to crop yield and reduced the presence of both insect pests and fungal pathogens. This additional phase was termed the 'sweating out process' and although this procedure became standard practice with some growers, its significance in producing homogeneity in composts was not fully understood until Lambert (1941) identified the major physical and environmental variables associated with the composting process. Lambert outlined the regions of the compost stack with differing temperature profiles and demonstrated that composts most suitable for mushroom culture came from parts of the stack where temperature had been maintained between 50–60°C and where there had been adequate aeration. He also identified that if compost taken from an anaerobic zone was subjected to further fermentation within this temperature range, it also became more productive. Lambert suggested that the 'sweating-out' period be regarded as an integral part of composting, by 'conditioning' the compost prior to mushroom inoculation and not merely a means to eradicate insect pests and competing fungi. This technique was later to become variously known as the 'peak heat', 'pasteurization' or 'phase II' process and this operation is now practised, in purpose-built buildings, on all major farms around the world today.

10.1.3 International production

Since France maintained the monopoly of the mushroom industry until the late 1920s, the white button mushroom (*A. bisporus*) is understandably referred to in most of the early literature as the 'Champignon de Paris'. During the Second World War (1939–45), mushroom production developed only in those countries not geographically involved. This explains why in countries like the USA, Canada, Sweden and Switzerland, mushroom production steadily increased while most other countries decreased or ceased production. By 1950, world mushroom production of *A. bisporus* was estimated at around 66 400 tons, 52% from Europe and 48% from North America (Delcaire, 1978). Major advances in composting technology, i.e. shorter composting durations for Phase I and Phase II composting (Sinden and Hauser, 1950, 1953) and axenic spawn preparation (Sinden, 1952; Elliott, 1985), i.e. growing mushroom mycelium on sterile cereal grains (Figure 10.2), resulted in a considerable improvement in crop yields. It was not until the 1960s that countries like Taiwan, Australia, New Zealand, Spain, Czechoslovakia, Hungary, Poland and Rumania began producing mushrooms in any great quantities (Delcaire, 1978). Today, total production has been assessed to be in excess of 1.42 million tonnes (Chang, 1991), the major countries in

Figure 10.2 Mushroom rye grain spawn (inset—individual grain covered with surface mycelium). Reproduced by permission of Horticulture Research International, Littlehampton, West Sussex.

Figure 10.3 Shelf-bed system of growing. Reproduced by permission of Chesswood Produce Ltd, Pulborough, Sussex.

descending order being USA, China, France, Holland, UK, Italy, Canada, Spain, Germany, and Taiwan. While many growing systems have evolved during the course of recent years, e.g. trays, shelves (Figure 10.3), bags, blocks, and troughs for economic as well as practical reasons (Gaze, 1985), there is a now a universally accepted two-phase composting procedure adopted by mushroom growers which ensures that the substrate is highly selective for *A. bisporus* at the time of inoculation. A concise step-wise list of procedures for growing *A. bisporus* is outlined in Table 10.1.

10.2 THE ROLE OF MICROORGANISMS IN COMPOSTING

The main purpose of the composting process is to prepare a substrate in which the growth of the mushroom is promoted to the practical exclusion of other microorganisms. The **physical qualities of the substrate** are that it must be freely permeable to the air, have a high moisture content (70–75%), be free of ammonia, and have a pH between 7.0–8.5. The chemical qualities must be such that while the compost provides an adequate nutrient status, such nutrients are in a form readily available to the mushroom and yet not so readily available to invading fungal contaminants. Above all, to produce high yielding good quality crops, the substrate must be free of insect pests, fungal pathogens and mushroom viruses.

The traditional substrate on which *A. bisporus* grows has been, until recent years, prepared from stable bedding, i.e. cereal straw, normally wheat straw, rich in horse manure. However, with the limited availability of such materials in some areas, many large farms today supplement their stable bedding supply with fresh wheat straw and deep litter chicken manure is used as an alternative nitrogen source. Other alternative nitrogen sources that have proved successful include pig and bullock manure. It is also now common practice to add to compost formulations, a proprietary activator (generally rich in soluble carbohydrate and nitrogen) to stimulate microbial activity in the early stages of composting and gypsum (Pizer, 1937) which improves the physical structure of the compost by flocculating the colloidal particles built up during the composting process, making it less greasy.

Today, the process for producing a **selective medium** for *A. bisporus* has evolved into a highly sophisticated commercial operation incorporating two distinct composting phases. Phase I is performed as **windrow stacks**, normally 2m × 2m cross section, and generally protected from the elements by a covered barn (Figure 10.4). Depending on the physical condition of the starting ingredients, Phase I composting is normally completed within 8–10 days, during which time the materials are mixed every 2–3 days and restacked using a turning machine. It is not uncom-

Table 10.1. Procedures for *Agaricus bisporus* cultivation

Operation	Procedure	Duration (days)
A. *Composting*		
Pre-wet	Gathering, mixing and wetting raw ingredients	7–10
Phase 1	Assembly of windrow stacks. Further mixing, wetting and application of manures/activators. Temperatures up to 75°C attained during this phase.	7–14
Phase 2	Transfer of compost to controlled environment room for pasteurization at 58–60°C (peak-heat). No mixing. Temperatures maintained between 45–55°C after pasteurization period (conditioning phase). Cool when compost is free of ammonia.	7–10
	Total days	21–34
B. *Growth of mycelium, fruit-body initiation, crop harvesting*		
Spawning	Inoculation of compost with *Agaricus* mycelium. Compost temperature 25°C.	<1
Spawn-run	Colonization of substrate by fungal mycelium. Compost temperature 25–28°C.	10–14
Casing	Covering colonized substrate with a layer of soil (normally peat/chalk mixture).	<1
Pre-cropping		
(a)	Continuation of both compost and casing layer colonization. Compost temperature 25°C.	6–8
(b)	Watering of beds to 'check' mycelial growth at surface of casing (knock-back) and reduction of room temperature to 16–18°C to initiate fruit-bodies.	8–10
Cropping	Harvesting of 3–5 crops (flushes) at 7–10-day intervals. Room temperature 16–20°C.	28–42
Cook-out	Steam treatment of growth rooms. Room temperature in excess of 60°C.	1–2
Emptying	Removal of compost, cleaning of chambers	1–3
	Total days	56–81

Figure 10.4 Phase I composting undercover. Reproduced by permission of Chesswood Produce Ltd, Pulborough, Sussex.

mon for stacks to reach temperatures as high as 80°C during the initial stages of composting.

The **phase 2** of composting commences after the compost is filled into shelves, trays or, as on most modern farms today, into long rectangular-shaped buildings normally referred to as tunnels, where it is initially subjected to a pasteurization temperature around 58–60°C (Fermor *et al.*, 1985). This is normally achieved by raising the air temperature with steam. While a pasteurization period of at least three hours is sufficient to kill insect pests, fungal pathogens and viruses, resident populations of thermotolerant and thermophilic microorganisms remain unaffected. On completion of the pasteurization phase, fresh filtered air is introduced into the facility to reduce the compost temperature to around 45 –50°C. This temperature, which is more favourable for development of thermotolerant and thermophilic microbial populations, is maintained for a further 3–4 days until the compost is sweet smelling and free of ammonia. Phase 2 composting is normally completed within 7–10 days and mushroom spawn is normally applied when the compost temperature has fallen to around 25°C.

Table 10.2. Major groups and succession of microorganisms occurring during compost preparation. (After Fermor *et al.*, 1985.)

Mesophiles	Thermotolerant	Thermophiles
	Succession of microorganisms →	
Bacteria		
Flavobacterium spp.	*Pseudomonas* spp.	*B. coagulans*
Pseudomonas spp.	*Bacillus licheniformis*	*B. stearothermophilis*
Serratia marcescens		*B. subtilis*
Actinomycetes		
Streptomyces spp.		*Thermoactinomycetes* spp.
Nocardia spp.		*Thermomonospora* spp.
Faeni spp.		*Saccharomonospora viridis*
		Streptomyces
Fungi		
Mucor spp.	*Aspergillus fumigatus*	*Torula thermophila* (syn.
		Scytalidium thermophilum)
Aspergillus spp.		*Chaetomium thermophile*
Penicillium spp.		*Humicola insolens*
		Rhizomucor pusillis
		Talaromyces thermophilus
		Thermomyces lanuginosa

10.2.1 Microbial succession

General microbiological surveys of organisms present in mushroom composts have been made by several workers including Hayes (1968), Laborde and Delmas (1969), Fordyce (1970), Imbernon and Leplae (1972), Lacey (1973), Chanter and Spencer (1974), Rosenburg (1975, 1978), Fermor *et al.* (1979), Kleyn and Wetzler (1981), Amner *et al.* (1988) and Derikx (1991). All these reports give quantitative assessments of the populations of fungi, bacteria and actinomycetes predominating within the substrate at different stages of its preparation.

Composting begins when readily available carbon and nitrogen sources are used by a mixed population of resident microorganisms, mainly mesophiles. Ammonia, also produced by microbial activity, aids in the decomposition of the straw by softening its cell walls enhancing further microbial activity (Bels-Koning, 1962). As the compost temperature increases, a **selection pressure** is exerted on the microbial community which results in a progressive reduction in species type. Mesophilic bacteria are quickly replaced by thermophilic spore formers and mesophilic fungi are gradually succeeded by more thermotolerant types which grow and sporulate at 45°C and above. As composting progresses, there is also a notable change in the actinomycete community, the high compost temperatures favouring the development of more thermotolerant types. Some of the major groups of microorganisms occurring in mushroom composts during its preparation are outlined in Table 10.2, but it must be emphasized that there is no clear dividing line between the groups. While all three microbial groups play an important role in preparing a selective medium for the mushroom, no prepared microbial 'cocktail' is used in commercial composting operations since no significant beneficial effects have yet been recorded. In recent years much attention has been paid to thermotolerant fungi which become predominant in mushroom composts in the latter stages of composting (the so-called 'conditioning phase' immediately following pasteurization). It is thought that the presence of fungi such as *Torula thermophila*; syn. *Scytalidium thermophilum* (Ross and Harris, 1983; Straatsma *et al.*, 1989) play a key role in compost selectivity favouring the mushroom during the early stages of substrate colonization.

10.2.2 Chemical changes occurring in mushroom composts due to microbial action

The organic constituents of 'mushroom compost' usually consist of a cereal straw (normally wheat), together with a manure source (normally horse or chicken), together with compost activators (generally rich in soluble carbon and nitrogen) and gypsum. For convenience of discussion, the chemical components of the starting ingredients can be sub-divided into three main fractions (Gerrits *et al.*, 1967; Hayes and Randle, 1968):

1. Carbohydrates consisting of cellulose, hemicellulose and simple sugars;
2. Organic and inorganic nitrogen sources;
3. Lignin.

An up-to-date schematic representation of the chemical transformations occurring in mushroom composts during its preparation (Smith, 1990), incorporating knowledge of humic acid synthesis (Stevenson, 1982) is shown in Figure 10.5.

During the initial stages of composting, the resident microflora utilize simple short-chain carbohydrates with the ultimate production of carbon dioxide, water and energy (Figure 10.5). The nitrogen requirement for these organisms is much less than its requirement for carbon, with only one part of nitrogen needed per 10–15 parts of carbon. Both organic and inorganic nitrogen sources are utilized by the microflora. Inorganic forms, such as ammonium or nitrate have to be converted into ammonia before being reconstituted into microbial metabolism, while organic forms such as urea, amino acids or protein are subjected to deamination reactions also resulting in the formation of ammonia.

On the depletion of the readily utilizable carbon and nitrogen sources, there is a notable change in the microbial groups. Thermophilic actinomycetes, which produce cellulases capable of degrading cellulose and hemicellulose, increase in number producing colonies that sporulate on the straw surface to give distinctive white flecks and a condition commonly termed 'fire fang' by mushroom growers. While this condition is a good indicator of the microbial succession taking place it cannot be solely used as an indicator that the compost is selective for *A. bisporus*.

Much less is known about the degradation of **lignin** during composting, although there are some notable structural changes. Waksman and McGrath (1931) were the first to pursue in depth studies on the changes that lignin undergoes during composting. They observed an association between lignin and protein and the formation of a 'lignin humus complex' which is resistant to further attack by microorganisms. Further studies to estimate the lignin content of commercial composts using exhaustive chemical techniques (e.g. 72% sulphuric acid), have confirmed the close association that occurs between lignin and protein complexes during the composting process (Gerrits, 1969; Grabbe, 1972). Although it is an accepted fact that lignin is devoid of nitrogen (Crawford, 1981), lignin determinations made using strong acids have revealed fixed nitrogen levels of between 2.5 and 4.0% of the lignin dry matter. This complex that remains insoluble in strong acid has been conveniently termed the 'nitrogen-rich lignin humus complex' (Gerrits, 1969), and as lignin is most degraded by the mushroom during the colonization period (spawn-running) it is assumed that the closely associated nitrogen is utilized along with it. It has been estimated that between 40 and 70% of the nitrogen utilized by mushroom fruit bodies comes from this complex.

Microbial biomass or the dark brown material which accumulates on straw surfaces during composting had long been inferred by research workers to be associated with mushroom nutrition (Hayes and Randle, 1968; Stanek, 1972; Smith and Spencer, 1976), but it was Eddy and

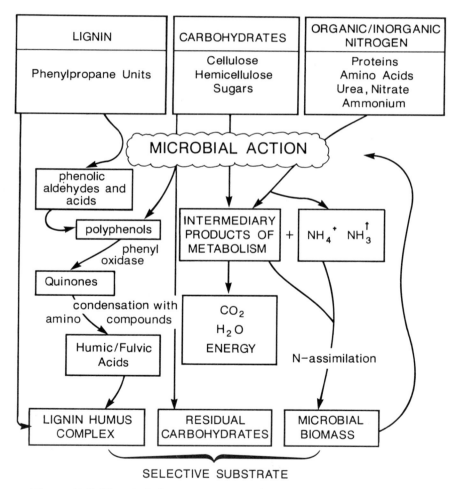

Figure 10.5 Chemical changes occurring during the composting cycle.

Figure 10.6 Electron micrograph of microbial communities present on straw surface at the completion of composting. Reproduced by permission of Horticulture Research International, Littlehampton, West Sussex.

Jacobs (1976) and later Wain (1981) who gave clear indications that the mushroom derives a large proportion of its nutriment from the accumulated biomass on the internal and external surfaces of the straw. This dark material, which builds up during the composting process, although heavily populated with microorganisms (Figure 10.6), has an amorphous matrix with the staining characteristics of bacterial polysaccharide. This material was shown to disappear rapidly during mycelial establishment by the mushroom, effecting a change in compost colour from dark to light brown. Investigations by Fermor and Wood (1981) and more recently Fermor (1988) have shown that mushroom mycelium produces a number of extracellular enzymes capable of attacking the cell walls and contents of bacteria, fungi and actinomycetes embedded in this matrix. This work demonstrates the importance of microbial protein as well as microbial polysaccharide to the nutrition of the mushroom. It is generally held that microbial protein together with nitrogen complexed to lignin are major sources of insoluble nitrogen for mushroom development.

Microorganisms colonizing mushroom compost during the composting process can therefore be regarded as **active agents** in the preparation of a nutrient substrate. They fashion the chemical composition, change the physical nature of the ingredients, contribute directly to the ultimate nutrition of the mushroom and probably the most important factor of all, they produce a selective medium on which competitor fungi cannot readily grow.

10.3 MICROORGANISMS AND FRUITBODY INITIATION

While mushroom mycelium will colonize a selective compost within 10 –14 days, no fruitbodies will be forthcoming unless the colonized compost is covered with a layer of soil (casing layer). Normally, mushroom fruitbodies appear on the beds 18–21 days after casing and four to five crops (flushes) can be expected in a 5–6-week cropping period. Flegg (1956) broadly defined the function of the casing layer as 'to induce sporophore production in quantity' and early research effort was directed at the physical and chemical characteristics of casing soils (Edwards and Flegg, 1953). Water-holding capacity, structure, porosity, salinity, pH, carbon dioxide gradients and nutrient status were all factors shown to influence fruitbody initiation, and after many years work the mushroom industry in the UK was encouraged to transfer from a traditional clay-loam subsoil to mixtures based on sphagnum peat. In addition to having properties appropriate for sporophore formation, the casing layer must also serve to protect the compost from desiccation, pest and disease attack, and to provide physical support for the developing fruit-bodies. It is also desirable that the nature of the casing material is such that no discolouration of the crop occurs, that it has a pH close to neutral, does not adhere to the harvested fruitbodies and has a physical nature able to withstand repeated watering without structural breakdown.

It was the work of Eger (1961) which changed the emphasis of study from the physical/chemical to the microbiological. She demonstrated for the first time a relationship between fruitbody initiation and the presence of bacteria, although both Mader (1943) and Stoller (1952) had proposed that fruiting was inhibited by volatile substances produced either by microorganisms or the mushroom mycelium itself. Eger (1961, 1962) demonstrated the inability of the mushroom to fruit in the absence of microorganisms. She also demonstrated that in axenic culture, the function of the microorganisms could be replaced by mixing activated charcoal into sterile casing soil mixtures. This work stimulated Hayes *et al.* (1969) to look more closely at the microorganisms involved. Populations of microorganisms isolated from unsterile casing soil were shown to contain bacteria and these bacteria when added to pure cultures of *A. bisporus* were shown to stimulate fruitbody formation. These stimulating bacteria were also shown to increase in number when cultured on a carbon-free liquid medium which had been exposed to atmospheres containing ethanol, ethyl acetate and acetone, known metabolites of mushroom. The ability to utilize these volatile chemicals was then exploited in a selective technique for isolating fruitbody stimulators which were subsequently identified as bacteria closely related to *Pseudomonas putida*. Work by Park and Agnihotri (1969) also confirmed that mushroom fruiting is stimulated by the presence of bacteria.

Later work by Hume and Hayes (1972) demonstrated that cultures of *Pseudomonas putida* stimulated primordium formation in cultures of *A. bisporus* grown on malt extract agar plates and primordium formation could also be induced using iron-containing compounds such as ethylene diamine tetra-acetic acid (Hayes, 1972). Wood (1976 a,b), failed to reproduce the stimulatory effect of *Pseudomonas putida* or iron-containing compounds on mycelial cultures of *A. bisporus*, but concluded that the role played by casing soil microorganisms was a negative one and he suggested that the mycelium produces substances that maintain the vegetative condition and the function of the resident microflora in the casing soil is to remove these inhibitory compounds. This would explain why **activated charcoal**, a known adsorbent of low molecular weight organic materials, has the apparent 'stimulatory' effect on fruitbody formation. Several compounds binding to activated charcoal during axenic culture of mushroom fruitbodies have been removed with solvents and identified by coupled gas chromatography and mass spectroscopy (Wood 1976b, 1982). Certain of these, xylene, toluene, benzaldehyde, benzyl alcohol, hexanol, iso-valeraldehyde, 1,2,4,5-tetrachlor-3,6-dimethoxybenzene, 1-octen-3-ol, octanol, 3-octanone, phenol, phenyl acetaldehyde and ethyl acetate, act as inhibitors of fruiting initiation in agar cultures but none of these have yet been positively identified as self-inhibitors since their concentrations in the casing layer are difficult to determine.

More recent work by Reddy and Patrick (1990) has shown that following spawning, stimulatory bacteria in the compost are found in their greatest numbers at the compost/casing interface. Bacterial populations in the casing layer were shown to increase beyond that found in the compost and the addition of spawned compost to the casing layer also resulted in a significant increase in bacterial numbers, and the number of fruitbodies formed and harvested.

10.4 FUTURE TRENDS

10.4.1 Disease and pest control

As mushrooms are grown on a continuous cycle throughout the year in warm protected environments with high humidities, it is not surprising that a number of pests (mainly flies, Sciarids, Phorids and Cecids) and diseases (fungal, bacterial and viral) will occur from time-to-time and cause major crop losses. While there are established practices for the control of pests and diseases (Fletcher *et al.*, 1989) public concern over the use of chemicals as control measures has led to increasing research worldwide into bio-control systems. With considerable attention to hygiene on modern farms today most fungal pathogens are well control-led and crop losses are generally minor. Of the bacterial pathogens *Pseudomonas tolaasii*, which causes brown blotches on the caps of mushrooms, is the most serious and is reported to cause crop losses between 5–10% each year in the UK (Fermor *et al.*, 1991). Unfortunately none of the commercial strains of mushroom is resistant to bacterial blotch and the casing soil is the most likely source of the pathogen. While **chemical control** measures have been identified to keep bacterial populations down in the casing material, none is registered for use within the UK.

Biological control agents using bacterial antagonists (fluorescent *Pseudomonas* spp.) have been researched in Australia (Nair and Fahy, 1976) and a commercial product 'Conquer' is now available to growers in Australasia. More recently, Fermor *et al.* (1991) in the UK have identified bacterial antagonists which have reduced blotch incidence in experimental trials by up to 50%.

Viral diseases

Virus diseases are less common today due to strict hygiene programmes operating on all farms but there are occasional outbreaks. One simple method of restricting the build-up of such a pathogen is to grow another species of mushroom, *Agaricus bitorquis* which is resistant to all the five known mushroom viruses (Atkey, 1985). A surge in popularity of this mushroom (called a virus-breaker) occurred in the 1970s when many farms throughout Europe suffered major crop losses from this disease. As the mushroom was morphologically similar to *A. bisporus* it suffered no consumer resistance and it could be grown for short periods to overcome the disease. Unfortunately, *A. bitorquis* requires an extended cropping time to produce commercially acceptable yields (Fritsche, 1978) and, as off-flavours were occasionally experienced, the popularity of this species was short-lived. Nevertheless, there are new good flavoured, high yielding strains of *A. bitorquis* re-appearing on the commercial market (Smith, 1991; Smith and Love, 1991) which could be grown as alternative crops in their own right and not solely as a virus-breakers.

Insecticides

While there are identified insecticides for the treatment of flies (White, 1985), over-use can lead to insect resistance. There is also much public concern associated with the use of pesticides on food products and this, together with the deleterious effect that such chemicals can have on ultimate yield of mushrooms, has encouraged research into alternative methods of pest control. Insect-parasitic nematodes have already proved successful against all *Diptera* larvae (Richardson and Grewal, 1991) and pathogenic bacteria *Bacillus thuringiensis* (White and Jarrett, 1990), while predatory mites (Al-Amidi *et al.*, 1989) have shown promise against sciarid larvae.

10.4.2 Advances in composting processes

The most likely major change that will affect the mushroom industry in the next decade is the transition of the composting process from basically a two-phase system to a single-phase operation completed totally indoors. Single-phase composting in controlled environments has been researched by many workers (Laborde and Delmas, 1969; Randle and Hayes, 1972; Smith and Spencer, 1976). The main objective has been to reduce the composting time thereby conserving time, energy and materials while maintaining commercially acceptable mushroom yields. More recent studies in single phase composting (Laborde *et al.*, 1986; Gerrits, 1987; Perrin and Gaze, 1987; Nair and Price, 1991) have been prompted by proposed legislation to control the release of noxious odours to the atmosphere and toxic run-offs into water courses. The long rectangular-shaped buildings used on most modern farms today for bulk pasteurization (normally 2.5 m × 2.5 m in cross section and referred to as 'tunnels'), are ideal structures in which singlephase composting can be investigated and such studies are being actively pursued in the UK, France, Holland, Italy, Switzerland and Australia. With the development of such a technology, it is highly likely that more defined compost formulations will be used and the addition of specific microorganisms, shown to have little or no beneficial effect in the traditional two-phase system of composting, can be tested under more ideal conditions. The traditional procedure of gathering partly composted manures as a base material will discontinue as mushroom composters will demand a high quality starting material and chopping and blending machines will replace the traditional turning machine. The use of more defined compost ingredients will significantly increase raw material costs but, as less dry matter is consumed by compost microorganisms during controlled environment composting, the total cost of materials utilized to prepare a given weight of compost for spawning is likely to remain unaltered.

10.4.3 Market trends

Commercial production in the UK is almost entirely composed of the white button mushroom *A. bisporus* and an efficient system of substrate preparation, environmental control and marketing has resulted in an industry providing in excess of 113 000 tonnes of mushrooms per annum (MAFF, 1988). While there has been a considerable interest in many parts of the world in other edible mushrooms, most notably *Pleurotus ostreatus* (oyster mushroom) and *Lentinus edodes* (Shiitake), these species have created only minor interest on the retail market in the UK. Supermarket chains, while failing to capture new markets with these species, have now broadened the commodity grouping by encouraging brown strains of *A. bisporus*, commonly called 'chestnut' mushrooms, to be grown and the consumption of these types is steadily increasing. This is very encouraging as there are a number of other species of *Agaricus* under investigation most notably the horse mushroom, *A. arvensis* (Fritsche, 1989), *A. silvicola* (the wood mushroom) and *A. silvaticus* (Elliott and Smith, 1988) and tropical strains of *A. bitorquis* (Smith and Love, 1991). All these new varieties, if successfully cultivated at a commercial level, would offer a greater diversity of mushrooms, i.e. shape, colour, texture and flavour, to the more demanding and adventurous cuisine of today's consumer.

REFERENCES

Abercrombie, J. (1779) in *The Garden Mushroom, Its Nature and Cultivation.* Lockyer Davis, London, pp.54.

Al-Amidi, A., Dunne, R. and Downes, M.J. (1989) Control of mushroom pests using a predatory mite. *Mushroom Journal*, **194**, 65.

Amner, W., McCarthy, A.J. and Edwards, C. (1988) Quantitative assessment of factors affecting the recovery of indigenous and released thermophilic bacteria from compost. *Applied and Environmental Microbiology*, **54**, (12), 3107–12.

Atkey, P.T. (1985) Viruses, in *The Biology and Technology of the Cultivated Mushroom*, (eds P.B. Flegg, D.M. Spencer and D.A. Wood), Wiley, Chichester, pp.241–60.

Atkins, F.C. (1966) *Mushroom Growing Today.* Faber and Faber, London.

Atkins, F.C. (1972) *Mushroom Growing Today.* 6th revised edn. republished by the Mushroom Growers Association, 1982.

Bels-Koning, H.C. (1962) Preliminary note on the analysis of the composting process. *Mushroom Science*, **5**, 30–8.

Callow, E. (1831) *Observations on methods now in use for the artificial growth of mushrooms, with full explanation of an improved mode of culture.* Fellowes, London. (Reprinted, 1965, by W.S. Maney and Son, Ltd., Hudson Road, Leeds.)

Cayley, D.M. (1938) The history of the cultivated mushroom. *Gardener's Chronicle*, **104**, (2690), 42–3.

Chang, S.T. (1991) Recent trends in world production of cultivated edible mushrooms. *Mushroom Journal*, **504**, 15–18.

Chanter, D.P. and Spencer, D.M. (1974) The importance of thermophilic bacteria in mushroom compost fermentation. *Scientia Horticulturae*, **2**, 249–56.

Crawford, R.L. (1981) *Lignin Biodegradation and Transformation.* Lockyer Davis, London.

Delcaire, J.R. (1978) Evolution de la consommation des champignons comestibles dans le monde. *Mushroom Science*, **10**, (2), 863–93.

Derikx, P.J.L. (1991) Gaseous compounds and microbial processes involved in the preparation of the substrate of *Agaricus bisporus*. Ph. D. Thesis, University of Nijmegen, Holland.

Duggar, B.M. (1905) The principles of mushroom growing and mushroom spawn making. *Bulletin of US Department of Agriculture, Bureau of Plant Industry*, **85**, 1–60.

Eddy, B.P and Jacobs, L. (1976) Mushroom compost as a source of food for *Agaricus bisporus*. *Mushroom Journal*, **38**, 56–67.

Edwards, R.L. and Flegg, P.B. (1953) Experiments with artificial mixtures for casing mushroom beds. *Mushroom Science*, **2**, 143–9.

Eger, G. (1961) Untersuchungen uber die Function der Deckschicht bei der Fruchtkorperbildung des Kulterchampignons, *Psalliota bispora* Lange. *Archiv fur Mikrobiologie*, **39**, 313–14.

Eger, G. (1962) Untersuchungen zur Frucht-korperbildung des Kulterchampignons. *Mushroom Science*, **5**, 314–20.

Elliott, T.J. (1985) Spawn-making and spawns, in *The Biology and Technology of the Cultivated Mushroom*, (eds P.B. Flegg, D.M. Spencer and D.A. Wood), Wiley, Chichester, pp.131–9.

Elliott, T.J. and Smith, J.F. (1988) Exploiting wild mushrooms, in *Science and Change in Agriculture*, (ed J.E.Y. Hardcastle), London: AFRC, pp.8–9.

Ferguson, M. (1902) A preliminary study of the germination of spores of *Agaricus campestris*. *Bulletin of US Department of Agriculture, Bureau of Plant Industry*, **16**, 1–40.

Fermor, T.R. (1988) Significance of micro-organisms in the composting process for cultivation of edible fungi, in *Treatment of Lignocellulosics with White Rot Fungi*, (eds F. Zadrazil and P. Reiniger), Elsevier Applied Science, Barking, Essex, pp.21–30.

Fermor, T.R. and Wood, D.A. (1981) Degradation of bacteria by *Agaricus bisporus* and other fungi. *Journal of General Microbiology*, **126**, 377–87.

Fermor, T.R, Randle, P.E. and Smith, J.F. (1985) Compost as a substrate and its preparation, in *The Biology and Technology of the Cultivated Mushroom*, (eds P.B. Flegg, D.M. Spencer and D.A. Wood), Wiley, Chichester, pp.81–109.

Fermor, T.R., Smith, J.F. and Spencer, D.M. (1979) The microflora of experimental mushroom composts. *Journal of Horticultural Science*, **54**, 137–47.

Fermor, T.R., Henry, M.B., Fenlon, J.S., *et al.* (1991) Development and application of a biocontrol system for bacterial blotch of the cultivated mushroom. *Crop Protection*, **10**, 271–8.

Flegg, P.B. (1956) The casing layer in the cultivation of the mushroom (*Psalliota hortensis*). *Journal of Soil Science*, **7**, 168–76.

Fletcher, J.T., White, P.F. and Gaze, R.H. (1989) *Mushrooms: Pest and Disease Control*, (eds J.T. Fletcher, P.F. White and R.H. Gaze), Intercept Ltd., Andover, Hants, England, pp.174.

Fordyce, C. (1970) Relative numbers of certain microbial groups present in compost used for mushroom (*Agaricus bisporus*) propagation. *Applied Microbiology*, **20**, 196–9.

Fritsche, G. (1978) Breeding work, in *The Biology and Cultivation of Edible Mushrooms*, (eds S.T. Chang and W.A. Hayes), Academic Press, New York, pp.371–92.

Fritsche, G. (1989) Ontwikkelingswerk met de Akkerchampignon (*Agaricus arvensis*, Schaeffer ex. secr.). *De Champignoncultur*, **33** (1), 7–13.

Gaze, R.H. (1985) Cultivation systems and their evolution, in *The Biology and Technology of the Cultivated Mushroom*, (eds P.B. Flegg, D.M. Spencer and D.A Wood), Wiley, Chichester, pp.23–41.

Gerrits, J.P.G. (1969) Organic compost constituents and water utilised by the cultivated mushroom during spawn run and cropping. *Mushroom Science*, **7**, 111–26.

Gerrits, J.P.G. (1987) Biedt composteren in tunnels perspectief? *Der Champignon-cultur*, **31** (7), 357–65.

Gerrits, J.P.G., Bels-Koning, H.C. and Muller, F.M. (1967) Changes in compost constituents during composting, pasteurisation and cropping. *Mushroom Science*, **6**, 225–43.

Grabbe, K. (1972) Vergleicherde Untersuchungen zur Spezifitat von kompostier-tem und nicht kompostiertem Champignonkultursubstrat. *Mushroom Science*, **8**, 533–52.

Hayes, W.A. (1968) Microbiological changes in composting wheatstraw/horse manure mixtures. *Mushroom Science*, **7**, 173–86.

Hayes, W.A. (1972) Nutritional factors in relation to mushroom production. *Mushroom Science*, **8**, 663–74.

Hayes, W.A. and Randle, P.E. (1968) The use of water soluble carbohydrates and methyl bromide in the preparation of mushroom composts. *MGA Bulletin*, **218**, 81–2, 87–92, 95–7.

Hayes, W.A., Randle, P.E. and Last, F.T. (1969) The nature of the microbial stimulus affecting sporophore stimulation in *Agaricus bisporus* (Lange) Sing. *Annals of Applied Biology*, **64**, 177–87.

Hume, D.P. and Hayes, W.A. (1972) The production of fruitbody primordia of *Agaricus bisporus* (Lange) Sing. in agar. *Mushroom Science*, **8**, 527–32.

Imbernon, M. and Leplae, M. (1972) Microbiologie des substrats destines à la culture du champignon de couche: dénombrements, groupements fonction-nels. *Mushroom Science*, **8**, 363–81.

Kleyn, J.G. and Wetzler, T.R. (1981) The microbiology of spent mushroom compost and its dust. *Canadian Journal of Microbiology*, **27**, 748–53.

Laborde, J. and Delmas, J. (1969) La preparation rapide de substrats. *Bulletin de la federation nationale des syndicats agricoles des cultivateurs de champignons*, **184**, 2093–109.

Laborde, J., Olivier, J.M., Houdeau, G., *et al.* (1986) Indoor static composting for mushroom (*Agaricus bisporus* [Lange] Sing.) cultivation, in *Cultivating Edible Fungi*, (eds P.J Wuest, D.J. Royse and R.B. Beelman), Elsevier, pp.91–100.

Lacey, J. (1973) Actinomycetes in soils, composts and fodders, in *Actinomycetales: Characteristics and Practical Importance*, (eds G. Sykes and F.A Skinner), Acade-mic Press, London, pp.231–51.

Lambert, E.B. (1941) Studies on the preparation of mushroom compost. *Journal of Agricultural Research*, **62**, 415–22.

Mader, E.O. (1943) Some factors inhibiting the fructification and production of the cultivated mushroom *Agaricus campestris*. *Phytopathology*, **33**, 1134–45.

MAFF (1988) *Basic Horticultural Statistics for the United Kingdom – Calendar and Crop Years 1979–1988*. Ministry of Agriculture, Fisheries and Food Statistics (Agricultural Commodities) Division, Branch B.

Nair, N.G. and Fahy, P.C. (1976) Commercial application of biological control of mushroom bacterial blotch. *Australian Journal of Agricultural Research*, **27**, 415–22.

Nair, N.G. and Price, G. (1991) A composting process to minimise odour pollution. *Mushroom Science*, **13**, 205–6.

Park, J.Y. and Agnihotri, V.P. (1969) Bacterial metabolites trigger sporophore formation in *Agaricus bisporus*. *Nature (Lond.)*, **222**, 984.

Perrin, P.S. and Gaze, R.H. (1987) Controlled environment composting. *Mushroom Journal*, **174**, 195–7.

Pizer, N.H. (1937) Investigations into the environment and initiation of the cultivated mushroom. Part 1. Some properties of composts in relation to growth of mycelium. *Journal of Agricultural Science*, **27**, 349–76.

Randle, P.E. and Hayes, W.A. (1972) Progress in the experimentation on the efficiency of composting and compost. *Mushroom Science*, **8**, 789–95.

Reddy, M.S. and Patrick, Z.A. (1990) Effect of bacteria associated with mushroom compost and casing materials on basidiomata formation in *Agaricus bisporus*. *Canadian Journal of Plant Pathology*, **12**, 236–42.

Rettew, G.R., Gahm, O.E. and Floyd, W.D. (1941) *Manual of Mushroom Culture*, Chester County Mushroom Laboratories, West Chester, PA.

Richardson, P.N. and Grewal, P.S. (1991) Comparative assessment of biological (Nematoda: *Steinernema feltiae*) and chemical methods of control of the mushroom fly *Lycoriella auripila* (Diptera: Sciaridae). *Biocontrol Science Technology*, **1**, 217–28.

Rosenburg, S.L. (1975) Temperature and pH optima for 21 species of thermophilic and thermotolerant fungi. *Canadian Journal of Microbiology*, **21**, 1535–40.

Rosenburg, S.L. (1978) Cellulose and lignocellulose degradation by thermophilic and thermotolerant fungi. *Mycologia*, **70**, 1–13.

Ross, R.C. and Harris, P.J. (1983) The significance of thermophilic fungi in mushroom compost preparation. *Scientia Horticulturae*, **28**, 37–45.

Sinden, J.W. (1952) *Grain Spawn, its Nature, Advantages and Use*. Technical leaflet, Champignon Laboratory, Gossau, Zurich.

Sinden, J.W. and Hauser, E. (1950) The short method of mushroom composting. *Mushroom Science*, **1**, 52–9.

Sinden, J.W. and Hauser, E. (1953) The nature of the composting process and its relation to short composting. *Mushroom Science*, **2**, 123–31.

Smith, J.F. (1990) Composted substrates used in the cultivation of the white button mushroom, *Agaricus bisporus*. *Aspects of Applied Biology*, **24**, 131–43.

Smith, J.F. (1991) A hot weather mushroom AGC W20. *Mushroom Journal*, **501**, 20–1.

Smith, J.F. and Love, M.E. (1991) Identification of environmental requirements for the commercial culture of a tropical *Agaricus* strain. *Mushroom Science*, **13**, 601–10.

Smith, J.F. and Spencer, D.M. (1976) Rapid preparation of composts suitable for the production of the cultivated mushroom. *Scientia Horticulturae*, **5**, 197–205.

Stanek, M. (1972) Micro-organisms inhabiting mushroom compost during fermentation. *Mushroom Science*, **8**, 797–811.

Stevenson, F.J. (1982) Biochemistry of the formation of humic substances, in *Humus Chemistry: Genesis, Composition, Reactions*, Wiley, New York, pp.195–220.

Stoller, B.B. (1952) Studies on the function of the casing for mushroom beds. *MGA Bulletin*, **34**, 289–97.

Straatsma, G., Gerrits, J.P.G., Augustijn, M.A.M., *et al.* (1989) Population dynamics of *Scytalidium thermophilum* in mushroom compost and stimulatory effects on growth rate and yield of *Agaricus bisporus*. *Journal of General Microbiology*, **135**, 751–9.

Tounefort, J de. (1707) Observations sur la naissance et sur la culture des champignons. *Memoires de l'Academie Royale des Sciences*, **1707**, 58–66.

Wain, D.I. (1981) Investigation of the nutrition of the mushroom *Agaricus bisporus* (Lange) Singer in compost. Ph.D. Thesis, University of Bath.

Waksman, S.A. and McGrath, J.M. (1931) Preliminary study of chemical processes involved in the decomposition of manure by *Agaricus bisporus*. *American Journal of Botany*, **8**, 573–81.

White, P.F. (1985) Pests and Pesticides, in *The Biology and Technology of the Cultivated Mushroom*, (eds P.B. Flegg, D.M. Spencer and D.A. Wood), Wiley, Chichester, pp.279–93.

White, P.F. and Jarrett, P. (1990) Laboratory and field tests with *Bacillus thuringiensis* for the control of the mushroom sciarid *Lycoriella auripila*. *Proceedings of the 1990 Brighton Crop Protection Conference – Pest and Diseases*, pp.373–8.

Wood, D.A. (1976a) Primordium formation in axenic cultures of *Agaricus bisporus* (Lange) Sing. *Journal of General Microbiology*, **95**, 313–23.

Wood, D.A. (1976b) Sporophore initiation in axenic cultures. *Annual Report of the Glasshouse Crops Research Institute, 1975*, 115.

Wood, D.A. (1982) Sporophore initiation in axenic cultures. *Annual Report of the Glasshouse Crops Research Institute 1981*, 140.

Wright, J. (1894) *Mushrooms for the Million*, 7th edn, 'Journal of Horticulture' Office, Fleet Street, London.

11

The exploitation of microorganisms in the processing of dairy products

A.H. Varnam

11.1 INTRODUCTION

The manufacture of cheese, yoghurt and other fermented milks and some types of butter depends on the activity of starter microorganisms. The most important of these, are species of *Lactobacillus*, *Lactococcus*, *Leuconostoc* and *Streptococcus* which form part of a group commonly referred to as lactic acid bacteria (LAB). Recently, use of a fifth member, *Pediococcus*, has been proposed (Tzanetakis *et al.*, 1991). A further recent development, stimulated by increased interest in the therapeutic properties of fermented milks is the use of the intestinal organism *Bifidobacterium* in starter cultures. Properties of starter bacteria are summarized in Table 11.1.

Yeasts are used in conjunction with LABs in production of two fermented milks, koumiss and kefir. *Kluyveromyces marxianus* var. *marxianus* and *K. marxianus* var. *lactis* are used as starter cultures for koumiss, while the kefir grain (refer to page 283) contains *Candida kefyr* together with one or more other yeasts.

Microorganisms other than starter species are involved in the ripening of many species of cheese. It should be appreciated that while these are important in imparting characteristic organoleptic properties, their role is strictly secondary to that of starter organisms.

Exploitation of Microorganisms Edited by D.G. Jones
Published in 1993 by Chapman & Hall, London ISBN 0 412 45740 7

Table 11.1. Bacteria used as starter microorganisms

Lactic acid bacteria	
Lactobacillus	Rod-shaped, grow at 45°C but not 10°C. Do not metabolize citrate.
L. delbrueckii ssp. *bulgaricus*	Does not produce ammonia from arginine; does not ferment galactose.
ssp. *lactis*	Some strains produce ammonia from arginine; some ferment galactose.
L. helveticus	Does not produce ammonia from arginine. Ferments galactose.
Lactococcus	Coccal-shaped, grow at 10°C but not at 45°C. Ferment galactose.
L. lactis ssp. *cremoris*	Does not produce ammonia from arginine or grow at 40°C. Does not metabolize citrate.
ssp. *lactis*	Produces ammonia from arginine and grows at 40°C. Does not metabolize citrate.
ssp. *lactis* biovar *diacetylactis*	Some strains produce ammonia from arginine and some grow at 40°C. Metabolizes citrate.
Leuconostoc mesenteroides	Coccal-shaped. Does not produce ammonia from arginine. Grows at 10°C but not at 40°C.
ssp. *cremoris*	Metabolizes citrate and ferments galactose.
Streptococcus salivarius ssp. *thermophilus*	Coccal-shaped, grows at 45°C but not at 10°C. Does not produce ammonia from arginine or metabolize citrate. Does not ferment galactose.
Bifidobacterium	Irregular rod-shaped, occasional branching in some species. Possess unique enzyme fructose-6-phosphate phosphoketolase. Require specfic growth factors ('bifidogenic factors'). Metabolize fructo-oligosaccharides.

Notes: 1. Bacteria other than those described may occasionally be used as starter cultures.
2. Non-starter members of the genera described may have different properties.

11.2 STARTER MICROORGANISMS IN MANUFACTURE OF CHEESE AND FERMENTED MILK

They key role of starter microorganisms is the production of **lactic acid** by fermentation of lactose. Lactic acid is responsible for the fresh acidic flavour of fermented milks and is of importance in the formation and texturizing of the curd during cheese manufacture. There are major differences in lactose metabolism between the different lactic acid bacteria used as starter cultures, while metabolism in *Bifidobacterium* is via the unique fructose-6-phosphate fructoketolase pathway. In *Lactococcus*, lactose is transported into the cell by the phosphoenol-pyruvate phosphotransferase system, involving the simultaneous formation of lactose-phosphate. Lactose-phosphate is hydrolysed by phospho-β-galactosidase to glucose and galactose-6-phosphate which are metabolized by the glycolytic and tagatose pathways respectively.

In *Lactobacillus, Leuconostoc* and *S. salivarius* sub-sp. *thermophilus*, lactose enters the cell intact and is hydrolyzed by β-galactosidase to glucose and galactose. *Lactobacillus* and *S. salivarius* ssp. *thermophilus* metabolize glucose by the glycolytic pathway and *Leuconostoc* by the phosphoketolase pathway. In most cases galactose is metabolized via glucose-1-phosphate in the Leloir pathway. Galactose cannot, however, be metabolized by *Lb. bulgaricus*, most strains of *S. salivarius* ssp. *thermophilus*, or some strains of *Lb. delbrueckii* ssp. *lactis* and is excreted by these bacteria.

Starters play important roles in addition to lactic acid production. These include the production of volatile flavour compounds such as diacetyl from citrate and acetaldehyde from threonine or sugars, the synthesis of proteolytic and lipolytic enzymes (important in the ripening of cheese), the production of body and texture in some fermented milks, the production of compounds characteristic of particular products such as ethanol in kefir and koumiss, and the suppression of pathogenic and some spoilage microorganisms. In addition some starter microorganisms may have probiotic or therapeutic properties.

11.2.1 Genetics of starter bacteria

Many starter bacteria contain **plasmid** as well as chromosomal genes. *Lactococcus lactis* sub-sp. *lactis* and *L. lactis* ssp. *cremoris*, for example, both contain a large and complex plasmid complement, typically ranging from four to seven. In contrast, *S. salivarius* ssp. *thermophilus* contains only a small number of plasmids or is plasmid-free (Gasson and Davies, 1984). The plasmid complement of *Lactobacillus* varies considerably, many strains containing none or only one plasmid, although at least six different plasmids are present among strains of *Lb. delbrueckii* ssp. *lactis* (Casey and Jimeno, 1989). *Leuconostoc* has two plasmids and multiple plasmids have also been reported in *B. longum* and *B. breve* (Iwata and Morishita, 1989).

Plasmid-borne genes are of particular importance because they code for many properties essential for successful starter activity (Table 11.2). Loss of plasmids partly explains the instability of some of these essential properties (Gasson and Davies, 1984). Molecular analysis of plasmid-borne genes is also relatively straightforward and it is now possible to

Table 11.2. Plasmid-borne, technologically important properties of starter bacteria.

Lactose fermentation	Lactose plasmids identified in most starter bacteria. Possible that lactose genes may be either plasmid or chromosomal in *L. lactis* ssp. *cremoris*.
Proteinase	Cell-wall proteinases recognized as plasmid mediated. Probable that both lactose and proteinase genes are encoded by same plasmid in *L. lactis* ssp. *lactis* but genetic linkage often broken by deletion.
Citrate metabolism	Citrate transport by *L. lactis* ssp. *lactis* biovar. *diacetylactis* plasmid mediated.
Antagonistic properties	Some, but not all, antagonistic properties plasmid mediated.
Phage resistance	At least three resistance mechanisms in *Lactococcus* plasmid mediated. Some may be encoded on same plasmid.

apply the knowledge of the genetics of lactic acid bacteria to improve existing or produce new strains. Conventional techniques such as conjugation are used alongside recombinant DNA technology. Progress is also being made in the directed integration of genetic material into the bacterial chromosome, a technique which may improve the stability of genetically manipulated cultures (Gasson, 1990). Work is naturally concentrated in areas offering the greatest technological benefit, including bacteriophage resistance, proteolysis and flavour generation and production of antagonistic compounds. A further important area is the stabilization of properties of fundamental importance, such as lactose metabolism.

The use of genetically manipulated starter cultures requires that special conditions are fulfilled. Vectors must use only DNA that is natural to lactic acid bacteria approved for food use and a selectable marker that does not compromise human drug therapy. Such a vector developed by Froseth and McKay (1991) uses nisin resistance as selectable marker.

11.2.2 Antagonistic properties

Suppression of undesirable microorganisms is an important function of starter cultures and, while a major role is played by lactic acid, it is known that a number of specific **antagonists** are also produced. The best known is the polypeptide antibiotic **nisin**, produced by strains of *L. lactis* ssp. *lactis*. In recent years there has been considerable interest in the use of starter microorganisms which produce antagonists and which may be targeted against pathogens or against spoilage microorganisms. Doubts persist as to the efficacy of antagonists in food systems but, in at least some circumstances, appear to be effective in controlling the spoilage microflora of more perishable products such as cottage cheese. Recent years have also seen the development of Microgard$^{(TM)}$, produced by starter microorganisms growing in skim milk, which may be effectively used as an additive to control the spoilage microflora of cottage cheese and yoghurt (Salih *et al.*, 1990). The nature of the inhibitor is not fully understood, but is not now believed to be a bacteriocin.

Problems can arise in mixed starter cultures if one component produces bacteriocins to which other components are sensitive. The consequences are likely to be particularly serious if the dominant strain is phage-sensitive and could result in starter failure (Piord *et al.*, 1990).

11.2.3 Probiotic and therapeutic properties

In recent years there has been a revival of interest in starter microorganisms as probiotic and therapeutic agents, with a number of products with alleged health-promoting properties having been developed in Japan and continental Europe. Despite a considerable degree of scepticism among both the public and the scientific community, there is evidence of beneficial effects under some circumstances. These involve different mechanisms, the benefit of which is restricted to certain individuals rather than to the population as a whole:

1. Maintenance and restoration of normal intestinal microflora balance.
2. Suppression of specific enteric pathogens such as *Salmonella*.
3. Alleviation of lactose maldigestion.
4. Anti-carcinogenic activity.
5. Reduction of serum cholesterol level.
6. Nutritional enhancement.

11.2.4 Types of starter cultures

Starter cultures containing lactic acid bacteria may be classified in a number of ways. The microorganisms themselves are referred to as either **mesophilic**, which have an optimal growth temperature of *ca.* 30°C, or **thermophilic**, which have an optimal growth temperature of 40 to 45°C. *Lactococcus* and *Leuconostoc* are mesophilic, while *S. salivarius* ssp. *thermophilus* and the more widely used *Lactobacillus* species are thermophilic.

Both mesophilic and thermophilic starters may be used as mixed cultures in which the number of strains is unknown, or as defined cultures which contain a known number of strains. Mesophilic mixed cultures, traditionally used in northern Europe, may be further classified on the basis of the presence, or absence, of the flavour-producing organisms *Leuconostoc* species and *L. lactis* ssp. *lactis* biovar *diacetylactis*.

Mesophilic defined cultures may consist of single, paired, or multiple strains. Multiple starters originally consisted of six·strains, but in recent years it has been found that the number may be reduced to two or three without reducing the effectiveness. This obviously facilitates starter handling. Traditional thermophilic starters were also of the mixed strain type but, in large-scale usage, these have been largely replaced by single or multiple defined strains.

11.2.5 Production of starter cultures

In modern large-scale usage, cultures are usually supplied by a commercial laboratory in either dried or liquid forms, or as a frozen concentrate. In some cases these cultures are supplied as stocks and propagated through mother and feeder stages, but it is more convenient to use cultures which can be inoculated direct into the bulk tank, or direct into the cheese milk (direct-in-vat inoculation).

Bacteriophage infection

The most important single aspect of the production of bulk starters is the protection of the culture from bacteriophage. Two main approaches are taken, namely cultivation in **phage-inhibitory media** (PIM) and the use of mechanically protected systems.

The development of PIM followed observations that phage require free calcium ions for multiplication (Reiter, 1956). Most PIM are whey-based and used with pH control which may be applied internally using buffer salts or externally using NH_4OH. Whey-based, pH-controlled PIM are cheap to produce, effective in control of phage and may be used with both mesophilic and thermophilic cultures (Cogan and Accolas, 1990). The use of pH-controlled media has further advantages – daily fluctuations in acid development in the bulk culture are minimized and cellular damage caused by low pH reduced. This latter factor means that the starter remains active for a longer period and is of higher cell count (Tamime, 1990).

Mechanically protected systems are of various types, but all are based on the principles of semi, or fully aseptic inoculation of the bulk starter and protection of the tank by maintaining a slight positive pressure and air filtration or sterilization. Antibiotic-free milk or reconstituted skim milk powder is most commonly used as growth medium in mechanically protected systems.

The use of either PIM or mechanically protected systems can only be effective if **good practice** and **strict control of hygiene** is employed in all aspects of starter handling and, indeed, throughout the factory. Lactococcal phages are of markedly greater heat resistance than their host cells (Cogan and Daly, 1987) and bulk starter media must be heated at 90°C for a minimum of 30 minutes. High heat treatment has the additional advantage of minimizing phage development due to reduction in the level of available Mg^{2+} ions.

Direct-in-vat inoculation minimizes the risk of phage infection and variations in starter performance. The process requires the concentration of starter cells, separation from their metabolites and preservation. Concentrated freeze-dried cultures are most commonly used and have been successfully applied to the production of cheese and yoghurt. Cultures preserved in liquid nitrogen are also available, but are less convenient in use.

11.2.6 Starter failure

Starter failure occurs when growth of and acid production by the starter microorganisms is significantly slower than the accepted norm. The consequences are often discussed in economic terms, but starter failure is also important from the public health viewpoint and can result in growth and enterotoxin production by *Staphylococcus aureus* and enhanced survival of *Salmonella*.

Starter failure has many causes of which bacteriophage infection is currently of greatest importance, earlier problems due to use of insufficient or inactive starter culture having largely been eliminated from modern industrial production.

Bacteriophage activity

The majority of phages isolated from starter cultures of LAB are of Bradley group B (Bradley, 1967), with a smaller number of group A. Phages have been isolated from all starter genera, although studies have tended to concentrate on *Lactococcus*.

Phage–host relationships for LAB are the same as for other bacteria. Two types of phage exist, the **virulent** (lytic) and the **temperate**, although meaningful distinction between the two is not always possible (Gilmour and Rowe, 1990). Virulent and temperate phages have different relationships with their hosts, the lytic cycle and lysogeny respectively.

The **lytic cycle** is commercially of greater significance and can result in total starter failure. Following infection, multiplication takes place in four sequential stages: adsorption, DNA injection, intracellular phage multiplication and phage release. The latent period, between infection and phage release is usually *ca.* 60 minutes and the number of phages released as high as 200 per cell (Cogan and Accolas, 1990). Infected starters may thus lose activity very quickly.

Temperate phages produce two responses in host cells. In most, the phage behaves like a virulent phage causing cell lysis but, in a small number, the phage is integrated into the host chromosome (prophage) or is maintained as a plasmid. The two latter states are very stable, the phage replicating with the host cell without causing lysis. Prophage integration also confers immunity (lysogenic immunity) from infection with virulent phage of the same, or closely related type. The stability of the prophage is dependent on the presence of a **repressor protein** which blocks the expression of phage growth genes. The repressor protein is inactivated by ultra-violet light or mitomycin C which may be used to induce the prophage and detect lysogeny. Spontaneous induction occurs at a low level and most lysogenic cultures contain free phage. **Lysogeny** is widespread among LAB and it is likely that many virulent phages currently present in factories evolved from temperate phage present in earlier starter cultures. In French factories, lysogenic artisan cultures are considered to be a continuing source of new virulent phage (Sechaud *et al.*, 1989).

A third condition, the **carrier state**, occurs when a culture is in equilibrium with the phages present in its own environment. Free phages are present and phage DNA is not incorporated into the host chromosome, but only a proportion of host cells are infected (de Vos, 1989). Cultures are still capable of fast acid production (Lodics and Steenson, 1990) and the carrier state appears to be common in the mixed strain cultures traditionally used in the Netherlands. Such cultures may themselves be a source of virulent phage.

Resistance is promoted by the presence of virulent phage (Sandine, 1989) and bacteriophage-insensitive mutants (BIMs), of major importance in many modern cheese making systems, may be selected by treatment of sensitive strains by exposure to phage. Three main mechanisms exist: prevention of phage adsorption, abortive infection and restriction/modification. Lysogenic immunity is also an important control mechanism (Cogan and Accolas, 1990).

Many BIMS are slow acid producers (prt$^-$), although phage resistance and the prt$^-$ phenotype are not necessarily genetically linked (Stadhouders *et al.*, 1988). Fast acid producing (prt$^+$) BIMs have also been isolated (Huggins and Sandine, 1984).

Other causes of starter failure

Agglutinins were previously recognized as a problem in cottage cheese manufacture only, but it is possible their importance in manufacture of other products has been underestimated. Agglutinins cause starter microorganisms and casein micelles to aggregate, leading to uneven distribution, slow acid production and, in cheese, a poor quality, 'grainy' curd. Homogenization of the bulk starter has been recommended to minimize problems (Milton *et al.*, 1990).

Antibiotic residues were previously a major cause of starter failure but, in most countries, the problem has been largely eliminated. Sanitizer residues are a continuing problem which can be eliminated by ensuring adequate rinsing of equipment after cleaning. Sensitivity to sanitizers varies greatly according to strain and species but, in practice problems are potentially greatest with quaternary ammonium compounds (Guirguin and Hickey, 1987a).

Starter failure may occur in spring or late lactation milk due to enhanced activity of the **lactoperoxidase system**. Inhibitory effects are strain-dependent, Guirguin and Hickey (1987b) suggesting that resistance to the lactoperoxidase system should be a factor in the selection of starter strains.

11.3 STARTER TECHNOLOGY IN MANUFACTURE OF YOGHURT AND OTHER FERMENTED MILKS

A very wide range of fermented milks are made but in most cases the technology is similar, differences lying in the nature of the starter microorganisms (Table 11.3) and the total solids content of the milk (Robinson and Tamime, 1990).

11.3.1 Yoghurt

Depending on the system of manufacture and the nature of the coagulum, yoghurts may be classified as being of two main types, **set** or **stirred**. Yoghurt may be made from whole or reduced fat milk but it is common commercial practice to fortify the solids content to *ca.* 15%. This improves the body of the final yoghurt and slightly reduces acid production during fermentation, resulting in the less acid product now generally preferred. Milk is homogenized and heat treated, the latter being of importance in control of pathogens.

Fermentation

Fermentation almost invariably involves the combined growth of *Lb. bulgaricus* and *S. salivarius* ssp. *thermophilus*. The relationship

Table 11.3. Starter microorganisms used in different fermented milks

Acidophilus milk	*Lb. acidophilus*
Buttermilk (acid)	*Lb. bulgaricus*
Buttermilk (cultured)	*L. lactis* ssp. *lactis* with some, or all, of *L. lactis* ssp. *lactis* biovar *diacetylactis*, *L. lactis* ssp. *cremoris*, *Leuc. mesenteroides* ssp. *cremoris*
Sour (cultured) cream	*S. lactis* ssp. *lactis* biovar *diacetylactis* with *S. lactis* ssp. *lactis* or ssp. *cremoris*
Kefir (traditional)	Kefir grains (refer to page 283)
Koumiss	*Lb. acidophilus* or *Lb. bulgaricus* and *K. marxianus* var *marxianus* or var. *lactis*
Scandinavian slimy milks (Filmjolk, villi, etc.)	*S. lactis* ssp. *lactis*, *S. lactis* ssp. *lactis* biovar *diacetylactis*, *Leuc. mesenteroides* ssp. *cremoris*
Yoghurt (coventional, including labneh, etc.)	*S. salivarius* ssp. *thermophilus* and *Lb. bulgaricus*
Yoghurt (therapeutic, various types)	*Lb. acidophilus*, *Lb. bulgaricus*, *Lb. casei*, *B. bifidum*, *B. longum* in various combinations or, in some cases, singly.

between these organisms is synergistic, *Str. salivarius* ssp. *thermophilus* being stimulated by amino acids and peptides released from casein by *Lb. bulgaricus* which, in turn, is stimulated by formic acid produced by the streptococci. Mixed strain cultures are used in traditional manufacture but, in modern factories, the starters are usually single or multiple strain defined cultures used in rotation (Cogan and Accolas, 1990). Bacteriophage-insensitive mutants are used in Australia (Hull, 1983) although problems due to phage are relatively rare. A fermentation temperature of 40 to 42°C is used and an addition of starter at 2% (v/v) permits the fermentation stage to be completed within four hours, at which point the acidity will be 0.90 to 0.95%. In production of set yoghurt, the fermentation is carried out in the final retail container, but stirred yoghurt is fermented in bulk, the coagulum being broken before cooling and packaging. Starter microorganisms have an important role in producing a desirable smooth, viscous consistency through the production of slime. This is composed of amino-sugar containing extracellular polysaccharide (Doco *et al.*, 1991), the structure of which may vary from strain to strain (Robinson, 1988).

Fruit, fruit flavourings and **sweeteners** are often incorporated into yoghurts. In the set type, additional ingredients must be incorporated into the milk and sweeteners can significantly lower the water activity level. This favours *S. salivarius* ssp. *thermophilus* over *Lb. bulgaricus* and can lead to quality problems resulting from unbalanced growth (Larsen and Anon, 1990).

The production of **therapeutic yoghurts** involves the same basic technology as the conventional product. *Lb acidophilus, B. bifidum* or *B. longum* and, less commonly, *Lb. casei* are used and may be grown alone in milk or in the presence of normal starter bacteria. Acid production is slow in the absence of the normal starter bacteria and strict precautions against contamination and overgrowth by undesirable bacteria must be taken. Criteria for selecting strains of *Lactobacillus* and *Bifidobacterium* for use in therapeutic yoghurt differ from the criteria used for selection of conventional starters. It is essential that such strains are able to survive transit through the stomach, remain active in the presence of bile and are able to colonize the intestine (Hughes and Hoover, 1991). It is considered that the minimum number of cells which must be present to have therapeutic effects is 10^6 viable cells per ml and that products should not be described as having health-promoting properties unless this number of cells are present at point of sale. Considerable strain variation has been demonstrated in survival attributes and some commercial products contain strains of *Bifidobacterium* which neither survive gastric transit, nor produce acidity during storage (Berroda *et al.* 1991).

In addition to yoghurt, a number of *Bifidobacterium*-containing dairy products have been developed in which the organism is cultured separately and incorporated into the non-fermented product before final processing. Examples of such 'bifid-amended' products include milk, milk powder and frozen desserts.

11.3.2 Other fermented milks

The starter cultures used in two types, kefir and Scandinavian ropy milks are of particular interest and are discussed in further detail below.

In traditionally made **kefir**, the starter consists of **kefir grains**. These are gelatinous granules some 2 to 15 mm in diameter which consist of a mixture of microorganisms grouped in a highly organized manner. The microorganisms present vary but may include *L. lactis* ssp. *lactis* and ssp. *cremoris, Lb. acidophilus, Lb. kefir, Lb. kefiranofaciens, Lb. casei, Candida kefyr, Kluyveromyces marxianus* var. *marxianus* and species of *Saccharomyces* including *Sacch. cerevisiae*. Providing sub-culturing is on a regular basis, the grains are able to proliferate in milk over many generations, without detectable changes in character or properties. Kefir grains appear to consist of a matrix of which *ca.* 50% is the carbohydrate, kefiran. Kefiran is produced by *L. kefiranofaciens* in the centre of the grain where growth is favoured by anaerobic conditions and the presence of ethanol (Arihara *et al.*, 1990). *L. kefiranofaciens* is thus responsible for propagation of the grains which does not occur in the absence of this organism, although non-propagable grains retain kefir-producing capacity.

Scandinavian ropy milks have a characteristic texture which is both sticky and yet easily cut by a spoon. Capsule-forming strains of *L. lactis* sub-species are responsible for slime production, the chemical nature of which appears to vary but includes lipotechoic acid and phospho-polysaccharides (Toba *et al.*, 1991). It is unlikely that capsular slime alone is responsible for the characteristic texture of ropy milks and it is probable that the slime forms a network which enmeshes the cells of the starter microorganisms and links protein micelles to form casein conglomerates (Toba *et al.*, 1990). One type of ropy milk, the Finnish villi, is the only mass produced fermented milk which is ripened by a non-starter organism, the mould *Geotrichum candidum*.

11.4 STARTER TECHNOLOGY IN PRODUCTION OF SOFT CHEESE

Soft cheese is a large group characterized by a high moisture content (50 to 80%). Cream, whole milk and partially, or fully, skimmed milk may be used according to the variety. Milk concentrated by reverse osmosis and ultrafiltration is also used for some types. Curds are subject to little, or no, cutting or scalding and are unpressed, drainage of the whey being by gravity. Soft cheese is of two basic types, **acid-set**, in which the coagulum results entirely from lactic acid produced by starter microorganisms and **renneted**, in which the coagulum results from the combined effects of the low pH value and rennet (or rennet substitute). Mixed strain starters are usual, but defined starters have been developed for large-scale production of cheeses such as Quarg. Manufacture of cheese of all types is summarized in Figure 11.1.

11.4.1 Acid-set cheeses

Acid-set cheeses are technologically simple and are typified by lactic cheese. Cottage cheese made by the commonly used short-set process is also an acid-set cheese, although rennet is added in the long-set. Cottage cheese differs from other soft cheese in that the curd particles are left separate and retain their identity. The curd is blended with a cream dressing made from cream ripened with *L. lactis* sub-sp. *lactis* biovar *diacetylactis* and *Leuc. cremoris*. Separate fermentation of cheese and dressing is required to prevent defects due CO_2 production by *L. lactis* sub-sp. *lactis* biovar *diacetylactis*.

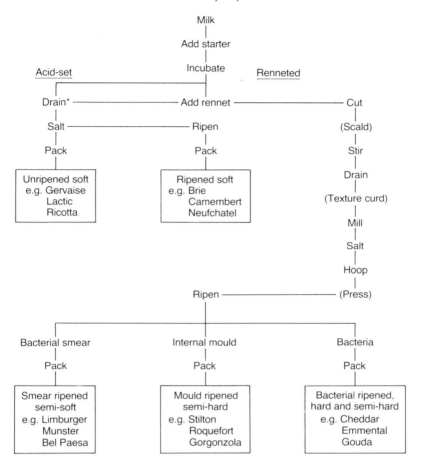

Figure 11.1 Simplified procedure for the manufacture of various types of cheese.

*Manufacture may involve light cutting and scalding. Note: Stages in parentheses are not involved in the manufacture of some varieties.

11.4.2 Renneted cheeses

The manufacture of renneted cheese depends entirely on the unique sensitivity of the Phe_{105}-Met_{106} bond of κ-casein to hydrolysis by the active enzymes present, acid proteinases (Fox, 1987). The major active ingredient of rennet, chymosin, is secreted in the bovine abomasum (fourth stomach) which is traditionally the main source of supply. Alternative sources of suitable proteolytic enzymes have been investigated and while many problems have been encountered, proteinases from species of *Mucor*, especially *M. miehei*, have found widespread application. More recently, chymosin produced by genetically engineered microorganisms has been introduced. Acid production by starter microorganisms is important in determining the properties of the coagulum and hence the quality of the final cheese.

Ripening of cheeses

Renneted soft cheeses may be further sub-divided into **unripened** and **ripened**. Unripened cheeses are ready for consumption immediately after manufacture and quality deteriorates relatively quickly on storage. In these cheeses, the sharp lactic acid flavour is predominant. Ripening of cheese is a complex process which involves enzymes derived from starter microorganisms, non-starter microorganisms, the milk and rennet. Non-starter microorganisms may either be added as part of the cheese-making process or be adventitious contaminants. Soft cheese made with unpasteurized milk contains a complex adventitious microflora which can contribute to both desirable and undesirable processes during ripening, but the role of this microflora and of milk enzymes is much reduced where the cheese milk is pasteurized before starter inoculation. Cheese made with unpasteurized milk is considered to be an inherently high risk product.

Added non-starter microorganisms are of major importance in the ripening of soft cheese, *Penicillium camembertii* being used to ripen the two major types, Camembert and Brie. Both types are surface ripened, *P. camembertii* developing following initial growth of adventitious film yeasts (*Mycoderma*) and *Geotrichum*. Ultimately, the mould develops over the whole cheese and is largely responsible for the characteristic body and flavour. Following mould growth, *Brevibacterium linens* may develop, especially in the high moisture content Brie, and impart a reddish colour to the curd. The major roles of *P. camembertii* are the production of free fatty acids and their resultant oxidation to methyl ketones such as 2-nonnone and 2-undecanone, which are responsible for correct flavour balance (Law, 1984), and the development of texture. Texture development does not directly involve proteolytic mould enzymes, but the deacidification of the cheese surface coupled with chymosin-mediated casein degradation (Noomem, 1983). There is probably also a contribution from other proteolytic enzymes present (Law, 1987). In addition to textural changes, amino acids are produced, catabolism of which *via* decarboxylation, deamination, desulphurylation or demethiolation produces flavour impact compounds, including ammonia, and their precursors.

11.5 STARTER TECHNOLOGY IN MANUFACTURE OF HARD, SEMI-HARD AND SEMI-SOFT CHEESE

These types of cheese have a water content in the range 26 to 50%, there being considerable overlap between the categories. Many varieties of each type exist, but in all cases differ from soft cheeses by the extent of heating, cutting and pressing the curd receives (Figure 11.1).

11.5.1 Hard cheeses

Cheddar is by far the most important variety and is produced on a world-wide basis. Whole milk is normally used, although low-fat cheddar is now available; starter cultures are of the mesophilic type, most containing *L. lactis* ssp. *cremoris* alone, or in combination with *L. lactis* ssp. *lactis*. In many countries, manufacture is on a very large scale and this has necessitated the development of special starter management systems to overcome problems due to phage. The initial approach to solving the problems of slow acid production due to phage involved the use of pairs of phage-unrelated strains in a four-day rotation; a different pair being used each day. This system worked well until the 1970s when the development of intensive, automated cheese making, often requiring the vats to be filled twice a day, led to much greater problems with phage. The problem has been much reduced by the development of defined multiple strain starters. Following pioneering work in New Zealand these are now used in Australia, the United States and Eire. Systems of use vary somewhat, but in all cases development of multiple strain starters is dependent on the use of a simple test for predicting phage resistance (Heap and Lawrence, 1976). Potential strains are then screened, over several growth cycles, for resistance to as large a number of phage as practical. Whey from previous production cycles is also included. On the basis of phage-resistance and satisfactory technological performance these strains are placed in a panel from which those for starter use are selected. Originally, multiple strain starters consisted of six strains used without rotation but the number may be reduced to two or three strains which may be used with, or without rotation (Prentice and Neaves, 1986). Daily testing for phage-sensitivity is required and while a simple acid inhibition test may be adequate, Timmons *et al.* (1988) found it necessary to use several different test methods and several incubation temperatures. As soon as a strain is found to be phage-sensitive it is removed and replaced with a resistant strain from the panel. The number of available strains can be rapidly depleted by phage infection and it is now usual practice to isolate BIMs from sensitive strains and, where possible, to re-introduce these to the panel. The isolation of BIMs may, however, be difficult and further problems arise from acid production and reversion to phage sensitivity (Timmons *et al.*, 1988).

In the Netherlands and other north European countries, a system of starter management evolved whereby mixed strain starter cultures are continuously challenged by phage with the resulting selection of resistant strains (Stadhouder and Leenders, 1982). Strains are not rotated but are used day after day (Prentice and Neaves, 1986).

Proteinase negative starters

In recent years much interest has been taken in the use of proteinase-negative strains (prt⁻) of starter microorganisms. These have a number of potential advantages including a high level of phage resistance, improved cheese yield due to reduced casein hydrolysis, and improved flavour due to reduction in the extent of bitter peptide formation (Muset *et al.*, 1989). Results of investigations into the use of prt⁻ strains have been variable and, while it appears that Cheddar cheese can be successfully made using prt⁻ strains only, Gouda cannot (Farkye *et al.*, 1990). Use of prt⁻ strains requires a high level of starter culture (or supplementation of milk with yeast extract) and a longer production cycle and Stadhouders *et al.* (1988) considered that 10 to 20% of the starter population should be prt⁺ to provide sufficient proteinase activity. Subsequently, it has been demonstrated that prt⁻ strains differ in their proteinase profiles and that some offer no advantage in terms of improved yield (Oberg *et al.*, 1990). Analysis before use is therefore considered essential.

Fermentation and 'cheddaring'

A fermentation temperature of 30°C is initially used in Cheddar cheese manufacture, but when the acidity reaches 0.10 to 0.14% lactic acid, the coagulum is cut into cubes of *ca.* 5 mm and stirred into the whey. The temperature is then raised to 39–40°C (**scalding** or **cooking**) and stirred until sufficiently dry and ready for separation from the whey. The curd is then 'cheddared', a laborious manual procedure which is now mechanized in large plants. During cheddaring, individual particles coalesce under their own weight, drainage of whey continues and finally the curd attains the characteristic 'chicken breast texture'. At this point the curd is milled, salted and pressed in moulds. The cheddaring process is unique to Cheddar and related cheeses and most other varieties are moulded immediately after whey separation. An exception is the pasta filata type (Mozzarella, Provolone) which is heated to *ca.* 70°C, stretched and kneaded prior to moulding.

Starter activity and thus acidification continues throughout scalding and cheddaring, where acidification is fastest. The rate of acidification affects the degree of syneresis and thus the moisture content of the final cheese, while the extent of acidification has other important effects. The calcium content of cheese falls with reducing pH value and has a major effect on texture. Low-calcium curds such as those of Cheddar-type cheeses have a crumbly texture in contrast to the rubbery, elastic texture of high-calcium Emmental or Edam types. pH value also determines the quantity of chymosin, but not of fungal pepsins or rennins, retained by the curd. Larger quantities are retained at low pH value with a resulting higher rate of proteolysis during ripening (Fox, 1987)

Control of acidification

Control of acidification is thus an important part of cheese manufacture, desired levels for cheddar being 0.15 to 0.19% lactic acid at the point of draining, 0.20 to 0.22% lactic acid at commencement of cheddaring and 0.60 to 0.80% lactic acid at the end of cheddaring. Excess acid production leads to harsh bitter flavours and where paired starter strains are used, it is common practice to use a 'fast' acid-producer which continues to grow during cooking with a 'slow' producer which does not grow but continues to produce acid at a slower rate. Excess activity can also result from the high efficiency of multiple strain starters containing BIMs and it may be necessary to modify manufacturing procedures to avoid over-production of lactic acid (Sandine, 1989). In the case of the related Cheshire variety, however, a fast rate of acid production is an essential part of manufacture and specially selected, highly active strains of *L. lactis* ssp. *lactis* are used as starters.

It is important for the ultimate quality of cheese that conversion of lactose to lactic acid should be completed, as far as possible by starter microorganisms. This is to minimize heterofermentative metabolism by non-starter lactobacilli and pediococci which results in over-production of formic acid, ethanol and acetic acid (Thomas *et al.*, 1979). Salting the curd to give a salt-in moisture level of 4% permits starters to continue metabolizing lactose in the freshly-pressed curd, conversion of lactic acid being complete within 24 hours An alternative method involves rapid cooling to 10°C at which temperature non-starter LAB metabolize lactose homofermentatively.

Thermophilic starters are used in the manufacture of very hard Italian cheeses such as Parmesan, Emmental and other Swiss-types and smear-ripened types such as Limburger. *S. thermophilus* is used in combination with *Lb. helveticus* and *Lb. delbrueckii* ssp. *lactis* or ssp. *bulgaricus*. The inclusion of strains capable of metabolizing galactose is essential to discourage the growth of heterofermentative spoilage bacteria (Benateya *et al.*, 1991). Manufacture of very hard cheese is characterized by the use of low-fat milk, very high cooking temperatures, a long immersion in brine and an extended ripening period. A high cooking temperature is also used in the manufacture of Swiss-type cheese, the curd temperature being raised from 35°C to 45°C at the rate of 1°C every two minutes and then to *ca*. 55°C at the rate of 1°C every minute. This protocol is essential to produce a texture sufficiently elastic to permit eye formation by the non-starter *Propionibacterium* and yet sufficiently firm to maintain shape. It is also essential that the cooking is not of sufficient severity to inactivate the starter microorganisms which grow in the cooling curd and produce sufficient acidity to mat the curd particles into a dense mass.

Ripening

All types of hard and semi-hard cheese are ripened before consumption. Factors involved in ripening are the same as those for soft cheese but, with certain notable exceptions, ripening does not involve the deliberate introduction of non-starter microorganisms. Ripened hard cheese contain a wide range of compounds which contribute to flavour and aroma and organoleptic characteristics have been defined in terms of a 'component balance' theory. More recently, much attention has been given to the pivotal role of **methanethiol** in Cheddar cheese, a correlation having been found between presence and absence of this compound and presence or absence of characteristic aroma.

It is now considered that methanethiol does not contribute directly to flavour, but that reactions and interactions between methanethiol and other compounds results in formation of flavour compounds (Law, 1987). The origin of methanethiol is obscure but it seems unlikely that microorganisms are directly involved. It has been postulated that methanethiol is produced by non-enzymic chemical reactions involving addition or substitution reactions between H_2S and casein or methionine (Manning, 1979; Green and Manning, 1982). Starter microorganisms are, however, involved to the extent of producing the reducing conditions which both favour the production of methanethiol and are necessary for its stability.

Although the degree of **proteolysis** in Cheddar-type cheese is very low compared with that in mould-ripened species, there is an appreciable effect on the body and texture during ripening. The effect on texture is modified by the pH value of the cheese. In Edam-type cheese, the high pH value permits proteins to exist in a matrix which softens with increasing proteolysis. Proteolysis also proceeds more rapidly in the centre of the cheese where the NaCl content is lowest. In contrast, the low pH value Cheddar-type cheese becomes crumbly with increasing proteolysis as interstitial water is bound by ionic groups (Creamer and Olson, 1982).

For a number of years, there has been discussion concerning the relative contribution of the starter and non-starter microflora to the ripening of hard cheese and it has been generally assumed that the role of non-starters such as *Lactobacillus* and *Pediococcus*, which are able to grow during ripening, at least equalled that of the starter organisms. This notion was dispelled by a series of experiments which showed that cheese made with starter organisms only was of normal organoleptic quality. Non-starter LAB may play a secondary role (Khalid and Marth, 1990), although this may be of less significance than the role in production of defects or, in some varieties, production of biogenic amines.

Ripening of **Swiss-type cheese** such as Emmental follows a different course from that of Cheddar-type and other varieties made with mesophilic starter cultures. This stems both from the technology of manufacture and the presence of *Propionibacterium freudenreichii* added with the starter culture but playing no part in the initial acidification. The role of *P. freudenreichii* is the metabolism of lactic acid to propionic and acetic acids and CO_2, the latter being responsible for the characteristic eye formation. Propionibacteria develop to the greatest extent in the centre of the cheese where NaCl and lactic acid concentrations are lowest and where the environment is anaerobic. The rate of growth and CO_2 production governs the size and distribution of eyes and thus the perceived quality. The characteristic sweet flavour of Swiss cheese is attributed to the formation of proline, although other compounds may be involved (Law, 1987). Propionibacteria are generally considered to be the source of proline-releasing peptidases, but other flavour compounds are produced by *Lb. helveticus* (Kowalewska *et al.*, 1985) and probably by other starter lactobacilli.

11.5.2 Blue vein cheeses

Blue vein cheeses such as Stilton, Roquefort and Gorgonzola have a moisture content of 40 to 45% and have been classified as both semi-hard and semi-soft. The unripened cheese is made from a high acid curd, acidity developing slowly over a long draining period (Chapman and Sharpe, 1990). *Leuconostoc* is included in the starter culture and produces CO_2 which imparts an open structure to the curd which aids subsequent mould growth. The curd is neither cooked nor pressed but allowed to consolidate under its own weight. Ripening involves the mould *Penicillium roquefortii* which may be added either to the cheese milk or to the curds. During ripening, the cheeses are pierced with needles to admit air to the interior. Mould growth follows the lines of piercing and under normal conditions (*ca.* 10°C; 96%RH) reaches a maximum after one to three months. In surface-salted varieties (Roquefort and Gorgonzola), mould growth is greatest in the mid-zone of the cheese where the NaCl concentration is an optimal 1 to 3%. The effects of *P. roquefortii* proteolytic enzymes complement those of residual chymosin, which initiates ripening, and those of starter microorganisms. The intracellular proteinases of *P. roquefortii* are of greater significance than the extracellular and the rate of proteolysis increases markedly between weeks 10 and 16 of storage, suggesting release of enzymes due to mycelial lysis and/or leakage (Hewedi and Fox, 1984). In addition to proteolytic enzymes, *P. roquefortii* has strong lipolytic activity and produces fatty acids, ketones and other flavour compounds.

11.5.3 Smear-ripened cheeses

Smear-ripened cheeses are semi-soft varieties such as Limburger and Port du Salut. *Brevibacterium linens* is the major non-starter microorganism involved in the ripening process, the organism being present adventitiously in cheese cellars and spread across the surfaces of the young cheese by wiping. The relationship between depth and surface diameter of the cheese is important for ensuring that ripening proceeds at the correct rate (Chapman and Sharpe, 1990). Salt-tolerant yeasts, *Geotrichum candidum* and micrococci are also present and contribute to the final properties of the cheese. Lactate utilization and consequent raising of the pH value by *G. candidum* is important in initiating growth of *B. linens*.

Brevibacterium linens has both proteolytic and amino acid catabolic activity. Proteolytic activity is responsible for texture modification and the extent to which proteolysis is allowed to proceed depends on the cheese variety. In the case of Limburger, for example, proteolysis proceeds to the extent that the consistency may resemble that of warm butter (Chapman and Sharpe, 1990). Flavour compounds are largely the products of amino acid catabolism and include 3-methyl-1-butanol, phenylethanol and 3-methylthiopropanol derived from leucine, phenylalanine and methionine respectively (Law, 1987). *B. linens* is also an established producer of methanethiol which in smear-ripened cheeses probably contributes to flavour as thioesters (Cuer *et al.*, 1979).

11.5.4 Accelerated ripening of cheese

In recent years, there has been considerable commercial pressure towards accelerated ripening of cheese. Accelerated ripening may be achieved by increasing the maturation temperature but this is unacceptable due to the possible growth of pathogens and the development of off-flavours. Of the various other approaches that have been taken, addition of enzymes related to ripening has attracted most attention and 'enzyme-modified' cheese is in small scale commercial production. Lipases from *Aspergillus* spp. have been used for strongly flavoured mould-ripened cheese, but most interest lies with hard cheeses which require a long, and expensive, ripening period. In this case proteinases and/or peptidases have been used from a number of sources including mammalian tissues, the moulds *Aspergillus* and *Penicillium*, and the bacteria *Bacillus subtilis, Lb. casei, L. lactis, Micrococcus* species and *Pseudomonas fluorescens*. The success of earlier work was limited by flavour and textural defects, but subsequently greater success has been achieved by combining bacterial or fungal endopeptidases with exopeptidases which degrade the small bitter peptides (Law, 1987).

Practical difficulties arise with the means of introducing enzymes, the most satisfactory solution being offered by the use of microencapsulation to separate enzymes and substrates until the cheese is made. This may be achieved by the use of liposomes – artificial vesicles with an aqueous, enzyme-containing core surrounded by concentral layers of lipid lamellae (Law and King, 1985). Liposomes added to milk are retained in the curd matrix at levels of up to 90% at the end of cheddaring (Kirby and Law, 1987), but are disrupted in the maturing cheese releasing the enzymes.

The possibility of using elevated levels of starter microorganisms to accelerate ripening has been investigated by a number of workers. Excess acid production is an inherent problem and while this may be overcome by in-vat neutralization with NaOH (Chapman and Sharpe, 1990), this is seen as being technically crude. Heat-shocking has been proposed to inactivate acid-producing capacity while retaining proteinase and peptidase activity of starter microorganisms. Results have, however, been variable with respect to organoleptic quality (Castenada *et al.*, 1990). Heat-shocking is also an expensive procedure and a more satisfactory approach may be to use lac⁻ strains of starters such as *Lactococcus* as a means of obtaining the necessary numbers without excess acid production (Sinha, 1990).

REFERENCES

Arihara, K., Toba, S. and Adachi, S. (1990) Immunofluoresence and microscopic studies on distribution of *Lactobacillus kefiranofaciens* and *Lactobacillus kefir* in kefir grains. *International Journal of Food Microbiology*, **11**, 127–34.

Benateya, A., Bracquart, P. and Linden, G. (1991) Galactose-fermenting mutants of *Streptococcus thermophilus*. *Canadian Journal of Microbiology*, **37**, 136–40.

Berroda, N., Lemeland, J.-F., Laroche, G., *et al.* (1991) *Bifidobacterium* from fermented milks: survival during gastric transit. *Journal of Dairy Science*, **74**, 409–13.

Bradley, D.E. (1967) Ultrastructure of bacteriophages and bacteriocins. *Bacteriological Reviews*, **31**, 230–7.

Casey, M.E. and Jimeno, J. (1989) *Lactobacillus delbrueckii* subsp. *lactis* plasmids. *Netherlands Milk and Dairy Journal*, **43**, 279–86.

Castaneda, R., Vassal, L., Rousseau, M. and Gripon, J.-C. (1990) Accelerated ripening of a Saint-Paulin cheese variant by addition of heat-shocked lactobacillus suspensions. *Netherlands Milk and Dairy Journal*, **44**, 49–62.

Chapman, H.R. and Sharpe, M.E. (1990) Microbiology of cheese, in *Dairy Microbiology, Volume 2, The Microbiology of Milk Products*, 2nd edn, (ed R.K. Robinson), Elsevier Applied Science, London, pp.203–89.

Cogan, T.M. and Daly, C. (1987) Cheese starter cultures, in *Cheese: Chemistry, Physics and Microbiology*, (ed P.F. Fox), Elsevier Applied Science, London, pp.179–249.

Cogan, T.M. and Accolas, J.-P. (1990) Starter cultures: Types, metabolism and bacteriophage, in *Dairy Microbiology, Volume 1. The Microbiology of Milk*, 2nd edn, (ed R.K. Robinson), Elsevier Applied Science, London, pp.77–114.

Creamer, L.K. and Olson, N.F. (1982) Rheological evaluation of maturing cheddar cheese. *Journal of Food Science*, **47**, 631–6.

Cuer, A., Dauphin, G., Kergomard, A., *et al.* (1979) Production of S-methylthioacetate by *Micrococcus* cheese strains. *Agricultural and Biological Chemistry*, **43**, 1783–4.

de Vos, W.M. (1989) On the carrier state of bacteriophages in starter lactococci: an elementary explanation involving a bacteriophage resistance plasmid. *Netherlands Milk and Dairy Journal*, **43**, 221–7.

Doco, T., Corcano, D., Ramos, P., *et al.* (1991) Rapid isolation and estimation of polysaccharide from fermented skim milk made with *Streptococcus salivarius* subsp. *thermophilus* by coupled anion exchange and gel-permeation high-performance liquid chromatography. *Journal of Dairy Research*, **58**, 147–50.

Farkye, N.Y., Fox, P.F., Fitzgerald, G.F. and Daly, C. (1990) Proteolysis and flavour development in Cheddar cheese made exclusively with single strain proteinase-positive and proteinase-negative starters. *Journal of Dairy Science*, **73**, 874–80.

Fox, P.F. (1987) New developments in cheese production. *Food Technology International Europe*, 112–15.

Froseth, B.R. and McKay, L.L. (1991) Development and application of *pfMO11* as a possible food-grade cloning vector. *Journal of Dairy Science*, **74**, 1445–53.

Gasson, M.J. (1990) Genetic manipulation of starter cultures. *Dairy Industries International*, **55**(4), 30–5.

Gasson, M.J. and Davies, F.L. (1984) The genetics of dairy lactic acid bacteria, in *Advances in the Microbiology of Cheese and Fermented Milks*, (eds F.L. Davies and B.A. Law), Elsevier Applied Science, London, pp.99–126.

Gilmour, A. and Rowe, M.T. (1990) Micro-organisms associated with milk, in *Dairy Microbiology, Volume 1. The Microbiology of Milk*, 2nd edn, (ed R.K. Robinson), Elsevier Applied Science, London, pp.37–57.

Green, M.L. and Manning, D.J. (1982) Development of texture and flavour in cheese and other fermented products. *Journal of Dairy Research*, **49**, 737–48.

Guirguin, N. and Hickey, M.W. (1987a) Factors affecting the performance of thermophilic starters 1. Sensitivity to dairy sanitizers. *Australian Journal of Dairy Technology*, **42**, 11–13.

Guirguin, N. and Hickey, M.W. (1987b) Factors affecting the performance of thermophilic starters 2. Sensitivity to the lactoperoxidase system. *Australian Journal of Dairy Technology*, **42**, 14–18.

Heap, H.A. and Lawrence, R.C. (1976) The selection of starter strains for cheese making. *New Zealand Journal of Dairy Science and Technology*, **11**, 16–20.

Hewedi, M.M. and Fox, P.F. (1984) Ripening of blue cheese: Characterisation of proteolysis. *Milchwissenschaft*, **39**, 198–201.

Huggins, A.R. and Sandine, W.E. (1984) Differentiation of fast and slow milk-coagulating isolates in studies of lactic streptococci. *Journal of Dairy Science*, **67**, 1674–9.

Hughes, D.S. and Hoover, D.G. (1991) Bifidobacteria: their potential for use in American dairy products. *Food Technology*, **45**(4), 74–83.

Hull, R.R. (1983) Factory derived starter cultures for the control of bacteriophage in cheese manufacture. *Australian Journal of Dairy Technology*, **38**, 149–54.

Iwata, M. and Morishita, T. (1989) The presence of plasmids in *Bifidobacterium breve*. *Letters in Applied Microbiology*, **9**, 165–8.

Khalid, N.M. and Marth, E.H. (1990) Lactobacilli – Their enzymes and role in ripening and spoilage of cheese: a review. *Journal of Dairy Science*, **73**, 2669–84.

Kirby, C.J. and Law, B.A. (1987) Recent developments in cheese flavour technology: application of enzyme microencapsulation. *Dairy Industries International*, **52**(2), 19–21.

Kowalewska, J., Zelazowska, H., Babuchowski, A., *et al.* (1985) Isolation of aroma-bearing material from *Lactobacillus helveticus* and cheese. *Journal of Dairy Science*, **68**, 2165–71.

Larsen, R.F. and Anon, M.C. (1990) Effect of water activity of milk upon growth and acid production by mixed cultures of *Streptococcus thermophilus* and *Lactobacillus bulgaricus*. *Journal of Food Science*, **55**, 708–10.

Law, B.A. (1984) Flavour development in cheeses, in *Advances in the Microbiology and Biochemistry of Cheese and Fermented Milk*, (eds F.L. Davies and B.A. Law), pp.187–208.

Law, B.A. (1987) Proteolysis in relation to normal and accelerated cheese ripening, in *Cheese: Chemistry, Physics and Microbiology, Volume 1. General Aspects*, (ed P.F. Fox), Elsevier Applied Science, London, pp.365–91.

Law, B.A. and King, J.S. (1985) Use of liposomes for proteinase addition to Cheddar cheese. *Journal of Dairy Research*, **52**, 183–7.

Lodics, T.A. and Steenson, L.R. (1990) Characterization of bacteriophages and bacteria indigenous to a mixed-strain cheese starter. *Journal of Dairy Science*, **73**, 2685–96.

Manning, D.J. (1979) Cheddar cheese flavour studies, II Relative flavour contributions of individual volatile compounds. *Journal of Dairy Research*, **46**, 523–9.

Milton, K., Hicks, C.L., O'Leary, J. and Langlois, B.E. (1990) Effect of lecithin addition and homogenization of bulk starter on agglutination. *Journal of Dairy Science* **73**, 2259–68.

Muset, G., Monnet, V. and Gripon, J.-C. (1989) Intracellular proteinase of *Lactococcus lactis* subsp. *lactis* NCDO 763. *Journal of Dairy Research*, **56**, 765–78.

Noomem, A. (1983) The role of the surface flora in the softening of cheese with a low initial pH. *Netherlands Milk and Dairy Journal*, **37**, 229–32.

Oberg, C.J., Khayat, F.A. and Richardson, G.H. (1990) Proteinase profiles of *Lactococcus lactis* ssp. *cremoris* using high performance liquid chromatography. *Journal of Dairy Science*, **73**, 1465–71.

Piord, J.C., Delorme, F., Giraffa, G., *et al.* (1990) Evidence for a bacteriocin produced by *Lactococcus lactis* CNRZ 481. *Netherlands Milk and Dairy Journal*, **44**, 143–58.

Prentice, G.A. and Neaves, P. (1986) The role of micro-organisms in the dairy industry. *Journal of Applied Bacteriology*, **61** (suppl.), 43–57.

Reiter, B. (1956) Inhibition of lactic streptococcus bacteriophage. *Dairy Industries*, **21**, 877–9.

Robinson, R.K. (1988) Cultures for yogurt- their selection and use. *Dairy Industries International*, **53**(7), 15–19.

Robinson, R.K. and Tamime, A.Y. (1990) Microbiology of fermented milks, in *Dairy Microbiology, Volume 2, The Microbiology of Milk Products*, 2nd edn., (ed R.K. Robinson), Elsevier Applied Science, London, pp.291–343.

Salih, M.A., Sandine, W.E. and Ayres, J.W. (1990) Inhibitory effects of Microgard$^{(TM)}$ on yogurt and cottage cheese spoilage organisms. *Journal of Dairy Science*, **73**, 887–97.

Sandine, W.E. (1989) Use of bacteriophage-resistant mutants of lactococcal starters in cheese making. *Netherlands Milk and Dairy Journal*, **43**, 211–19.

Sechaud, L., Callegari, M.L., Rousseau, M., *et al.* (1989) Relationship between temperate bacteriophage 0241 and virulent bacteriophage 832-B1 of *Lactobacillus helveticus*. *Netherlands Milk and Dairy Journal*, **43**, 261–77.

Sinha, R.P. (1990) Effect of growth media and extended incubation on the appearance of lactose-nonfermenting variants in Lactococci. *Journal of Food Protection*, **53**, 583–7.

Stadhouders, J. and Leenders, G.J.M. (1982) *The Effect of Phage Contamination on the Activity of Mixed Strain Starters in Relation to Process Control*, N.O.V.-890, Netherlands Institute for Dairy Research (NIZO), Ede.

Stadhouders, J., Toepol, L. and Wouters, J.T.M. (1988) Cheesemaking with prt⁻ and prt⁺ variants of N-streptococci and their mixtures. Phage sensitivity, proteolysis and flavour development during ripening. *Netherlands Milk and Dairy Journal*, **42**, 183–93.

Tamime, A.Y. (1990) Microbiology of 'starter cultures', in *Dairy Microbiology, Volume 2. The Microbiology of Milk Products*, 2nd edn., (ed R.K.Robinson), Elsevier Applied Science, London, pp.131–201.

Thomas, T.D., Ellwood, D.C. and Longyear, V.M.C. (1979) Change from homo- to heterolactic fermentation by *Streptococcus lactis* resulting from glucose limitation in anaerobe chemostat cultures. *Journal of Bacteriology*, **138**, 109–12.

Timmons, P., Hurley, M., Drinan, F., *et al.* (1988) Development and use of a defined strain starter system for Cheddar cheese. *Journal of the Society of Dairy Technology*, **41**, 49–53.

Toba, T., Nakajina, H., Tobitani, A. and Adachi, S. (1990) Scanning electron microscopic and texture studies on characteristic consistency of Nordic ropy sour milk. *International Journal of Food Microbiology*, **11**, 313–20.

Toba, T., Kotari, T. and Adachi, S. (1991) Capsular polysaccharide of a slime-forming *Lactococcus lactis* subsp *cremoris* LAPT 3001 from Swedish fermented milk 'langfil'. *International Journal of Food Microbiology*, **12**, 167–73.

Tzanetakis, N., Litopoulou-Tzanetaki, E. and Vafopoulou-Mastrojiannaki, A. (1991) Effect of *Pediococcus pentosaceus* on microbiology and chemistry of Teleme cheese. *Food Science and Technology*, **24**, 173–6.

12

The exploitation of microorganisms in the manufacture of alcoholic beverages

A.H. Varnam

12.1 INTRODUCTION

The key and definitive manufacturing stage in the manufacture of all types of alcoholic beverage is the fermentation of carbohydrates to ethanol, and in all but a small minority the microorganism used is a species of *Saccharomyces*.

For obvious reasons, most studies concerning the role of microorganisms in production of alcoholic beverages have concentrated on *Saccharomyces* and its properties of alcoholic fermentation. Other microorganisms do, however, play an important role in the manufacture of specific beverages and should not be overlooked. The roles played by microorganisms are summarized in Table 12.1 and discussed below.

12.2 THE KEY ROLE OF *SACCHAROMYCES*

12.2.1 Taxonomy and nomenclature

For many years it was considered that two species of *Saccharomyces* were involved in the production of alcoholic beverages, *Sacch. cerevisiae* in ale-

Exploitation of Microorganisms Edited by D.G. Jones
Published in 1993 by Chapman & Hall, London ISBN 0 412 45740 7

Table 12.1. Role of microorganisms in the production of alcoholic beverages

Bacteria	
Acetic acid bacteria	Flavour development: light rum
Clostridium spp.	Fermentation: heavy rum
Enterobacteriaceae	Fermentation: lambic beer
Lactic acid bacteria	Malo-lactic fermentation: wine
	Fermentation: whisky mash
	Fermentation: palm wine
	Fermentation: lambic beer
	Moto acidification: sake
Zymomonas mobilis	Fermentation: pulque
Moulds	
Aspergillus oryzae	*Koji* preparation: sake
Yeasts*	
Saccharomyces carlsbergensis	Fermentation: lager-type beer
Saccharomyces cerevisiae	Fermentation: ale-type beer, wine, etc.
Schizosaccharomyces pombe	Fermentation: heavy rum

*A number of other species are involved in 'spontaneous' fermentation of wine, cider, etc.
Note: Enzymes derived from microorganisms are used for a wide range of purposes in production of alcoholic beverages.

type beers and wine and *Sacch. carlsbergensis* in lager-type beers. A subsequent change in nomenclature led to *Sacch. carlsbergensis* being renamed *Sacch. uvarum*, but separate species status was retained until Kreger-van Rij (1984), on the basis of DNA homology studies, placed a number of former species including *Sacch. uvarum* into *Sacch. cerevisiae*. Further studies, however, favoured the earlier view that *Sacch. cerevisiae* and *Sacch. carlsbergensis* (*uvarum*) are separate species (Vaughan-Martini and Kurtzman, 1985; Vaughan-Martini and Martini, 1987). The species name *Sacch. pastorianus* has now been proposed in place of *Sacch. carlsbergensis* (Deak, 1991), but in view of historical precedence and common usage, *Sacch. carlsbergensis* is retained in this chapter.

Strains of *Saccharomyces* currently in industrial use in production of alcoholic beverages have been selected over many years on the basis of technologically desirable characteristics. For this reason, strains used for producing different beverages may be differentiated from each other and from 'laboratory' strains and strains used in processes such as baking (Figure 12.1).

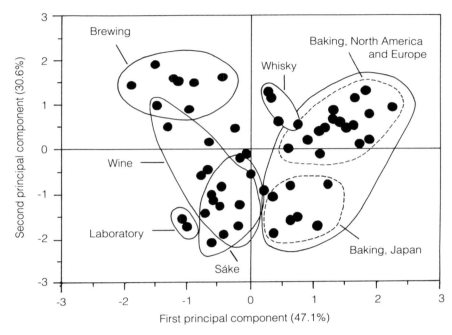

Figure 12.1 Principal-component analysis of industrial and laboratory strains of yeast. (After Oda and Ouchi, 1989.)

12.2.2 The alcoholic fermentation

Alcoholic fermentation by *Saccharomyces* involves two stages, the aerobic formation of pyruvate *via* the Embden–Meyerhof pathway, and the anaerobic decarboxylation of pyruvate to form acetaldehyde, followed by reduction of acetaldehyde to ethanol with concomitant reoxidation of NADH (Figure 12.2). Both *Sacch. cerevisiae* and *Sacch. carlsbergensis* are capable of fermenting a wide range of sugars including sucrose, glucose, fructose, galactose, mannose, maltose and maltotriose.

Glycerol, formed by the glycerol-3-phosphate dehydrogenase-catalysed conversion of dihydroxyacetone phosphate to glycerol-3-phosphate, is the major fermentation end product after ethanol and CO_2, and significant amounts of acetic acid may also be formed through the action of pyruvate decarboxylase. The relative quantities of end products depend to a large extent on conditions pertaining during the fermentation, but there is also variation according to the yeast strain used (Eustace and Thornton, 1987).

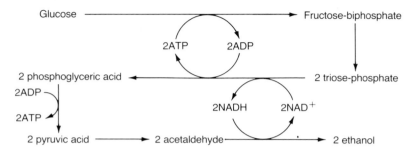

Figure 12.2 Simplified pathway of alcoholic fermentation.

12.2.3 Other technologically important properties

Ethanol tolerance

It is obviously necessary that alcoholic fermentations should continue to the desired ethanol content, and most industrially used strains of *Saccharomyces* have an ethanol tolerance significantly higher than their wild counterpart. Ethanol tolerance is under polygenic control and involves a number of possible mechanisms. Two basic hypotheses exist, **end-product inhibition** of glycolytic enzymes and **damage to the cell membrane**. These factors must be considered in the perspective of environmental conditions (Stewart and Russell, 1986) and the possibility that minor fermentation end products such as octanoic and decanoic acids may contribute to inhibition (Viegas *et al.*, 1989).

Flocculation

Flocculation is an important property which permits ready separation of the yeast from the product at the end of fermentation and minimizes off-flavours due to an excessively long contact period with yeast. At the same time, a too-rapid flocculation can result in failure to complete the fermentation. Flocculation is also an undesirable property in brewing processes using centrifugation to remove yeast from the immature beer. Although flocculation has, in the past, been largely associated with brewing, its importance is increasingly recognized in yeasts used for making wine and spirits.

Flocculation is a complex phenomenon which is a consequence both of the genetic make-up and of physical and chemical features of the environment operating at the level of the yeast cell wall (Lyons and Hough, 1970). The process, which involves the formation of an open agglomeration of cells, occurs in the absence of cell division and involves divalent ions, usually Ca^{2+}, forming bridges between anionic groups at the cell surface. Differences in flocculation characteristics between strains result primarily from differences in cell wall structure.

A number of factors affect flocculation and it may be broadly stated that flocculation is inhibited by sugars and promoted by salts. There are also differences between *Sacch. cerevisiae* and *Sacch. carlsbergensis* which may reflect the higher surface charge and lower cell hydrophobicity of *Sacch. carlsbergensis* (Raspor *et al.*, 1990).

Resistance to killer activity

A number of strains of *Saccharomyces* produce extra-cellular toxins, **zymocins**, which are lethal for other strains (killer-sensitive strains). Fermentations can be seriously disrupted if wild yeasts with killer activity are present.

Killer activity is not restricted to wild yeasts, but has also been reported in wine yeasts from some areas (Heard and Fleet, 1987). Killer activity is beneficial in suppressing wild yeasts and a number of attempts have been made to introduce the property using genetic engineering techniques (Hara *et al.*, 1981; Seki *et al.*, 1985; Tubbs, 1987). The benefits are, however, limited by the specificity of zymocins which are active against wild strains of *Saccharomyces*, but not other genera (Heard and Fleet, 1987). Furthermore, the use of both a killer and sensitive strain in the same premises could lead to fermentation problems with the latter (Thornton and Barker, 1989)

12.2.4 Genetics of *Saccharomyces*

Industrial strains of *Saccharomyces* differ from 'laboratory' strains in being usually polyploid or aneuploid. Polyploidy has advantages with respect to industrial use since the multiple gene structure is genetically more stable and less prone to mutation. The inate stability, however, means that industrial strains are not amenable to easy genetic manipulation; a mating type characteristic is usually lacking, the yeasts sporulate poorly, if at all, and tetrad analysis of any spores that do form is difficult (Stewart and Russell, 1986).

Plasmids have been demonstrated in *Sacch. cerevisiae* and both zymocin production and resistance are plasmid-encoded (Seki *et al.*, 1985). Most industrially important properties, however, are chromosomally mediated.

In recent years there has been increasing interest in either breeding, or **constructing**, yeasts in order to maximize performance in production of alcoholic beverages as well as in other industrial uses. A range of techniques have been used including conventional selective hybridiza-

Table 12.2. Examples of the genetic modification of *Saccharomyces*

Introduction of glucoamylase activity in brewing of low carbohydrate beer
Elimination of phenolic off-flavour in brewing of beer
Introduction of killer activity and killer resistance to both wine and beer yeasts
Reduction of vicinal diketone production during brewing
Reduction of higher alcohol production during wine making
Improvement of wine yeast by introducing flocculation and reducing H_2S
   production
Improvement of fermentation efficiency in wine yeast
Reduction of foaming during wine making
Increasing yield of glycerol during wine making
Introduction to wine yeast of ability to effect malo-lactic fermentation
Elimination of urea production (and thus ethyl carbamate) by sake yeast

---

tion, rare mating, protoplast fusion, mutagenesis and recombinant DNA technology and a wide range of properties have been introduced to, or eliminated from, yeasts (Table 12.2). Work is still largely at an experimental stage but it seems likely that most future emphasis will be placed on the more precise genetic engineering techniques which minimize problems encountered during yeast breeding experiments, with co-transfer of undesirable properties or loss of desirable properties.

12.3 MICROORGANISMS IN THE BREWING OF BEER

Beer is generally considered to be of two types, the **ale** type, brewed with *Sacch. cerevisiae* and the **lager** type, brewed with *Sacch. carlsbergensis*. The basic process for the two types is summarized in Figure 12.3.

Beer is brewed according to a recipe and is, at least in theory, of highly consistent character. Furthermore, the same recipe may be used in plants in different parts of the world to produce a beer which is at least of recognizable character.

12.3.1 The ingredients of beer

The main ingredients of beer are malted barley, the source of fermentable carbohydrates, proteins, polypeptides, minerals, etc., hops, the primary purpose of which is to impart bitterness and the hop characteristic, but which also have anti-microbial properties valuable in draught beer, yeast and water (liquor). In some countries, no other ingredients are permitted but elsewhere use may be made of adjuncts, which partly replace malted barley as a source of fermentable carbohydrates, and various enzyme preparations.

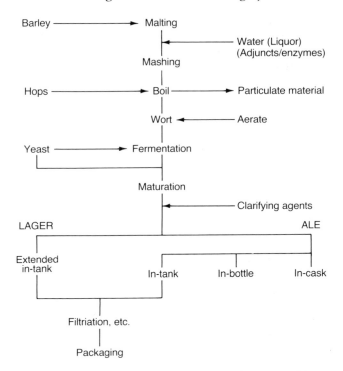

Figure 12.3 Simplified procedure for the brewing of beer.

12.3.2 Malting

Malting involves the mobilization and development of the enzymes formed during germination of the barley grain. The grain is permitted to germinate under controlled conditions of moisture and temperature, the starch/enzyme balance then being fixed by kilning at drying temperatures as high as 104°C. Fungal **gibberellic acid** is used in a number of countries to regulate and accelerate germination. Conditions during malting affect the amino acid levels in the wort and thus the progress of fermentation. Levels fall markedly as kilning temperatures increase from 82° to 104°C (Pierce, 1987) but are increased by use of gibberellic acid.

12.3.3 Mashing

During mashing, ground malt is mixed (mashed) with hot water in the mash tun. This serves both to extract existing soluble compounds from the malt and to reactivate malt enzymes which complete the breakdown of starch and proteins. The temperature used varies from 45° to 70°C, depending on the type of beer and the required sugar and nitrogen spectrum (Marchbanks, 1989), lower temperatures being generally used for lager-type beers than for ale-type.

Adjuncts are of two types, sugars such as enzymically prepared glucose syrup and starches derived from unmalted cereals. Where starchy adjuncts exceed 25% in the mash it is necessary to augment malt enzymes with commercial enzyme preparations.

12.3.4 Wort boiling

Wort is drained from the mash tun into a copper and boiled to inactivate malt enzymes. In traditional brewing, hops are added at this stage, the humulones (α-acids) being extracted and chemically isomerized. The resulting iso-humulones have a greater solubility and contribute the characteristic bitter flavour to beer, while the 'hop character' is derived from essential oils. In recent years there has been a tendency to replace hop cones with various types of hop pellets, powders or extracts including pre-isomerized hop products which may be added after fermentation (Westwood, 1989)

Boiling serves two other functions; reducing the potential for microbiological problems by effectively sterilizing the wort and coagulation of proteins followed by their removal as 'trub'. Inadequate coagulation may adversely affect the subsequent fermentation due to interference with yeast: substrate exchange processes (membrane blocking) and lead to poor quality beer (Miedaner, 1986).

12.3.5 Fermentation

In traditional brewing, fermentations are considered to be of two distinct types: the **top** fermentation used in production of ales, in which CO_2 carries flocculated *Sacch. cerevisiae* to the surface of the fermenting vessel, and the **bottom** fermentation used in production of lagers, in which *Sacch. carlsbergensis* sediments to the bottom of the vessel. Differentiation on the basis of the behaviour of the yeast is, however, becoming less distinct with the increasing use of cylindro-conical fermenters (page 306) and centrifuges.

Yeast handling

In commercial practice, yeast cells grown during one fermentation cycle are used as inoculum in subsequent cycles. It is usually necessary to store yeast, in water or beer, before re-use and it is important that a satisfactory physiological condition is maintained. In this context, glycogen levels within the cell are of prime importance, glycogen serving as an energy and carbon store to support the cells under the conditions of effective starvation during storage and providing an immediate source of energy for lipid synthesis during the lag phase. Good practice involves handling procedures which avoid inclusion of oxygen and rapid cooling to 4° to 6°C. It is also important to recognize a yeast of low glycogen content and to increase the inoculum size (pitching rate) accordingly (Stewart and Russell, 1986).

Pitching yeast is recognized as the major reservoir of beer spoilage microorganisms, especially *Lactobacillus, Pediococcus* and *Obesumbacterium* (Ogden, 1987) and decontamination procedures have become widely used. Those currently used involve an acid treatment with dilute phosphoric acid or, more effectively, acidified ammonium persulphate (Simpson, 1987a). Acid washing can have an adverse effect on yeast performance and an alternative treatment using nisin has been proposed (Ogden, 1987). However, Simpson and Hammond (1989) showed that acid washing is an effective procedure providing precautions are taken to ensure that low (<5°C) temperatures are maintained and the yeast is pitched immediately. The effectiveness of phosphoric acid is enhanced by the synergistic action of hop acids, while the post-fermentation use of pre-isomerized hop extracts in place of copper hopping may lead to problems of spoilage by lactic acid bacteria (Simpson, 1987b).

Wort aeration

The main role of wort aeration is the promotion of the biosynthesis of lipids required for yeast growth (Ohno and Takahashi, 1986a). For optimal fermentation efficiency the level of aeration should be set according to lipid levels in the early yeast rather than to the oxygen levels in the wort (Ohno and Takahashi, 1986b).

Design of fermentation vessels

In traditional practice, ale-type beers were fermented in relatively shallow, circular or rectangular vessels from which the 'top' yeast was removed by skimming. Similar, but rather deeper vessels, were used for lager-type although the use of a 'bottom' yeast obviated the need for skimming. Since the 1960s, however, the Nathan cylindro-conical design (Figure 12.4) has become predominant. This type has a number of advantages including the ability to produce either ale-type or lager-type beers, a high ratio of fermenting capacity to total vessel capacity, rapid fermentation due to effective agitation by CO_2 and efficient temperature control. The sedimentation of the yeast into a compact plug in

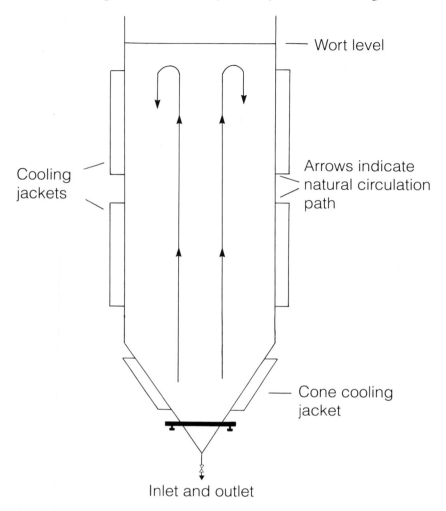

Wort level

Cooling jackets

Arrows indicate natural circulation path

Cone cooling jacket

Inlet and outlet

Figure 12.4 Cylindro-conical brewery fermenter.

the cone simplifies yeast handling and reduces the filtration required, while the design is also well suited to cleaning-in-place systems.

The development of inexpensive microprocessors now permits the control of fermentation using interactive systems. Application in breweries has, however, been limited by sensor technology failing to match information handling technology (Hammond, 1988).

Continuous fermentation, first developed in the early 20th century, has a number of potential advantages but, despite a revival of interest in the 1960s, the process is in commercial use in only two breweries,

both situated New Zealand. This situation may partly reflect technological inertia but more fundamental objections to continuous fermentation
lie in its inflexibility and the fact that the benefits fail to outweigh the
disadvantages in terms of the greater complexity of the plant and its
operation. It seems likely that widescale adoption of continuous fermentation must await a continuous wort-producing process (Atkinson,
1987) and that, ultimately, processes involving immobilized enzymes
may be most acceptable (Stewart and Russell, 1986).

Behaviour of yeast during fermentation

Fermentations for production of ale-type beer are conducted at 12° to
24°C for four to seven days where traditional fermenting vessels are
employed, or two to three days in the case of cylindro-conical vessels.
Temperatures in the range 3° to 14°C are used for fermentation of lager-
type beer, the eight to ten days required in traditional practice being
reduced to five to seven days by use of cylindro-conical vessels. Fermentations proceed more rapidly at higher temperatures, but the risk of
flavour-defects is greater. In the case of lager, attempts have been made
to overcome this problem by brewing under moderate CO_2 pressures,
but this process has adverse effects on the viability of the yeast crop
(Knatchbull and Slaughter, 1987).

During the initial stages of fermentation, oxygen uptake by the yeast
is rapid and glycogen is mobilized, serving as a source of intracellular
glucose to fuel the synthesis of the lipids required for growth. There is
no significant uptake of wort glucose until five hours, or more, after
pitching, at which stage the oxygen dissolved in the wort is almost
depleted. Excessive yeast growth is undesirable, a five-fold increase
over the initial inoculum level being usual.

Brewery strains of *Saccharomyces* are able to ferment a wide range of
sugars including sucrose, glucose, fructose, galactose, mannose, maltose and maltotriose. Glucose represses uptake of maltose and maltotriose and this imposes a restriction on the fermentation rate. The use of
derepressed mutants is a possible future development and offers the
potential for increased control of fermentation rate (Stewart and Russell,
1986). At present, however, failure of yeast to take up and metabolize
maltotriose remains an important cause of a 'stuck' fermentation. The
uptake of fructose, which is used as an adjunct, is also repressed by
glucose and its presence at the end of fermentation can lead to a sweet
off-flavour (Cason *et al.*, 1987). Conversely, however, the presence of
fructose is desirable in some types of sweet beer.

Amino acids are the major source of assimilable nitrogen in wort and
the concentration is usually kept fairly low to prevent excessive yeast
growth. It is desirable that uptake of amino acids is complete since this

limits the growth of any contaminating bacteria. The uptake of amino acids is regulated by a small number of transport systems, different amino acids being absorbed at different points in the fermentation cycle.

Excretion products of yeast

The major excretion product of wort fermentation by brewery strains of *Saccharomyces* is, of course, ethanol. In addition a large number of other products are excreted (Table 12.3), the concentration and type of which determine to a large extent the final flavour, and thus perceived quality, of the beer. The relationship between excretion products and flavour is not straightforward since, in many cases, products contribute to the overall flavour spectrum rather than having a direct flavour impact. In a number of cases, excretion products are considered desirable when present in limited concentrations as part of the flavour spectrum, but undesirable if present as flavour impact compounds at high concentrations. In the case of **esters**, for example, over-production results in flavour defects in high alcohol beers, while under-production results in lack of flavour in low alcohol beers (Peddie, 1990). The situation is

Table 12.3. A summary of flavour compounds excreted by *Sacch cerevisiae* during brewery fermentations

Product	Conditions favouring production
Higher alcohols (Fusel oils) e.g. propanol, butanol, 2-phenylethanol	Certain yeast strains; High fermentation temperature (especially 2-phenylethanol)
Esters e.g. ethyl acetate	High fermentation temperature; Low pitching rate; Wort aeration; Continuous fermentation.
Carbonyl compounds e.g. acetaldehyde	Certain brewery designs? Certain yeast strains? High pitching rate.
Vicinal diketones e.g. diacetyl, pentane-2,3-dione	Produced in all fermentations but subsequently removed
Hydrogen sulphide	High level of threonine or glycine in wort; Low level of methionine in wort; Low level of pantothenate in wort (some yeast strains); Few yeast cells budding.

further complicated by the influence of product type on the perceived effect of excretion products. The flavour of diacetyl and pentane-2,3-dione is readily detectable in lager and light Canadian ales and invariably regarded as a defect, whereas in English ales the taste threshold concentration is higher (Stewart and Russell, 1986) and the presence considered less objectionable and, in the case of some 'brown' ales, even desirable.

Fermentation of high gravity wort

The use of high gravity wort (high gravity brewing) involves fermentation of a concentrated wort and consequent dilution with deoxygenated water at a later stage. The process has a number of technological advantages and also produces a beer of improved flavour stability and smoother taste. Brewhouse material efficiency is, however, reduced.

The use of high gravity wort influences the behaviour of yeast which become more flocculent and sedimentary as the gravity is increased. Markedly higher quantities of acetate esters are produced, although this tendency can be controlled by either introducing oxygen into the fermenting vessel as oxygen-saturated water, or by increasing the levels of fermentable sugars in the wort (Stewart and Russell, 1986). It is also necessary to use a strain of yeast which can tolerate the higher level of ethanol produced.

Fermentation of wort in production of low carbohydrate, high alcohol beer

Low carbohydrate beers have become increasingly popular in recent years. Several methods of manufacture may be used, but the most common involves the use of amyloglucosidase from *Aspergillus niger* to hydrolyse dextrins to sugars which are then fermented by brewery strains of *Saccharomyces*. Although neither *Sacch. cerevisiae* nor *Sacch. carlsbergensis* are capable of fermenting dextrins directly a separate species, *Sacch. diastaticus* does possess this property. *Saccharomyces diastaticus* is not, however, suitable for use in beer production since it possesses the gene for phenolic off-flavour production (*POF1*), a serious flavour defect. Attempts have been made to introduce the amyloglucosidase gene from either *Sacch. diastaticus* or *A. niger* into brewery yeast, the use of the *A. niger* gene showing the greater promise.

12.3.6 Maturation (conditioning; secondary fermentation)

Maturation may be considered to include all transformations between the end of primary fermentation and the final filtration of the beer (Masschelein, 1986). These include carbonation by fermentation of residual sugars, removal of excess yeast, adsorption of various non-volatiles onto the surface of the yeast and progressive change in aroma and flavour. Fermentation and maturation are usually considered to be two separate stages, but the distinction between the two is becoming increasingly blurred.

In **traditional practice**, ale-type beer is conditioned for one to two weeks at 12° to 20°C, while lager-type beers are conditioned at *ca*.4°C for several months. Ale-type beers may be conditioned either in bulk tanks or, especially in the case of high quality beers, in individual casks. In-bottle maturation is also used for a few types of beer including stout and high alcohol light ale. In each case **priming sugar** may be added or amyloglucosidase used to hydrolyse dextrins. Sugars are not fully metabolized by the secondary fermentation and the quantity remaining determines the sweetness of the beer.

Traditional lagers are conditioned in bulk, although the process is occasionally completed in-bottle. Partly fermented wort may be added to initiate the secondary fermentation (krausening).

Chemical changes during maturation are complex and, in some cases remain poorly understood. It is generally accepted that the removal of acetaldehyde, vicinal diketones such as diacetyl, and sulphur compounds such as hydrogen sulphide is of major importance in the maturation of beer, although, in some cases, the actual participation of secondary fermentation has been doubted (Masschelein, 1986). However, other changes result from the release of compounds from the yeast during the maturation process. Compounds involved include amino acids, peptides, nucleotides and both inorganic and organic phosphates. Participation of these compounds in flavour improvement depends upon their intracellular accumulation, changes in cell permeability and consequent exchange reactions between the yeast and the beer. These factors in turn depend on many variables including the yeast, its physiological condition and behaviour, and the length and temperature of maturation. The release of compounds from yeast may also have a deleterious effect on quality; medium chain length fatty acids, for example, can impart autolytic and yeasty flavours to beer (Masschelein, 1986).

It is current practice to reduce the maturation period by producing **'green beer'** containing minimal levels of undesirable flavour compounds. This is usually achieved by control of fermentation, although some progress has been made in the genetic engineering of yeasts which do not produce vicinal diketones. Maturation periods may also be shortened by washing out volatiles with carbon dioxide. The use of Nathan-type cylindro-conical fermenters permits fermentation and maturation to be combined in a single operation. This is achieved by introducing a warm maturation period towards the end of the main maturation. The use of immobilized yeast to speed maturation has been investigated and offers considerable promise for future exploitation (Lommi *et al.*, 1990).

12.4 MICROORGANISMS IN WINE-MAKING

Although wine may be made from any fruit, or indeed vegetable, which contains sufficient fermentable sugars, the grape (*Vitis vinifera*) is of overwhelming importance in commerce and is the major concern of the present discussion. Wine is also made on a limited commercial scale from fruits such as gooseberry and peach. **Mead** is a type of wine, now little produced, made from fermented honey and water, while **palm wine** is made by mixed lactic-alcoholic fermentation of the sap of various species of palm tree.

12.4.1 Production of the must

A large number of grape cultivars have been used in commercial wine making, it being recognized that the type of grape, together with factors such as the soil type and overall climate, is important in determining the character of the finished wine. The quality of the grape, however, and its suitability for fermentation is affected by the climate during each individual growing season, cold, wet weather leading to a grape of low sugar content and high volatile acidity and an increased possibility of mould infection. The composition of the natural yeast micro-flora is also likely to be affected (see below).

Grapes are picked by hand, but stemming and crushing is usually mechanical, the colour of the wine being determined at this stage by the colour of the grape and the length of contact between the juice and the skin. It is usual practice to add sulphur dioxide to the must to discourage growth of indigenous (wild) yeasts and spoilage bacteria. The use of pasteurization in place of SO_2 treatment has been re-proposed (Mallet-roit *et al.*, 1991) but seems unlikely to be accepted.

The sugar content of the must may be increased by addition, although this process is not permitted in some areas. In the case of rich, sweet wines such as Tokay the high sugar content of the must is dependent on the growth of the mould *Botrytis cinerea* (the 'noble rot') on the grapes before harvesting.

12.4.2 Fermentation

Yeasts and other microorganisms present during fermentation

In traditional European practice, fermentation depends on the presence of naturally occurring yeasts in the must. *Saccharomyces* is not significant at this stage, dominant yeasts including *Aureobasidium pullulans*, *Candida stellata, Hanseniaspora uvarum, Issatchenkia orientalis, Kloeckera javanica, Metschnikowia pulcherrima* and *Pichia anomala* (Holloway *et al.*, 1990, Querol *et al.*, 1990; Deak, 1991). Some variation occurs according to the geographic location of the vineyard, *H. uvarum* being dominant in the early stages of fermentation in California and mid-Europe, while *H. osmophila* dominates elsewhere. *Metschnikowia pulcherrima* is of particular prevalence in Europe (Deak, 1991). The yeast microflora of a particular vineyard is affected by climatic conditions during growth. In a Spanish vineyard, for example, damage to grapes during cold, wet years led to increased numbers of oxidative yeasts (Querol *et al.*, 1990).

The addition of SO_2 to musts reduces both the total numbers of yeast and the number of genera. *Saccharomyces cerevisiae* and *Saccharomycoides ludwigii* are both of high SO_2 resistance, while *K. apiculata* is highly sensitive (Herraiz *et al.*, 1989).

Saccharomyces cerevisiae is derived from equipment and becomes of increasing dominance as the fermentation proceeds and other yeast die-off with increasing ethanol content. Temperatures in excess of 25°C, however, may result in domination by *Sacch. exiguus* and *Zygosaccharomyces bailii* (Mauricio *et al.*, 1986), while at 10°C, *K. apiculata, C. stellata* and *C. krusei* are present in large numbers in addition to *Sacch. cerevisiae* (Heard and Fleet, 1988). A primary fermentation at *ca.* 24°C for three to five days is used for red wines and 7 to 21°C for seven days to several weeks for white. Over-heating can be a significant problem and the introduction of cooled vats in the Californian wine industry has been considered to have a greater beneficial effect on quality than the use of pure yeast cultures.

The addition of cultures of *Sacch. cerevisiae* to the must is common practice in the non-traditional wine making regions such as California and is being increasingly adopted in the old established regions. The addition of *Sacch. cerevisiae* results in a more rapid and even rate of fermentation and a more consistent wine quality. Indigenous yeasts are still present, however, and the pattern of fermentation is similar whether or not *Sacch. cerevisiae* is added. It has been suggested that different strains of *Sacch. cerevisiae* may develop at different stages of the natural fermentation and that when cultures of *Sacch. cerevisiae* are added to induce fermentation, the main effect is to influence the development of *Saccharomyces* rather than to inhibit non-*Saccharomyces* (Heard and Fleet, 1985).

Non-growing cells of *Sacch. cerevisiae* are responsible for completing the alcoholic fermentation and problems can arise, especially in natural fermentations, from premature death of cells as a result of ethanol intolerance. Premature death is a consequence of the physiological state of the yeast and may be prevented by adding sterol, a readily assimilable nitrogen source, or yeast hulls to the must in the early stages of fermentation. Oxygenation of the must provides an alternative which has met with some success in California (Munoz and Ingledew, 1989)

Moulds die quickly during fermentation, while lactic acid bacteria (Pondo *et al.*, 1989), which are unable to compete effectively for fermentable sugars (Kennes *et al.*, 1991), decline during the alcoholic fermentation but increase in numbers during any subsequent malo-lactic fermentation.

Role of yeast in determining the organoleptic properties of wine

The flavour and aroma of wine is derived from a number of sources including the grape, products of yeast metabolism and, where used, the wooden barrels in which the wine is matured. Organoleptic properties other than those derived from the grape are commonly referred to as the 'complexity' of the wine. As in brewing, many excretion products of yeast are involved in determining the organoleptic properties of wine, the most important being higher alcohols, esters, fatty acids and carbonyl compounds. The overall contribution of excretion products varies considerably according to the type of wine. In some types of wine, yeast autolysis products are also involved in determination of organoleptic properties.

12.4.3 The malo-lactic fermentation

The malo-lactic fermentation (MLF) involves the decarboxylation of malic to lactic acid by strains of lactic acid bacteria (Kunkee, 1974). *Leuconostoc oenos* is most commonly involved, but some strains of *Lactobacillus* and *Pediococcus* are also capable of the conversion. In the majority of wines, the MLF follows the alcoholic fermentation, but during wine making using 'carbonic maceration', where whole grapes are fermented under a CO_2 atmosphere, the MLF and alcoholic fermentation proceed simultaneously.

Lactic acid is a weaker acid than malic acid and the MLF thus deacidifies the wine, the extent depending on the proportion of the grape acids malic and tartaric present in the wine (Rodriguez *et al.*, 1990). Other effects include the stabilization of wine to microbial spoilage by removal of malic acid and an increase in the complexity of the wine. The effects of the MLF may be considered desirable or undesirable according to the type of wine. A reduction of acidity is desirable, for example, in the highly acidic Italian red wines, but undesirable in the low acid wines of California and other warm climates. The aroma and flavour modification associated with the MLF is usually considered undesirable in high acid white wines and, with the exception of wines fermented from Chardonnay juice, chemical means of acidity reduction are employed.

To some extent, the MLF may be induced, or inhibited, by manipulating winemaking techniques to change the environment of the wine (Pilone, 1975). It is, however, paradoxical that the MLF is most difficult to induce when desirable, and most difficult to inhibit when undesirable. In recent years, use has been made of pure culture inocula of *L. oenos*, which have the advantage of more reliably initiating an effective MLF and reducing quality problems associated with some indigenous lactic acid bacteria. The use of pure cultures introduces potential problems with bacteriophage (Davis *et al.*, 1985), while the suitability of strains differs according to the nature of the wine (Rodriguez *et al.*, 1990; Wibowo *et al.*, 1988). The use of added cultures does not entirely eliminate problems due to indigenous lactic acid bacteria and it has been proposed that nisin should be used to control these strains, the added cultures being selected for nisin-resistance (Daeschel *et al.*, 1991). Nisin is also seen as being preferable to SO_2 or pasteurization for preventing the MLF.

The use of **immobilized microorganisms** as a means of improving the reliability and efficiency of the malo-lactic fermentation has been investigated (Crapisi *et al.*, 1987; Magyar and Panyik, 1991) but further work is required before industrial application is possible.

The development of **genetic engineering techniques** for yeasts has led to attempts to introduce the malo-lactic gene into *Sacch. cerevisiae*. Transfer of the malo-lactic gene of *Lb. delbrueckii* was successful, but the gene was expressed to only a limited extent (Williams *et al.*, 1984) and further work is required in this area.

12.4.4 Special types of wine

Champagne

Champagne and similar wines are made using a secondary in-bottle fermentation. In the case of champagne itself, the wine remains in contact with the secondary yeast growth for several years, the bottles being manipulated during storage so that the yeast forms a plug of sediment in the neck of the bottle from where it is removed by freezing. The yeast must be capable of initiating a secondary fermentation under increased pressure, at a relatively low temperature of *ca*. 10°C and in the presence of 10 to 12% alcohol, of a substrate which has been deprived of some of its growth factors. The yeast must also settle out rapidly and completely die, or become inactive, before the 'dosage', the final addition of sugar syrup in Champagne brandy (Kunkee and Amerine, 1970). The long period of contact with the yeast cells is considered an important factor in development of the Champagne character and in some accelerated processes autolysed yeast cells or yeast extract is added at the end of the secondary fermentation. In some other sparkling wines, however, the yeast is removed as soon as the active stage of the secondary fermentation is complete, prolonged contact with yeast being considered to introduce off-flavours.

Sherry and similar wines

Sherry proper is made only in a small area of southern Spain around Jerez de la Frontera, but similar wines are made world-wide. The wine, fortified with neutral grain spirit, is characterized by its maturation under a film (*flor*) of yeast which largely consists of strains of *Sacch. cerevisiae* capable of growth in 15 to 16% ethanol and of film formation. There is a considerable utilization of acetic acid and, to a lesser extent, ethanol and production of acetaldehyde, acetal and other flavour compounds including esters, isopentanol and hexanol. Although *Sacch. cerevisiae* is dominant in the *flor*, other film yeasts are present and probably contribute to flavour development. Sugars are largely depleted during fermentation and maturation and it is necessary to add special sweetening wines, the quantity added varying according to the desired 'dryness'.

Madeira is similar to sherry but the characteristic flavour is achieved by heating the wine to *ca*. 41°C for up to three months (Goswell and Kunkee, 1977).

12.5 MANUFACTURE OF OTHER NON-DISTILLED ALCOHOLIC BEVERAGES

12.5.1 Cider and perry

Cider and perry are similar to wine both in terms of manufacture and microbiology. In traditional manufacture, the apple juice is fermented by naturally occurring microorganisms. The general pattern of microbial succession and the course of the fermentation is similar to that of wine, although there tends to be considerably more variation both between plants and between batches produced by the same plant (Cabranes *et al.*, 1990). A malo-lactic fermentation often occurs concurrently with the alcoholic fermentation (Salih *et al.*, 1990).

Large-scale **cider** production is conducted on a year-round basis which involves the use of heat-treated and concentrated juice and pure cultures of yeast. In recent years, yeasts have been selected on the basis of producing desirable flavours as well as their growth and fermentation characteristics (Beech and Carr, 1977). Cider yeasts should also produce polygalacturonase to degrade de-esterified pectins and permit clarification of the cider.

A small quantity of cider is still carbonated in-bottle but in the vast majority of cases the Charmat system is used in which the secondary fermentation takes place under pressure in a temperature controlled tank.

12.5.2 Sake

Sake is a traditional Japanese drink, similar products being made in China and other countries of south-east Asia. The raw material for fermentation, the *koji*, is steamed rice, which provides a substrate for the mould *Aspergillus oryzae* (Kodoma, 1970). Sake is fermented by a strain of *Sacch. cerevisiae* which is tolerant of high concentrations of ethanol, acid and sugar and can initiate an effective fermentation at relatively low temperatures (<10°C). The yeast is prepared as a starter culture, the *moto*, which contains lactic acid, either added or produced by indigenous lactic acid bacteria, which favours the sake yeast and inhibits undesirable microorganisms. The main fermentation, the *moromi*, takes 20 to 25 days, during which saccharification by *Aspergillus* amylase and fermentation by yeast occurs simultaneously. Sake is fermented in an open system and no attempt is made to exclude microorganisms. It is necessary therefore that conditions should favour the growth of the starter strain of *Sacch. cerevisiae*. This is achieved by preparing the *moromi* in three successive stages during which the temperature is gradually lowered to less than 10°C. At the end of fermentation the ethanol content is 17.5 to 19.5%, the *moromi* is filtered and held for *ca.* four weeks before pasteurization and bottling.

12.6 MANUFACTURE OF DISTILLED ALCOHOLIC BEVERAGES

Distilled beverages may be considered to be of two types. The first, represented by vodka and gin, are **rectified** and have no requirement for congeners. The raw material is spirit which may be obtained from any source. The second, is the **congeneric** or self-flavoured type, represented by whisky and rum, where a fermentable starting material is traditional and where conditions of fermentation and distillation are prescribed (Simpson, 1977).

Whisky varies in nature according to the raw materials and the method of manufacture. Scotch malt whisky is made from a mash of malted barley dried in peat-fired kilns, though maize starch may also be present as an additional source of carbohydrate. In contrast to brewery practice, the wort is not boiled and thus the malt enzymes remain active throughout the fermentation. Pure cultures of a strain of

Sacch. cerevisiae with a high ethanol tolerance are added to the wort at a relatively high concentration (Harrison and Graham, 1970). The fermentation is not conducted under sterile conditions and lactic acid bacteria, predominantly *Leuconostoc*, grow in the late stages of fermentation at the expense of yeast autolysis products. This 'late fermentation' is considered to enhance product quality (Priest and Pleasants, 1988).

Growth earlier in the fermentation may occur where hygiene is poor and very large numbers of lactic acid bacteria are present. This leads to an unacceptable reduction in alcohol yield (Dolan, 1976) and, on some occasions, the production of undesirable compounds such as hydrogen sulphide (Geddes, 1986).

Two types of whisky are produced in the United States, **rye** and **bourbon**. In neither case is the fermentation carried out under sterile conditions and lactic acid bacteria are of importance in determining flavour. In the case of bourbon, the wort may be acidified by addition of pure cultures of *Lb. delbrueckii* before the main alcoholic fermentation (Lyons and Rose, 1977).

Rum is usually made from molasses, although cane juice and other cane by-products may also be used. Small-scale production in Jamaica still involves 'spontaneous fermentations', the resulting rum having a highly characteristic and individual flavour. Acetic acid bacteria, regarded as spoilage organisms in other alcoholic beverages, contribute acetic acid, an important flavour compound in light rum. In heavy rum fermentations, species of *Clostridium* grow synergistically with yeast producing butyric acid, *Schizosaccharomyces pombe* being considered to produce a higher quality product than *Sacch. cerevisiae*.

In large-scale rum production, increasing use is made of pure culture fermentations in which the molasses are sterilized and essentially sterile conditions maintained throughout the fermentation.

REFERENCES

Atkinson, B. (1987) The recent advances in brewing technology. *Food Technology International Europe*, 142–5.

Beech, F.W. and Carr, J.G. (1977) Cider and perry, in *Alcoholic Beverages*, (ed A.H. Rose), Academic Press, London, pp.139–313.

Cabranes, C., Moreno, J. and Mongas, J.M. (1990) Dynamics of yeast populations during cider fermentation in the Asturian region of Spain. *Applied and Environmental Microbiology*. **56**, 3881–4.

Cason, D.T., Reid, G.C. and Gatner, E.M.S. (1987) On the differing rates of fructose and glucose utilisation in *Saccharomyces cerevisae*. *Journal of the Institute of Brewing*, **93**, 23–5.

Crapisi, A., Spettoli, P., Nuti, M.P. and Zamorani, A. (1987) Comparative traits of *Lactobacillus brevis*, *Lact. fructivorans* and *Leuconostoc oenos* immobilized cells for the control of malo-lactic fermentation in wine. *Journal of Applied Bacteriology*, **63**, 513–21.

Daeschel, M.A., Jung, D.-S. and Watson, B.T. (1991) Controlling wine malolactic fermentation with nisin and nisin-resistant strains of *Leuconostoc oenos*. *Applied and Environmental Microbiology*, **57**, 601–3.

Davis, C., Silveira, N.F.A. and Fleet, G.H. (1985) Occurrence and properties of bacteriophages of *Leuconostoc oenos* in Australian wines. *Applied and Environmental Microbiology*, **50**, 872–6.

Deak, T. (1991) Foodborne yeasts, in *Advances in Applied Microbiology, Volume 36*, (eds S.L. Neidleman and A.I. Laskin), Academic Press, San Diego, pp.179–278.

Dolan, T.C.S. (1976) Some aspects of the impact of brewing science on Scotch malt whisky production. *Journal of the Institute of Brewing*, **82**, 177–81.

Eustace, R. and Thornton, R.J. (1987) Select hybridization of wine yeasts for higher yields of glycerol. *Canadian Journal of Microbiology*, **33**, 112–17.

Geddes, P.A. (1986) The production of hydrogen sulphide by *Lactobacillus* spp. in fermenting wort. *Proceedings of the Second Aviemore Conference on Malting, Brewing and Distilling*, (eds I. Campbell and F.G. Priest), Institute of Brewing, London, pp.364–70.

Goswell, R.W. and Kunkee, R.E. (1977) Fortified wines in *Alcoholic Beverages*, (ed A.H. Rose), Academic Press. London, pp.477–535.

Hammond, J.R.M. (1988) Brewery fermentation in the future. *Journal of Applied Bacteriology*, **65**, 169–77.

Hara, S., Iimura, Y., Oyuma, H., *et al.* (1981) The breeding of cryophilic killer yeasts. *Agricultural and Biological Chemistry*, **45**, 1327–34.

Harrison, J.S. and Graham, J.C.J. (1970) Yeasts in distillery practice, in *The Yeasts, Volume 3, Yeast Technology*, (eds J.S. Harrison and A.H. Rose), Academic Press, London, pp.283–348.

Heard, G.M. and Fleet, G.H. (1985) Growth of natural yeast flora during the fermentation of inoculated wines. *Applied and Environmental Microbiology*, **50**, 727–8.

Heard, G.M. and Fleet, G.H. (1987) Occurrence and growth of killer yeasts during fermentation. *Applied and Environmental Microbiology*, **53**, 2171–4.

Heard, G.M. and Fleet, G.H. (1988) The effect of temperature and pH on the growth of yeast species during the fermentation of grape juice. *Journal of Applied Bacteriology*, **65**, 23–8.

Herraiz, T., Martin-Alvarez, J., Reglero, G., *et al.* (1989) Differences between wines fermented with and without sulphur dioxide using various selected yeasts. *Journal of the Science of Food and Agriculture*, **49**, 249–58.

Holloway, P., Subden, R.E. and Lachance, M.-A. (1990) The yeasts in a riesling must from the Niagara grape-growing region of Ontario. *Canadian Institute of Food Science and Technology Journal*, **23**, 212–16.

Kennes, C., Veiga, M.C., Dubourguier, H.C., *et al.* (1991) Trophic relationships between *Saccharomyces cerevisiae* and *Lactobacillus plantarum* and their metabolism of glucose and citrate. *Applied and Environmental Microbiology*. **57**, 1046–51.

Knatchbull, F.B. and Slaughter, J.C. (1987) The effect of low CO_2 pressure on the absorption of amino acids and production of flavour-active volatiles by yeast. *Journal of the Institute of Brewing*, **93**, 420–4.

Kodoma, K. (1970) Sake yeast, in *The Yeast, Volume 3, Yeast Technology*, (eds J.S. Harrison and A.H. Rose), Academic Press, London, pp.225–82.

Kreger-van Rij, N.J.W. (1984) *The Yeast, a Taxonomic Study*, 3rd edn, Elsevier Science Publishers, Amsterdam.

Kunkee, R.E. (1974) Malo-lactic fermentation and wine making, in *Chemistry of Wine Making*, (ed A.D. Webb), American Chemical Society, Washington, pp.151–70.

Kunkee, R.E. and Amerine, M.A. (1970) Yeasts in wine making, in *The Yeasts, Volume 3, Yeast Technology*, (eds J.S. Harrison and A.H. Rose), Academic Press, London, pp.5–71.

Lommi, H., Gronqvist, A. and Pajuner, E. (1990) Immobilized yeast reactor speeds beer production. *Food Technology*, **44** (4), 128–33.

Lyons, T.P. and Hough, J.S. (1970) Flocculation of brewer's yeast. *Journal of the Institute of Brewing*, **76**, 564–71.

Lyons, T.P. and Rose, A.H. (1977) Whisky, in *Alcoholic Beverages*, (ed A.H. Rose), Academic Press, London, pp.635–92.

Magyar, I. and Panyik, I. (1991) Reducing the malic acid content of wines by a combined biological process. *Acta Alimentaria Hungarica*, **20**, 80–1.

Malletroit, V., Guinard, J.-X., Kunkee, R.E. and Lewis, M.J. (1991) Effect of pasteurisation on microbiological and sensory qualities of white grape juice and wine. *Journal of Food Processing and Preservation*, **15**, 19–29.

Marchbanks, C. (1989) Enzymes and the brewing process. *Food Technology International Europe*, 207–10.

Masschelein, C.A. (1986) The biochemistry of maturation. *Journal of the Institute of Brewing*, **92**, 213–19.

Mauricio, J.C., Moreno, J., Medina, M. and Ortega, J.M. (1986) Fermentation of 'Pedro Ximinez' must at various temperatures and different degrees of ripeness. *Belgian Journal of Food Chemistry and Biotechnology*, **41** (3), 71–6.

Miedaner, H. (1986) Wort boiling today – old and new aspects. *Journal of the Institute of Brewing*, **92**, 330–5.

Munoz, E. and Ingledew, W.M. (1989) Effect of yeast cells on stuck and sluggish wine fermentations: importance of the lipid component. *Applied and Environmental Microbiology*, **55**, 1560–4.

Oda, Y. and Ouchi, K. (1989) Principal-component analysis of the characteristics desirable in baker's yeasts. *Applied and Environmental Microbiology*, **55**, 1495–9.

Ogden, K. (1987) Cleaning contaminated pitching yeast with nisin. *Journal of the Institute of Brewing*, **93**, 302–7.

Ohno, T. and Takahashi, T. (1986a) Role of wort aeration in the brewing process. Part I. Oxygen uptake and biosynthesis of lipid in the final yeast. *Journal of the Institute of Brewing*, **92**, 84–7.

Ohno, T. and Takahashi, T. (1986b) Role of wort aeration in the brewing process. Part II. The optimal aeration conditions for the brewing process. *Journal of the Institute of Brewing*, **92**, 88–92.

Peddie, H.A.B. (1990) Ester formation in brewery fermentations. *Journal of the Institute of Brewing*, **96**, 327–31.

Pierce, J.S. (1987) The role of nitrogen in brewing. *Journal of the Institute of Brewing*, **93**, 322–4.

Pilone, G.J. (1975) Control of malo-lactic fermentation in table wine by addition of fumaric acid, in *Lactic Acid Bacteria in Beverages and Food*, (eds J.G. Carr, C.V. Cutting and G.C. Whiting), Academic Press, London, pp.121–38.

Pondo, I., Garcia, M.J., Zuniga, M. and Urubura, F. (1989) Dynamics of microbial populations during fermentation of wine from the Utiel-Requera region of Spain. *Applied and Environmental Microbiology*, **55**, 539–41.

Priest, F.G. and Pleasants, J.G. (1988) Numerical taxonomy of some leuconostocs and related bacteria isolated from Scotch whisky distilleries. *Journal of Applied Bacteriology*, **64**, 379–87.

Querol, A., Jiminez, M. and Huerta, T. (1990) Microbiological and enological parameters during fermentation of musts from poor and normal grape-harvests in the region of Alicante. *Journal of Food Science*, **55**, 1603–6.

Raspor, P., Russell, I. and Stewart, G.G. (1990) An update of zinc as an effector of flocculation in Brewer's yeast strains. *Journal of the Institute of Brewing*, **96**, 303–5.

Rodriguez, S.B., Amberg, E., Thornton, R.J. and McLellan, M.R. (1990) Malolactic fermentation in Chardonnay: growth and sensory effects of commercial strains of *Leuconostoc oenos*. *Journal of Applied Bacteriology*, **68**, 139–44.

Salih, A.G., Le Quere, J.M., Drilleau, J.-F. and Moreno Fernandez, J. (1990) Lactic acid bacteria and malolactic fermentations in Spanish cider. *Journal of the Institute of Brewing*, **96**, 369–72.

Seki, T., Choi, T.-H. and Ryu, D. (1985) Construction of a killer wine yeast strain. *Applied and Environmental Microbiology*, **49**, 1211–15.

Simpson, A.C. (1977) Gin and Vodka, in *Alcoholic Beverages*, (ed A.H. Rose), Academic Press, London, pp.537–93.

Simpson, W.J. (1987a) Kinetic studies of the decontamination of yeast slurries with phosphoric acid and added ammonium persulphate and a method for the detection of surviving bacteria involving solid medium repair in the presence of catalase. *Journal of the Institute of Brewing*, **93**, 313–18.

Simpson, W.J. (1987b) Synergism between hop resins and phosphoric acid and its relevance to the acid washing of yeast. *Journal of the Institute of Brewing*, **93**, 405–6.

Simpson, W.J. and Hammond, J.R.M. (1989) The response of brewing yeasts to acid washing. *Journal of the Institute of Brewing*, **95**, 347–54.

Stewart, G.G. and Russell, I. (1986) One hundred years of yeast research and development in the brewing industry. *Journal of the Institute of Brewing*, **92**, 537–58.

Thornton, R.J. and Barker, A. (1989) Characterization of yeast for genetically modifiable properties. *Journal of the Institute of Brewing*, **95**, 181–4.

Tubbs, R.S. (1987) Gene technology for industrial yeasts. *Journal of the Institute of Brewing*, **93**, 91–6.

Vaughan-Martini, A. and Kurtzman, C.P. (1985) Deoxyribonucleic relatedness among species of the genus *Saccharomyces sensu stricto*. *International Journal of Systematic Bacteriology*, **35**, 508–11.

Vaughan-Martini, A. and Martini, A. (1987) The newly delimited species of *Saccharomyces sensu stricto*. *Antonie van Leeuwenhoek*, **53**, 77–84.

Viegas, C., Rosa, M.F., Sa-Correia, I. and Navais, J.M. (1989) Inhibition of yeast growth by octanoic and decanoic acids produced during ethanolic fermentation. *Applied and Environmental Microbiology*, **55**, 21–8.

Westwood, K.T. (1989) Hops and hop products in the brewing process. *Food Technology International Europe*, 281–4.

Wibowo, D., Fleet, G.H., Lee, T.H. and Eschenbruch, R.E. (1988) Factors affecting the induction of the malolactic fermentation in red wines with *Leuconostoc oenos*. *Journal of Applied Bacteriology*, **64**, 421–8.

Williams, S.A., Hedges, R.A., Srike, T.L., *et al.* (1984) Cloning the gene for malolactic fermentation of wine from *Lactobacillus delbrueckii* in *Escherichia coli* and yeast. *Applied and Environmental Microbiology*, **47**, 288–93.

13

The exploitation of moulds in fermented foods

Robert A. Samson

13.1 INTRODUCTION

The impact of bacterial or fungal contamination of food is significant. Waites and Arbuthnott (1990) stated that a large portion of food-related illness is caused by bacterial food-borne infections and intoxications. It has also long been recognized that consumption of foods contaminated with moulds may cause serious health problems which could ultimately result in death. Toxins produced by moulds (mycotoxins) have been attracting global attention in view of their adverse health effects and negative economic impact (Smith and Moss, 1985; Champ *et al.*, 1991; Pitt, 1993; Samson, 1992). The presence of microorganisms in foods and beverages is, therefore, mostly regarded as negative and the beneficial aspects of moulds as fermenting agents are often little understood by the layman. The acceptance of unknown fermented foodproducts will be difficult.

The West is just becoming aware of the possibilities of fermented foods while in Asia, and to a lesser extent Africa and South and Central America, the fermentation of food is regarded beneficial and the attitude of the people is positive. Traditional fermentations may have originated centuries ago and there is a steadily growing list of food products and associated fermenting organisms. Several of these organisms have been studied extensively, demonstrating an increasing array of possibilities for their application as agents other than in food fermentation.

Exploitation of Microorganisms Edited by D.G. Jones
Published in 1993 by Chapman & Hall, London ISBN 0 412 45740 7

In this chapter, the role as fermenting organisms is focused on fungi. No attempt is made to list all different food and associated fermenting flora and the reader is referred to excellent reviews by Steinkraus (1983), Hesseltine (1983 a, b, 1985, 1989), Beuchat (1987), Campbell-Platt (1987) and Ko (1988).

13.2 FUNGAL FLORA INVOLVED IN FERMENTATION OF FOOD

In comparison with the estimated number of known fungal taxa, the number of species involved in food fermentation is relatively small. The important species involved in food fermentation are listed in Table 13.1.

There is a noticeable difference in fungal species used in the West and those from traditional Oriental foods. In the West where food fermentation mostly takes place at temperatures of a moderate climate, the mould flora is exclusively represented by species of *Penicillium*: *P. camemberti* and *P. roqueforti* for the fermentation of soft cheese and *P. nalgiovense* for the preparation of cured hams and sausages. The usage of fungi in the Orient where fermentation takes place at the higher temperatures of (sub)tropical climates, the flora consists of zygomycetous species, such as *Rhizopus*, *Amylomyces* and *Mucor* and *Aspergillus* (*A. oryzae*, *A. sojae*, *A. awamori*) (Figure 13.1) (Yokotsuka, 1986, 1991 a, b; Hesseltine, 1991). In addition, species of *Monascus* are widely used in Asian countries for the production of pigments.

Several **yeasts** are involved in the fermentation of food products. However, their usage is often in beverages or semi-liquid substances. In recent years, a fermented product called 'tea fungus', kambucha or Kargasok tea (Stadelmann, 1957) has become very popular in Europe and other countries. Although the name suggests that filamentous fungi are involved, the fermentation of tea with sugar, a mixture of yeasts species *Saccharomyces cerevisiae*, *Pichia membranaefaciens*, *Wickerhamiella domercqiae* could be isolated from this drink.

13.2.1 Taxonomic aspects

In applied mycology, the identity and names of species involved is often a serious problem. One of the main difficulties of confusing nomenclature of food fermenting moulds is the occurrence and persistent use of 'old' and incorrect names (e.g. *P. glaucum*, *P. candidum*, *P. caseicolum*, *Monilia*, etc.) by the food industry. However, in general, the taxonomy of moulds found on foods and feeds including the species involved in food fermentation has now been elucidated and proposed nomenclature has gained acceptance (Pitt and Hocking, 1985; Samson and Van Reenen-Hoekstra, 1988; Samson *et al.*, 1991).

Table 13.1. Fungal species found as food fermenting agents. (Partly compiled from Batra and Miller (1974), Yokotsuka (1991 a,b), Hesseltine (1983a))

Zygomycetes

Rhizopus Ehrenberg	*R. arrhizus, R. chinensis, R. oligosporus, R. oryzae*
Mucor Mich ex L.:Fr.	*M. circinelloides, M. javanicus, M. prainii, M. racemosus, M. rouxii*
Antinomucor Schotakowitsch	*A. elegans*
Absidia v. Tieghem	*A. corymbifera*
Amylomyces Calmette	*A. rouxii*

Ascomycetes

Monascus v. Tieghem	*M. purpureus, M. pilosus, M. ruber*
Neurospora Shear and Dodge	*N. intermedia*
Eurotium Link	

Deuteromycetes

Aspergillus Mich ex Fr.	*A. oryzae, A. sojae, A. awamori*
Penicillium Link ex Fr.	*P. camemberti, P. roqueforti, P. nalgiovense*

Yeasts

Endomycopsis Dekker	*E. burtonii* (ragi)
Hansenula Sydow	*H. anomala* (ragi, murcha, Kauji)
Zygosaccharomyces Barker	*Z. rouxii* (miso, shoya)
Saccharomyces Meyen ex Hansen	*S. cerevisiae* (various foods and beverages)
Torulopsis Berlese	*T. etchellsii* (shoya, cucumber brine)
	T. versatilis (shoya)
	T. candida (idli)
	T. pullulans (idli)
Trichosporon Behrend	*S. fibuligera* (Chinese yeast, ragi, tape)
Saccharomycopsis Schionn.	*S. malanga* (Chinese yeast, ragi)

Figure 13.1 *Aspergillus oryzae*, an isolate from commercial Japanese starter culture for the production of koji. (*a*), (*c*), (*d*): Normarski's interference micrographs of conidiophores and conidia ((*a*), × 128; (*c*), × 512; (*d*) × 512). (*b*): scanning electron micrograph of a conidiophore showing the typical phialides and catenulate, globose, roughened conidia (× 1152).

In a few genera such as *Aspergillus* and *Rhizopus*, the taxonomy of food fermenting species and related taxa remain problematical. For example, from all recent morphological, biochemical and genetical studies on *Aspergillus* section *Flavi*, it is apparent that the distinction between the toxinogenic species *A. parasiticus*, *A. flavus* and the two taxa traditionally used in food fermentation, *A. oryzae* and *A. sojae*, is very little. However, Klich and Pitt (1988) argued that a very strong case exists for maintaining separate species for the food fermentation Aspergilli, regardless of the closeness of those relationships. This practical approach of nomenclature and taxonomy of such an important group as the section *Flavi* is recommended (Samson, 1991; Samson and Frisvad, 1991).

Also, in the genus *Rhizopus*, species delimitation and nomenclature is still a matter of discussion. According to the revised nomenclature of Schipper (1984) and Schipper and Stalpers (1984), the genus *Rhizopus* consists of the groups *R. stolonifer*, *R. oryzae* and *R. microsporus*. The latter species is subdivided into four varieties: *microsporus*, *rhizopodiformis*, *oligosporus* and *chinensis*. *Rhizopus arrhizus*, a common name, has been placed in the *R. oryzae* group. The taxonomy of nomenclature of this genus is also becoming important because species have been reported as pathogens of man and animals and can produce toxic substances (Rabie *et al.*, 1985).

13.3 GROWTH CHARACTERISTICS OF FOOD-FERMENTING MOULDS

Most species used in food fermentation are characterized by fast growth. This is certainly true for the species of *Neurospora*, *Rhizopus*, *Mucor* and other zygomycetous species where the strains can reach a radial growth of one to more centimetres per day. Also, *P. roqueforti* and *P. camemberti* are fast growing in comparision with most species of the genus. Optimal growth of *Penicillium* species occurs at 20–25°C while, for species used in oriental foods, optimal and maximal growth can occur between 30–40°C. Several *Rhizopus* species are even **thermotolerant**, showing development between 40–45°C. Other species which are **thermophilic**, *Rhizomucor miehi* and *R. pusillus* are not known to be involved in food fermentation processes, but these species are used extensively for the industrial production of lipases, proteases and other enzymes (Bigelis, 1991).

During submerged fermentation, sporulation rarely occurs, although some species, e.g. *P. roqueforti* exhibits sporulation soon after inoculation in the small cavities of the soft blue cheese (Figure 13.2 b-c). Both *P. nalgiovense*, *P. camemberti* and *Neurospora intermedia* sporulate soon after inoculation, but on the surfaces of the product. Although the ascomycete *N. intermedia* (Figure 13.3) and species of *Monascus* (Figure

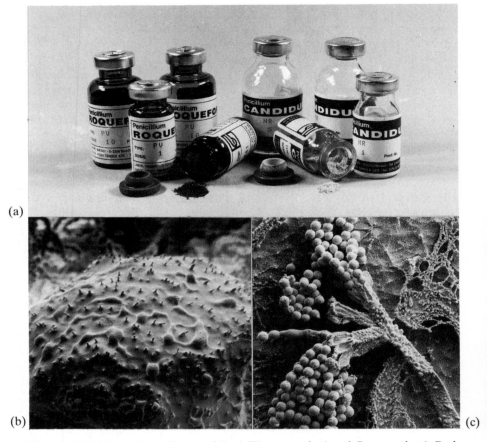

(a)

(b)

(c)

Figure 13.2 (*a*): starter cultures of *Penicillium roqueforti* and *P. camemberti*. Both commercial products are available as lyophilized preparations of pure conidia. (*b*), (*c*): low temperature scanning electron microscopy of *P. roqueforti* growing in soft blue cheese. (*b*): young mycelial growth which result in sporulating structures in small cavities in blue cheese (× 396). (*c*): conidiophore of *P. roqueforti* showing the typical penicillate structure with conidia in chains and warted conidiophore stipe (× 792).

13.4) readily produce teleomorphs in pure culture, the moulds produce only sporulating structures of the *Chrysonilia* and *Basipetospora* anamorph during fermentation. During the tempe fermentation, the growth of *Rhizopus* mycelium must take place without sporulation, giving the product its characteristic white appearance. However, the production of the *Rhizopus* starter is aimed at maximum sporulation, which implies the requirement of aerobic conditions.

Figure 13.3 *Neurospora intermedia* used for oncom fermentation in Indonesia. (*a*): blocks of fermented peanut press cake (× 0.34) which are sold on local markets after 24 hours' incubation. These cakes are covered by a dense, pink, powdery layer of sporulating structures and are sliced and cooked or fried for consumption. (*b*): Normarski's interference micrograph of the *Chrysonilia* anamorph (× 544). (*c*), (*d*): asci of the *Neurospora* produced in pure cultures on oatmeal agar ((*c*), × 136; (*d*), × 544).

There has been some discussion about the use of the word '**fermentation**' because it has been used with different meanings. Woolford (1990) defines it as an anaerobic process, which is particularly true for the natural process of microbial deterioration in silage, for example. However, in food fermentation, most moulds show aerobic growth, although *P. roqueforti* is one of the rare fungi which have the lowest

(a)

(b)

(c)

Figure 13.4 Anka or angkak fungus *Monascus purpureus*. (*a*): commercial product readily available in Asia and Chinese shops in the west, depicting fermented rice kernels which are deeply red-stained by *Monascus* pigments. (*b*): scanning electron micrograph of the conidia and conidiophores of the *Basipetospora* anamorph growing on the rice kernels (× 960), (*c*): Normarski's interference micrograph of a young *Monascus*, typically-stalked ascoma produced in pure culture on malt extract agar (× 704).

requirement for oxygen for growth (Golding, 1940; Pitt and Hocking, 1985). This explains its abundant growth during the blue-cheese fermentation. Hesseltine *et al.* (1985) found that many starters used in Asia contain species of *Mucor, Rhizopus* and *Amylomyces*, which could also grow under anaerobic conditions. Anaerobic growth of fungi has been known (Tabak and Cooke, 1968), but is extremely rare. Hesseltine *et al.* (1985) suggested that the development of anaerobic conditions during the preparation of the starters may be selective and exclude unwanted contaminating microorganisms.

13.4 TWO EXAMPLES OF TRADITIONAL FOOD-FERMENTING PRODUCTS

There are many food products fermented by means of moulds and the reader is referred to the introduction of this chapter and reviews which detail aspects of preparation process, etc. In the following section, two products are briefly described to exemplify the role of the moulds and possibilities for their use.

13.4.1 Tempe fermentation

Tempe is a traditional Indonesian fermented food in which moulds, particularly *Rhizopus* spp., play an essential role (Figure 13.5). Yellow-seeded soya beans are the most common raw material; the resulting 'tempe kedele' is usually referred to as 'tempe'. Most cultivars of yellow-seeded soya beans are suitable for tempe, in contrast to black-seeded ones (Sharma and Sarbhoy, 1984). Several examples of tempe and their raw materials are known (Nout and Rombouts, 1990). Generally speaking, fresh tempe of good quality is a compact and sliceable mass of cooked particles of raw material covered, penetrated and held together by dense non-sporulating mycelium of *Rhizopus* spp. The major desirable aspects of tempe are its attractive flavour and texture, certain nutritional properties, and 50% reduced cooking time compared with whole soya beans (Shurtleff and Aoyagi, 1979, 1980).

With its high protein content, tempe serves as a tasty **protein complement** to starchy staples, and can substitute for meat or fish. In 1986, the annual production of tempe in Indonesia was estimated as between 154 000 and 500 000 tonnes, mainly by small-scale industries (the largest Indonesian factory produces 800 kg tempe/day (Nout and Rombouts, 1990). Tempe consumption is highest on the island of Java, where it can range from 19–34 g/day/person (Karyadi, 1985).

Outside Indonesia, tempe is well known in the Netherlands and is becoming increasingly popular as a nutritious **non-meat protein food**, e.g. in the USA and Japan (Soewito, 1985) and Africa (Djurtoft, 1985).

The largest-scale tempe factories are found in Japan (6.6 tonnes/week), The Netherlands (4 tonnes/week) and the USA (3 tonnes/week) (Hesseltine, 1989).

Most fermented food products such as tempe are not consumed raw, but are first heated. In tempe, heating develops meat-like flavours, e.g. by frying spiced and salted slices in oil, by boiling with coconut milk in soups, by stewing, by roasting spiced kebabs, and in peppered ground

(a)

(c)

(b)

(d)

Figure 13.5 *Rhizopus oligosporus* and tempe. (*a*): typical appearance of tempe cake produced from cooked soyabean fermented for 24 hours by *R. oligosporus* (× 0.32). (*b*)–(*d*): scanning electron micrographs of the sporulating structures and spores ((*b*), × 230; (*c*), × 704; (*d*), × 1408). Note that the spores are slightly striate and that they vary in shape.

pastes (Shurtleff and Aoyagi, 1979; Soewito, 1985). Tempe has also been tested for use as a meat extender (Kuo *et al.*, 1989) and in the production of mixed potato/tempe snackfoods (Grzeskowiak and Berghofer, 1989).

Because of microbial enzymatic activities, fresh tempe has a **short shelf-life** which also limits its keeping quality in refrigerators. Irrespective of storage temperature, fresh tempe eventually turns brown, the beans become visible due to senescence of the fungal mycelium, the material softens and ammoniacal odours emerge (Nout *et al.*, 1985). In Indonesia, while such tempe is unacceptable for frying or stewing purposes, it is used to produce strongly-flavoured cookies which require an habituated palate in order to be appreciated.

Several **zygomycetous species** have been isolated from traditional tempe (Hesseltine, 1991), but *Rhizopus* spp. are considered to be essential for tempe making. Other genera are considered to be contaminants which may not even have grown but are present as spores.

13.4.2 *Monascus* pigments

Natural colourants are gaining importance as food additives and can replace synthetic compounds (Spears, 1988). Among the fungal species which could be good alternatives, the yeasts *Pfaffia rhodozyma* and *Monascus* spp. have been used. Particularly, *Monascus* pigments or the commonly-called **anka** or **angkak** pigment, which is a mixture of at least six different molecules, have been used in drinks and foods in South East Asia (Su and Huang, 1980; Lin and Hzuka, 1982; Kiyohara *et al.*, 1990).

Monascus purpureus and related species have been traditionally grown on dehulled rice in the Orient since ancient times, and impart a brilliant purple red colour to the product. This can be added directly to foods and beverages (e.g. rice wine) or dried to a powder and used as needed. The four main *Monascus* pigments (monascin, ankaflavin, rubropunctatin and monascorubrin) are heat stable and unchanged over a pH range of 2–10, besides having solubility properties adaptable to water- or oil-based systems. The colours range from yellow to orange to red and even purple, depending on culture conditions, species and strain, means of pigment preparation and storage, and possible chemical substitution on the polyketide structure. In the last 15 years, 38 patents pertaining to *Monascus* colourants have appeared and have been proposed for use with meat products, meat substitutes, vegetable protein, fish, candies, cakes, milk products, and various beverages. In cured meats, their addition yields a colour similar to that developed by nitrite treatment.

Recent research has focused on the improved fermentatative production of *Monascus* pigments using UV mutants (Kim and Kim, 1990), roller bottle cultures on articifial media (Mak *et al.*, 1990) or food-processing waste such as sugar cane bagasse (Chiu and Chan, 1992).

13.5 STARTER CULTURES

In traditional oriental food fermentation, natural **starters** are still used for some products such as tempe (Ko and Hesseltine, 1979; Nout and Rombouts, 1990). These natural starters are made with plant leaves (e.g. *Hibiscus* spp.) and soybeans, and known as 'usar'. For the production of most fermented food products, starter cultures are now commercially available. Starters are mostly available as pure cultures, containing **dry conidia** which are often preserved in lyophilized conditions (Figure 13.2a). With the exception of *Amylomyces rouxii*, most species produce numerous spores. The production of a pure conidial product is therefore easy to achieve.

In the production of various foods, purposely mixed cultures containing moulds, yeasts and bacteria are often used as starters in Asia (Nout, 1992a). Although these starters are made under unsanitary conditions, the products contain few contaminating microorganisms. Hesseltine (1983a) and Hesseltine *et al.* (1985) found that these starters have a uniform occurrence of certain species of *Mucor*, *Rhizopus* and *Amylomyces*, and suggested that the development of anaerobic conditions during the preparation of the starters may be selective and exclude unwanted contaminating microorganisms.

Most starter organisms used commercially are considered to be non-pathogenic. Exceptions may be *Staphylococcus saprophyticus* and *Staph. xylosus*, which are part of certain meat-curing inocula (Hammes, 1988). As the latter organisms have also been isolated from human infections, their pathogenicity merits further study.

13.6 SAFETY

When microorganisms are used for the fermentation of food, their safety is often a major concern (Nout, 1991). Even in the case of well-known and controlled manufactured products such as soft-cheeses (Camembert, Brie), outbreaks of toxic strains of *Listeria monocytogenes* and *Escherichia coli* can occur (Roberts, 1990).

Fermented food products and especially those from the Orient are made and sold fresh the same day, where it will be immediately prepared for consumption. In the West however, these products may be stored for several weeks before they are purchased and, in addition, will be kept for some time in a refrigerator by the consumer prior to preparation. Most of the products have a **high protein content** and are, therefore, a good substrate for particular food-poisoning bacteria. Samson *et al.* (1987) found, besides the starter cultures, several species of *Candida* spp. in 110 samples of Dutch tempe. In addition, lactic acid bacteria and Enterobacteriaceae were sometimes predominantly found. At unacceptable levels, *Staphylococcus aureus*, *Bacillus cereus* and *Escherichia coli* were found in several samples.

The presence of **spoilage bacteria**, **moulds** or **yeasts** often results in off-flavours, unattractive appearance and enzymatic changes. For many food products, a high salt and low pH can have an effect of preventing undesirable microbial growth. In addition, the high population of fermenting moulds prevents low numbers of spoilage microorganisms from increasing and, in some cases, may reduce the numbers because of population pressure. This has been shown for the Indonesian product tempe 'bongkrek', which can contain bongkrek acid and toxoflavin, produced by the bacterium *Pseudomonas cocovenenans*. A large *Rhizopus* inoculum as starter will prevent the growth and toxin production of this toxic contaminant (Ko, 1985).

Many species of **filamentous fungi** are able to produce mycotoxins and particular taxa of the genera *Penicillium* and *Aspergillus* are known toxin producers (Frisvad and Samson, 1991 a, b). *P. roqueforti*, used for the fermentation of blue-veined cheese is not only known as a starter-culture but is now reported as a common contaminant of foods and feeds. Strains of *P. roqueforti* can grow under microaerobic conditions and may therefore be a contaminant in airtight commodities, e.g. corn silage used for feedstuff. It is tolerant of alkaline conditions, with a pH range for growth between 3 and 10.5, and also shows resistance to food preservatives (Samson, 1989). Strains of *P. roqueforti* and *P. camemberti* can produce a number of mycotoxins (refer to Table 13.2) and many researchers have raised the question as to what extent consumption of mould-ripened cheese presents a risk to health. Engel and Teuber (1989) stated that the PR toxin produced by *P. roqueforti* is not stable in cheese and that the reaction products formed are relatively non-toxic. Other compounds which are produced in commercial blue cheese also have a relatively low toxicity. **Cyclopiazonic acid**, which is produced by *P. camemberti* in the rind of Brie and Camembert cheese may present a health risk when this product is consumed regularly, but Engel and Teuber (1989) believe that the doses of this common mycotoxin ingested by consumers are very low compared with the oral LD_{50} dose in rats. In cured meat, strains of *P. nalgiovense* contribute to the aroma, the quality of the skin, and product safety by suppressing wild strains and their metabolites. However, more than 50% of *P. nalgiovense* and *P. chrysogenum* isolated from fermented meats are toxinogenic when tested on laboratory media (Leistner and Eckhardt, 1979). Mycotoxins isolated from starter cultures of moulds used in food fermentation are listed in Table 13.2.

Table 13.2. Some mycotoxins produced by species used in food fermentation. (Partly from Frisvad and Samson, 1991b.)

Aspergillus oryzae	Cyclopiazonic acid; Kojic acid; 3-nitropropionic acid
Penicillium camemberti	Cyclopiazonic acid
Penicillium commune	Cyclopaldic acid; Cyclopiazonic acid; Isofumigaclavine A; Palitantin; Rugulovasine A
Penicillium nalgiovense	Penicillin
Penicillium roqueforti	Isofumigaclavine A and B; Marcfortines; Mycophenolic acid; PR-toxin; Roquefortine C; Botryodiploidin; Patulin; Penicillic acid

13.6.1 Ioxins in raw materials

A number of raw materials naturally contain toxic substances (e.g. cyanogenic glycosides); in addition, environmental contaminants such as pesticides, herbicides and hormones may be present. There are no indications that food fermentation has a diminishing effect on such residues. In the field and during storage, plant foods especially may become contaminated with mycotoxins.

Groundnut presscake or **maize** used as a raw material for the production of fermented products, e.g. Indonesian oncom and Ghanaian kenkey are sometimes contaminated with **aflatoxins**. The possible contamination with aflatoxin B_1 of fermented food has been investigated in a variety of products. Fungi involved in food fermentations, for instance *Rhizopus oryzae* and *Rhiz. oligosporus*, are able to reduce the cyclopentanon moiety which results in aflatoxicol A, which is a reversible reaction. Under suitable medium conditions (e.g. the presence of organic acids), aflatoxicol A is irreversibly converted into its stereoisomer aflatoxicol B (Nakazato *et al.*, 1990). Aflatoxicol is approximately 18 times less toxic than aflatoxin B_1. Furthermore, Bol and Smith (1989), Nout (1989) and Bol and Knol (1990) have found that *Rhizopus* and *Neurospora* strains have the ability to reduce fluorescence in an aflatoxin B_1 medium. They observed that certain strains were able to degrade 87% of aflatoxin B_1 into non-fluorescent substances of, as yet, unknown nature and toxicity. For detoxification of food and feed in solid substrate, the role of the food-fermenting moulds could have a significant biotechnological application.

13.7 ALTERNATIVE SUBSTRATES OF FOOD-FERMENTING FUNGI

Fermented foods are products which have been subjected to the biochemical activities of microorganisms, and particular enzymes can lead to improved modifications of the original raw materials. The final products may be more **nutritious, palatable** and **digestible** and, in many cases, are **safer** to eat. Because these moulds are excellent producers of enzymes, their industrial usage is now widely applied. Many enzymes are used in other food-processing applications (Bigelis, 1991) and enzyme-producing, food-fermenting species are listed in Table 13.3.

Besides the industrial production of enzymes, much attention has been given to the utilization of moulds for the bioconversion of **waste**. Several species which are originally known from food fermentation have been used: orange finisher pulp as substrate for *Rhizopus oryzae* (Hart *et al.*, 1991); *Aspergillus awamori* growing in waste water of a soysauce factory (Morimura *et al.*, 1991); *Aspergillus niger* for the bioconversion of olive waste water (Hamdi *et al.*, 1991); *Monascus purpureus* on sugarcane bagasse (Chiu and Chan, 1992) and *Rhizopus oligosporus* on okara (tofu residue) (Matsuo, 1990). *Aspergillus oryzae* and *Aspergillus terreus* were investigated for their application on agricultural and food wastes (Aladashvili *et al.*, 1990).

The fermentation extract of *Aspergillus oryzae* has also been used to increase milk production of lactating cows. Kellems *et al.* (1990) found that the addition of this fermentation extract had an increased effect during the early stages of the lactation cycle and subsequent milk production was probably a result of the higher initial production. An *A. oryzae* fermentation extract could also increase the initial rate of straw degradation during rumen fermentation of sheep fed with chopped barley straw (Fondevila *et al.*, 1990).

13.8 CONCLUSIONS

The interest and consumption of food products fermented by moulds is increasing. Several fermented products such as soysauce produced in many Asian countries are now readily available in the west and are becoming increasingly popular. In addition, while there are trends to the belief that animal fat and protein consumption should be reduced, mould fermented products may be regarded as potential replacements. In future, moulds can play a significant role for the worldwide need for protein which although available from plant sources is currently not readily consumable. A further area of interest is the role of moulds, produced from fermented food, and used for the bioconversion of commodities for animal feeding.

Fungal species which have been known for centuries have also been used and studied for their excellent enzyme systems, and their role will increase and prove further valuable contributions to the biotechnology industry. However, since many of the species used in food fermentation

Table 13.3. Example of commercially used enzymes with applications in food produced by fungal species known as food-fermenting organisms. (Modified from Anon., 1989.)

Fungal origin	Principal enzymatic activity
Aspergillus niger Including strains *A. aculeatus, A. awamori, A. ficuum, A. foetidus, A. japonicus, A. phoenicis, A. saitoi* and *A. usamii.*	α-Amylase
	α-L-Arabinofuranosidase
	Catalase
	Cellobiase (β-Glucosidase)
	Cellulase
	α-D-Galactosidase
	endo-1,3(4)-β-D-Glucanase
	Glucoamylase
	Glucose oxidase
	Hemicellulase
	Inulinase
	Invertase (β-Fructofuranosidase)
	Lactase (β-Galactosidase)
	Lipase, triacylglycerol
	Maltase (α-Glucosidase)
	Pectinase (Polygalacturonase)
	Pectinesterase
	Pectin lyase
	Protease (Proteinase)
	Tannase
	endo-1,3-β-D-Xylanase
Aspergillus melleus	Protease (Proteinase)

Aspergillus oryzae
Including strains: *A. sojae* and *A. effuses.*

α-Amylase
Cellulase
endo-1,3(4)-β-D-Glucanase
Glucoamylase
Hemicellulase
Lactase (β-Galactosidase)
Lipase, triacylglycerol
Maltase (α-glucosidase)
Pectinase (Polyglactturonase)
Protease (Proteinase)
Tannase

Rhizopus arrhizus

Glucoamylase
Lipase, triacylglycerol

Rhizopus delemar

α-Amylase
Cellulase
endo-1,3(4)-β-D-Glucanase
Glucoamylase
Hemicellulase

Rhizopus niveus

Glucoamylase
Lipase, triacylglycerol

Rhizopus oryzae

α-Amylase
Cellulase
endo-1,3(4)-β-D-Glucanase
Glucoamylase
Hemicellulase
Maltase (α-Glucosidase)
Pectinase (Polygalcturonase)

are known to produce toxic metabolites, their safety should be carefully evaluated.

REFERENCES

Aladashvili, N.V., Tkeshelashvili, M.Y., Metrevili, E.M., Berikashvili, V.S. and Dolidze, D.A. (1990) Preparation of feed additives through microbiological conversion of agricultural and food wastes. *Izvestija Akadeiia Nauk Gruzinskoj SRR Serija Biologiceskaja*, **16**, 325–9.
Anon. (1989) *Regulatory Aspects of Microbial Food Enzymes*. Third Edition. The Association of Microbial Food Enzyme Producers.
Batra, L.R. and Millner, P.D. (1974) Some Asian fermented foods and beverages and associated fungi. *Mycologia*, **66**, 942–50.
Beuchat, L.R. (1987) *Food and Beverage Mycology*. Second edition. Van Nostrand Reinhold, New York.
Bigelis, R. (1991) Fungal enzymes in food processing, in *Handbook of Applied Mycology, Vol. 3. Foods and Feeds*, (eds D. Arora, K. Mukerji and E. Marth), Marcel Dekker, New York, pp.445–98.
Bol, J. and Smith, J.E. (1989) Biotransformation of aflatoxin. *Food Biotechnology*, **3**, 127–44.
Bol, J. and Knol, W. (1990) Biotechnological methods for detoxification of mycotoxins. *Proceedings of the IFF Symposium, Braunschweig*, 30–31 October, 1990.
Campbell-Platt, G. (1987) *Fermented Foods of the World. A Dictionary and Guide*. Butterworths, London.
Champ, B.R., Highley, E., Hocking, A.D. and Pitt, J.I. (1991) Fungi and mycotoxins in stored products. *Proceedings of an international conference, Bangkok, Thailand*, 23–26 April 1991, Canberra ACIAR Proceedings, No. 36.
Chiu, S.W. and Chan, S.M. (1992) Production of pigments by *Monascus purpureus* using sugar-combagasse in roller bottle cultures. *World Journal of Microbiology and Biotechnology*, **8**, 68–70.
Djurtoft, R. (1985) Tempe from cowpeas introduced in Nigeria. *Proceedings Asian Symposium on Non-salted soyabean fermentation.*, Tsukuba, Japan, 1985, pp.144–62.
Engel, G. and Teuber, M. (1989) Toxic metabolites from fungal cheese starter cultures (*Penicillium camemberti* and *Penicillium roqueforti*), in *Mycotoxins in Dairy Products*, (ed H.P. van Egmond), Elsevier Applied Sciences, London and New York, pp.163–92.
Fondevila, M., Newbold, C.J., Hotten, P.M. and Orskov, E.R. (1990) A note on the effect of *Aspergillus oryzae* fermentation extract on rumen fermentation of sheep given straw. *Animal Production*, **51**, 422–5.
Frisvad, J.C. and Samson, R.A. (1991a) Filamentous fungi in foods and feeds: ecology, spoilage and mycotoxin production, in *Handbook of Applied Mycology, Vol. 3, Foods and Feeds*, (eds D.K. Arora, K.G. Mukerji and E.H. Marth), Marcel Dekker, New York, pp.31–68.
Frisvad, J.C. and Samson, R.A. (1991b) Mycotoxins produced by species of *Penicillium* and *Aspergillus* occurring in cereals, in *Cereal Grain, Mycotoxin, Fungi and Quality in Drying and Storage*, (ed J. Chelkowski), Elsevier, Amsterdam, pp.441–76.
Golding, N.S. (1940) The gas requirements of molds II. The oxygen requirements of *Penicillium roqueforti* (three strains originally isolated from blue

veined cheese) in the presence of nitrogen as diluent and the absence of carbon dioxide. *Journal of Dairy Science*, **23**, 879–89.

Grzeskowiak, B. and Berghofer, E. (1989) Production of a stable, fried snack food containing tempeh. *Abstracts International Conference on Biotechnology and Food*, Stuttgart, Hohenheim University, February, 1989.

Hamdi, M., Khadir, A. and Arcia, J. (1991) The use of *Aspergillus niger* for the bioconversion of olive waste waters. *Applied Microbiology and Biotechnology*, **34**, 828–31.

Hammes, W.P. (1988) Health hazards due to the use of starter cultures in the food industry. *Alimenta*, **7**, 55–9.

Hart, H.E., Parish, M.E., Burns, J.K. and Wicker, L. (1991) Orange finisher pulp as substrate for polygalacturonase production by *Rhizopus oryzae*. *Journal of Food Science*, **56**, 480–3.

Hesseltine, C.W. (1983a) Microbiology of oriental fermented food. *Annual Review for Microbiology*, **37**, 575–601.

Hesseltine, C.W. (1983b) The future of fermented foods. *Nutrition Reviews*, **41**, 293–301.

Hesseltine, C.W. (1985) Fungi, people and soybeans. *Mycologia*, **77**, 505–25.

Hesseltine, C.W. (1989) Fermented products, in *Legumes: Chemistry, Technology and Human Nutrition*, (ed R.H. Matthews), Marcel Dekker, New York, pp.161–85.

Hesseltine, C.W. (1991) Zygomycetes in food fermentations. *The Mycologist*, **5**, 162–9.

Hesseltine, C.W., Featherston, C.L., Lombard, G.L. and Dowell, V.R. (1985) Anaerobic growth of molds isolated from fermentation starters used for foods in Asian countries. *Mycologia*, **77**, 390–400.

Karyadi, D. (1985) Nutritional implications of tempe in Indonesian rural community. *Proceedings Asian Symposium on Non-salted soyabean fermentation*. Tsukuba, Japan, 1985, pp.112–24.

Kellems, R.O., Lagerstedt, A. and Wallentine, M.V. (1990) Effect of feeding *Aspergillus oryzae* fermentation extract or *Aspergillus oryzae* plus yeast culture plus mineral and vitamin supplement on performance of Holstein cows during a complete lactation. *Journal of Diary Science*, **73**, 2922–3.

Kim, S. and Kim, J. (1990) Pigment production in *Monascus auka*. *Journal of Korean Agricultural Chemical Society*, **33**, 239–46.

Kiyohara, H., Watanabe, T., Imai, J., Takizawa, N., *et al.* (1990) Intergeneric hybridization between *Monascus anka* and *Aspergillus oryzae* by protoplast fusion. *Applied and Microbiological Biotechnology*, **33**, 671–6.

Klich, M.A. and Pitt, J.I. (1988) Differentiation of *Aspergillus flavus* from *A. parasiticus* and other closely related species. *Transactions of the British Mycological Society*, **91**, 99–108.

Ko, S.D. (1985) Growth and toxin production of *Pseudomonas cocovenenans*, the so-called 'Bongkrek Bacteria'. *Asian Food Journal*, **1**, 78–84.

Ko, S.D. (1988) Useful role of moulds in foods, in *Introduction to Food-borne Fungi*, Third edition, (eds R.A. Samson and E.S. Van Reenen-Hoekstra), Centraalbureau voor Schimmelcultures, Baarn, pp.274–83.

Ko, S.D. and Hesseltine, C.W. (1979) Tempeh and related foods, in *Economic Microbiology, vol. 4, Microbial Biomass*, (ed A.H. Rose), Academic Press, London, pp.115–40.

Kuo, J.C., Wang, S.Y., Peng, A.C. and Ockerman, H.W. (1989) Effect of tempeh on properties of hams. *Journal of Food Science*, **54**, 1186–9.

Leistner, L. and Eckhardt, C. (1979) Vorkommen toxinogener Penicillien bei Fleischerzeugnissen. *Fleischwirtschaft*, **59**, 1892–6.

Lin, C.F. and Hzuka, H. (1982) Production of extra-cellular pigment by a mutant of *Monascus kaoliang* sp. nov. *Applied and Environmental Microbiology*, **43**, 671–6.

Mak, N.K., Fong, W.F. and Wong-leung, Y.L. (1990) Improved fermentative production of *Monascus* pigments in roller bottle culture. *Enzyme Microbiological Technology*, **12**, 965–8.

Matsuo, M. (1990) Suitability of Okara tempe as foodstuff. *Nippon Noquekagaku Kaishi*, **64**, 1235–6.

Morimura, S., Kida, K., Yakita, Y., *et al.* (1991) Production of saccharifying enzyme using the waste water of a shochu distillery. *Fermentation Bioengineering*, **71**, 329–34.

Nakazato, M., Morozumi, S., Saito, K., *et al.* (1990) Interconversion of aflatoxin B_1 and aflatoxicol by several fungi. *Applied and Environmental Microbiology*, **56**, 1465–70.

Nout, M.J.R. (1989) Effect of *Rhizopus* and *Neurospora* spp. on growth of *Aspergillus flavus* and *A. parasiticus* and accumulation of aflatoxin B_1 in groundnuts. *Mycologigal Research*, **90**, 518–20.

Nout, M.J.R. (1991) Fermented foods and food safety. *Proceedings Eighth World Congress of Food Science and Technology*, 29 September–4 October 1991, Toronto, Canada.

Nout, M.J.R. (1992) Ecological aspects of mixed culture food fermentations, in *The Fungal Community, its Organization and Role in the Ecosystem*, Second Edition, (eds G.C. Carroll and D. Wicklow), Marcel Dekker, New York, pp.817–51.

Nout, M.J.R. and Rombouts, F.M. (1990) Recent developments in tempe research. *Journal of Applied Bacteriology*, **69**, 609–33.

Nout, M.J.R., Bonants-van Laarhoven, T.M.G., De Drue, R. and Gerats, A.A.G.M. (1985) The influence of some process variables and storage conditions on the quality and shelf-life of soyabean tempeh. *Antonie van Leeuwenhoek*, **51**, 532–4.

Pitt, J.I. (1993) The most significant toxigenic fungi, in *Toxigenic Microorganisms*. Blackwell, Oxford, in press.

Pitt, J.I. and Hocking, A.D. (1985) *Fungi and Food Spoilage*. Academic Press, Sydney. pp.413.

Rabie, C.J., Lubben, A., Schipper, M.A.A., *et al.* (1985) Toxinogenicity of *Rhizopus* species. *International Journal of Food Microbiology*, **1**, 263–70.

Roberts, D. (1990) Sources of infection: food. *Lancet*, **336** (No. 8719), 859–61.

Samson, R.A. (1989) Filamentous fungi in food and feed. *Journal of Applied Bacteriology, Symposium Supplement*, **1989**, 27S–35S.

Samson, R.A. (1991) Identification of food-borne *Penicillium, Aspergillus* and *Fusarium* species, in *Fungi and Mycotoxins in Stored Products: Proceedings of an international conference*, (eds B.R. Champ, E. Highley, A.D. Hocking and J.I. Pitt), Bangkok, Thailand, 23–26 April, 1991. Canberra ACIAR Proceedings, no. 36, pp.39–46.

Samson, R.A. (1992) Mycotoxins: a mycologist's perspective. *Journal of Medical and Veterinary Mycology*, **30**, 9–18

Samson, R.A. and Frisvad, J.C. (1991) Taxonomic species concepts of Hyphomycetes related to mycotoxin production. *Proceedings of the Japanese Association of Mycotoxicology*, **32**, 3–10.

Samson, R.A. and Van Reenen-Hoekstra, E.S. (1988) *Introduction to Food-borne Fungi*, Third edition. Centraalbureau voor Schimmelcultures, Baarn, pp.299.

Samson, R.A., Van Kooiji, J.A. and De Boer, E. (1987) Microbiological quality of commercial tempeh in the Netherlands. *Journal of Food Protection*, **50**, 92–4.

Samson, R.A., Frisvad, J.C. and Arora, D. (1991) Taxonomy of filamentous fungi in foods and feeds, in *Handbook of Applied Mycology, Vol. 3, Foods and Feeds,* (eds D. Arora, K. Mukerji and E. Marth), Marcel Dekker, New York, pp.1–29.

Schipper, M.A.A., (1984) A revision of the genus *Rhizopus*. I The *Rhizopus stolonifer*–group and *Rhizopus oryzae. Studies in Mycology,* Baarn **25**, 1–19.

Schipper, M.A.A. and Stalpers, J.A. (1984) A revision of the genus *Rhizopus*. II The *Rhizopus microsporus*–group. *Studies in Mycology,* Baarn, **25**, 20–34.

Sharma, R. and Sarbhoy, A.K. (1984) Tempeh – a fermented food from soybean. *Current Science,* **53**, 325–6.

Shurtleff, W. and Aoyagi, A. (1979) *The Book of Tempeh, a Super Soyfood from Indonesia.* Harper and Row, New York.

Shurtleff, W. and Aoyagi, A. (1980) *Tempeh Production. The Book of Tempeh: Volume*
II. New-Age Foods, Lafayette, California.

Smith, J.E. and Moss, M.O. (1985) *Mycotoxins: Formation, Analysis and Significance.* John Wiley, Chichester.

Soewito, A. (1985) The cooking of tempeh – Indonesia. *Proceedings Asian Symposium on Non-salted soyabean fermentation.* Tsukuba, Japan, 1985, pp.270–2.

Spears, K. (1988) Developments in food colourings: the natural alternatives. *Trends in Biotechnology,* **6**, 283–8.

Stadelmann, E. (1957) Der Teepilz. Eine Literaturzusammenstellung. *Sydowia,* **11**, 380–8.

Steinkraus, K.H. (1983) *Handbook of Indigenous Fermented Foods.* Marcel Dekker, New York.

Su, Y.C. and Huang, J.H. (1980) Fermentation production of anka pigment. *Proceedings of National Science Council, Republic of China,* **4**, 201–5.

Tabak, H.H. and Cooke, W.B. (1968) The effects of gaseous environments on the growth and metabolism of fungi. *Botanical Review (Lancaster),* **34**, 126–252.

Waites, W.M. and Arbuthnott, J.P. (1990) Foodborne illness: an overview. *Lancet,* **336** (No. 8717), 722–5.

Woolford, M.K. (1990) A review. The detrimental effects of air on silage. *Journal of Applied Bacteriology,* **68**, 101–16.

Yokotsuka, T. (1986) Traditional fermented soybean foods, in *Comprehensive Biotechnology, vol. 3, The Practise of Biotechnology,* (ed M. Moo-Young), Pergamon Press, Oxford, pp.395–427.

Yokotsuka, T. (1991a) Nonproteinaceous fermented foods and beverages, in *Handbook of Applied Mycology, Vol. 3, Foods and Feeds,* (eds D. Arora, K. Mukerji and E. Marth), Marcel Dekker, New York, pp.293–328.

Yokotsuka, T. (1991b) Proteinaceous fermented foods and condiments, in *Handbook of Applied Mycology, Vol. 3, Foods and Feeds* (eds D. Arora, K. Mukerji and E. Marth), Marcel Dekker, New York, pp.329–76.

14

The exploitation of microorganisms in the developing countries of the tropics

K.A. Jones, A. Westby, P.J.A. Reilly and M.J. Jeger

14.1 INTRODUCTION

Many chapters in this book have outlined the potential and commercial use of microorganisms in applied biology. Microorganisms have been exploited at all stages of food production from the use of microbial inoculants in crop production (Chapters 6–8) to the varied uses of microorganisms in food processing (Chapters 11–13). All of the applications described have potential for use in developing countries, and many are already in use. Some applications are, however, particularly suited to the developing countries and can play a significant role in the development process. This is because different criteria for success or failure are often used in the developed and developing world. For example, agricultural production requiring high value inputs, particularly those based on petroleum products, is often difficult to sustain in developing countries. In contrast, low input technologies while often resulting in a lesser increase in productivity can, nevertheless, be sustainable. In many cases, the exploitation of microorganisms offers a low input option to local needs, across a whole range of applications, and often based on, or adapted from, local indigenous knowledge.

Exploitation of Microorganisms Edited by D.G. Jones
Published in 1993 by Chapman & Hall, London ISBN 0 412 45740 7

This chapter reviews two important areas where microorganisms are being exploited in developing countries. These are the microbial **control of crop pests** and the uses of microorganisms in **fermentation of food products**, both areas of current activity in research and developmental aid funded by the Overseas Development Administration of the UK Government. They provide illustrations of how, and to what extent, microorganisms are being exploited in the developing countries of the tropics, given the very different environmental and socio-economic conditions in these regions. In both of these areas, the present state of exploitation is reviewed and the future role of each in development is considered.

14.2 USE OF MICROBIALS FOR PEST CONTROL

Microbial pesticides offer a safe and environmentally benign means of pest control. In many cases, they can also be produced locally in developing countries and are therefore cost-effective, particularly where production techniques are labour intensive (Jones, 1988). Such production also relies mainly on locally-produced items and avoids the need to import chemical pesticides, saving the countries involved scarce foreign exchange.

Microbial pesticides have been used for many years in a number of developing countries, but recently, reflecting the situation in the developed world, there has been an increase in research and interest in their use.

Microorganisms have been used to control insects, weeds, nematodes and plant pathogens (Jeger *et al.*, 1990). The increased interest in microbial pesticides is a result of several factors. A major factor has been the massive upsurge in the use of synthetic chemicals in recent years. Taking Thailand as an example, pesticide use was 20 000 tonnes in 1981/82 and rising at an annual rate of 20% each year (Oka, 1988). Increasing pesticide use, and often misuse, has led to increased problems of resistance. In 1980, it was estimated that world-wide there were 392 species of resistant insects, mites and ticks, compared with only 25 in 1955 (Youdeowei and Service, 1983). Coupled with this has been the well publicized environmental effects, such as toxic residues on produce, destruction of beneficial insects and other non-target organisms, and human poisoning. The World Health Organization (WHO) estimates that world-wide over a million people are poisoned with pesticides each year, and up to 2% of cases may prove fatal (Ambridge and Haines, 1987). These problems are particularly severe, and often unaddressed by legislation in developing countries. Microbial pesticides offer an alternative method of control that do not seem to promote the rapid development of resistance in the field (Briese, 1981, 1986a, b), although recent developments with bacterial insecticides suggest considerable caution should be exhibited, especially with regard to mode of action (refer to Section 14.2.1). Little or no toxic residues are left by microbial pesticides and they are generally harmless to beneficial insects and other non-target organisms, including humans (Laird *et al.*, 1990).

14.2.1 Microbial control of insect pests

Microbial insecticides have been reviewed in this book (Chapters 2 and 5) and by, among others, Burges (1981), Kurstak (1985), Payne (1988) and Roberts *et al.* (1991). Bacteria, fungi, viruses and protozoa have all been used experimentally or commercially for insect control in developing countries. Examples of their use are given below.

Bacteria

The most widely used microbial pesticides are those based on *Bacillus thuringiensis* (Bt), although other species such as *B. sphaericus* (Bs) have also been used. In developing countries, a major success in recent years has been the large-scale use of Bt and Bs to control disease vectors. In the Chinese province of Hubei, 10 tonnes of Bt var. *israelensis* (Bti) and 14 tonnes of Bs are produced annually, enough to treat 12 000 ha of mosquito breeding sites each year. Between 1986 to 1989 this control programme resulted in a decrease in malaria from 5.6 cases/10 000 people to 0.8 cases/10 000 people (Becker, 1990). Bti is also successfully used to control larvae of the blackfly, *Similium damnosum*, the vector of Onchocerciasis. In response to increasing resistance of the insect to chemical insecticides, the WHO embarked on a programme to develop Bti as a control measure. Some 50 000 river kilometres in 11 West African countries are now treated with 700 000 litres of Bti per annum (Walsh, 1986; Becker, 1990, 1991).

Several important agricultural pests are controlled with Bt. This has also often been stimulated by problems resulting from widespread resistance to chemical insecticides. An example is the **diamondback moth**, *Plutella xylostella*, in Malaysia. In this country, 50% of farmers spray crucifers two–three times a week (Ooi and Sudderuddin, 1978). This has resulted in high levels of resistance to all pesticides, for example a 2096-fold resistance to malathion (Lim, 1988), as well as high levels of toxic residues on fresh produce. Unless alternative control measures were to be found, crucifer production in the region was threatened. In response, commercial preparations of Bt var *kurstaki* were imported and used extensively from 1979. In some areas this is the only insecticide used to control this pest (H. Kadir, personal communication); a situation that has now resulted in the development of resistance to the

Bt strain and its substitution with commercial preparations of Bt var. *aizawai*. A similar situation developed in Thailand: in 1985, 46 tonnes of Bt were imported, mainly for use against *P. xylostella* on vegetables in the cabbage growing areas west of Bangkok (Tryon and Litsinger, 1988). Again, constant use has led to resistance development of 25-fold (Zoebelein, 1990).

Bt has also been used on a commercial scale in a number of other developing countries. In China, where Bt was first introduced in 1953 and the first production plant was set up in 1966 (Zhong-yun, 1984), some 880 tonnes of Bt was produced at the Microbial Pilot Plant, Hubei in 1990 for control of 14 agricultural pests (Tianjian *et al.*, 1991); this rose to more than 1000 tonnes in 1991. The main target crops were rice (38.4% of uses), followed by corn (34%), vegetables (15.9%) and cotton (11%). Hussey and Tinsley (1981) reported that the total production of Bt in China in 1977 was 2500 tonnes. Liangsheng (1991) reported the total area then treated with Bt to be approximately 320 000 ha. Significant amounts of Bt were also exported to other South east Asian countries. In Mexico, some 60 tonnes of Bt were used in 1989, representing 4% of the insecticide market and worth some US$1.5 million.

Wider use of Bt in developing countries not only depends on effective control and environmental safety, but also on **reduced cost**. Imported Bt products cost more than many chemical insecticides. This has resulted in many countries investigating cheap methods of local production. In Egypt, Bt var. *galleriae*, used in field trials against *Spodoptera littoralis*, is produced in a pilot-scale submerged fermentation unit located at a sugar factory near Cairo (Foda and Salama, 1991) where agriculture by-products are used in the fermentation process. The cost of the Bt, however, is still two–three times that of chemical insecticides (Zaki, 1991). The cost of Bt var. *morrisoni* produced in the Philippines for mosquito control is 50% more than the cost of DDT. However, research in improved local production of *Bacillus* spp. is continuing using products such as legumes, fermented cassava, cowpea and maize in Africa (Ejiofor and Okafor, 1988, 1989; Obeta and Okafor, 1988) and by-products from monosodium glutamate production in south east Asia (Dharnsthiti *et al.*, 1985). In China, one liquid fermentation medium is based on 4% peanut bran, 0.5% yeast and 0.5% silkworm pupal powder; solid fermentation is also employed using a mixture of wheat-bran and corn meal or soybean and cotton seed cake (Hussey and Tinsley, 1981). The successful control of pests with Bt, particularly mosquitoes, and the possibilities of local production has led to the preparation of UNDP/WHO guide-lines for local production (Dulmage, 1982).

Formulation of Bt products is generally simple; in China, for example, the spores are dried, ground and diluted with chalk powder (Hussey and Tinsley, 1981).

Viruses

Insect viruses have not been commercially developed on the scale of Bt (see Chapter 5 for a description of viral insecticides), but there are several examples of their large-scale use in developing countries, the majority based on local production of viruses. Insect viruses are particularly suited to low-input, local production. At present, they can only be effectively produced *in vivo* which, although labour intensive, does not need the high cost capital equipment required for Bt fermentation. Labour, of course is readily available and comparatively cheap in many developing countries.

The largest field use of viruses world-wide is the use of a **nuclear polyhedrosis virus** (NPV) to control the velvetbean caterpillar, *Anticarsia gemmatalis*, on soya in Brazil (Moscardi and Ferreira, 1985). In 1990, approximately 1 million ha were sprayed with this virus, a single application providing season-long control, whereas chemical insecticides required repeated applications (Moscardi, 1990, 1991). Production is achieved at a cost of US\$2 ha^{-1} (40% of chemical insecticide costs) by collecting infected larvae from the field. In 1990, a plot of approximately 30 ha was sprayed with virus and over a six-week period, three tonnes of infected larvae were collected, enough to spray more than 1 million ha (Moscardi, 1990, Moscardi, personal communication). The virus is air dried and simply formulated with clays to give a wettable powder. Farmers are also recommended to collect infected insects for later application in the field. This successful control measure has also been adopted in Paraguay where 40 000 ha were treated in 1989/90. Other insect pests for which NPVs are being used for control in Brazil include the cassava hornworm, *Errynis ello* (annual application to more than 5000 ha) and the fall armyworm, *Spodoptera frugiperda* (annual application to some 20 000 ha).

In China, **insect viruses** were used to control *Pieris rapae* on 16 000 ha in the early 1980s in more than 30 provinces (Huber, 1986). The NPVs of *Spodoptera litura*, *Helicoverpa armigera*, *Mythimna seperata*, *Euproctis similis* and *E. pseudoconspersa* have also been developed for use on crops including cotton, mulberry and tea (Hussey and Tinsley, 1981) In Colombia, large-scale application of NPV against *Trichoplusia ni* on cotton was so successful that it replaced all other control measures, with the result that the insect no longer was an important pest of that crop (Belloti and Reyes, 1980). Other examples include the control of *Spodoptera littoralis* on cotton in Egypt (Topper *et al.*, 1984); control of *Helicoverpa armigera* and *S. exigua* on vegetables and cotton in Thailand (Huber, 1986; Jones, 1988; Ketunuti and Prathomrut, 1989; Ketunuti and Tantichodok, 1990); control of *H. armigera* on vegetables and cotton in Senegal, Cameroon and Ivory Coast (Anon., 1986); control of *Darna trima* on oil palm in Sarawak (Entwistle, 1986) and control of several lepidopterous pests, including *S. exigua*, *P. xylostella*, *Autographa californica*, *T. ni*, *Bacculatrix thurberiella* and *Pseudoplusia includens* in Guatemala (Cherry, NRI, personal communication, 1992).

The previous examples illustrate the use of viruses as an inundative approach to pest control. The alternative is to adopt **classical biological control**; that is, sustainable control through the introduction, transmission and spread of the disease among the pest population. Such control has been achieved with the non-occluded baculovirus of the rhinoceros beetle, *Oryctes rhinoceros*. This virus, first reported in Malaysia in 1966, is the most important control agent for control of *O. rhinoceros* in the Southern Pacific region (Bedford, 1981). Control is achieved by release of infected adult beetles in the field, which return to breeding sites and infect larvae. The virus was introduced into Western Samoa in 1967 (Mohan, 1985) and, subsequently, reduction in damage levels led to the revival of the copra industry (Marshall and Ioane, 1982). The disease, and hence population suppression, continues from generation to generation. Young and Longworth (1981) found that 84% of beetles were infected with the virus seven years after its introduction into Tonga. Successful control has also been demonstrated in the Maldives, Indonesia, Fiji, Mauritius, Wallis Island, American Samoa, Takelau Islands and Papua New Guinea (Bedford, 1976; Mohan, 1985; Huber, 1986; Zelazny *et al.*, 1989; Zelazny *et al.*, 1992). The closely related *O. monoceros* was also controlled with the virus in the Seychelles (Lomer, 1986). Costs of control are favourable; between 1970 and 1974 10 000 infected beetles were released in Fiji, costing an estimated US$0.9 per beetle, whereas the combined cost of traditional and chemical control measures and crop losses resulting from the insect was estimated to be US $114 000 (Huber, 1986).

One approach to production of insect virus inoculum, through replication in the field, was mentioned above. A similar approach has been adopted in Thailand where a **seed inoculum** of *S. exigua* NPV is supplied to farmers with instruction on how to multiply the virus in the field and store for later use (Huber, 1986; Jones, 1988). Most production, however, involves rearing infected insects in a factory system. This can be achieved using automated machinery, as is done in France by the Calliope company to produce *Mamestra brassicae* virus (Mamestrin) for control of *H. armigera* and other lepidopterous pests (Anon., 1986; Payne, 1988; Campion and Jones, 1991). However, in developing countries it is cost effective to adopt a labour intensive system. Pilot scale production of NPV has been developed in Egypt for control of *Spodoptera littoralis* on cotton and in Thailand for control of *Helicoverpa armigera* and *S. exigua* on a number of crops including cotton and vegetables (Jones, 1988, 1990; McKinley *et al.*, 1989). In Egypt, production, formulation and application costs of *S. littoralis* NPV were estimated to be US$18.06 ha^{-1}

(including the capital costs of constructing a production facility for 25 000 ha) compared with a range of US$17.80–42.00 for the standard control measures (Anon., 1991; Bickersteth, 1990). The 'insect factory' approach has been used, at least on experimental or pilot-scales, in a number of other countries: for example in India for *S. litura* and *H. armigera* NPVs; in Guatemala and Nicaragua for *A. californica* NPV (Cherry, NRI, personal communication, 1992; Mulock *et al.*, 1990); and in Zimbabwe for production of a seed inoculum for control of *Trichoplusia oricalsia* (Kunjeku, personal communication, 1990).

Formulations of insect viruses in **developing countries** are simple or non-existent. Generally, it is too expensive to purify the virus from dead insects and therefore unpurifed virus is used (Topper, 1984; Jones, 1988). This itself can provide considerable environmental protection to the virus, removing the need for some additives (Jones and McKinley, 1986; Jones and Grzywacz, 1990). Taking the formulation of *S. littoralis* NPV in Egypt as an example, this is formulated as a wettable powder by combining freeze-dried, virus-infected insects with china clay and a mixture of synthetic silica and wetting agent (McKinley *et al.*, 1989). Locally available materials are often added to unpurified virus suspensions to improve persistence or activity, for example molasses, cotton seed oil, jaggery or whole milk (Topper, 1984; Topper *et al.*, 1984; Dhandapani *et al.*, 1987; Jones, 1990).

Fungi

Insect pathogenic fungi have also been exploited on a reasonably large scale in some developing countries, as with entomopathogenic viruses predominantly through local production. An example of large scale use is again Brazil, where *Metarhizium anisopliae* is used to control the froghopper (*Mahanarva posticata*) in sugar-cane. In North east Brazil, over 50 tonnes of a powder formulation of the fungus have been produced since 1977 for treatment to more than 600 000 ha, reported to be the largest use of fungi for insect control in the world (Mendonça, 1992). Chemical treatments are now applied to only 18.4% of the area requiring protection (Zimmerman, 1986). This control measure is also being used in Honduras (Cherry, NRI, personal communication 1992). *M. anisopliae* has also been used in combination with the baculovirus to control *O. rhinoceros* in Indonesia (Munaan and Wikardi, 1986), as well as being effective against the Hispine beetle, *Brontispa longissinia* and *Pleisispa* spp. In China, 400 000 ha of corn were treated in 1977 with *Beauvaria bassiana* for control of the stem-boring caterpillar *Ostrinia nubilalis*; and use of this fungus involving some 1000 production units and up to 20 000 people (Hussey and Tinsley, 1981). *B. bassiana* has also been used to control *Dendrolinus punctitus* on pine and *Nepholettix* spp.

on rice and tea (Hussey and Tinsley, 1981). *Verticillium lecanii* has been used to control the green coffee bug, *Coccus viridis*, in India reducing the pest population to negligible proportions (Mohan Naidu, 1985). This fungus has also been applied on a commercial farm in Venezuela to control *Bemisia tabaci*. Field trials with fungi to control major pests such as *Schistocerca gregaria* in Niger and *Nilapavata lugens* in Philippines have also been reported (Gillespie *et al.*, 1987; Bateman, 1992), but treatments have not yet reached commercial use.

Local production of fungus normally utilizes agricultural products. *M. anisopliae* used to control *O. rhinoceros*, was grown on bran, brewer's grain, corn/maize or rice (Kaske, 1988). Rice or bran is used to produce *M. anisopliae* in Brazil and rice bran or corn stalks to produce *B. bassiana* in China (Aquino *et al.*, 1975; Prior 1989). In Venezuela, several fungal strains, including *B. bassiana*, *V. lecanii*, *Nomuraea rileyi* and *Metarhizum* spp. are produced at a small factory using rice as a substrate. In Columbia, *B. bassiana* is produced in a simple liquid culture (Prior, IIBC, personal communication, 1992). In Taiwan, *B. bassiana* is cultured on wine derivatives (Hou and Chiuo, 1991). With all of these examples, simple equipment is used and normally the fungus is inoculated into bottles or plastic bags containing the substrate (Prior, 1989; Goettel and Roberts, 1992; Mendonça, 1992). In some cases, e.g. Venezuela and Columbia, sporulation takes place in shallow, open trays.

Formulation of fungal products is also simple. In Brazil and Venezuela, rice supporting the fungus is sold to farmers who remove the fungus by washing in water which is then applied in the field. In Brazil, the fungus and substrate is also ground to a powder. This is sold under the tradename Metaquino (Mendonça, 1992). In Columbia the fungus is dried and ground (Prior, IIBC, personal communication, 1992). In Taiwan, experimental formulations are simply prepared by mixing the fungus with sand (Hou and Chiou, 1991). In China, the fungus plus growth medium (rice/wheatbran) is dried and diluted 1:10 with sand (Hussey and Tinsley, 1981). Experimental preparations of mycopesticides for locust and grasshopper control are formulated in a mixture of vegetable oil and mineral kerosene (Prior *et al.*, 1992).

Protozoa

No commercial use of protozoa has been reported in developing countries. Some trials have been undertaken, most notably the large-scale trials to control grasshoppers in Cape Verde using bran-bait formulations of *Nosema locustae* (Lobo Lima *et al.*, 1992).

As with insect viruses, protozoans are produced *in vivo* by inoculating susceptible insects (Wilson, 1982); this technology should therefore be suitable for use in developing countries.

14.2.2 Microbial control of nematodes

The potential for microbial control of nematodes has been reviewed by Kerry (Chapter 4) and by Davies *et al.* (1991). In 1985, van Gundy reported that over 200 pathogens, parasites and predators were known to have potential for biological control of nematodes, of these 75% were fungi (van Gundy, 1985). At that time, only two commercial products were available, and therefore used on any scale; both of these were in France and consisted of nematode trapping fungi (Mankau, 1981). Some research has been done in developing countries to assess the potential of microbial control of nematodes. Field trials in Pakistan and Malawi with the mycelium-forming bacterium, *Pasteuria penetrans*, against root-knot nematodes, *Meloidogyne* spp., have demonstrated that the nematode populations can be suppressed (Gowen and de Silva, 1989; Anon., 1991), and that this can be achieved by re-incorporation into the soil of roots which contain nematodes infected with *P. penetrans* (Daudi *et al.*, 1990; Gowen and Ahmed, 1990). Production of *P. penetrans* is carried out *in vivo* but present systems are inconsistent (Gowen and Channer, 1988; Anon., 1991). Davies *et al.* (1991) concluded that the potential for microbial control of nematodes in tropical agriculture has not been realized. A major reason for this is that limited studies have been carried out to date, particularly on fundamental aspects of host–pathogen interactions and virulence, topics that will need to be addressed.

14.2.3 Microbial control of plant pathogens

Control of **soil-borne** and **foliar pathogens** with microorganisms has been studied experimentally and a few cases exist of commercial exploitation, the majority in developed countries (for example, Lindow, 1985; Martin *et al.*, 1985; Schroth and Hancock, 1985; Deacon, 1988; Rishbeth, 1988; Wong, 1986; Buck, 1988). Jeffries and Jeger (1990) reviewed the control of post-harvest diseases of fruit using microbes and concluded that there was good potential for biocontrol, but wide-scale adoption probably required legislation for withdrawal of chemical fungicides. As with chemical insecticides, environmental and resistance concerns may well lead to this.

Some use in developing countries is reported. In Chile, the fungus *Trichoderma* spp. was registered in 1987 to control silver leaf on Japanese plums caused by *Chondrostereum purpureum*. Control was superior to chemical measures (Poblete, 1988). *Trichoderma* was also reported to be used in South America for control of chlorotic vine leaf curl (Ricard and Highley, 1988). Successful control of mango anthracnose on an experimental level has been achieved in the Philippines by a post-harvest dip in a *Pseudomonas* suspension (Koomen *et al.*, 1990). Use of antagonistic microorganisms has also shown potential for control of anthracnose on banana (Kanapathipillai and Jantan, 1986). Trials have been undertaken on disease suppression by the use of mild or avirulent strains of pathogens. In Thailand, the introduction of mild strain, nitrous acid-induced mutants of papaya ringspot virus resulted in a reduction in disease incidence (Amaritsut *et al.*, 1988).

Local **production** and **formulation** methods for commercial use of these microorganisms in developing countries have not been studied in detail. Simple media for production of fungi, such as potato dextrose agar has been used, as well as simple fermentation technology. Application of some of these microorganisms lends itself to low input technology. For example, *Trichoderma* used to control soil pathogens has been delivered to the soil on wheat bran or inactivated rye and sunflower seeds, as well as molasses-enriched meals and clay granules (Martin *et al.*, 1985). Such an approach would be suitable for developing countries.

14.2.4 Microbial control of weeds

Control of weeds is also an area where there is potential for microbial control (see Chapter 1; Cullen and Hasan, 1988: Templeton and Trujillo, 1991), but to date there are very few examples of use in developing countries. Evans (1987) listed the pathogens of 18 important tropical or subtropical weeds and concluded that the prospects for microbial control of eight species seemed to be good, either through a classical or inundative approach. Hassan (1986) listed 42 examples of projects using pathogens to control weeds. Of these, only five were from developing countries – two from Chile, one from Indonesia and two from South Africa. Examples of successful use in developing countries include the introduction of *Phragmidium violacaeum* into Chile for classical biological control of *Rubus* spp. (Cullen and Hasan, 1988) and the use of a mycoherbicide based on *Colletotricum gloeosporioides* to control dodder, *Cuscuta* spp., in China (Wang, 1986). Trials are also currently underway in Thailand on the use of a mycoherbicide to control *Rottboellia cochinchinensis* (Ellison and Evans, 1992). Formulation studies in this case have included the addition of vegetable gums, use of invert emulsions and addition of low doses of chemical herbicides. Production is carried out on a small laboratory scale only using agar plates.

14.2.5 The future role of microbial pesticides in development

These examples provide an illustration of the widespread use of microbial pesticides in developing countries. They are by no means an exhaustive list and undoubtedly many more examples of microbial control exist. In two cases, that of insect viruses and fungi, they represent the largest use of these products world-wide. In both cases, the products are produced locally, and local commercial companies have shown an interest in producing them. In a number of countries, for example Brazil, Guatemala and Venezuela, small private companies producing microbial pesticides are well established. Large-scale use of Bt has been adopted where local production has been used, or when commercial products are easily available and their use is the only, if often more expensive, option for control. At present, predictions of Bt use put the total market for the Far East, Middle East, Africa and Central and South America at 45% (US $45 million) of the total world market, a figure that includes developed countries in the region such as Japan (Lambert and Peferoen, 1992). The authors conclude that this is a limited market. However, the rapid rise in interest and use of microbial pesticides, as well as the considerable research effort now being directed toward commercial and small-scale local production, will undoubtedly lead to an increase in their use in developing countries. Indeed, in many cases, it is the developing countries that are leading the way in the entrepreneurial development and large-scale use of these pesticides.

Registration procedures

A further push to the wider uptake of microbial pesticides will be the adoption of appropriate registration procedures by developing countries. This process is already underway. The FAO has issued guidelines for registration (FAO, 1988) and countries like India have already drawn up their own procedures for some microbial pesticides (Jeger *et al.*, 1990). These procedures recognize the inherent safety of microbial pesticides and are simpler and therefore less expensive to enact compared with the procedures for chemical insecticides. Lisansky (1984) estimated the cost of safety tests for microbial pesticides in UK and the USA to be £40 000 compared with £3 million for chemical active ingredients. It is currently being suggested that generic registration of microbial pesticides should be adopted. Registration procedures should also consider appropriate means of efficacy testing. Microbial pesticides behave differently from most chemical pesticides, for example they often take longer to kill the pest. **Standard efficacy tests** by local agriculture ministries generally have been developed for chemical pesticides and may not be suitable for microbial pesticides. To register an insecticide for use on cotton in Egypt, data normally are required on the level of mortality obtained 48 hours after feeding field-sprayed leaves to third instar larvae. Microbial pesticides, in general, will show little mortality after this time, but results of field trials have shown that they can be effective control agents (e.g. Campion and Jones, 1991). This is an area that obviously needs to be addressed.

Other developments required for increased uptake of microbial pesticides in developing countries include **dissemination of information** and
training of producers and users, including appropriate mechanisms for
the transfer of technology from developed to developing countries and
between developing countries. It is also important that the nature and
mode of action of microbial pesticides be explained to farmers and that
efficacy and cost-effectiveness be demonstrated in the field. In the
developing country context, this training and technology transfer is
likely to require the investment of public rather than private funds, at
least in the initial stages; these funds need to be available and well
targeted.

Finally, policy-makers and users should be aware that indiscriminate
and excessive use of microbial pesticides may lead to the same problems
of resistance encountered with conventional chemical pesticides. This is
already starting to be seen with Bt. Microbial pesticides should primarily
be developed for use within an integrated strategy for sustainable pest
control.

14.3 FERMENTATION – A LOW-COST, APPROPRIATE
TECHNOLOGY FOR LESSER DEVELOPED COUNTRIES

Lesser developed countries require food processing technologies that
are technologically appropriate, suitable for tropical regions and have a
low cost. Fermentation is one such technology that has been developed
indigenously for a wide range of food commodities including cereals,
root crops, legumes, fruits, vegetables, milk, fish and meat. This section
of this chapter considers the advantages of fermentation as a processing
operation in developing countries, gives examples of food fermentations
that demonstrate these advantages, and considers the role of fermentation technologies and fermented foods in development.

The term **'fermentation'** is often used with imprecision when referring
to foods (Adams, 1990). Strictly, it describes a type of energy yielding
microbial metabolism in which an organic substrate is incompletely
oxidized and an organic compound acts as an electron acceptor; examples are the production of ethanol by yeasts or the production of organic
acids by lactic acid bacteria. However, the term fermented food is
applied to any foods that have been subjected to the action of microorganisms or enzymes so that desirable biochemical changes cause
significant modification to the food (Campbell-Platt, 1987).

Foods such as **tempe** (from soybean), involving mould growth, and
dawadawa (from African locust bean), involving growth of *Bacillus* spp.,
are therefore considered as fermented foods, although the metabolism
of the organisms involved may not strictly be fermentative.

There is a wide range of fermented foods in the world and to describe each in detail would be impossible here. Several excellent books and reviews (Steinkraus, 1983, 1989; Wood, 1985; Campbell-Platt, 1987; Uzogara *et al.*, 1990) have been devoted to describing and cataloguing the production, use, nutritional changes in and microbiology of the many fermented foods.

14.3.1 Advantages of using fermentation

As a unit operation in food processing, fermentation offers a large number of advantages. These are: food **preservation**, improved food **safety**, enhanced **flavour** and **acceptability**, increased **variety** in the diet, improved **nutritional value**, reduction in anti-nutritional compounds and in some cases improved functional properties. Each of the advantages is discussed below, together with relevant examples. Often, a particular type of fermentation offers more than one advantage and the advantage under which it is discussed may not be the primary reason for carrying out the fermentation; for example, the lactic fermentation of fish may offer a means of preservation, but may also impart desirable flavour to the products.

Low cost preservation technique

Lactic acid fermentation is a common means of preservation in tropical countries (Cooke *et al.*, 1987). Preservation techniques used in developed countries (such as refrigeration, freezing, canning, modified atmosphere packaging, irradiation) have a high cost and certain infrastructural requirements which greatly restrict their use in the developing world. The high ambient temperatures of most tropical countries heighten the need for low cost, low technology preservation techniques such as salting, drying and traditional fermentations.

There is a wide range of perishable substrates that can be preserved by lactic fermentation, including meat (for example, nham from Thailand, chorizo from Mexico), fish (burong isda from Philippines, pla-som and som-fak from Thailand), leafy vegetables (kimchi from Korea) and cassava (akpu from Nigeria). Cereals (maize, sorghum, rice) can also be lactic fermented but preservation is not the primary reason for fermentation.

The dominance of lactic acid bacteria in food fermentations is a consequence of their dominance in carbohydrate-rich anaerobic environments and their rapid suppression of food spoilage and pathogenic bacteria. This antagonistic property can be a result of organic acid production, reduction in pH value, hydrogen peroxide production, carbon dioxide production, nutrient depletion or production of antibiotics (Andersson, 1986).

Sea foods

Lactic acid fermented sea foods are an example of fermentation as a low cost preservation technique. Fish are highly perishable proteinaceous foods which spoil rapidly when exposed to tropical ambient temperatures. The keeping time of fresh fish which are stored under these conditions is less than one day (Barile *et al.*, 1985a, b). In many developing countries, particularly in rural coastal areas, usual techniques for extending the shelf-life of fish such as chilling, freezing and canning are too expensive for domestic markets. In many poorer communities, fish processing depends entirely on traditional processing techniques which are low-cost and rely upon the reduction of water activity (salting, drying and smoking) or on a fermentation process. Traditional fermented fishery products are particularly popular in south east Asia, a range of products exist and are reviewed by Lee *et al.* (1991).

The term **fermented fish** commonly applies to two categories of products (Adams *et al.*, 1985); fish and salt formulations such as fish sauce; and fish, salt and carbohydrate mixtures, which involve a lactic fermentation. These products are not well defined and there are many variations in processing methods and raw materials used (Lee, 1990). Manufacture of fish sauce occurs on a commercial scale and although these products could be described as fish hydrolysates, bacteria which can tolerate high salt concentrations seem to be involved in flavour and aroma development (Thonghtai *et al.*, 1990). In the second category of products, lactic acid fermentation occurs and contributes to an extended shelf-life. There are many examples of this type of product where the storage life at ambient tropical temperatures is extended by a combination of increased acidity and the addition of salt (Lee *et al.*, 1991).

The lactic fermentation involves a succession of bacteria where lactic streptococci and leuconostocs dominate in the early stages (1–3 days). These eventually give way to homofermenters (pediococci and lactobacilli) which are more acid tolerant (3–10 days) (Solidum and Acedevo, 1983). The heterofermenters produce both acetic and lactic acids and production of acetic acid, which has a greater anti-microbial effect (Adams, 1990), is important in the early stages of fermentation to ensure the inhibition of competing spoilage and pathogenic bacteria.

Lactic fermented seafoods are produced mainly by women on a household scale as an income-generating activity. They are consumed as main courses, rather than as condiments or side-dishes, as is the role of fish sauces and fish pastes.

Improved food safety

Food safety is an important issue in developing countries. Processing techniques that ensure food safety are required at both the rural and urban levels, particularly in view of the frequently poor sanitary conditions and high ambient temperatures. Lactic fermentation as a food preservation technique has been discussed above; in addition there are many reports of the ability of lactic acid bacteria to inhibit the growth of food-borne pathogenic bacteria such as *Salmonella* spp., *Shigella* spp., *Escherichia coli* and *Staphylococcus aureus* (Nout *et al.*, 1989; Mbugua and Njenga, 1992) in a range of fermented foods used for weaning infants.

It was estimated that based on 1980 data, 4.6 million deaths of children aged under five years occurred in developing countries as a result of **acute diarrhoeal disease**. In this light, weaning foods have been a particular focus of attention for research on the beneficial effects of fermentation (for example, Alnwick *et al.*, 1987; Nout *et al.*, 1989; Mensah *et al.*, 1991). Recently, there have been attempts to correlate the use of fermented weaning gruels in villages with reduced instances of diarrhoea. Design of such studies is difficult because of the problem in identifying suitable controls not using fermented gruels. Lorri and Svanberg (1992) selected two groups of about 100 children from two villages in Tanzania of less than five years of age based on the use or not of fermented gruels. Bi-monthly follow-ups on the occurrence of diarrhoea over a nine-month study period were carried out. The mean number of diarrhoea episodes over the study period was 2.1 for children using fermented porridges, compared with 3.5 for those using non-fermented porridges, a significant difference at $P<0.0001$. It was concluded that lactic fermentation is an important factor in the preparation of weaning gruels. However, further studies of a similar nature are required to verify this conclusion.

Improved flavour, variety and acceptability

Fermentation is well known for its ability to improve the flavour and acceptability of food commodities in developed countries, for example the production of wine from grape juice and cheese from milk. Similar improvements have been reported for food commodities in developing countries where bland raw materials are turned into foods with enhanced flavour. With the imparting of flavour, it is also possible to introduce variety into the diet.

Tempe is a white mould-covered cake produced by the fungal fermentation of dehulled, hydrated (soaked) and partially cooked soyabean cotyledons (Steinkraus, 1983) that is produced mainly in Indonesia. Van Veen and Steinkraus (1970) noted that tempe has little of the initial soyabean flavour, making it more acceptable than the raw material.

Cassava is an important staple root crop for many people in Sub-Saharan Africa. Cassava stores poorly after harvest and has to be processed into a form that can be stored. Many different fermented

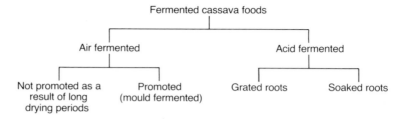

Figure 14.1 Classification of fermented cassava foods, based on primary (first) method of fermentation.

foods can be produced and hence fermentation serves as a means of introducing variety in the diet. NRI has contributed to the recent Collaborative Study of Cassava in Africa (COSCA), which is a study of all aspects of cassava production and processing in six African countries (Ivory Coast, Ghana, Nigeria, Uganda, Tanzania and Zaire) (Nweke, 1990). Information obtained at the village level (233 villages) on the classification of 488 processed cassava products showed that more than 90% were fermented, according to the definitions used. Fermentation is therefore confirmed as being an important technique in the processing of cassava in Africa. Fermented cassava foods can be divided into a number of different classes based upon the primary (first) method of fermentation (Figure 14.1).

Grated acid fermented products include roasted granules such as West African gari, steamed products such as Ivorian attieke and Ghanaian yakayake, and fermented pastes such as Ghanaian agbelima. Products prepared by soaking roots in water include acidic pastes such as Zairian chickwangue, Nigerian fu-fu and akpu, and dried products such as Nigerian lafun. Air fermented cassava products, where fresh cassava can be covered with leaves for a number of days to encourage mould growth prior to sun drying, include products such as dark moulded cassava flour (Essers and Nout, 1989) and Tanzanian ugada. The other class of air fermented products become mouldy as a result of long drying times and include products such as Ghanaian and Ivorian kokonte. Fermentation in this latter case may be seasonal, i.e. in the dry season the product may dry so quickly that fermentation does not occur.

Reduction in anti-nutritional factors

Anti-nutritional factors are of concern with the consumption of certain foods, for example lactose in milk, cyanogenic glucosides in cassava, phytate in cereals and legumes, and oligosaccharides in legumes. Fermentation has been reported to be responsible for the removal of these anti-nutritional factors (Chevan and Kadam, 1989; Adams, 1990; Liew and Buckle, 1990).

Although fermentation has been associated with the removal of anti-nutritional factors, there are examples of foods where fermentation is not directly responsible for the observed phenomenon. The hydrolysis of cyanogenic glucosides during the processing of grated fermented cassava products has been attributed to fermentation. Recent work (Vasconcelos *et al.*, 1990) has shown that it is the release of an endogenous enzyme during grating that is responsible for hydrolysis of the glucoside. The level of endogenous enzyme activity and the efficiency of grating are therefore key parameters in glucoside hydrolysis (Westby and Twiddy, 1992a).

Improved nutritional value

There are many examples of fermentation improving the nutritional value of foods. Chevan and Kadam (1989) have reviewed the situation in relation to **cereals**. Improvements include reduction in anti-nutritional factors such as tannins leading to better protein digestibility (Svanberg and Lorri, 1992) and hydrolysis of phytate leading to improved iron availability. Vitamin production during some fermentations has been reported (for example, tempe fermentation of soybeans and cowpeas; Djurtoft and Nielsen, 1983).

Modified functional properties

Fermentation can bring about changes in functional properties. One example of this is the production of fermented cassava starch in South America, notably in Brazil and Colombia. In Brazil, the state of Minas Gerais is the main production area with smaller amounts produced in São Paulo, Parana and Santa Catarina States. Fermented starch is produced because it has modified functional properties, the most important of which is expansion during cooking which allows the production of textured products. No chemical or biological raising agent is required during cooking when fermented cassava starch is used. The most important Brazilian products from fermented cassava starch are biscoitos de polvilho, an expanded snack food, and pão de queijo, a cheese bread.

The technology for processing fermented starch is very simple (Cereda, 1987). Starch is extracted from the cassava roots using water. After removal of the cellulosic fibre, the starch is sedimented and transferred into tanks to ferment under water. Fermentation takes between 20 and 40 days after which time the product is sun dried.

The mechanism by which the change in functional properties is brought about has not been elucidated, but fermentation and sun drying are key factors. Camargo *et al.* (1988) associated the production of acids with the ability to expand during cooking. A high content of lactic acid has been associated with good quality starch.

14.3.2 The future role of fermentation

The nature of food processing and marketing chains in rural and urban populations in developing countries is related to the stage of development, level of income and socio-cultural characteristics of different population groups. To make generalizations on the future role of fermented foods in development is therefore difficult. This discussion has been centred on a number of important developmental issues.

Indigenous foods in developing countries

Rural–urban migration is one of the major problems facing many countries, particularly in Africa. Traditional food processing systems will have to adapt to the consequences of urbanization. The type of adaptation necessary will be dictated by the nature of the raw material; for example commodities that can be stored such as maize and rice can be transported to urban centres and processed, whereas root crops such as cassava are difficult to transport and store poorly, and as such are better processed close to production.

People in urban centres still demand **traditional foods** made from locally grown crops. Such consumption patterns should be encouraged as they are sustainable and avoid dependence on imported materials. The processing of many traditional foods involves fermentation and urban demand may result in increased industrialization of traditional food processing. For some indigenous fermented foods, for example: mageu (Holzapfel, 1992), soy sauce (Fukushima, 1989) and gari (Onyekwere *et al.*, 1989), some degree of industrialization has been achieved. Urban demand will probably result in a requirement for foods of a consistent quality that are more convenient to use in the home. The economics of modified processing systems to meet these requirements will have to be carefully evaluated to ensure products are competitive in the marketplace.

In cases where rural production of traditional foods supplies urban areas, the income generated may help stem migration to urban centres. In such cases, the needs of rural processors (predominantly women) need to be addressed. This may involve reducing their work load or increasing turnover of materials and releasing finance by, for example, speeding up fermentations.

Contribution of fermentation to food safety

Food safety problems are usually associated with the contamination of raw materials by pathogenic or toxin-forming bacteria or fungi. Equally important are the preparation and storage of foods in the home, where poor standards of hygiene and sanitation may exist and water supplies are contaminated. Fermentation is clearly one low cost, low technology processing operation that can contribute to food safety.

In the future, it will be important to identify situations where fermentation may be successfully used to contribute **safe food delivery**. One example of where this principle has been employed is in the application of an accelerated souring (lactic fermentation) to the rehydration step for tempe preparation. A promoted lactic fermentation during soaking of beans in water was shown to inhibit the growth of *S. aureus* (Nout *et al.*, 1988) and *B. cereus* (Nout *et al.*, 1987), whereas soaking without a promoted fermentation allowed the organisms to grow.

As discussed above, fermentation, should have an important role to play in the development of weaning foods. Tomkins *et al.* (1988) identified several important areas of research in this field centring on the lack of basic definitive information rather than anecdotal experience. There is still insufficient information on the role of fermentation in preventing contamination of weaning foods and the role of fermented foods in the management of diarrhoea.

Role of starter cultures

The provision of **starter cultures** is an issue commonly raised in relation to fermented foods in developing countries. There are three main simple, low cost, low technologies: natural inoculation, transfer of an amount of old batch of fermented product to a new batch ('back-slopping'), and indigenously derived starter cultures. Some fermentations (for example gari processing; Westby and Twiddy, 1992b) do not currently need starter cultures because the processing environment provides a good natural inoculum. Higher levels of technology should only be used for these products if the need can be demonstrated. In some cases, back-slopping is a suitable technology to ensure a reproducible fermentation and has been applied to new situations (Nout *et al.*, 1988). Several indigenous starter cultures exist, such as those for tempe (Steinkraus, 1983), attieke and the ragi starter culture for tapé (Steinkraus, 1983).

The development of specific starter cultures with desirable properties (such as antagonistic activity, production of vitamins, removal of anti-nutritional factors) should be combined with the development of appropriate technologies for disseminating and propagating the culture.

Higher levels of starter culture technologies (such as freeze-dried inocula) are only currently suitable for large-scale industrial applications.

Research on fermented foods

Research work on indigenous fermented foods has concentrated on the characterization (microbiological, chemical and biochemical) of changes that occur during the fermentation. This work has provided a strong basis on which to build, and clearly more work has to be done. To be effective from a development standpoint, however, research will have to focus more on addressing the needs of developing countries, both in terms of indigenous processes and future developments in biotechnology. In the section on fermentation and its use in developing countries, the discussion has centred upon indigenous fermented foods because this is the state-of-the-art in many developing countries. Indigenous fermented foods are an application of biotechnology at its most basic level and the point from which we believe progress on biotechnology will, by way of commercialization and industrialization, be made in the future. Research is also being done on more advanced forms of biotechnology for use in the food/feed industries of developing countries (for example, Doelle, 1989) such as production of enzymes, protein enrichment of starchy staples, utilization of processing wastes and production of food components. Such technologies could have wide applications in the future.

REFERENCES

Adams, M.R. (1990) Topical aspects of fermented foods. *Trends in Food Science and Technology*, December 1990, 140–4.

Adams, M.R., Cooke, R.D. and Rattagool Pongpen (1985) Fermented fish products of S. E. Asia. *Tropical Science*, **25**, 61–73.

Alnwick, D., Moses, S. and Schmidt, O.G. (1988) *Improving Young Child Feeding in Eastern and Southern Africa*. Ottawa, Canada: International Development Research Centre.

Amaritsut, W., Prasartsri, W. and Craig, I.A. (1988) Crop protection and IPM in rainfed cropping systems in Northeast Thailand, in *Pesticide Management and Integrated Pest Management in Southeast Asia*, (eds E.H. Teng and K.L. Heong), Consortium for International Crop Protection, Maryland, pp.135–43.

Ambridge, E.M. and Haines, I.H. (1987) Some aspects of pesticide use and human safety in Southeast Asia, in *Proceedings, 11th International Congress of Plant Protection, Vol.II*, Manila, Philippines, October 5–9, 1987 pp.219–27.

Andersson, R. (1986) Inhibition of *Staphylococcus aureus* and spheroplasts of Gram-negative bacteria by an antagonistic compound produced by a strain of *Lactobacillus plantarum*. *International Journal of Food Microbiology*, **3**, 149–60.

Anon. (1986) Mamestrin[R] a new biological insecticide, *Technical leaflet*, Calliope S.A., Beziers, France.

Anon. (1991) Biorational control technologies, in *Report on Operational Programmes: 1989–91*, Natural Resources Institute, Chatham, UK, pp.162–71.

Aquino, M. de L., Cavalcanti, V.A.L.B., Sena, R.C. and Queiroz (1975) Novo tecnologia de multiplicacao do fungo *Metarhizium anisopliae*. *CODECAP Technical Bulletin (Recife, Pernambuco)*, **4**, 1–31.

Barile, L.E., Milla, A.D., Reilly, A. and Villafsen, A. (1985a) Spoilage patterns of Mackerel (*Rastrelliger faughni* Matsui). 1. Delays in icing. *ASEAN Food Journal*, **1**, 70–7.

Barile, L.E., Milla, A.D., Reilly, A. and Villafsen, A. (1985b) Spoilage patterns of Mackerel (*Rastrelliger faughni* Matsui). 2. Mesophilic and psychrophilic spoilage. *ASEAN Food Journal*, **1**, 121–7.

Bateman, R. (1992) Controlled droplet application by myco-insecticides: an environmentally friendly way to control insects. *Antenna*, **16**, 6–13.

Becker, N. (1990) Microbial control of mosquitoes and black flies, in *Proceedings, Vth International Colloquium on Invertebrate Pathology and Microbial Control*, Adelaide, Australia, 20–24 August 1990, Society of Inverebrate Pathology, pp.84–9.

Becker, N. (1991) Microbial control of blood sucking diptera. In *Abstracts, Bt1991 First International Conference on* Bacillus thuringiensis. Oxford, UK, July 28–31 1991, Institute of Virology and Environmental Microbiology, Oxford, (unnumbered).

Bedford, G.O. (1976) Use of virus against the coconut palm rhinoceros beetle in Fiji. *PANS*, **22**, 11–25.

Bedford, G.O. (1981) Control of the rhinoceros beetle by baculovirus, in *Microbial Control of Pests and Plant Diseases 1970–1980*, (ed H.D. Burges), Academic Press, London, pp.409–26.

Bellotti, A.C. and Reyes, J.A. (1980) Progress in microbial control (1975–1980): World activities: N. South and Central America, in *Proceedings, Worshop on Insect Pest Management with Microbial Agents: recent Achievemnts, Deficiencies and Innovations*, Ithaca, N.Y., p.20.

Bickersteth, S. (1990) A socio-economic assessment of the use of viruses for control of cotton leafworm in Egypt. *NRI Overseas Assignment Report R1634 (R)*, NRI, Chatham, pp.37.

Briese, D.T. (1981) Resistance of insect species to microbial pathogens, in *Pathogenesis of Invertebrate Microbial Diseases*, (ed E.W. Davidson), Allanheld Osmum, New Jersey pp.511–45.

Briese, D.T. (1986a) Host resistance to microbial control agents, in *Biological Plant and Health Protection, Biological Control of Plant Pests and Vectors of Human and Animal Diseases, Progress in Zoology, Vol. 32*, (ed J.M. Franz), Fischer, Stuttgart, pp.233–56.

Briese, D.T. (1986b) Insect resistance to baculoviruses, in *The Biology of Baculoviruses, Volume II, Practical Application for Insect Control*, (eds R.R. Granados and B.A. Federici), pp.237–63.

Buck, K.W. (1988) Control of plant pathogens with viruses and related agents. *Philosophical Transactions of the Royal Society of London Series B*, **318**, 295–317.

Burges, H.D. (ed.) (1981) *Microbial Control of Pests and Plant Diseases, 1970–1980*, Academic Press, London, pp.949.

Camargo, C., Colonna, P., Buleon, A. and Richard-Molard, D. (1988) Functional properties of sour cassava starch (*Manihot utilissima*) starch: Polvilho azedo. *Journal of the Science of Food and Agriculture*, **45**, 273–89.

Campbell-Platt, G. (1987) *Fermented Foods of the World* London: Butterworths.

Campion, D.G. and Jones, K.A. (1991) Pheromones and microbial insecticides for the control of cotton pests, in *Growing Cotton in a Safe Environment, 50th Plenary Meeting of the ICAC*, Antalya, Turkey, September 1991, pp.10–18.

Cereda, M.P. (1987) Tecnologia e qualidade do polvilho azedo. *Inf. Agropec., Belo Horizonte*, **13**, 63–8.

Chevan, J.K. and Kadam, S.S. (1989) Nutritional improvement of cereals by fermentation. *CRC Critical Reviews in Food Science and Nutrition*, **28**, 349–400.

Cooke, R.D., Twiddy, D.R. and Reilly, P.J.A. (1987) Lactic acid fermentation as a low cost means of food preservation in tropical countries. *FEMS Microbiology Reviews*, **46**, 369–79.

Cullen, J.M. and Hasan, S. (1988) Pathogens for the control of weeds. *Philosophical Transactions of the Royal Society of London Series B*, **318**, 213–24.

Daudi, A.T., Channer, A.G., Ahmed, R. and Gowen, S.R. (1990) *Pasteuria penetrans* as a biocontrol agent of *Meloidogyne javanica* in the field in Malawi and in microplots in Pakistan, in *Proceedings, Brighton Crop Protection Conference – Pest and Diseases, Vol. 1*, November 19–22, 1990, Brighton, British Crop Protection Council, Farnham, pp.253–7.

Davies, K.G., de Leij, A.A.M. and Kerry, B.R. (1991) Microbial agents for the biological control of plant-parasitic nematodes in tropical agriculture. *Tropical Pest Management*, **37**, 303–20.

Deacon, J.W. (1988) Biocontrol of soil-borne plant pathogens with introduced inocula. *Philosophical Transactions of the Royal Society of London Series B*, **318**, 249–64.

Dhandapani, N., Jayaraj, S. and Rabindra, R.J. (1987) Efficacy of ULV application of a nuclear polyhedrosis virus with certain adjuvants for control of *Heliothis armigera* (Hbn) on cotton. *Journal of Biological Control*, **1**, 111–17.

Dharnsthiti, S.C., Pantuwatana, S. and Bhumiratana, A. (1985) Production of *Bacillus thuringiensis* and *Bacillus sphaericus* 1593 on media using a byproduct from a monosodium glutamate factory. *Journal of Invertebrate Pathology*, **46**, 231–8.

Djurtoft, R. and Nielsen, J.P. (1983) Increase in some B vitamins, including B_{12}, during fermentation of tempe produced from cowpeas or soy bean. *Journal of Plant Foods*, **5**, 135–41.

Doelle, H.W. (1989) Socio-ecological biotechnology concepts for developing countries. *MIRCEN Journal*, **5**, 391–410.

Dulmage, H.T. (1982) Guidelines on the local production for operational use of *Bacillus thuringiensis*, especially serotype H-14, in *Guidelines for Production of Bacillus thuringiensis H-14*, (eds M. Vandekar and H.T. Dulmage), UNDP/World Bank/WHO, Geneva, pp.31–124.

Ejiofor, A.O. and Okafor, N. (1988) The production of *Bacillus sphaericus* using fermented cowpea (*Vichua unguiculata*) medium containing mineral substitutes from Nigeria. *Journal of Applied Microbiology and Biotechnology*, **4**, 455–62.

Ejiofor, A.O. and Okafor, N. (1989) Production of mosquito larvicidal *Bacillus thuringiensis* H-14 on raw material from Nigeria. *Journal of Applied Bacteriology*, **67**, 5–9.

Ellison, C.A. and Evans, H.C. (1992) Present status of the biological control programme for the graminaceous weed *Rottboellia cochinchinensis*, in *Proceedings, 8th International Symposium on Biological Control of Weeds*, (eds E.S. Delfosse and R.R. Scott) Canterbury, New Zealand, February 2–7, 1992 DSIR/CSIRO, Melbourne, (in press).

Entwistle, P.F. (1986) Viruses for insect pest control, in *Biological Control in the Tropics*, (eds M.Y. Hussein and A.G. Ibrahim), Penerbit University, Pertanian, Malaysia, pp.237–54.

Essers, A.J.A. and Nout, M.J.R. (1989) The safety of dark, moulded cassava flour – a comparison of traditionally dried cassava pieces in North East Mozambique. *Tropical Science*, **29**, 261–8.

Evans, H.C. (1987) Fungal pathogens of some subtropical and tropical weeds and the possibilities for biological control. *Biocontrol News and Information*, **8**, 7–30.

FAO (1988) *Guidelines on the registration of biological pest control agents*. Food and Agriculture Organisation of the United Nations, Rome, pp.7.

Foda, M.S. and Salama, H.S. (1991) Local production of *Bacillus thuringiensis* in Egypt: advantages and constraints, in *Abstracts, International Workshop on Bacillus thuringiensis and its Applications in Developing Countries*, Cairo, Egypt, 4–6 November 1991, National Research Center, Cairo, p.19.

Fukushima, D. (1989) Industrialization of fermented soy sauce production centering around Japanese shoyu, in *Industrialization of Indigenous Fermented Foods*, (ed K.H. Steinkraus), Marcel Dekker, New York, pp.1–88.

Gillespie, A.T., Collins, M.D., Atienza, A. and Entwistle, J.C. (1987) Use of entomogenous fungi to control brown planthopper (*Nilaparvata lugens*) on rice, in *Proceedings, 11th International Congress of Plant Protection, Vol II*, Manila, Philippines, October 5–9, 1987 pp.303–9.

Goettel, M.S. and Roberts, D.W. (1992) Mass production, formulation and application of entomopathogenic fungi, in *Biological Control of Locusts and Grasshoppers*, (eds C.J. Lomer and C. Prior), CAB International, Wallingford, UK, pp.230–8.

Gowen, S.R. and Ahmed, R. (1990) *Pasteuria penetrans* for control of pathogenic nematodes. *Aspects of Applied Biology*, **24**, 25–32.

Gowen, S.R. and Channer, A.G. (1988) The production of *Pasteuria penetrans* for control of root-knot nematodes, in *Proceedings, Brighton Crop Protection Conference – Pest and Diseases*, November, 1988, Brighton, British Crop Protection Council, Farnham, pp.1215–20.

Gowen, S.R. and de Silva, P. (1989) Infectivity of *Pasteuria penetrans* on *Meloidogyne* spp. in roots and the soil environment. *Aspects of Applied Biology*, **22**, 437–8.

Hassan, S. (1986) Recent developments in the biocontrol of weeds using pathogens, in *Biological Control in the Tropics*, (eds M.Y. Hussein and A.G. Ibrahim), Penerbit University, Pertanian, Malaysia, pp.399–414.

Holzapfel, W.H. (1992) Industrialisation of mageu (mahewu) and sorghum beer fermentation, in *Proceedings of a Regional Workshop on Traditional African Foods – Quality and Nutrition*, (ed. A. Westby, and P.J.A. Reilly) 25–29 November 1991, International Foundation for Science, Stockholm, (in press).

Hou, R.F. and Chiuo, W.C. (1991) Control of the Asian corn borer, *Ostrinia furnacalis*, using three entomopathogens, in *Proceedings Oral and Poster Sessions Programs and Abstracts, XII International Plant Protection Congress*, Rio de Janeiro, Brazil, August 11–16, 1991, (unnumbered).

Huber, J. (1986) Use of baculoviruses in pest management programs, in *The Biology of Baculoviruses, Volume II, Practical Application for Insect Control*, (eds R.R. Granados and B.A. Federici), CRC Press, Boca Raton, FL, pp.181–202.

Hussey, N.W. and Tinsley, T.W. (1981) Impressions of insect pathology in the Peoples Republic of China, in *Microbial Control of Pests and Plant Diseases, 1970–1980*, (ed H.D. Burges), Academic Press, London, pp.785–95.

Jeffries, P. and Jeger, M.J. (1990) The biological control of postharvest diseases of fruit. *Postharvest News and Information*, **1**, 365–8.

Jeger, M.J., Jones, K.A. and Hall, D.R. (1990) Microbial control of pests: recent progress and prospects for developing countries. *Aspects of Applied Biology*, **24**, 263–70.

Jones, K.A. (1988) The use of insect viruses for pest control in developing countries. *Aspects of Applied Biology*, **17**, 425–33.

Jones, K.A. (1990) Control of *Spodoptera littoralis* in Crete and Egypt with NPV, in *Pesticides and Alternatives: Innovative Approaches to Pest Control*, (ed J.E. Casida) Elsevier, Amsterdam, pp.131–42.

Jones, K.A. and Grzywacz, D. (1990) Formulation of baculoviruses for pest control in developing countries, in *Proceedings, Vth International Colloquium on Invertebrate Pathology and Microbial Control*, Adelaide, Australia 20–24 August 1990, Society of Invertebrate Pathology, p.494.

Jones, K.A. and McKinley, D.J. (1986) UV inactivation of *Spodoptera littoralis* nuclear polyhedrosis virus in Egypt: assessment and protection, in *Fundamental and Applied Aspects of Invertebrate Pathology*, (eds R.A. Samson, J.M Vlak and D. Peters), Foundation of the Fourth International Colloquium of Invertebrate Pathology, Wageningen, The Netherlands, p.155.

Kanapathipillai, V.S. and Jantan, R. (1986) Approach to biological control of anthracnose fruit rot of bananas, in *Biological Control in the Tropics*, (eds M.Y. Hussein and A.G. Ibrahim), Penerbit University, Pertanian, Malaysia, pp.387–98.

Kaske, R. (1988) IPM activities in coconut in Southeast Asia, in *Pesticide Management and Integrated Pest Management in Southeast Asia*, (eds E.H. Teng and K.L. Heong), Consortium for International Crop Protection, Maryland, pp.181–6.

Ketunuti, U. and Prathomrut, S. (1989) Cotton bollworm larvae control by *Heliothis armigera* nuclear polyhedrosis virus, in *Abstracts, First Asia-Pacific Conference of Entomology*, November 8–13, 1989, Chang Mai, Thailand, p.16.

Ketunuti, U. and Tantichodok, A. (1990) The use of *Heliothis armigera* nuclear polyhedrosis virus to control *Heliothis armigera* (Hubner) on okra, in *Proceedings, Vth International Colloquium on Invertebrate Pathology and Microbial Control*, Adelaide, Australia 20–24 August 1990, Society of Invertebrate Pathology, pp.257.

Koomen, I., Dodd, J.C., Jeger, M.J. and Jeffries, P. (1990) Postharvest biocontrol of anthracnose disease of mangoes. *Journal of the Science of Food and Agriculture*, **50**, 137–8.

Kurstak, E. (ed) (1985) *Microbial and Viral Pesticides*, Marcel Dekker, New York, pp.720.

Laird, M., Lacey, L.A. and Davidson, E.W. (eds) (1990) *Safety of Microbial Insecticides*, CRC Press, Boca Raton, pp.259.

Lambert, B. and Peferoen, M. (1992) Insecticidal promise of *Bacillus thuringiensis*: facts and mysteries about a successful biopesticide. *BioScience*, **42**, 112–22.

Lee, C.-H. (1990) Fish fermentation technology – a review, in *Post-Harvest Technology, Preservation and Quality of Fish in Southeast Asia*, (eds P.J.A. Reilly, R.W.H. Parry and L.E. Barile), Stockholm: International Foundation for Science, pp.1–13.

Lee, C.-H., Steinkraus, K.H. and Reilly, P.J.A. (eds) (1991) *Fish Fermentation Technology*, Tokyo: United Nations University.

Liangsheng, Z. (1991) Application of Bt insecticides in China, in *Abstracts, Bt1991 First International Conference on Bacillus thuringiensis*, Oxford, UK, July 28–31 1991, Institute of Virology and Environmental Microbiology, Oxford.

Liew, C.C.V. and Buckle, K.A. (1990) Oligosaccharide levels in pigeonpea and pigeonpea tempe. *ASEAN Food Journal*, **5**, 79–81.

Lim, G.S. (1988) IPM of *Plutella xylostella* (L.) on vegetables in Southeast Asia: rationale, need, and prospects, in *Pesticide Management and Integrated Pest Management in Southeast Asia*, (eds E.H. Teng and K.L. Heong), Consortium for International Crop Protection, Maryland, pp.145–55.

Lindow, S.E. (1985) Foliar antagonists: status and prospects, in *Biological Control in Agricultural IPM Systems*, (eds M.A. Hoy and D.C. Herzog), Academic Press, London, pp.395–413.

Lisansky, S.G. (1984) Biological alternatives to chemical pesticides. *World Biotechnology Report*, **1**, Online Publications, Pinner, pp.455–66.

Lobo Lima, M.L., Brito, J.M. and Henry, J.E. (1992) Biological control of grasshoppers in the Cape Verde Islands, in *Biological Control of Locusts and Grasshoppers*, (eds C.J. Lomer and C. Prior), CAB International, Wallingford, UK, pp.287–310.

Lomer, C. (1986) Baculovirus oryctes: *Biochemical studies and field use*, Overseas Development Administration, London, UK, pp.136.

Lorri, W. and Svanberg, U. (1992) The potential role of fermented cereal gruels in reduction of diarrhoea among young children, in *Proceedings of a Regional Workshop on Traditional African Foods – Quality and Nutrition*, (eds Westby, A. and Reilly, P.J.A.), 25–29 November 1991, International Foundation for Science Stockholm, (in press).

Mankau, R. (1981) Microbial control of nematodes, in *Plant Parasitic Nematodes, Vol. III*, (eds B.M Zuckerman, W.F. Mai, and R.A. Rohde), Academic Press, London, pp.475–94.

Marshall, K.J. and Ioane, I. (1982) The effect of re-release of *Oryctes rhinoceros* baculovirus in the biological control of rhinoceros beetles in Westerm Samoa. *Journal of Invertebrate Pathology*, **39**, 267–76.

Martin, S.B., Abawi, G.S. and Hoch, H.S. (1985) Biological control of soilborne pathogens with antogonists, in *Biological Control in Agricultural IPM Systems*, (eds M.A. Hoy and D.C. Herzog), Academic Press, London, pp.433–54.

Mbugua, S.K. and Njenga, J. (1992) Anti-microbial properties of fermented uji as a weaning food, in *Proceedings of a Regional Workshop on Traditional African Foods – Quality and Nutrition*, (eds A. Westby, and P.J.A. Reilly), 25–29 November 1991, International Foundation for Science Stockholm, (in press).

McKinley, D.J., Moawad, G., Jones, K.A., Grzywacz, D. and Turner, C. (1989) The development of nuclear polyhedrosis virus for the control of *Spodoptera littoralis* (Boisd.) in cotton, in *Pest Management in Cotton*, (eds M.B. Green and D.J. de B. Lyon), Ellis Horwood, Chichester, pp.93–100.

Mendonça, A.F. (1992) Mass production, application and formulation of *Metarhizium anisopliae* for control of sugarcane froghopper, *Mahanarva posticata*, in Brazil, in *Biological Control of Locusts and Grasshoppers*, (eds C.J. Lomer and C. Prior), CAB International, Wallingford, UK, pp.239–44.

Mensah, P., Drasar, B.S., Harrison, T.J. and Tomkins, A.M. (1991) Fermented cereal gruels: Towards a solution of the weanling's dilemma. *Food and Nutrition Bulletin*, **13**, 50–7.

Mohan, K.S. (1985) Baculovirus of *Oryctes rhinoceros*, in *Microbial Control and Pest Management*, (ed S. Jayaraj), Tamil Nadu Agricultural University, Tamil Nadu, pp.102–6.

Mohan Naidu, K. (1985) Microbial control and pest management, in *Microbial Control and Pest Management*, (ed S. Jayaraj), Tamil Nadu Agricultural University, Tamil Nadu, pp.1–6.

Moscardi, F. (1990) Development and use of soybean caterpillar baculovirus in Brazil, in *Proceedings, Vth International Colloquium on Invertebrate Pathology and Microbial Control*, Adelaide, Australia 20–24 August 1990, Society of Invertebrate Pathology, pp.184–7.

Moscardi, F. (1991) Use of viral insectcides in Brazil, in *Proceedings Plenary Lectures and Symposia Programs and Abstracts, XII International Plant Protection Congress*, Rio de Janeiro, Brazil, August 11–16, 1991, (unnumbered).

Moscardi, F. and Ferreira, B.S.C. (1985) Biological control of soybean caterpillars, in *World Soybean Research Conference III*, (ed R. Shibles), Westview Press, London, pp.703–11.

Mulock, B., Swezey, S.S., Narvaez, C., Castillo, P. and Rizo, C.M. (1990)

Development of baculoviruses as a contribution to biological control of Lepidopterous pests of basic grains in Nicaragua, in *Proceedings, Vth International Colloquium on Invertebrate Pathology and Microbial Control*, Adelaide, Australia, 20–24 August 1990, Society of Invertebrate Pathology, pp.179–83.

Munaan, A. and Wikardi, W.A. (1986) Toward the biological control of coconut insect pest in Indonesia, in *Biological Control in the Tropics*, (eds M.Y. Hussein and A.G. Ibrahim), Penerbit University, Pertanian, Malaysia, pp.149–57.

Nout, M.J.R., Beernink, G. and Bonants-van Laarhoven, T.M.G. (1987) Growth of *Bacillus cereus* in soya-bean tempeh. *International Journal of Food Microbiology*, **4**, 293–301.

Nout, M.J.R., Notermans, S. and Rombouts, F.M. (1988) Effect of environmental condition during soya-bean fermentation on the growth of *Staphylococcus aureus* and production and thermal stability of enterotoxins A and B. *International Journal of Food Microbiology*, **7**, 299–309.

Nout, M.J.R., Rombouts, F.M. and Havelaar, A. (1989) Effect of accelerated natural lactic fermentation of infant food ingredients on some pathogenic bacteria. *International Journal of Food Microbiology*, **8**, 351–61.

Nweke, F.I. (1990) Collaborative Study of Cassava in Africa (COSCA) Project Description. Working paper no. 1. International Institute of Tropical Agriculture, Ibadan, Nigeria.

Obeta, J.A. and Okafor N. (1988) Medium for the production of primary powder of *Bacillus thuringiensis* subsp. *israelensis*. *Applied and Environmental Microbiology*, **47**, 863–7.

Oka, I.N. (1988) Future needs for pesticide management in Southeast Asia, in *Pesticide Management and Integrated Pest Management in Southeast Asia*, (eds E.H. Teng and K.L. Heong), Consortium for International Crop Protection, Maryland, pp.1–11.

Onyekwere, O.O., Akinrele, I.A., Koleoso, O.A. and Heys, G. (1989) Industrialisation of gari fermentation, in *Industrialization of Indigenous Fermented Foods*, (ed K.H. Steinkraus), Marcel Dekker, New York, pp.363–410.

Ooi, A.C.P. and Sudderuddin, K.I. (1978) Control of diamondback moth in Cameron Highlands, Malaysia, in *Proceedings, Plant Protection Conference*, Kuala Lumpur, Malaysia, 1978, Rubber Research Institute of Malaysia, Kuala Lumpur, pp.193–227.

Payne, C.C. (1988) Pathogens for control of insects: where next? *Philosophical Transactions of the Royal Society of London Series B*, **318**, 225–48.

Poblete, J. (1988) Experience with *Trichoderma* ATCC 20476 preparations in top quality fruit production. *Trichoderma Newsletter*, **4**, 6.

Prior, C. (1989) Biological pesticides for low external-input agriculture. *Biocontrol News and Information*, **10**, 17–22.

Prior, C., Lomer, C.J., Herren, H., Paraiso, A., Kooyman, C. and Smit, J.J. (1992) The IIBC/IITA/DFPV collaborative research programme on the biological control of locusts and grasshoppers, in *Biological Control of Locusts and Grasshoppers*, (eds C.J. Lomer and C. Prior), CAB International, Wallingford, UK, pp.8–18.

Ricard, J.L. and Highley T.L. (1988) Biological control of pathogenic fungi in wood and trees, with particular emphasis on the use of *Trichoderma*. *Biocontrol News and Information*, **9**, 133–42.

Rishbeth, J. (1988) Biological control of air-borne pathogens. *Philosophical Transactions of the Royal Society of London Series B*, **318**, 265–81.

Roberts, D.W., Fuxa, J.R., Gaugler, R., Goettel, M., Jacques R. and Maddox R. (1991) Use of pathogens in insect control, in *CRC Handbook of Pest Management on Agriculture*, 2nd ed. *Vol. II*, (ed D. Pimental), CRC Press, Boca Raton, pp.243–78.

Schroth, M.N. and Hancock, J.G. (1985) Soil antagonists in IPM systems, in *Biological Control in Agricultural IPM Systems*, (eds M.A. Hoy and D.C. Herzog), Academic Press, London, pp.415–31.

Solidum, M.H.T. and Acedevo, T.P. (1983) Lactic acid fermentation of balaobalao. *Philippine Journal of Food Science and Technology*, **7**, 56–86.

Steinkraus, K.H. (1983) *Handbook of Indigenous Fermented Foods*, Marcel Dekker, New York.

Steinkraus, K.H. (1989) *Industrialization of Indigenous Fermented Foods*, Marcel Dekker, New York.

Svanberg, U. and Lorri, W. (1992) Lactic fermentation of cereal-based weaning gruels and improved nutritional quality, in *Proceedings of a Regional Workshop on Traditional African Foods – Quality and Nutrition*, (eds A. Westby, and P.J.A. Reilly) (25–29 November 1991), International Foundation for Science, Stockholm, (in press).

Templeton, G.E. and Trujillo, E.E. (1991) The use of plant pathogens in the biological control of weeds, in *CRC Handbook of Pest Management in Agriculture, 2nd ed., Vol. II*, (ed D. Pimental), CRC Press, Boca Raton, pp.353–8.

Thonghtai, C., Panbangred, W., Khoprasert, C. and Dhaveetiyanond, S. (1990) Protease activities in the traditional process of fish sauce fermentation, in *Post-Harvest Technology, Preservation and Quality of Fish in Southeast Asia*, (eds P.J.A. Reilly, R.W.H. Parry and L.E. Barile). International Foundation for Science, Stockholm, pp.61–5.

Tianjian, X., Bingao, H., Liansen, Z. and Gixin, W (1991) Commercialization and utilization of *Bacillus thuringiensis* for crop protection in China, in *Abstracts, International Workshop on* Bacillus thuringiensis *and its Applications in Developing Countries*, Cairo, Egypt, 4–6 November 1991, National Research Center, Cairo, p.18.

Tomkins, A., Alnwick, D. and Haggerty, P. (1988) Fermented foods for improving child feeding in eastern and southern Africa: a review, in *Improving Young Child Feeding in Eastern and Southern Africa*, (eds D. Alnwick, S. Moses and O.G. Schmidt), International Development Research Centre, Ottawa, Canada, pp.136–67.

Topper, C. (1984) Integrated pest management in Egypt. *Report on the research and development of a nuclear polyhedrosis virus of* Spodoptera littoralis *1979–1983*, Overseas Development Administration, London, 3 volumes.

Topper, C., Moawad, G., McKinley, D., Hosny, M., Jones, K., Cooper, J., El-Nagar, S. and El-Sheik, M. (1984) Field trials with a nuclear polyhedrosis virus against *Spodoptera littoralis* on cotton in Egypt. *Tropical Pest Management*, **30**, 372–8.

Tryon, E.H. and Litsinger, J.A. (1988) Feasibility of using locally produced *Bacillus thuringiensis* to control tropical insect pests, in *Pesticide Management and Integrated Pest Management in Southeast Asia*, (eds E.H. Teng and K.L. Heong), Consortium for International Crop Protection, Maryland, pp.73–81.

Uzogara, S.G., Agu, L.N. and Uzogara, E.O. (1990) A review of traditional fermented foods, condiments and beverages in Nigeria: their problems and possible benefits. *Ecology of Food and Nutrition*, **24**, 267–88.

van Gundy, S.D. (1985) Biological control of nematodes: status and prospects in agriculture IPM systems, in *Biological Control in Agricultural IPM Systems*, (eds M.A. Hoy and D.C. Herzog), Academic Press, London, pp.467–78.

Van Veen, A.G. and Steinkraus, K.H. (1970) Nutritive value of and wholesomeness of fermented foods. *Journal of Agricultural and Food Chemistry*, **18**, 576–9.

Vasconcelos, A.T., Twiddy, D.R., Westby, A. and Reilly, P.J.A. (1990) Detoxification of cassava during gari preparation. *International Journal of Food Science and Technology*, **25**, 198–203.

Walsh, J. (1986) River blindness, a gamble pays off. *Science*, **232**, 922–5.

Wang, R (1986) Current status and perspectives of biological weed control in China. *Chinese Journal of Biological Control*, **2**, 173–7.

Westby, A. and Twiddy, D.R. (1992a) Role of microorganisms in the reduction of cyanide during traditional processing of African cassava products, in *Proceedings of a Regional Workshop on Traditional African Foods – Quality and Nutrition*, (eds A. Westby and P.J.A. Reilly) 25–29 November 1991, International Foundation for Science, Stockholm, (in press).

Westby, A. and Twiddy, D.R. (1992b) Characterisation of the gari and fu-fu preparation procedures in Nigeria. *World Journal of Microbiology and Biotechnology*, **8**, 175–82.

Wilson G.G. (1982) Protozoans for insect control, in *Microbial and Viral Pesticides*, (ed E. Kurstak), Marcel Dekker, New York, pp.587–600.

Wong, P.T.W. (1986) Biological control of fusarium wilts and other soilborne diseases by soil bacteria and fungi, in *Biological Control in the Tropics*, (eds M.Y. Hussein and A.G. Ibrahim), Penerbit University, Pertanian, Malaysia, pp.333–47.

Wood, B.J.B. (1985) *Microbiology of Fermented Foods, Volumes 1 & 2*, Elsevier Applied Science Publishers, London.

Youdeowei, A. and Service, M.W. (1983) *Pest and Vector Management in the Tropics*, Longman, Harlow.

Young, E.C. and Longworth, J.F. (1981) The epizootiology of the baculovirus of the coconut palm rhinoceros beetle (*Oryctes rhinoceros*) in Tonga. *Journal of Invertebrate Pathology*, **38**, 362–9.

Zaki, F.N. (1991) Utilization of *Bacillus thuringiensis* for crop protection in Egypt, emphasizing constraints, in *Abstracts, International Workshop on Bacillus thuringiensis and its Applications in Developing Countries*, Cairo, Egypt, 4–6 November, 1991, National Research Center, Cairo, p.24.

Zelazny, B., Lolony, A. and Pattang, B. (1992) *Orcytes rhinoceros* Coleoptera: Scarabaeidae populations suppressed by a baculovirus. *Journal of Invertebrate Pathology*, **59**, 61–8.

Zelazny, B., Lolony, A. and Crawford, A.M. (1989) Introduction and field comparison of baculovirus strains against *Oryctes rhinoceros* in the Maldives. *Industrial Crops Research Journal*, **2**, 50–6.

Zhong-yun, P. (1984) Study and utilization of *Bacillus thuringiensis* Berlinger in China, in *Proceedings of the Chinese Academy of Sciences–United States National Academy of Sciences Joint Symposium on Biological Control of Insects*, (eds P.L. Adkisson and S. Ma), Beijing, China, 25–28 September 1982, Science Press, Beijing, pp.235–46.

Zimmerman, G. (1986) Insect pathogenic fungi as pest control agents, in *Biological Plant and Health Protection, Biological Control of Plant Pets and Vectors of Human and Animal Diseases, Progress in Zoology, Vol. 32*, (ed J.M. Franz), Fischer, Stuttgart, pp.217–31.

Zoebelein, G. (1990) Twenty-three-year surveillance of development of insecticide resistance in diamondback moth from Thailand (*Plutella xylostella* L., Lepidoptera, Plutellidae). *Mededelingin van de Faculteit Landbouwwetenschappen Rijksuniversiteit Ghent*, **55**, 313–22.

15

Microorganisms and detoxification of industrial waste

M. Talaat Balba

15.1 INTRODUCTION

The problems caused by the increases in the total volumes of wastes generated globally are compounded by the presence of an expanding array of pollutants. As a consequence, all waste streams are being viewed as potential threats to the environment, thus necessitating the development of innovative waste disposal strategies.

The role of microorganisms in maintaining the carbon cycle at balance is evident from the lack of major accumulations of biological products in aerated environments. The absence of any vast build-up of naturally-occurring compounds is testimony to the catabolic versatility of microbial communities. On the other hand, the persistence in nature of some chemicals, synthesized by man, their apparent recalcitrance to biodegradation, and the frequent inability to show biological decomposition in the laboratory argue for the innate stability of many synthetic products.

15.2 FACTORS AFFECTING BIODEGRADATION

Many factors are concerned in the degradation of organic compounds, irrespective of whether they are of natural or synthetic origin. They are all exposed to physico-chemical forces and biological catalysts capable of causing chemical changes in the biosphere. Photo-chemical and other non-enzymatically-effected reactions undoubtedly cause some transformation in many substrates formed or introduced into the environ-

Exploitation of Microorganisms Edited by D.G. Jones
Published in 1993 by Chapman & Hall, London ISBN 0 412 45740 7

ment. However, by contrast, degradative reactions catalysed by the natural microbial population usually result in the conversion of the substrate into harmless end products such as carbon dioxide, water and other organic products. Whereas, non-microbial processes rarely lead to total conversion of a complex compound to the elemental state, dilution processes and time eventually ensure that conditions are created for these physico-chemical and biological forces to operate.

There are many chemical structures introduced continually into the biosphere that are, to varying degrees, resistant to microbiological degradation. Such compounds are often spoken of as being **recalcitrant** or **refractile**. These biologically resistant molecules enter the biosphere either through the activity of industrial societies or via natural, biological processes. Generally, the chemicals which are resistant to microbial attack in nature can be classified into four categories:

1. Very refractory compounds;
2. Those which are only slowly biodegradable but never at a rapid rate;
3. Compounds that can be utilized rapidly in microbial cultures, but are not known to be destroyed readily under natural conditions;
4. Chemicals that are suitable substrates for populations in axenic cultures or in one or more microbial habitat but which are occasionally quite persistent.

Many **synthetic polymers** would seem to fit into the first category. **Lignin** is an excellent example of the second category of molecules, because it is a substrate for many fungi although not dissimilated readily. Certain **chlorinated hydrocarbon insecticides** fit into the third category in as much as they are metabolized in culture, but rapid conversion to carbon dioxide, water and chloride has yet to be demonstrated in nature. In the last category are some **petroleum hydrocarbons** and many **plant constituents**.

The factors contributing to the biodegradation of organic compounds are in fact numerous and inter-related, but can be divided for convenience into those relating to the contaminant and those relating to the environment, which may be physical, chemical or biological in origin.

15.2.1 Factors relating to the contaminants

A significant number of man-made chemicals will be biodegradable if the necessary enzymes have been acquired by microorganisms during the course of evolution. This depends on two major factors:

1. The structure of the compounds and the ability of microbial enzymes to accept these as substrate compounds;

2. The ability of the compounds to induce or suppress the necessary catabolic enzymes in the microorganism.

Biodegradation is less likely in a molecule having features never encountered in natural products because the required degradative enzyme system may simply not exist.

Effects of concentration

The concentration of the pollutants may be either too low to support microbial growth or too high, causing biodegradation inhibition. For example, **marine bacteria** have been reported not to grow at low substrate concentrations (Jannasch, 1967), and a pseudomonad able to grow at glucose levels of 18 ng ml^{-1} or higher had almost no effect on lower levels of the sugar (Boethling and Alexander, 1979). Many xenobiotic compounds that are mineralizable under laboratory conditions persist in sewage, natural water, and soils because they may exist at toxic levels. The toxicity can be exacerbated by the presence of specific organic pollutants or toxic heavy metals. **Phenols**, for example, are common pollutants which are known to persist only under environmental adverse conditions or when present at elevated concentrations. Toxicity of such ionizable molecules is also expected to be greatly affected by the predominant pH of the environment. This is due to the effect of pH on the solubility/permeability of the resultant form of the molecule.

Relationship between chemical structure and biodegradability

The search for readily biodegradable replacements for persistent pesticides, detergents, polymers and surfactants requires a thorough understanding of the structural features that hinder or favour biodegradation. Such information is also necessary in order to be able to adopt a rational, rather than an empirical, approach to synthetic programmes in these industries.

Slight modifications in the structure of many small molecules and at least some polymers greatly alter their biological availability. The effect of the type of substituent is evident by the resistance imparted to:

1. Mono- and dicarboxylic acids, aliphatic alcohols and alkylbenzenesulphonates when one or more hydrogen atoms are replaced by methyl group(s).
2. Aliphatic acids when a hydrogen is replaced by chlorine.

3. Triazines or the insecticide methoxychlor, when the methoxy groups are replaced by chlorine atoms.
4. Benzoate derivatives when carboxyl or hydroxyl groups are replaced by nitro-, sulphonate or chloro- groups.
5. Triazine derivatives when a hydroxyl is replaced by an amino group, or a chloro- is replaced by a methylthio- or methoxy group.

An influence of the number of substituents is evident from:

1. The greater resistance of di- and tri-, as contrasted with the mono-chlorobenzoic acid.
2. The longer persistence of chloroacetic acids, -substituted propionic acid, isopropyl-*N*-phenyl carbamate, containing two rather than one or three rather than two chlorine atoms.
3. The lesser bioavailability of diamino benzene as contrasted with aniline.

The **position** of a substituent also governs its potential degradability. Thus, mono-, di-, and trichloro-phenols with a halogen atom on the *meta* position to the hydroxyl group, phenoxyalkanoic acids with a chlorine in the *meta* position to the ether- oxygen, and benzoic acids with amino, nitro or a methoxy group in the *meta* position are less readily metabolized than the corresponding isomers with the same substituent but in the *ortho* or *para* positions (Alexander and Aleem, 1961; Alexander and Lustigman, 1966). Conversely, it is the *ortho* isomers of nitrophenols, methylanilines, sulphonates of l-phenyldodecane, and chlorine-containing isopropylphenylcarbonate that show the greatest longevity (Kaufman, 1978). The position of a substituent is also of paramount importance among the fatty acids, and the presence of a halogen or phenyl group on the *beta* carbon markedly reduces the biodegradation potential as compared with the same group on the *omega*-carbon (Dias and Alexander, 1971).

Chemical treatment of **natural polymers** affects the biodegradability to some extent. For example, the acetylation and carboxylation of natural cellulose fibres protect them from biodeterioration.

Limited information is available on the effect of the **replacement** of carbon atoms with oxygen, sulphur or nitrogen on the biodegradability of organic molecules. Available data, however, suggest that this replacement reduces the bioavailability of the molecule to microorganisms; for example, many ethers are refractory. Chlorophenylcarbamates are usually more quickly decomposed than the corresponding chlorophenyl carbonates, whereas the corresponding phenylureas are quite resistant, meaning that the replacement of the two nitrogens in the phenylureas by one and then a second atom favours biodegradability.

Chemical recalcitrance is not limited to some xenobiotic compound, but many forms of natural structures such as coal, graphite, and diamond are also highly resistant to microbial metabolism.

15.2.2 Environmental factors

Despite their occurrence in concentrations which fall within the limits amenable to degradation, there is ample evidence that even quite labile compounds can persist at contaminated sites, e.g. phenols in the soils of disused coal gasification plants. The reasons for such persistence can be attributable to one or more factors relating to the micro-environment of which that contaminate forms an integral part. Just as environmental factors will mediate the effect of a pollutant on the organism, so will they determine whether, and to what extent, biodegradation will take place. Such factors may be **chemical**, **physical** or **biological** in origin.

Chemical and physical factors

Of all environmental factors which can influence biodegradation, probably most research has been performed on the effects of both organic and inorganic nutrients.

As with any other living organisms, bacteria and fungi require a wide range of macro and micronutrients for metabolic activity to take place. Natural environments, both aquatic and terrestrial, are often nutritionally poor, so that lack of sufficient nitrogen or phosphorus is very often the cause of the slow breakdown of carbonaceous substrates. Most of the interest in nutrient limitation to biodegradation has focused on hydrocarbon degradation, particularly oil, in aquatic environments (Atlas, 1977; Colwell and Walker, 1977). Concentrations of nitrogen and phosphorus are usually limiting for degradation of large oil spills at sea whereas, for low level discharges of hydrocarbons, the amounts of these elements may be adequate (Atlas, 1981).

Application of nitrogenous and phosphate fertilizers to contaminated **marine environments** clearly has to take account of their solubility, so that oleophilic fertilizers may be necessary for effective degradation to take place. A formulation based on paraffinized urea and octyl phosphate, for example, selectively enhance hydrocarbon-degrading microorganisms due to the affinity of such compounds for the oil phase (Atlas and Bartha, 1973a).

In **soil environments**, the effect of nutrients on hydrocarbon degradation has also been extensively researched (Raymond *et al.*, 1976; Blakebrough, 1978). Carbon : nitrogen and carbon : phosphorus ratios of 60:1 and 800:1 respectively were the most beneficial for oil sludge degradation to occur when other factors were optimized (Dibble and Bartha, 1979).

The **anionic composition** of the inorganic nutrient may also influence pollutant transformation, particularly if certain elements are limiting, e.g. ammonium sulphate addition to groundwater resulted in much greater growth of hydrocarbon-degrading flora than ammonium nitrate (Lee and Ward, 1985). Organic supplements may have conflicting effects on pollutant degradation. Yeast extract or brewer's yeast has been observed both to stimulate (Lehtomaki and Niemala, 1975) or retard (Dibble and Bartha, 1979) hydrocarbon degradation. Sewage sludge may also be a repressant to oil degradation but composted sewage sludge accelerated biodegradation of toxic compounds (Marinucci and Bartha, 1979). Addition of birch sawdust to oil-supplemented soil also tended to suppress hydrocarbon mineralization (Loynachan, 1978). These discrepancies are attributable to the different degrees to which provision of an alternate carbon source will repress xenobiotic-degrading enzymes. The degradation of many xenobiotic compounds for example, is usually inhibited by the presence of more readily biodegradable compounds (Liu *et al.*, 1981). On the other hand, the presence of an **alternate substrate** may be required for co-metabolism of the compound to be degraded, e.g. provision of biphenyl for degradation of PCBs (Baxter *et al.*, 1975; Furukawa *et al.*, 1978).

In addition to providing nutrient sources for degradation, the presence of **inorganic ions** may influence biological transformation of organic pollutants through effects on osmotic pressure and salinity. The presence of elevated concentrations of sodium chloride suppressed oil emulsification by bacteria known to be capable of hydrocarbon degradation (Zajic *et al.*, 1977; Ajisebutu, 1988).

The presence of toxic levels of **heavy metals**, could also significantly repress transformation of organic pollutants. Metal toxicity in soil is usually less than in an aquatic system due to reduction in bioavailability through adsorption on clay particles and organic fractions. Heavy metals in used motor oil had little effect on biodegradation however, due to their non-accessible form (Vazquez-Duhalt and Greppin, 1986). Other factors inhibitory to the degradation of a pollutant may result directly from the production of a toxic intermediate or the accumulation of the terminal end product. Fatty acids, for example can inhibit degradation of petroleum (Atlas and Bartha, 1973b). This is possibly due to the effect of inhibitory competition for key enzyme factors such as CoA.

The **pH** of the environment will also markedly influence biotransformations taking place. As with other microorganisms, most of the pollutant-degrading microflora will flourish best at pH values near to 7 (Boethling and Alexander, 1979), and raising the pH of soil from 5 to 7.8 with $CaCO_3$ progressively enhances degradation of oil sludge. Carbofuran was degraded seven to ten times faster than under acidic or neutral conditions (Getzin, 1973). Soil pH may also indirectly affect degradation by influencing the adsorption of organic pollutants to soil (Pancorbo and Varney, 1986).

The presence or absence of **oxygen** will clearly have a most important effect on the rate and extent of degradation, depending on whether the pollutant is metabolized through an aerobic or an anaerobic pathway. **Hydrocarbon biodegradation** is predominantly aerobic as the major degradation pathways involve oxygenases and molecular oxygen, whereas degradation of certain halogenated compounds, particularly certain pesticides such as BHC (benzene hexachloride) and lindane, can proceed more effectively under anaerobic environments (MacRae *et al.*, 1967, 1969).

Microbial populations from lake sediments or sewage sludge have been shown to mediate the anaerobic conversion of various chlorinated compounds to methane, given a suitable incubation time. More recently, reductive dechlorination of the higher chlorinated PCB isomers has been observed in aquatic sediments (Brown *et al.*, 1987). Under aerobic conditions, such congeners are particularly resistant to microbial attack. There are also examples of xenobiotics which may be degraded almost equally well under environments of contrasting redox potential. Biodegradation of nitrilotriacetate, for example, in subsurface soils was as rapid in the absence of molecular oxygen as it was under aerobic conditions (Ward, 1986).

Physical conditions will affect biodegradation both directly or indirectly through influences on chemical or biological parameters. Temperature may appear to be the most obvious direct influence on the rates of biological activity through microbial growth rate and enzyme reaction kinetics. However, a review of its influence on hydrocarbon degradation in natural environments has indicated that the relationship of temperature to degradation kinetics is more complex than revealed merely by Q_{10} values. Phsychotrophic, mesophilic and thermophilic hydrocarbon degrading microorganisms have all been isolated and it has been concluded that the natural environment temperature is very often not the major factor limiting such activities. Increase in temperature under mesophilic conditions does not necessarily bring about an increase in degradation, e.g. slower decomposition of 2,4-D took place at 30°C compared with 25°C (Ou, 1984). Temperature will also indirectly affect degradation by its influence on the solubility of xenobiotic compounds (Pancorbo and Varney, 1986). Volatile hydrocarbons with toxic properties will become more soluble in water under low temperature conditions (Colwell and Walker, 1977).

Physical properties of the soil will have a most important influence on biodegradation rates and all other aspects of microbial ecology (Alexander, 1977). Water and air in soil are mutually exclusive so that anaerobic conditions will prevail in waterlogged soil. Under arid conditions, however, the lack of available moisture may be the primary limitation to degradation. The relationship between moisture and aeration will be directly influenced by soil texture. Oxygen diffusion rates will decrease with finer textured soils so that aerobic degradation will be very limited in clay soil where the movement of microorganisms may also be restricted by pore size. Sunlight may photo-oxidize xenobiotics to produce degradation products which have a susceptibility to biodegradation differing from that of the parent molecule (Walker, 1985).

15.2.3 Biotic factors

Whereas the influence of physical and chemical factors on biodegradation of many xenobiotic chemicals has been well researched in different natural environments, far less is known about the interaction of pollutant-degrading microorganisms and other members of the soil, aquatic or sediment flora. Natural environments generally contain a wide variety of organic materials providing substrates for a diverse range of bacterial and fungal species. The organism capable of degrading a specific xenobiotic compound will have to compete with such species for nutrients, and space: both groups of organisms could also limit each other's growth through production of volatiles or direct antagonism in the form of antibiotics.

Such interactions may provide an explanation as to why some compounds, particularly those which are metabolized by a narrow range of organisms with a highly specific enzymic capability persist when soil conditions are otherwise favourable. Conversely, hydrocarbons with similar chemical structures to naturally-occurring residues, which are metabolized by a much wider range of the existing microflora are more liable to degradation through a general stimulation of microbial activity simply by optimizing the physico-chemical conditions for biodegradation.

Microorganisms are of course subject to predation by **protozoa**: the influence of the interaction on organic pollutant persistence is not known. The zone surrounding the roots of plants (the rhizosphere) has been shown to markedly stimulate various physiological groups of microorganisms (Lynch, 1982) due to the release of a vast range of organic materials from the plants, particularly carbohydrate, vitamins, amino acids and enzymes. Such input or organic material may have important direct and indirect influences on degradation of xenobiotics through the mechanisms illustrated above. Both diazinon and parathion were degraded to a great extent in rhizosphere compared with non-rhizosphere soil (Hsu and Bartha, 1979). Certain reeds (*Phragmites* spp.) can absorb oxygen through their pores above ground in air and transport it to the root zone where it enters the surrounding soil. The dense root network provides a complex mosaic of aerobic and anaerobic microenvironments, and the microbial activities taking place there have been commercially exploited for waste water treatment (Arthur, 1986).

15.3 STRATAGEMS USED BY MICROORGANISMS FOR THE DEGRADATION OF XENOBIOTICS

15.3.1 Use of constitutive enzymes

In all microorganisms, a certain minimal array of enzymes is necessary for maintenance and growth. These enzymes which are always present are **constitutive**, i.e. they are always formed irrespective of the nature of the environment (medium). If a synthetic chemical is sufficiently similar to a natural product in structure, then it is likely that biodegradation will be initiated by the effect of constitutive enzyme through the cleavage of common chemical bonds.

15.3.2 Enzyme induction

The xenobiotic compounds may be capable of **inducing** the necessary enzymatic apparatus. These inducible arrays of enzymes are synthesized only in response to the presence of specific substrate known as inducers. The mechanism for the induction of enzyme synthesis involves the regulatory genes on the DNA chromosome. These genes are responsible for the production of substances, each of which acts specifically on a system for synthesizing induced enzymes.

15.3.3 Co-metabolism

Many xenobiotic compounds were shown to be biodegradable via a mechanism known as **co-metabolism**, which refers to the ability of microorganisms to metabolize a compound which cannot be used as an independent source of energy or growth (Horvath, 1972). Co-metabolism is thus a fortuitous process in which microorganisms, while growing at the expense of one substrate, also have the capacity to transform another compound without deriving any benefit from its metabolism.

In many cases, complete mineralization of certain chemicals is linked to the co-metabolic activities of more than single organisms. This may explain the difficulty in isolating pure cultures capable of utilizing certain xenobiotics from active enrichments degrading these compounds. Examples are cycloalkane-degrading microorganisms. Co-metabolic steps may also be involved in the biotransformation of some chlorinated pesticides.

15.3.4 Transfer of plasmid coding for certain metabolic pathways

Some members of the genus *Pseudomonas* and others have been observed to possess unusual versatility in the degradation of synthetic organic compounds. This capacity has been shown to be **transmissible** at least between members of the same genera, and is mediated by the storing of genetic information in extra-chromosomal DNA structures called **plasmids**. The recent discoveries of inter-species transfer of catabolic plasmids highlight the significance of this phenomenon on the continuous evolution of microbial strains with novel catabolic capabilities.

15.3.5 Enhancement of pollutant bioavailability

Insolubility of some contaminants may often be the rate-limiting step to biodegradation. Some organisms have developed the capability to produce **surfactants** which can emulsify the pollutant and enhance its bioavailability. *Pseudomonas cepacia* for example, produces an agent which can emulsify 2,4,5-T and also has displayed some activity towards other chlorinated compounds such as chlorophenols (Chakrabarty, 1987). There are a number of reports of emulsification of hydrocarbons by extracellular agents, e.g. *Corynebacterium hydrocarboclastus* (Zajic *et al.*, 1974) and *Acinetobacter calcoaceticus* (Goldman *et al.*, 1982).

15.4 BIODEGRADATION OF MAJOR GROUPS OF ENVIRONMENTAL POLLUTANTS

15.4.1 Petroleum hydrocarbon compounds

The main forms of hydrocarbon contamination which give rise to most concern in the environment in terms of both their visual impact and ecological effects are **oils** and **tars**. Reports of oil spills on beaches have always attracted particular public concern, the spills from the Torrey Canyon, the Amoco and the Exxon Valdez tankers being among the most notable. Oil represents a highly complex mixture of hydrocarbons, which can be divided into crude oil, petroleum distillates, lubricating oils, black fuel oils, mixed oils and non-petroleum products. The main chemical fractions of oil include: naphthenes (cycloalkanes), normal alkanes, isoalkanes and aromatics, together with a variety of non-hydrocarbon compounds. The latter include sulphur, nitrogen and oxygen-containing compounds, porphyrins and asphaltenes and resins, these last two groups being of higher molecular weight material. Trace elements may also be complexed with these constituents.

There is a vast literature on the subject of oil breakdown by microorganisms with several major review papers (Atlas, 1977; Higgins and Gilbert, 1978; Bartha, 1986; Leahy and Colwell, 1990). No one species of microorganisms will completely degrade any particular oil (Colwell and Walker, 1977) and reviews of the literature suggest that biodegradation of both crude and refined oils involves a consortium of organisms, including both eukaryotic and prokaryotic forms. The principal bacterial genera responsible for oil degradation in both soils and aquatic environments comprise mainly *Pseudomonas, Achromobacter, Arthrobacter, Flavobacterium, Micrococcus,* and *Nocardia* (Atlas, 1981; Bossert and Bartha, 1984).

Of the various petroleum fractions, n-alkanes, n-alkylaromatic and aromatic compounds of the C_{10}–C_{22} range tend to be the most readily degradable, shorter chain compounds being rather more toxic. Despite their low toxicities, chain lengths of these groups of compounds which exceed C_{22} tend to be degraded more slowly due to a reduction in bioavailability. Branched chain alkanes are degraded more slowly than the corresponding normal alkanes. Cycloalkane degradation rates are somewhat variable but tend to be much slower, often including several microbial species. Components which exhibit the greatest resistance to biodegradation comprise the highly condensed aromatic and cycloparaffinic structures, together with the tars, bitumen and asphaltic materials which have the highest boiling points (Blakebrough, 1978; Atlas, 1981). Asphaltenes can also appear as the resistant microbial products of petroleum hydrocarbons in soil (Bossert and Bartha, 1984). It has been proposed that such residual material from oil degradation is analogous to, and could even be regarded as, humic material (Jobson *et al.*, 1972). One can speculate that due to its inert characteristic, insolubility and structural similarity to humic materials that it is unlikely to be an environmental hazard.

15.4.2 Coal tar waste

The manufactured gas industry produced gas for lighting and heating and generated by-products which served as fuel or feedstocks for the production of chemicals. The peak of production in Europe and the United States was from mid 1800s to the late 1940s and early 1950s. The manufactured gas industry began to decline rapidly in 1950s as transcontinental pipelines brought natural gas to more and more of the country. By the 1980s, all the manufactured gas plant (MGP) sites became inactive. During the operation of these plants, both organic and inorganic process residuals accumulated.

Organic contaminants primarily from coal tar and ammoniacal liquors are produced as by-products of the coal gasification process. The chemical composition of coal tar is extremely complex, but the principal constituents include saturated and unsaturated aliphatic and alicyclic hydrocarbons (including high-molecular-weight paraffin waxes), volatile aromatic hydrocarbons (including benzene, toluene, xylenes), polynuclear aromatic hydrocarbons, pyridine and other heterocyclic nitrogen compounds (e.g. tar bases), thiophene and related compounds, and charred material (carbon). The principal organic constituents of the ammoniacal liquor are **phenolic compounds**.

One of the principal sources of **inorganic contamination** in gas manufactured plant sites is spent oxide from the gas purification. Spent oxide contains typically high levels of free sulphur (up to 30%), sulphate (2–3%), and total cyanide (3–6%, mainly as thiocyanate and complex cyanide). Other inorganic species which may be encountered on a GMP site include ammonia and ammonium salts (e.g. sulphate, sulphide, cyanide, thiocyanate, ferro and ferricyanide, thiosulphate, etc.), and carbon, in the form of coal or coke residues.

Contamination by **heavy metals** is also typical of GMP sites; these metals include iron, titanium, manganese, arsenic, chromium, molybdenum and selenium which are found in coal, ash, and coke residues. Catalysts and corrosion inhibitors, which were used frequently on site, are another source of heavy metal contamination. Iron, nickel, copper, zinc, and vanadium are the principal metals associated with this source of contamination. Lead is usually present, since it was used in paint and in batteries.

Polynuclear aromatic hydrocarbons (PAHs)

The primary source of PAHs at MGP sites are typically tars and tar sludge. PAHs are neutral, non-polar organic compounds, consisting of two or more fused benzene rings in linear, angular, or cluster arrangements. Due to the acute and chronic toxicity associated primarily with the lower molecular-weight PAHs, the USA Environmental Protection Agency (USEPA) has designated sixteen PAHs as being environmentally important and representative of PAHs as a class of compounds. These sixteen compounds are known as the EPA's Priority 16 PAHs. Table 15.1 includes key characteristics of these compounds. It should be noted that the water solubility of the PAHs decreases significantly with the increase in molecular weight. This has a direct impact on their bioavailability to the microorganisms as a carbon source, with high ring structure meaning less bioavailability and high degree of recalcitrancy.

At present, many microorganisms are known to have the enzymatic capacity to catabolize PAHs that range in size from naphthalene to benzo(a)pyrene (Cerniglia, 1984; Gibson and Subramanian, 1984; Weis-

Table 15.1. General characteristics of key PAHs

Compound	Rings	Molecular weight	Melting point,°C	Boiling point,°C	Aqueous solubility (µg/mol)	Solubility Log_{10} Molar	Octanol-water partition coefficient
Naphthalene	2	128	82	218	31.7	-3.606	3.37
1-Methylnaphthalene	2	142	-222	245	—	-3.705	—
Acenaphthylene	3	152	92	265	3.93	—	4.07
Acenaphthene	3	154	95	278	3.47	-4.594	4.33
Fluorene	3	166	116	295	1.98	-4.925	4.18
Phenanthrene	3	178	101	340	1.29	-5.15	4.46
Anthracene	3	178	218	342	0.073	-6.377	4.45
Fluoranthene	4	202	110	393	0.26	-5.898	5.33
Pyrene	4	202	156	404	0.135	-0.6176	5.32
Benz(a)anthracene	4	228	160	400	0.014	-7.214	5.61
Chrysene	4	228	254	448	0.002	-8.057	5.61
Benzo(b)fluoranthene	5	252	167	481	0.0012	—	6.57
Benzo(k)fluoranthene	5	252	215	480	0.00055	—	6.84
Benzo(a)pyrene	5	252	178	495	0.0038	-7.82	6.04
Ideno(1,2,3-cd)pyrene	6	276	163	—	0.062	—	7.66
Benzo(g,h,i)perylene	6	276	220	500	0.00026	-9.018	7.23

senfels *et al.*, 1990). Given the requisite environmental conditions, microbial communities are able to readily degrade these chemicals (Morgan and Watkinson 1989; Mueller *et al.*, 1989). Experience gained from studies in biological soil remediation however, have shown that the **degree of PAH degradation** in different soils may differ significantly, even under the same optimum growth conditions concerning temperature, nutrients, oxygen supply and presence of sufficient population of PAH-degrading bacteria.

The soil physical and chemical characteristics play a major role in determining the rate at which PAHs are degraded. For example, the presence of elevated concentrations of cyanide or heavy metals may result in the resistance of PAHs toward biological attack due to microbial toxicity. More significant is the effect of tight PAH binding to the soil fine particles and clay content, resulting in a reduced substrate availability for microbial attack. This inhibitory mechanism of PAH degradation was reported also recently by Weissenfels *et al.* (1992). Based on our own experience in this field, we have found also that the soil organic carbon can cause a significant inhibitory effect on the degradation of PAHs. The mechanism of this effect is due to one or more of the following reasons:

1. Organic carbon may be utilized by the soil microorganisms in preference to the PAH compounds;
2. The adsorption of PAH compounds onto the soil organic content, which protects them from microbial attack and thus enhances their persistence;
3. Organic carbon may contain readily degradable substrates which inhibit the induction of PAH's metabolic pathways.

Metabolic pathways of PAHs

Most of the literature concerning microbial transformation of PAH has centred on the lower molecular weight compounds such as naphthalene, anthracene and phenanthrene (Evans *et al.*, 1965; Fewson, 1981). More recent studies have described capacities of other microorganisms to metabolize 3- or 4-ringed compounds such an benzo(a)pyrene, including the white rot fungus *Phanerochaete chrysosporium* (Bumpus, 1989).

One of the concerns which has arisen out of *in vitro* studies has been the possible production of certain intermediates from PAH degradation, particularly **dihydrodiols** which are of greater toxicity than the parent compound; however, studies on PAH degradation in sediments have suggested that accumulation of such compounds may not actually occur in the natural environment due to the rapidity with which they are fully transformed, and the small size of the steady-state intermediate pool (Herbes and Schwall, 1978). A similar absence of intermediate accumulation in the degradation of PAHs with three or more aromatic rings in laboratory microcosms tests has also recently been observed (Heitkamp *et al.*, 1988).

The microbial degradation of PAHs is initiated by **oxidation reactions**, catalysed by dioxygenase enzymes to form the *cis*-dihydrodiol derivatives as transient intermediates; these intermediates are subsequently oxidized by specific monooxygenase systems resulting in ring cleavage. The ring cleavage mechanisms, combined with series of oxidation and hydrolytic reactions, results eventually in the production of fatty acids. The fatty acids are then metabolized via the tricarboxylic acid (TCA) cycle to provide the necessary carbon and energy source for microbial growth.

Land farming
Land farming has been used with some success for the remediation of PAH-contaminated soil. This method of biological treatment involves batch addition of nutrients using physical mixing of the nutrients into the soils and passive introduction of oxygen (air). Nutrients consist of a **phosphorus source** such as phosphate and a **nitrogen source** such as urea, ammonium and a nitrate salt. Nutrient requirement can be estimated from the contamination concentrations and/or laboratory treatability tests. The benefits of adding microbial inocula to enhance PAH's degradation are not always clear. Despite a number of brief reports in the literature, there are only a few examples of detailed data from full-scale land farming of coal tar-contaminated soil (Bewley and Theile, 1988). Most of the results presented have referred to lab tests, pilot-scale or field trials. The results available so far suggest the PAH degradation in soil is somehow variable and end point analytical concentration is difficult to predict. The performance of the process is site specific and detailed laboratory treatability studies are required to assess the achievable final PAH concentration after treatment. Traditionally, these tests are accomplished with pan studies involving extensive sampling and analysis. Typically four to six months are required to approach the plateau level of PAH concentration. This length of time can be significantly reduced if slurry systems are used, due to the enhancement of substrate bioavailability. Coal tar-contaminated soils appear to contain PAHs portioned into two fractions, one readily available for biodegradation and the other more resistant. The pool size of each fraction is dependent on the soil physical and chemical characteristic such as soil particle size distribution, organic content, degree of soil weathering, etc. It is generally now believed that the chief factor limiting the rate and extent of PAH biodegradation in contaminated soil is **mass transfer** and,

when PAH's are rendered bioavailable in soil or water, microorganisms are able to biodegrade them extensively.

Soil slurry treatment

Another technology which is more efficient in treating PAH contamination in soil is **soil slurry treatment**. This technology uses a slurry-phase bioreactor in which the soil is mixed with water to form a slurry. After the contaminated soil is excavated and screened to remove oversized material, it is mixed with water to form a slurry (10–30%, w/v). The slurry is then passed through a milling process to achieve a slurry with grain size distribution suitable for charging to the bioreactor, which is supplemented with oxygen (air), nutrient and when necessary, a specific inoculum of microorganisms to enhance the biodegradation process. The residence time in the bioreactor varies depending on the soil type, PAH concentration and clean-up requirement. Once biodegradation of the contaminant is completed, the treated slurry is sent to a separation/dewatering system. The solids may be further treated if they still contain residual PAH. The processed water can be treated on an on-site treatment system prior to discharge, or it can be recycled to the front end of the system for slurrying. PAH degradation in slurry bioreactors proceeds much faster than in land farming due to the optimized mixing condition and the improved bioavailability of PAHs, particularly those of high molecular weight. Usually, the greatest decline in PAH concentration occurs in the first two weeks of treatment. Careful attention should be given to control air emission during operation.

The result of many *in vitro* studies are encouraging and suggest that: (i) biodegradation of residual PAHs in MGP soils is potentially viable and, (ii) that the slurry bioreactor can achieve significant improvements in the rate of degradation. Pilot studies and available field data so far have indicated that soils with high silt, clay and or organic content may be less susceptible to contaminant removal via biodegradation than soils with a high sand content.

For further enhancement of degradation of coal tar-contaminated soil, chemical pre-treatment may have to be considered; for example, cyanide toxicity can be eliminated with dilute alkali. Ozone and hydrogen peroxide pre-treatment may be used to initiate PAH oxidation and convert recalcitrant molecules into a readily biodegradable product. The application of surface active agents may assist also in improving the bioavailability of PAHs in certain cases.

Halogenated organic compounds

Whether or not a halogenated compound will be utilized by microorganisms in natural environments depends upon two factors: the structural similarity of the compound with a natural product, or the ability of such a compound to induce the synthesis of enzymes that may act on structurally analogous, naturally-occurring compounds. Thus DDT is rarely degraded by microorganisms, but if the *para* chlorine atoms of DDT are replaced by methoxy groups resembling natural products of similar structure, the resulting compound (methoxychlor) is much more amenable to microbial attack. Many other halogenated compounds, including some of the highly toxic molecules, were shown to undergo biological degradation in the natural environment. If concentrations are too high, biodegradation will not take place. For example, high concentrations of chlorinated aliphatic compounds may inhibit the microbial flora of sewage sludge.

Halogenated compounds vary considerably in their **biodegradability**, depending on their molecular configuration and the number and position of the halogens attached to the molecules. Examples of readily degradable substrate include simple chlorinated alkanoic acids such as 2,2′-dichloropropionate, which is commonly used as a herbicide. The microbial degradation of chlorinated aliphatic and chlorinated aromatic compounds involves two different types of dehalogenase systems. Highly chlorinated aliphatic and aromatic compounds appear to be persistent, these include some of the higher chlorinated PCB isomers, carbon tetrachloride, dioxin, etc.

The microbial transformation of the **aliphatic halogenated compounds** usually involves the elimination of the halogen atoms by halide hydrolases, dehalogenases, or even fortuitously. The dehalogenation of aromatic molecules takes place often after aromaticity has been abolished by the activities of dioxygenases, then the carbon–halogen bond becomes labilized. Initial dehalogenations of arylhalides have rarely been found under aerobic conditions, and these cases have turned out to be basically oxygenolytic elimination. Under **anoxic conditions**, many of the highly chlorinated aromatic compounds were shown to be transformed into **less** chlorinated compounds. Little information is available on the nature or the characteristics of these types of dehalogenase systems.

The majority of halogenated aromatic compounds are degraded via **halocatechol** which may be converted by a variety of different mechanisms to non-chlorinated ring cleavage products. For example, monochlorobenzoates may be degraded to terminal end products such as CO_2, H_2O, or methane and CO_2 depending on the available electron acceptor. Increasing information is now also becoming available on the ability of mixed populations from several environmental habitats (including sewage) to dehalogenate chlorinated aliphatic compounds. In addition, defined strains capable of utilizing haloalkanoic acids for growth have been isolated, but only a few have been examined in detail for their dehalogenase activity.

Volatile halogenated C_1 and C_2 hydrocarbons, in general, are poorly degradable in the environment and persist for long periods of time. The chlorinated phenols and benzoates are more biodegradable than chlorinated benzene and recalcitrance increases with increasing the degree of chlorine substitution. Highly chlorinated polychlorinated biphenyls (PCBs) and the chlorinated pesticides hexachlorocyclohexanes (HCHs) and the drins (aldrin, dieldrin and telodrin) are extremely recalcitrant and are hardly degraded by microorganisms in the environment.

15.4.3 Organophosphate compounds

Organosphosphate pesticides have recently been extensively used as replacement chemicals for the more persistent organochlorine and mercuric compounds. They are characterized as having the general structures shown in Figure 15.1. Several reviews have been published discussing the environmental fate of organophosphate pesticides and the role that microbes play in their degradation. Studies have concentrated mainly on parathion ($R_1 + R_2$ = ethyl, R_3 = *para*-nitrophenyl) and methylparathion ($R_1 + R_2$ = methyl; R_3 = *para*-nitrophenyl) (refer to Figure 15.1) due to their intensive agricultural use. Indirect evidence of microbial participation in the degradation of both pesticides was shown by comparing their stability in sterilized and non-sterilized soil and water samples.

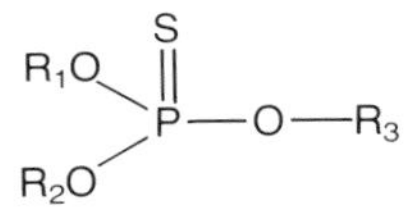

R_1 and R_2 = alkyl moiety
R_3 = substituted alkyl or alkyl moieties

Figure 15.1 General structures of organophosphate pesticides.

The **hydrolysis** of organophosphate compounds can be both chemical and biological in nature. The major pathway of malathion disappearance in soil, water, sediments and salt marsh environments is biologically-catalysed. Parathion degradation can proceed through different pathways depending on the environmental conditions and the type of microorganisms involved. For example, the nitro group can be reduced, resulting in amino parathion; alternatively, the molecule can be hydrolysed to *para*-nitrophenol and diethylthiophosphoric acid.

15.5 THE PROSPECTS OF MICROBIAL APPLICATION TO TOXIC WASTE TREATMENT

As wastes generated by society continue to increase in both volume and complexity, the need for novel treatment approaches and efficient recycling process also will substantially increase. The required improvement in biological waste treatment is likely to be dependent upon successful exploitation of gene manipulation technology. Some success has already been achieved in this area and a good example is the TCE-degrading microorganism which was developed by the scientists at Celgene recently and is currently being evaluated in the field.

One of the problems in developing highly specialized organisms in general, using genetic engineering techniques, is that the organisms produced do not usually compete favourably with the wild-type microorganisms. Potentially, this represents a major limitation in waste treatment processes, since most of these operate in non-sterile conditions and may even require the presence of more than one microorganism for the complete processing of the waste. Instability and loss of catabolic capabilities of genetically engineered microorganisms in a biological systems can arise in at least three different ways:

1. Loss of the structured plasmid, resulting in a loss of the catabolic properties.
2. Dilution of the active organism from the system by plasmid-free microorganisms.
3. Outcompetition by contaminating microorganisms.

Natural plasmids are maintained in bacterial populations because they contain special genes which control the segregation of the plasmid into the daughter cells produced at cell division. These are not always present in plasmids structured by gene manipulation, so plasmids can be lost by being segregated out at cell division. The presence of a plasmid also imposes extra nutritional requirements on the bacterial host cell, which places it in a selectively disadvantageous position unless the plasmid is of some other advantage. The problem of protecting genetically-engineered strains from contaminating bacteria is similar to that encountered in most fermentation processes. It can easily be eliminated if the process can be operated as a sterile process, but this is normally too complex and expensive for waste treatment processes. Problems of contamination may be minimized by the use of high inoculum density of the selective organism, frequent inoculation, and/or by using immobilized cell systems which retain the selected organisms in the treatment system at a high concentration. In general, due to the recognized problems associated with the use of genetically-engineered microorganisms, conventional approaches for selecting and using natural populations are likely to remain important in waste treatment for some time.

During the next few years, research is also likely to continue to focus on the isolation and genetic development of microbial strains with **novel catabolic capabilities**. These strains will be tailored towards dealing with specific chemicals which resist degradation by natural microbial activities, e.g. chlorinated solvents, PCBs, chlorinated pesticides, dioxins, etc.

15.5.1 Soil and groundwater bioremediation

Bioremediation of soil and groundwater appears to be among the most promising methods for dealing with a wide range of organic contaminants, particularly petroleum hydrocarbons. The most reliable way to achieve successful bioremediation is to ensure that the appropriate microorganisms are present in adequate numbers and that the physico-chemical environmental conditions are optimized for their growth and catabolic activities.

Probably the greatest value of adding relatively large numbers of a particular strain occurs in cases where such organisms will catalyse a key initial biochemical reaction which would normally be the rate-limiting step, e.g. ring cleavage or initial oxidation dehalogenation, etc. Such organisms are therefore the **initiators** of the degradation process. Given the diverse range of contaminants present at many abandoned industrial sites and their chemical complexity, it is unlikely that a 'superbug' can be employed which will completely mineralize all such pollutants to carbon dioxide and water. In many cases, effective degradation may be achieved entirely by a judicious optimization of the physico-chemical properties of the environment. The need for biological augmentation of the soil will need to be evaluated on a site-by-site basis, and will be dependent on a number of factors which include (i) the **length of time** the chemical spill has been present and whether this has been sufficient to allow the development of an appropriate pollutant-degrading microflora; (ii) the **nature** of the contamination and how similar the compound is to naturally-occurring organic substrates in the soil; (iii) the presence of **other carbon sources** which might be utilized in preference to the target compounds.

15.5.2 Bioremediation strategies

Microbial strains for site remediation may be obtained from a variety of sources. The obvious source is from the contaminated site itself, where it may be assumed that such species will have adapted to the prevailing ecological conditions. Samples of contaminated material can be inoculated into particular growth media using the target pollutant as the sole carbon or nitrogen source and, following incubation, suitable pollutant-degrading organisms may be isolated.

One of the key properties limiting the degradation of many contaminants is **insolubility**: it may be possible to enhance this using a suitable **surfactant**, provided that the latter is biodegradable and non-toxic at the concentrations used to both microorganisms and other flora and fauna. The success of surfactants in enhancing degradation has been so far, somewhat variable (Robichaux and Myrick, 1972; Mulkins-Philips and Stewart, 1974) and further investigation in this area is necessary to define their potential role in bioremediation. Several hydrocarbon-degrading microorganisms were shown to produce very active emulsifying agents. These extracellular polymers usually consist of high molecular-weight polysaccharides, synthesized from the growth on paraffinic substrates. The manipulation of soil conditions to enhance the proliferation of these types of microorganism is likely to be more cost effective than surfactant applications. Following selection of microorganisms, nutrients and surfactants based on liquid culture studies, such reagents can then be tested in microcosm experiments.

Microcosms

There are many definitions of a 'microcosm', a typical one being that of an intact, minimally disturbed piece of an ecosystem brought into the laboratory for study in its natural state (Pritchard and Bourquin, 1984). Microcosms can vary in complexity from simple columns or static jars of contaminated soil to highly sophisticated systems engineered to enable variations encountered in different environmental parameters on site to be more accurately simulated in the laboratory.

In the development of **bioremediation strategy**, the purpose of the microcosm systems is to select the most appropriate combination of biological and chemical reagents by modelling their effects on pollutant concentration. Such systems embody the 'community' approach to a bioremediation strategy, combining the effects of vanguard microorganisms with the naturally-occurring microflora (including all co-metabolic mechanisms) under optimal conditions. Obviously there have been a large number of more general uses of microcosms for studying microbe–pollutant interactions and these can be used to develop models for predicting the fate of xenobiotics introduced into water/sediment systems.

Mathematical equations can be formulated to describe the **kinetics** of each of the processes involving transformation of the specific xenobiotic under consideration. Concentrations of the xenobiotic and its degradation products can then be monitored in the various components of the microcosm to obtain data on adsorption rates, equilibrium partitioning, abiotic and biotic degradation. 'Spiked' systems have been used frequently in these studies and some caution should be exercised in the assumption that the behaviour of spiked systems necessarily reflects the corresponding microbe–pollutant interactions in the natural environment. The bioavailability and adsorptive properties of a specific xenobiotic in, for example a tarry or oil residue is likely to be very different from that of an introduced, pure, labelled compound in laboratory microcosm.

Having selected the most appropriate chemical and biological reagents in microcosm experiments, **physical conditions** required for effective biodegradation can be optimized in on-site field trials. These involve a careful synthesis of civil engineering procedures with the biological and chemical aspects of the treatment developed in the laboratory, to tailor the remediation strategy to the specific site in question.

There are a range of methods available for ensuring that **optimal site conditions** are maintained during the treatment period (Ellis *et al.*, 1990), all of which are dependent on the nature of the pollutant and the geotechnical properties of the site. Oxygen supply is usually maintained at adequate levels by regular soil rotovation or by passing forced air through perforated pipes. In some cases, hydrogen peroxide, ozone or nitrate may be injected into saturated zones to optimize the availability of electron acceptors. In addition to supplying electron acceptors, a **co-substrate** may be required in gaseous form. For example, certain halogenated aliphatic hydrocarbons, such as trichloroethylene, have been shown to be removed from groundwater by treatment with organisms which use mono-oxygenase enzyme systems. These organisms will require provision of low molecular weight hydrocarbons such as methane, ethane, propane or butane: natural gas is reported to be satisfactory for this purpose (Wilson and Wilson, 1987).

15.5.3 Types of bioremediation

There are basically five types of bioremediation: above-ground bioreactor, solid phase treatment, composting, land farming, and *in situ* treatment.

Above-ground bioreactors

Above-ground bioreactors have been used widely to treat industrial effluent and waste water. Bioreactors can also be designed to treat contaminated groundwater and soils (in a slurry form), and require much the same technology as those used in waste water treatment reactors. The preferred design of these reactors is based on the use of suspended microbial growth (Balba and Rees, 1987) or growth on solid support media (fixed film reactors). The solid support media can be activated carbon, plastic spheres, glass, diatomaceous earth, etc. The advantages of the fixed film reactors are the reduced rate of sludge generation and the biodegradation efficiency of contaminants, particularly in the cases of low strength effluents and groundwater.

The above-ground bioreactors can be operated in an **aerobic** or **anaerobic** mode depending on the nature of contaminants, though, to date, most large scale plants have been aerobic operations. It is also possible to design two bioreactors in series, with the first unit operated anaerobically (for example, to dehalogenate a compound) and the second reactor aerobically (to mineralize the resulting metabolic by-product). **Sequence batch reactors** are also frequently used because of their flexible design, which is particularly useful in dealing with effluents which have irregular flow and variable chemical characteristics.

Solid phase treatment

In this method, the contaminated material is usually constructed into treatment mounds in an enclosure, such as a greenhouse structure or a polytunnel, or on the top of a liner. The treatment mounds can be aerated regularly by mechanical tilling or via forced aeration through perforated plastic pipes installed at the base of the mounds. Fertilizer, microbial inocula and or surfactants may be applied to the treatment mounds to enhance biodegradation (Balba *et al.*, 1991). This system can be optimized if linked with vacuum extraction (biovacuum pile), to enhance the aeration of the soil matrix and in the same time assist in removing toxic volatile pollutants.

Composting

Composting techniques similar to those used widely in the farming industry can also be applied to degrade resistant chemicals such as TNT (2,4,6-trinitrotoluene), RDX (hexahydro-1,3,5-trinitro-1,3,5-triazine) and many other explosive chemicals. Such chemicals appear to be amenable to biodegradation under thermophilic conditions. These conditions prevail during the initial phase of the composting process (Senior and Balba, 1990). The process should also be investigated further to determine its potential application towards other groups of toxic chemicals.

Contaminated soil is usually mixed with compostable materials prior to treatment, there are several relevant processes:

1. Windrow system: soil is piled up in mounds which are periodically turned to facilitate aeration.
2. Beltsville system: as in the Windrow system, but aeration takes place by forced air through perforated pipes passing through the mound.
3. Dano system: a closed system in which soil is treated in horizontally-rotating drums.
4. Kneer system: a closed system in which the soil is passed downwards through a column and air is blown in counterflow through the column.
5. Schnorr system: similar to Kneer system, but the column is divided into stages by horizontal plates.

Landfarming

In landfarming, chemical waste is spread thinly over agricultural soil. Biodegradation is then promoted by ploughing, fertilizing and other standard farming procedures. Aerobic degradation of waste is generally faster than anaerobic degradation; it is thus of great importance to keep the soil **aerated** (e.g. by ploughing). Depending on the type of waste and on the composition of the soil, it may be necessary to add *nutrients* in the form of artificial fertilizer or animal manure. A high moisture content hinders oxygen transport, a low moisture content inhibits the biological activity. A pH value of 6–8 is normally required; the pH can be adjusted by addition of, e.g. $Ca(OH)_2$ or $(NH_4)_2SO_4$. An increase in temperature (up to 30°C) generally increases the rate of the biodegradation. In landfarming, however, the temperature depends on the climate, and can hardly be influenced. Addition of microorganisms, which are adapted to the biodegradation of the contaminants, can hasten the onset and possibly the decontamination process itself.

In situ *decontamination (without soil excavation)*

Shallow contamination
This may be treated by landfarming, the technique being suitable when the contamination is concentrated in a top soil layer. Sufficient oxygen is provided by regular ploughing and harrowing the soil layer. Nutrients, surfactants and/or microbes can be added by conventional methods to enhance the biodegradation of target compounds.

Saturated zone/aquifers

For the effective treatment of the plumes in the saturated zones it is essential to keep a supply of oxygen and nutrient in the aquifer for the added microbes to use. Oxygen has a limited solubility (7–10 mg/l, depending on temperature), therefore highly oxygenated water should be recirculated through the aquifer as fast as possible by pumping out the water from a central well and recirculating it through infiltration trenches. Nutrients can be applied to the recycled water beside the appropriate microorganisms. Hydrogen peroxide may be introduced as a supplementary oxygen source to satisfy oxygen demands of the subsurface soil when aeration becomes inadequate.

The water recharged to the aquifer is obviously not completely clean and may therefore contain bacteria and nutrients which can be considered contaminant down-flow in the aquifer. The recharge water should not be placed outside the treatment zone. To ensure that the water placed in the infiltration trenches does not get out of the treatment zone, it is necessary to place less water in the infiltration system than was drawn-down from the central well.

The unsaturated zone

The *in situ* clean-up technique can also be applied to the unsaturated zone. The design criteria are the same as those for the *in situ* clean-up of the saturated zones with two exceptions. First, the upper layers of the unsaturated zone may not require any water pumping and second, the objective in the unsaturated zone is usually flushing of the contaminants along with in-place destruction. In this case, the contaminants present in soil are biodegraded by adding biomass, nutrients, surfactants and/or hydrogen peroxide. Drainage pipes can be installed to allow the recirculation of leachate. Unsaturated zones can also be activated by flowing or infiltrating bioaugmented water through trenches filled with gravel. The biodegradation process can be significantly enhanced if it is linked with a soil vapour extraction system to provide a renewable source of oxygen.

15.5.4 Case studies of large-scale bioremediation

Despite a number of brief reports in the literature, there are few examples of detailed data from full-scale bioremediation. Most of the results presented have referred to pilot-scale field trials. There are probably two major reasons for this. Contaminated sites are often remarkably heterogeneous in nature, so that the initial starting data vary from very low to very high concentrations over a relatively small area. In addition, large volumes, generally of the order of thousands of cubic yards of soil, are involved. Under such circumstances, it is very difficult to obtain statistically meaningful data without analysing a massive number of samples which would be prohibitively expensive. The following projects present some examples of the use of biotechnology in the remediation of contaminated soil and groundwater.

Case study 1: Bioremediation of heavy petroleum hydrocarbon-contaminated soil

Numerous locomotive maintenance yards are operated worldwide by major railroad companies. In recent years, an increasing number of these yards are being identified as having soils contaminated with various concentrations of petroleum hydrocarbons. The source of the contamination is refuelling operations and general locomotive servicing where hydrocarbon-type products such as diesel fuel and heavy motor oil are routinely used. The current available means for cleaning up these oil-contaminated soils is off-site disposal at a secured land disposal facility. In addition to the significant cost of this approach, the inherent liabilities associated with waste transportation and land disposal make this a very unattractive remedial option.

In early 1988, TreaTek, Inc., was commissioned to develop and design a cost-effective biological treatment programme specifically for this type of heavy engine oil contamination. This project was partially funded by the Alternative Technology Section of California's Department of Health Services under the California Hazardous Waste Reduction Grand Programme.

Laboratory programme
A multi-step laboratory treatability programme was carried out using actual oil-contaminated soil samples collected from an operating railyard. The soil contained from 5000 ppm to 60000 ppm petroleum hydrocarbons. The programme was designed to focus primarily on microbial and chemical additives for enhancing the natural degradation of oil in the soil.

The oil extracted from the railyard soil was characterized and found to contain mostly linear and branched alkanes in the C_{22} plus range (Figure 15.2). The oil consists mainly of heavy engine oil with a boiling point of higher than 200°C. The experimental programme also involved several other preliminary tasks, including microbial assessment and inocula optimization. These initial steps were then followed by two actual benchscale treatment operations. The first involved soil microcosm tests, each containing 500 g of contaminated soil. These micro-

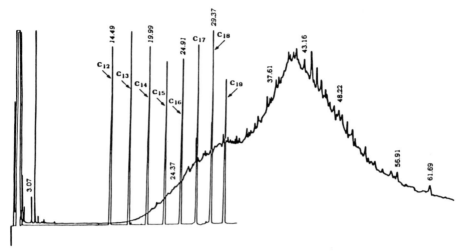

Figure 15.2 Gas chromatographic analysis of heavy engine oil.

cosms were treated with multiple combinations of different microbial and chemical reagents. Oil degradation in these microcosms was measured and compared over a 16-week period. The second set of treatment simulations involved a much larger quantity of soil (approximately 9 kg each). These scaled-up systems enabled actual soil aeration testing and provided means to measure potential oil removal via volatilization or leachate generation. The laboratory data produced at 15°C showed more than 85% of the total petroleum hydrocarbons in soil were degraded in less than 16 weeks. Rigorous aeration of the soil in the simulation units showed minor volatilization of the contaminant at high concentration (soil with >3% oil), while no volatilization was observed with soil below 1% oil.

The field demonstration phase involved more than 150 cubic yards of heavily contaminated soil (>10% oil) which was excavated from the site and placed on two enclosed rectangular beds (40 × 4 m each), resting on concrete liners to prevent leaching into subsurface soils. The beds drained into an adjacent sump from which leachate could be collected and analysed (Balba and Ying, 1991). The soil was kept aerated and mixed by using an agricultural spading machine. Nutrients and microbial inoculation of the soil were applied to the soil beds by spraying, followed by soil mixing. Biodegradation parameters such as pH, nitrogen, phosphorus, and microbial population count were monitored regularly and adjusted when required. The beds were sampled once every four weeks for total petroleum hydrocarbons (TPH) analysis using a sampling grid consisting of 13 samples from each bed. Routine TPH

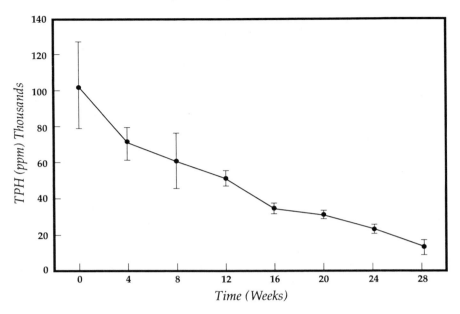

Figure 15.3 Bioremediation of heavy petroleum oil in soil at a railroad maintenance yard. (Figures represent the means of 13 samples.)

analyses were carried out using EPA Modified Method 418.1 (IR) and gas chromatographic analysis, using a flame ionization detector was performed only intermittently to confirm biodegradation.

In spite of the complex nature of the contaminant and its extremely high starting concentration of TPH in the soil ($>100\,000$ mg/kg), the treatment was successful in substantially reducing the contamination level. More than 85% biodegradation was achieved, within less than 28 weeks (Figure 15.3). These field data show that high molecular weight hydrocarbon can be successfully treated using bioremediation techniques.

Case study 2: Bioremediation of an oil-polluted refinery site

This study was designed to assess the feasibility of bioremediation within an oil-polluted refinery site. The site occupies several hectares and is situated near a major river in West Germany. Oil contamination penetrated the soil to depths in excess of 6 m, and therefore excavation and removal of the polluted area would prove very expensive and complicated by fluctuations in the groundwater level. Furthermore, certain surface layers of the refinery soil contain contaminated clay lattices not amenable to *in situ* processes. Therefore, a dual treatment process was designed, incorporating microbial remediation of heavy contaminated excavated clay soil, combined with bioaugmented *in situ* techniques for deeper layers. During the field trials, two areas of contamination were selected. Area A underwent *in situ* treatment and consisted of approximately 1600 cubic metres of oil-contaminated land (0–8 m depth). Area B consisted of approximately 200 cubic metres of more heavily contaminated subsurface soil layers (0–2 m depth).

Laboratory programme
The laboratory programme consisted of several tasks which included oil and groundwater characterization, contaminant characterization, microbial isolation and screening of oil-degrading microorganisms, surfactant screening for oil dispersants, and soil microcosm optimization. Detailed analysis of the oil extracted from the refinery soil showed that it had the composition shown in Table 15.2.

Axenic cultures capable of oil degradation were successfully isolated from the soil and screened for their oil degradation capabilities and survival in the contaminated soil. The results in Table 15.3 show that

Table 15.2. Analysis of oil extracted from polluted soil of refinery

Hydrocarbon	Proportion (%)
n-Alkane:	
C_5-C_{10}	Insignificant
$C_{11}-C_{13}$	0.22
$C_{14}-C_{25}$	1.59
$C_{26}-C_{27}$	0.04
Higher carbons	Insignificant
Branched-chain alkane	
$C_{11}-C_{13}$	0.08
$C_{14}-C_{25}$	1.10
$C_{26}-C_{27}$	0.03
Naphthenes	59.44
Total saturates	62.5
Aromatics	25.0
Resins/residue	6.25
Asphaltines	6.25

Table 15.3. Characteristics of oil-degrading microorganisms isolated from the refinery soil

Designated code	Microorganism species	Growth of oil fraction[a] Gas oil	Growth of oil fraction[a] Hexadecane	Asphaltene	Degradation of refinery oil(%)[b]	Survival of cells in contaminated soil(%)[c]
ETB001/2	Pseudomonas	+	+	–	81.0	55.9
ETB004	Rhodococcus	+	+	–	66.0	70.8
ETB004/2	Acinetobacter	+	+	–	56.0	17.2
ETB005/2	Pseudomonas	+	+	–	61.0	51.7
ETB007	Mycobacterium	+	+	+	92.0	ND
ETB015/3	Arthrobacter	+	+	–	82.0	55.6
Control	None	–	–	–	0	—

[a]Growth was monitored qualitatively (increase in absorbance of >0.5 compared with controls at 620 nm) in liquid culture after 10 days' incubation (25°C, 200 rpm). Medium consisted of basal salts containing the relevant oil fraction (0.2%, w/v).
[b]Refinery oil was determined quantitatively using infrared spectroscopy.
[c]Survival (viable cell counts as a percentage of the initial values) after 10 days' incubation in oil-contaminated soil.
ND = Not Determined.

not only did each species tested survive but they also proliferated within the soil, clearly indicating their tolerance to the soil contaminants.

Oil dispersants and surfactants were also screened and tested for their ability to mobilize oil contaminants in soil. More than 50 compounds were examined and the most promising of these agents were subjected to permeability tests using standard methods described elsewhere. Surface active compounds were further examined using soil column leach tests with and without recirculation of the leachate. These investigations focused attention on the ability of the various surfactants to effect oil movement from contaminated soil under *in situ* conditions and also monitored their influence on soil permeability. One compound, cyanamer P70 was selected because it removed a relatively high proportion (47 %) and in the same time did not adversely affect soil permeability. Another advantage of this compound was its descalent properties which alleviated the possibility of metal precipitation during groundwater extraction, aeration and infiltration. Metal precipitation, especially of iron and manganese, is a common problem in *in situ* remediation and can result in severe soil clogging problems, which can affect the success of the remediation process.

Toxicity of this surfactant was also examined toward three common soil microorganisms: *Arthrobacter*, *Rhodococcus*, and *Pseudomonas*, plus the oil degrading species isolated from the refinery site. Each species examined was found to tolerate up to 1000 mg/l surfactants in liquid culture, and the presence of 2000 mg/l did not affect growth on solid media.

The laboratory programme also included soil microcosm tests to optimize microbial treatment of contaminated soil; a series of soil microcosm tests consisting of more than 80 different treatment combinations were tested, with the objective of optimizing the nutrient status of the soil and mobilizing oil contaminants in order to enhance degradation by microbial inocula. The results showed that nutrient addition and moisture optimization alone resulted in approximately 79% oil reduction within 23 weeks, while the use of nutrient and microbial inocula plus the surfactants resulted in 95% oil reduction during the same period (Figure 15.4). These successful laboratory data led to field trials which were performed over a 6-month period to treat a portion of the refinery site; the area examined represented a volume of approximately 2000 cubic metres of soil.

Field trials
The field trials involved two separate trials; (a) *in situ* treatment and (b) solid phase treatment.

(a) *In situ* treatment: The treatment of groundwater and subsurface soil over an area measuring 10 × 20 m and to a depth of 8 m was achieved

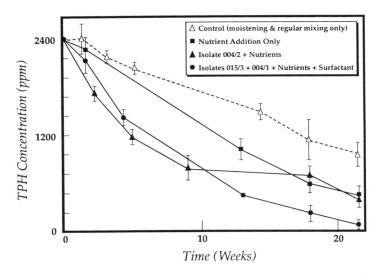

Figure 15.4 Effects of biological treatment on oil concentrations in soil micro-cosm test. (Mean values of four replicates.)

over a 15-week period. Treatment was effected using groundwater extraction and infiltration systems. *In-situ* biodegradation was enhanced by applying surfactant, nutrients, aeration, and inoculation with selected microorganisms. Groundwater was extracted from an 8 m-deep central well to an aerated 50 m³-capacity reactor tank and then piped back into gravel infiltration trenches at the perimeters of the treatment area (Figure 15.5).

Initially, **tracer dye tests** were performed using a series of well points around the *in situ* treatment area in order to follow the nature of flow during treatment. Approximately 5% of flow was diverted to an on-site waste water treatment facility to enhance groundwater movement and to maintain an effective zone of influence. This was also necessary to maintain the treatment zone under hydraulic containment. Organic and inorganic nutrients were applied to maintain an optimum C:N:P ratio. Microorganisms were used as inoculants on four occasions during the trial period.

Soil and groundwater samples were taken before and after treatment over a 15-week period. A 20-point sampling grid was used, with soil samples being taken at five depths per point (100 samples per sample time). Liquid samples were taken within the treatment area and at five points surrounding the site in order to monitor possible interference and changes in groundwater characteristics. Liquid entering the *in situ*

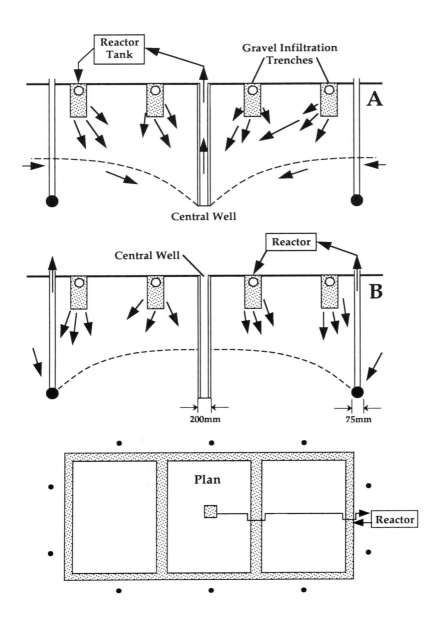

Figure 15.5 *In situ* treatment operation modes: (A) during normal pumping out through the central well; (B) when the system was reversed and liquid was pumped out using the perimeter wells.

treatment area was found to contain an oxygen concentration of 8.4 mg/l but only 2.4 mg/l in extracted leachate. This provided some indication of the rate of oxygen consumption within the zone of treatment during the remediation programme and confirmed biological degradation.

Results of oil analyses from the *in situ* treatment of the site are shown in Figure 15.6, including the mean oil concentrations at various depths. Clearly there were significant reductions at every depth tested, and the overall mean oil hydrocarbon concentration was reduced from 185 ± 49 mg/kg to 26 ± 6 mg/kg (86% reduction) within 15 weeks. Leachate analysis on the surrounding groundwater area resulted in no detectable oil hydrocarbons.

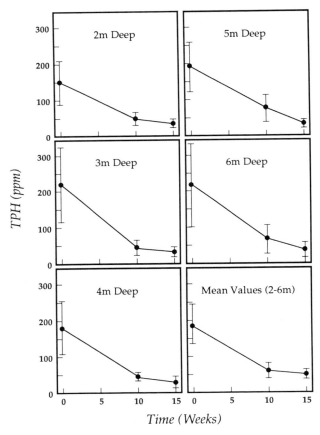

Figure 15.6 *In situ* treatment of refinery land. Figures indicate the mean concentrations of oil hydrocarbons at various depths (20 samples per depth.)

(b) Solid phase treatment: soil containing 10 000 to 30 000 mg/kg oil contaminants was excavated from the surface (2 m) layer of part of the refinery site and placed on an enclosed rectangular bed. The soil was kept aerated; it was mixed using an agricultural spading machine, and its moisture content was maintained at approximately 15% (w/w). Nutrients and surfactant were added to the soil as dilute solutions and applied by tractor-spraying onto the soil surface, followed by mixing. Microorganisms were inoculated into the soil as dilute suspensions; altogether, five separate inoculations of the soil were made over the trial period. Sampling involved 100 soil samples being taken at regular intervals using approximately a 2 m sampling grid (0.4 m depth). In addition, leachate in the sump was monitored.

The excavated treatment had consisted of soil initially containing an average of 12 980 mg/kg oil hydrocarbons which, after 34 weeks of treatment, were reduced by more than 90% to 1273 mg/kg (Figure 15.7).

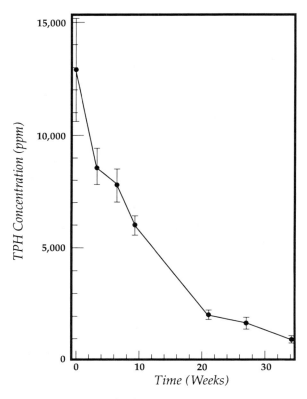

Figure 15.7 *Ex situ* treatment of refinery soil. Figures indicate the means of 100 samples.

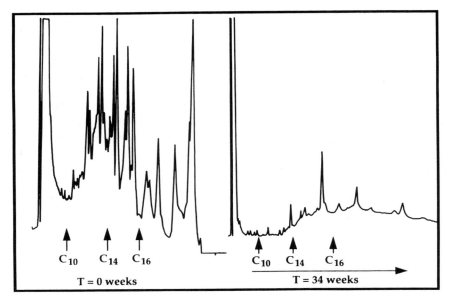

Figure 15.8 Gas chromatograms of refinery soil showing progressive oil degradation during the bioremediation programme.

The soil used in the field trials was also examined for hydrocarbon constituents before and after treatment by gas chromatography; the chromatographs in Figure 15.8 clearly show substantial reduction in oil compounds across the range of chain lengths. The residue remaining in the soil after this period was found to be immobile, as when subjected to repeat 24-hour leach tests (shaking the soil in water, 5:1 water:soil), no oil hydrocarbons were detected in the leachate.

15.6 GENERAL CONCLUSIONS

Biological degradation is a very effective treatment technology for remediating petroleum contamination in soils. It is cost effective and the process results in the on-site destruction of the pollutants. *In situ* bioremediation has the added advantage of minimizing the disruption to on-site plant operations. However, a successful remediation programme can only be properly instigated once the problems of a particular site are fully understood. The economics of the various bioremediation options must also be evaluated. Some issues which are relevant here include: time to identify the appropriate analytical protocols for validation length of time for the remediation, cost of construction and cost of maintenance and operation. Once such knowledge is established then the remedial system should be engineered so as to optimize the balance of both treatment and cost.

Like many other technologies, there are limitations to the application of biotechnology. These result from the heterogeneous nature of the waste to be treated. For example, the waste may contain not only contaminated soil and water, but also concrete, scrap metal, building debris, etc. This complex nature of waste highlights the problems encountered usually in developing adequate analytical methods for the validation of the treatment. This may also affect the economics of the process. The prospects for biotechnology will probably be dependent on the success of integrating it with other technologies such as thermal desorption, soil vapour extraction, chemical treatment, etc. Bioremediation alone will not always be the solution for all contamination problems.

REFERENCES

Alexander, M. (1977) *Introduction to Soil Microbiology*, Wiley & Sons, New York.
Alexander, M. and Aleem, M.I.H. (1961) Effect of chemical structure on microbial decomposition of aromatic herbicides. *Journal of Agriculture and Food Chemistry*, **9**, 44–7.
Alexander, M. and Lustigman, B.K. (1966) Effect of chemical structure on microbial degradation of substituted benzenes. *Journal of Agriculture and Food Chemistry*, **14**, 410–13.
Arthur, R.A.J. (1986) At the root of the matter. A profile of the Kickuth root zone system. *Water and Waste Treatment*, **48**, 34–5.
Ajisebutu, S.O. (1988) Effects of sodium chloride on biodegradation of crude oils by two species of *Aeromonas*. *Applied Microbiology and Biotechnology*, **28**, 203–8.
Atlas, R.M. (1977) Stimulated petroleum degradation. *Critical Reviews of Microbiology*, **5**, 371–86.
Atlas, R.M. (1981) Microbial degradation of petroleum hydrocarbons: An environmental perspective. *Microbiological Reviews*, **45**, 180–209.
Atlas, R.M. and Bartha, R. (1973a) Stimulated biodegradation of oil slicks using oleophilic fertilizers. *Environmental Science and Technology*, **7**, 538–41.
Atlas, R.M. and Bartha, R. (1973b) Inhibition by fatty acids of the biodegradation of petroleum. *Antonie van Leeuwenhoek. Journal of Microbiology and Serology*, **39**, 257–71.
Balba, M.T. and Bewley, R.F. (1991) Organic contaminants and microorganisms, in *Organic Contaminants in the Environment*, (ed K.C. Jones), Elsevier, Applied Science Publisher Ltd., London.
Balba, M.T. and Rees, J.F. (1987) The potential application of microbes to the detoxification of noxious effluent generated by pre-wash tank cleaning, *IMP Proceedings of the International Symposium on Reception Facilities for Noxious Liquid Substances*, London, pp.298–326.
Balba, M.T. and Ying, A.C. (1991) Enchanced biodegradation of heavy engine oil in soil from railroad maintenance yards – Phase II Field Demonstration. *Proceedings of the Sixth Annual Conference on 'Hydrocarbon Contaminated Soil'*. University of Massachusetts, Amherst.

Balba, M.T., Ying, A.C. and McNeice, T.G. (1991) Bioremediation of contaminated land: Bench-scale to field application, in *Proceedings of National Research and Development Conference on the Control of Hazardous Materials*, HMCRI, pp.145–51.

Bartha, R. (1986) Biotechnology of petroleum pollutant biodegradation. *Microbial Ecology*, **12**, 155–72.

Baxter, R.A., Gilbert, R.E., Lidgett, R.A., Mainprize, J.H. and Vodden, H.A. (1975) The degradation of PCBs by microorganisms. *Science of the Total Environment*, **4**, 53–61.

Bewley, R.J.F. and Theile, P. (1988) Decontamination of a coal gasification site through application of vanguard microorganisms, in *Contaminated Soil '88*, (ed K. Wolf, W.J. van den Brink and F.J. Colon), Kluwer Academic Publishers, Dordrecht, pp.739–43.

Blakebrough, N. (1978) Interactions of oil and microorganisms in soil, in *The Oil Industry and Microbial Ecosystems*, (eds K.W.A. Chater and H.J. Somerville), Heydon and Son Ltd, London, pp.28–40.

Boethling, R.S. and Alexander, M. (1979) Effect of concentration of organic chemicals on their biodegradation by natural microbial communities. *Applied and Environmental Microbiology*, **37**, 1211–16.

Bossert, I. and Bartha, R. (1984) The fate of petroleum in soil ecosystems, in *Petroleum Microbiology*, (ed R.M. Atlas), Macmillan, New York, pp.435–73.

Brown, J.F., Bedard, D.L., Brennan, M.J., Carnahan, J.C., Feng, H. and Wagner, R.E. (1987) Polychlorinated biphenyl dechlorination in aquatic sediments. *Science*, **236**, 709–11.

Bumpus, J.A. (1989) Biodegradation of polycyclic aromatic hydrocarbons by *Phanerochaete chrysosporium*. *Applied and Environmental Microbiology*, **55**, 154–8.

Cerniglia, C.E. (1984) Microbial metabolism of polycyclic aromatic hydrocarbons. *Advances in Applied Microbiology*, **30**, 31–71.

Chakrabarty, A.M. (1987) New biotechnological approaches to environmental pollution problems. *Biotec*, **1**, 67–74.

Colwell, R.E. and Walker, J.D. (1977) Ecological aspects of microbial degradation of petroleum in the marine environment. *Critical Reviews in Microbiology*, **5**, 423–45.

Dias, F.F. and Alexander, M. (1971) Effects of chemical structure on the biodegradability of aliphatic acids and alcohols. *Applied Microbiology*, **22**, 1114–18.

Dibble, J.T. and Bartha, R. (1979) Effect of environmental parameters on the biodegradation of oil sludge. *Applied and Environmental Microbiology*, **37**, 729–39.

Ellis, B., Balba, M.T., and Theile, P. (1990) Bioremediation of oil-contaminated land, in *Environmental Technology*, **11**, 443–55.

Evans, W.C., Fernley, H.N. and Griffiths, E. (1965) Oxidative metabolism of phenanthrene and anthracene by soil pseudomonads. The ring fission mechanisms. *Biochemical Journal*, **95**, 819–31.

Evans, W.C., Smith, B.S.W., Moss, P., Fernley, H.N. and Davis, J.I. (1971) Bacterial metabolism of 4-chlorophenoxyacetate. *Biochemical Journal*, **122**, 509–17.

Fewson, C.A. (1981) Biodegradation of aromatics with industrial relevance, in *Micro Degradation of Xenobiotics and Recalcitrant Compounds*, (eds A.M. Cook, R. Hutter, T.Leising, and H.Nuesch), Academic Press, London, pp.141–79.

Furukawa, K., Matsumara, F. and Tonomura, K. (1978) *Alcaligenes* and *Acinetobacter* strains capable of degrading polychlorinated biphenyls. *Agricultural and Biological Chemistry*, **42**, 543–8.

Getzin, L.W. (1973) Persistence and degradation of carbofuran in soil. *Environmental Entomology*, **2**, 461–7.

Gibson, D.T. and Subramanian, V. (1984) Microbial degradation of aromatic hydrocarbons, in *Microbial Degradation of Organic Compounds*, (ed D.T. Gibson), Dekker, New York, pp.181–282.

Goldman, S., Shabtai, Y., Rubinovitz, C., Rosenberg, E. and Gutnick, D.L. (1982) Emulsan in *Acinetobacter calcoaceticus* RAG-1: Distribution of cell-free and cell-associated cross-reacting material. *Applied and Environmental Microbiology*, **44**, 165–70.

Heitkamp, M.A., Freeman, J.P., Miller, D.W. and Cerniglia, C.E. (1988) Pyrene degradation by a *Mycobacterium* sp.: Identification of ring oxidation and ring fission products. *Applied and Environmental Microbiology*, **54**, 2556–65.

Herbes, S.E. and Schwall, L.R. (1978) Microbial transformation of polycylic aromatic hydrocarbons in pristine and petroleum-contaminated sediments. *Applied and Environmental Microbiology*, **35**, 306–16.

Higgins, I.J. and Gilbert, P.D. (1978) The Biodegradation of hydrocarbons, in *The Oil Industry and Microbial Ecosystems*, (eds K.W.A. Chater and H.J. Somerville), Heydon and Son Ltd., London, pp.80–117.

Horvath, R.S. (1972) Microbial co-metabolism and the degradation of organic compounds in nature. *Bacteriological Reviews*, **36**, 146–55.

Hsu, T.S. and Bartha, R. (1979) Accelerated mineralization of two organophosphate insecticides in the rhizosphere. *Applied and Environmental Microbiology*, **37**, 36–41.

Jannasch, H.W. (1967) Growth of marine bacteria at limiting concentrations of organic carbon in seawater. *Limnology and Oceanography*, **12**, 264–71.

Jobson, A., Cook, F.D. and Westlake, D.W.S. (1972) Microbial utilization of crude oil. *Applied Microbiology*, **23**, 1082–9.

Kaufman, D.D. (1978) Degradation of pentachlorophenol in soil and by soil microorganisms, in *Pentachlorophenol*, (ed K.R. Rae), Plenum Press, New York, pp.27–39.

Leahy, J.G. and Colwell, R.R. (1990) Microbial degradation of hydrocarbons in the environment. *Microbiological Reviews*, **54**, 305–15.

Lee, M.D. and Ward, C.H. (1985) Biological methods for the restoration of contaminated aquifers. *Environmental Toxicology and Chemistry*, **4**, 743–50.

Lehtomaki, M. and Niemala, S. (1975) Improving microbial degradation of oil in soil. *Ambio*, **4**, 126–9.

Liu, D., Thompson, K. and Strachan, W.M.J. (1981) Biodegradation of pentachlorophenol in a simulated aquatic environment. *Bulletin of Environmental Contamination and Toxicology*, **26**, 84–90.

Loynachan, T.E. (1978) Low temperature mineralization of crude oil in soil. *Applied and Environmental Microbiology*, **7**, 494–500.

Lynch, J.M. (1982) The Rhizosphere, in *Experimental Microbial Ecology*, (eds R.G. Burns and J.H. Slater), Blackwell Scientific Publications, Oxford, pp.395–411.

MacRae, I.C., Raghu, K., and Castro, T.F. (1967) Persistence and biodegradation of four common isomers of benzene hexachloride in submerged soils. *Journal of Agricultural and Food Chemistry*, **15**, 911–14.

MacRae, I.C., Raghu, K. and Bautista, E.M. (1969) Anaerobic degradation of the insecticide lindane by *Clostridium* spp. *Nature, London*, **221**, 859–60.

Marinucci, A.C. and Bartha, R. (1979) Biodegradation of 1,2,3-, and 1,2,4-trichlorobenzene in soil and in liquid enrichment culture. *Applied and Environmental Microbiology*, **38**, 811–17.

Morgan, P. and Watkinson, R.J., (1989) Hydrocarbon degradation in soils and methods for soil biotreatment. *Critical Reviews in Biotechnology*, **8**, 305–33.

Mueller, J.C., Chapman, P.J. and Pritchard, P.H. (1989) Creosote contaminated sites: Their potential for bioremediation. *Environmental Science and Technology*, **23**, 1197–201.

Mulkins Philips, G.J. and Stewart, J.E. (1974) Effects of four dispersants on biodegradation and growth of bacteria in crude oil. *Applied Microbiology*, **28**, 547–52.

Ou, L.T. (1984) 2,4-D-degrading microorganisms in soils. *Soil Science*, **137**, 100–7.

Pancorbo, O.C. and Varney, T.C. (1986) Fate of synthetic organic chemicals in soil-groundwater systems. *Veterinary and Human Toxicology*, **28**, 127–43.

Pritchard, P.H. and Bourquin, A.W. (1984) The use of microcosms for evaluation of interactions between pollutants and microorganisms, in *Advances in Microbial Ecology*, vol. 7, (ed K.C. Marshall), Plenum Press, New York, London, pp.133–215.

Raymond, R.L., Hudson, J.O. and Jamison, V.W. (1976) Oil degradation in soil. *Applied and Environmental Microbiology*, **31**, 522–35.

Robichaux, J.J. and Myrick, H.N. (1972) Chemical enhancement of the biodegradation of crude-oil pollutants. *Journal of Petroleum Technology*, **24**, 16–20.

Senior, E. and Balba, M.T. (1990) Refuse decomposition, in: *Microbiology of Landfill Sites*, (ed E. Senior), CRC Press, Boca Raton, FL, pp.17–57.

Vazquez-Duhalt, R. and Greppin, H. (1986) Biodegradation of used motor oil by bacteria promoting the solubilization of heavy metals. *Science of the Total Environment*, **52**, 109–21.

Walker, J.D. (1985) Chemical fate of toxic substances: Biodegradation of petroleum. *Marine Technology Society Journal*, **18**, 73–86.

Ward, T.E. (1986) Aerobic and anaerobic biodegradation of nitrilotriacetate in subsurface soils. *Ecotoxicology and Environmental Safety*, **11**, 112–25.

Weissenfels, W.D., Beyer, M. and Klein, J. (1990) Degradation of phenanthrene, fluorene, and fluoranthene by pure bacterial cultures. *Applied Microbiology and Biotechnology*, **32**, 479–84.

Weissenfels, W.D., Klewer, H. and Langhoff, J. (1992) Adsorption of polyaromatic hydrocarbons (PAHs) by soil particles; Influence on biodegradability and biotoxicity. *Applied Microbiology and Biotechnology*, **36**, 689–96

Wilson, J.T.J. and Wilson, B.H. (1987) Biodegradation of halogenated aliphatic hydrocarbons. *United States Patent*, No. 4, 713,343

Zajic, J.E., Guignard, H. and Gerson, D.F. (1977) Properties and biodegradation of a bioemulsifier from *Corynebacterium hydrocarboclastus*. *Biotechnology and Bioengineering*, **19**, 1303–20.

Zajic, J.E., Suplisson, B. and Volesky, B. (1974) Bacterial degradation and emulsification of No. 6 fuel oil. *Environmental Science and Technology*, **8**, 664–8.

16

Formulation of biological control agents

D.J Rhodes

16.1 INTRODUCTION

Biological control, in the broadest sense, could be defined as the use of living agents to control pests, plant pathogens and weeds; in this context both the activities of man and the use of resistant plants might be included. This chapter addresses only the formulation requirements of microbial agents used for crop protection and public health. Although some success is being achieved with the use of parasitoids and predators, especially in the glasshouse industry, and some ingenious formulations have been developed for this purpose, these applications are beyond the scope of this book.

During the past decade, biological control has received increasing attention from the scientific community, the press and the public. This has been fuelled by a desire for non-chemical means of controlling plant pests, pathogens and weeds. The research effort expended in this area has increased dramatically in both public and private sectors over this period, and this is beginning to be reflected in the number of biological control products available in the marketplace. Although sales volume remains modest, these products are achieving substantial rates of sales growth (Wood Mackenzie, 1991) which, in general, outstrip those of the crop protection market as a whole, providing a further impetus for research.

Exploitation of Microorganisms Edited by D.G. Jones
Published in 1993 by Chapman & Hall, London ISBN 0 412 45740 7

Interest in microbial pest control has also benefited from the anticipated impact of recombinant DNA technology on the exploitation of
microorganisms as delivery vehicles for novel peptides and proteins,
with the expectation that genetic manipulation may allow inherent
deficiencies to be overcome, and greater levels of efficacy to be achieved.
The use of genetic manipulation appears likely to raise the development
costs of microbial products, costs which must be recouped by increased
sales and profits.

Industry, government, and the public, therefore, share high expectations of microbial control. If these expectations are to be fulfilled,
appropriate formulations will be required which are capable of exploiting and optimizing the intrinsic properties of biological control agents.
Until recently, the published research in this area is disappointingly
sparse, in comparison with the considerable volume of literature on
intrinsic activity, genetics and mode of action. This may, in part, be due
to the fact that much of the information exists as 'trade secrets'; it is
frequently difficult to protect formulations adequately through patents.
Nevertheless, the development of formulation technology which
addresses the requirements of microbial agents will be essential if
microbial control agents are to offer realistic alternative approaches to
crop protection.

16.2 PRINCIPLES OF FORMULATION

Almost all active ingredients used in crop protection and public health
are formulated prior to distribution and sale. Formulation is necessary
in order to present the product in a usable form and in order to optimize
the efficacy, stability, safety and ease of application of the product. The
type of formulation used in any situation will depend both on the
application technology which is available and on the active ingredients
which are registered for that purpose. Any one active ingredient is likely
to be sold in a variety of formulations, depending on local custom in the
area of sale, the purposes for which the product will be used, and on
the physicochemical properties of the active ingredient such as solubility, volatility, melting point and behaviour at the target site (Barlow,
1985). In addition, the formulations available in the marketplace are
constantly changing in response to demands from the user and availability of new formulation and packaging technology. Of prime concern,
especially in recent years, has been safety to the user and to the
environment, which has tended to favour the use of disposable or
degradable packaging, and has increased demand for solid and water-
based formulations.

A great variety of formulations are therefore available, and this has
led to a need for a standardized nomenclature. All formulation types

are therefore designated by a two-letter code, regardless of the active ingredient, although individual manufacturers may choose not to use this code in the trade name (GIFAP, 1984). The main formulation classes are summarized in Table 16.1.

16.2.1 Adjuvants

In addition to the active ingredient and the solvent or carrier, formulations may contain a number of **adjuvants**, defined as compounds which assist or modify the action of the active ingredient (Foy, 1989). The most important of these are the **surfactants**, compounds which alter the wetting and spreading properties of the formulation, but a number of other adjuvants may also be required in order to optimize a formulation. The role and properties of adjuvants have been reviewed by Foy (1989); the properties of the main classes of adjuvant are summarized in Table 16.2.

Adjuvants serve three main purposes. Firstly, they may be used to optimize the activity of the active ingredient, for example by enhancing the uptake of the compound. Secondly, they may improve the characteristics of the formulated product during application, in order to distribute and retain the active ingredient in the most efficient way. Thirdly, adjuvants may be used to retain the physical integrity and stability of the formulation during the application process. A particular adjuvant may serve a number of purposes simultaneously, so the terminology used to describe adjuvants is fairly imprecise. Adjuvants may be incorporated into the formulation itself or applied separately, for example in a tank mix.

It should be apparent that a wide range of formulation approaches and techniques is potentially available for the formulation of biological control agents. It should be stressed, however, that almost all of these techniques have been developed specifically in order to apply chemical active ingredients. The technology developed for application of chemicals has required significant adaptation in order to meet the needs of biological control.

16.3 PROPERTIES OF BIOLOGICAL CONTROL AGENTS

The most significant distinction between chemicals and biological control agents is that the latter are living organisms which are capable of replicating in the environment and, frequently, are required to multiply after application in order to control the target pest, pathogen or weed. The formulation must therefore provide conditions which retain viability during preparation, storage and application, and favour survival of the agent in the environment. An astonishing variety of living

Table 16.1. Classification of formulations

Formulation	Code	Properties
Formulations diluted in water		
Emulsifiable concentrate	EC	Emulsion formed when added to spray tank
Water-in-oil emulsion	EO	Pre-formed emulsion
Oil-in-water emulsion	EW	Pre-formed emulsion
Suspension concentrate	SC	Suspended insoluble AI
Capsule suspension	CS	AI contained in capsules
Soluble concentrate	SL	Used for water-soluble AI
Water soluble powder	SP	Powder soluble, but may contain inert ingredients
Water soluble granule	SG	Used for water-soluble AI
Tablet	TB	
Briquette	BR	Controlled-release formulation
Wettable powder	WP	Typically consist of AI, clay carrier and surfactants
Water dispersible granule	WG	AI dispersed, but not dissolved, in water
Formulations diluted		
with organic solvents		
Oil miscible liquid	OL	AI dissolved in organic solvent
Oil miscible flowable	OF	Suspension in organic liquid
Oil dispersible powder	OP	Powder to be applied in oil

Formulations applied undiluted

Dustable powder	DP	AI carried on free-flowing powder
Granule	GR	Applied to soil and water
Encapsulated granule	CG	Controlled-release granule
Microgranule	MG	Diameter below 0.6mm
Electrochargeable liquid	ED	Used with electrostatic spray equipment
Spreading oil	SO	Applied to water surface
Ultra low volume liquid	UL	Applied through ULV sprayers
Ultra low volume suspension	SU	As above

Seed treatments

Powder for dry seed treatment	DS	
Flowable concentrate	FS	Liquid suspension
Solution for seed treatment	LS	
Coated seed	PS	
Water dispersible powder	SS	
	WS	Applied as slurry

Miscellaneous

Bait concentrate	CB	Bait diluted before application
Bait	RB	
Smoke generator	FU	

Table 16.2. Classification of adjuvents

Adjuvant	Types	Purpose
Oils	Mineral oils Crop oils	Improve uptake, photostability
Surfactants	Wetting agents Spreaders Penetrants	Improve spreading, wetting or dispersion
Stabilizing agents	Emulsifiers Dispersants Antiflocculating agents Compatibility agents	Maintain stability during application
Solvents	Cosolvents Coupling agents	Maintain AI in solution
Hygroscopic agents	Humectants	Prevent premature drying of deposit
Deposit builders	Stickers Film formers	Improve adhesion
Foam modifiers	Foams Antifoam agents	Control foaming during application
Buffering agents		Stabilize pH

organisms have been proposed as potential biological control agents. These include viruses, fungi, bacteria, protozoa and nematodes, all of which are presently sold or distributed for use in agriculture or public health, and have been used to control insects, nematodes, mites, plant pathogens and weeds. Each presents a different set of constraints or problems in formulation. It is difficult to generalize, therefore, about biological control agents or their formulation requirements. Some features, however, are common to all.

Perhaps the most important distinction between biologicals and chemicals is that biological control agents consist of **discrete particles**, whether these are cells, spores, virus particles or multicellular structures. In general, the integrity of these structures cannot be disrupted without inactivating the agent. Biological agents may be suspended in fluid, but can rarely be dissolved without loss of activity. These properties place constraints on the formulation techniques which can be applied.

In order to function as a biological control agent, a microorganism must exert one or more specific effects, either directly or indirectly, on a target pest. The majority of biocontrol agents which are currently available accomplish this through production of toxic compounds, parasitism or competition. Each mode of action presents a different set of formulation requirements, but all are bound by some common principles.

Perhaps the simplest case is that of the *Bacillus thuringiensis* products, in which the **toxic component** (the protein endotoxin) accumulates during fermentation and is formulated along with spores, cells and cellular debris. The requirement of the formulation, therefore, is simply to deliver the material intact and in a form which can be ingested by the target insect. If a toxic compound (for example the bacteriocin secreted by *Agrobacterium radiobacter*) is produced only after application (Kerr, 1980), the formulation must favour not only the multiplication of the organism in the environment, but also the production of the toxic compound.

The same principles apply to those agents which are **parasitic** on the host. Organisms which multiply within the host, but not in the environment, for example the baculoviruses and plant-parasitic fungi, must simply be maintained in a viable and active form which favours entry into the host. Those which are required to increase their biomass in the environment before contacting the host, for example many of the biocontrol agents applied to soil, must be provided with the means to do so by provision of nutrients or placement where nutrients are readily available. Organisms which actively seek out the host, such as insect-parasitic nematodes, present a further level of complexity in that they must be released from the formulation in order to move freely in the environment.

Where **competition** is the principal mode of action, for example in the case of biocontrol fungi applied to pine stumps, the formulation must again provide whatever nutrients and environmental conditions are necessary for multiplication of the biocontrol agent, since rapid colonization is likely to be the key to success.

16.4 REQUIREMENTS FOR FORMULATION, DISTRIBUTION AND APPLICATION

16.4.1 Formulation

In general, formulation of a biological control agent fulfils the same purpose as formulation of a chemical; a formulation is required in order to deliver the active ingredient at the point of use in a form which is both cost-effective and convenient to apply. Many of the requirements for formulation of a biological are therefore identical to those of chemicals, but the biological nature of the active ingredient also poses several unique problems.

In all cases, the biological control agent must be formulated and maintained in a **viable** or **active** form. This presents a number of technical problems in that microorganisms and proteins, particularly in the absence of specialized resting structures, are easily killed or inactivated by unfavourable environmental conditions (Kirsop and Doyle, 1991). This may occur during the formulation process itself, for example if the organism is exposed to heat, shear stress, toxic compounds, uncontrolled desiccation or premature germination in the presence of moisture. The formulation process must therefore be carefully designed to reflect the particular requirements and limitations of each organism.

16.4.2 Distribution and storage

A more difficult problem is the need to maintain viability or activity during distribution and storage, since products may be stored for periods of two to four years, and the conditions of storage are usually out of the control of the producer. In some cases it may be possible to stipulate that the product must be refrigerated, used within a restricted period, or ordered in advance and used immediately upon delivery. However, such restrictions are usually only acceptable where the product is used in small quantities or possesses some unique property which justifies the inconvenience and expense on the part of the distributor and user. A number of biologicals have succeeded in gaining acceptance despite such restrictions, although the total market value of such products is limited and is likely to remain so.

An alternative approach is to formulate the microbial agent in a way which allows it to be handled through the normal channels of distribution and storage. This usually involves drying of the product and maintenance in a dry environment or suspended in oil. The exclusion of water slows the metabolic rate of cells, prevents depletion of nutrients and accumulation of toxic metabolites, and slows the denaturation of proteins. This approach has been used successfully to date with bacteria, viruses and fungi. Unfortunately, not all microorganisms have proved amenable to drying and many tend to lose viability both during the drying process and in storage (Lapage and Redway, 1974; Kirsop and Doyle, 1991).

16.4.3 Application

Formulations of biological control agents must be convenient to use and compatible with existing application techniques and equipment. Although users may be prepared to adapt conventional methodology in order to accommodate a novel product, and indeed have done so in practice in order to use biological control agents, a requirement for specialized application equipment or inconvenient application techniques is likely to limit the acceptance of any new product. The requirement for ease of **reconstitution** and **stability** during application is identical to that of chemical products, and the formulation techniques available are similar.

As with chemical products, the formulation must distribute the active ingredient in a way which maximizes cost efficacy. This is particularly important in the case of biologicals, since no redistribution of the active ingredient due to vapour action or systemicity can be expected. Propagules which fail to reach the site of action or a suitable environment for multiplication are unlikely to contribute to control of the target pest. The formulation must also maintain the biocontrol agent in a viable form in the environment for as long as is necessary for the mode of action to be expressed. In some cases, this period may be relatively short. For example, in the case of mycoherbicides the biocontrol agent must only be protected for a matter of hours until the penetration process is complete. In other cases, for example the control of insects which may come into contact with the biocontrol agent some time after application, greater persistence is necessary. The requirement for persistence and stability is, of course, not unique to biologicals; photostability, rainfastness and rate of degradation in the soil are also issues which must be addressed in formulation of many chemical products.

In addition to the requirements for efficacy and stability, formulations must also optimize the **safety** of the final product. Although the issue of toxicity is perhaps of less concern with biologicals, which are generally non-toxic, than with certain chemical active ingredients, other safety considerations such as crop safety, flammability, risk of explosion and allergenicity must be addressed in the design and registration of the formulation.

16.5 AVAILABLE TECHNOLOGY

Despite its importance in determining the success or failure of biological control strategies, formulation technology designed specifically for microorganisms has been surprisingly slow to develop. However, a number of valuable techniques are available for formulation of microbials, and a variety of formulations of biological control agents are now reaching the marketplace (Connick *et al.*, 1990; Lisansky, 1990; Rajnchapel-Messaï, 1990). The majority of formulations of biological control agents which are available commercially are listed in Table 16.3.

Many biological control products rely on formulation technology developed in the agrochemical industry; for this reason most biocontrol products physically resemble agrochemicals. Most of the carriers and adjuvants which have been proposed for use with biologicals are in common use in agrochemical formulations (Couch and Ignoffo, 1981; Ward, 1984). Other techniques developed originally for use with agrochemicals, but now used widely for formulation of biological control agents include the enhancement of photostability and palatability to insects by incorporation of photoprotectants and phagostimulants respectively (Couch and Ignoffo, 1981). Invert emulsions, which have recently been demonstrated to enhance the efficacy of mycoherbicides (Connick *et al.*, 1991) derive from agrochemical technology.

16.5.1 Formulating microbials

Some techniques have, however, been developed specifically for microbials. The majority of formulations which provide an extended shelf-life require cells to be maintained in a dry state. Drying can be accomplished by a number of means including freeze-drying (Kirsop and Doyle, 1991), drying on silica gel (Blachère *et al.*, 1973; Kirsop and Doyle, 1991) and spray drying (Dulmage *et al.*, 1990). Organisms vary in their ability to withstand desiccation and storage in the dry state (Lapage and Redway, 1974) but dry formulations can, in some cases, remain stable over a period of several years.

A number of attempts have been made to exploit the properties of **cross-linked alginate gels** to encapsulate biocontrol agents (Fravel *et al.*, 1985; Kaya and Nelsen, 1985; Lewis and Papavizas, 1985, 1987; Papa-

Table 16.3. Currently available formulations of biological control agents

Active ingredient	Formulation
Bacteria	
Bacillus thuringiensis	Wettable powder (WP)
	Emulsifiable concentrate (EC)
	Suspension concentrate (SC)
	Granule (GR)
	Oil miscible suspension (OF)
	Ultra low volume suspension (SU)
	Dustable powder (DP)
	Water dispersible granule (WG)
	Briquette (BR)
Bacillus popilliae	Dustable powder (DP)
	Granule (GR)
Bacillus subtilis	Water dispersible seed treatment (WS)
	Powder for seed treatment (DS)
Streptomyces spp.	Water dispersible seed treatment (WS)
	Wettable powder (WP)
Agrobacterium radiobacter	Wettable powder (WP)
	Gel
	Suspension concentrate (SC)
Pseudomonas fluorescens	Granule (GR)
Fungi	
Verticillium lecanii	Wettable powder (WP)
Colletotrichum gloeosporoides	Wettable powder (WP)
Phytophthora palmivora	Suspension concentrate (SC)
Metarhizium anisopliae	Dustable powder (DP)
	Granule (GR)
	Cockroach trap
Arthrobotrys spp.	Dustable powder (DP)
Trichoderma spp.	Wettable powder (WP)
	Dowels
Nematodes	
Nematodes	Wettable powder (WP)
	Soluble gel
	Sponge
Protozoa	
Nosema locustae	Bait
Viruses	
Baculoviruses	Wettable powder (WP)
	Ultra low volume suspension (SU)
	Suspension concentrate (SC)

vizas *et al.*, 1987; Knudsen *et al.*, 1991; Lewis, 1991). Cells of the biocontrol agent are suspended in sodium alginate, and drops of the suspension are then added to a solution of a calcium salt. Calcium ions cross-link the alginate matrix, forming a pellet which can then be dried. Substrates which favour the growth of the biocontrol agent can be added to the suspension and encapsulated together with the cells. The use of alginate encapsulation in biological control has been reviewed by Connick (1988).

Attempts have been made to utilize the desiccation-protective properties of **polysaccharides** such as methylcellulose and xanthan gum in preservation of cultures and in formulation of microbial cells (Kloepper and Schroth, 1981; Suslow and Schroth, 1981, 1982). Unfortunately, such formulations tend to lose viability during storage (Kloepper and Schroth, 1981; Suslow and Schroth, 1981).

Adsorption of cells on to peat is a technique commonly used for the formulation of *Rhizobium* inoculants which has also been applied to biological control agents such as *Pseudomonas fluorescens* (McIntyre and Press, 1991). The function of the peat is both as a dispersant and as a means of protecting the organisms from environmental stress. This approach has proved only partially successful since peat is not homogeneous and shelf-life on non-sterile peat is limited.

Attempts have been made to take advantage of the biological nature of microbial active ingredients, and to simplify the production and formulation process, by fermenting microorganisms on a **solid substrate** which later functions as a carrier, substrate and organism being applied together in the field. Examples of this approach include the direct fermentation of microorganisms on vermiculite or diatomaceous earth (Backman and Rodriguez-Kabana, 1974; Paau *et al.*, 1991). The material can be applied directly as soil or seed treatments, extruded into granules, or suspended in water to liberate the cells, which are applied as a drench for biological control of soil-borne plant pathogens. A similar approach is also being developed for production of fungi for insect control; the fungus *Beauveria bassiana* is grown on nutrient-impregnated clay microgranules which are then applied directly to the whorls of maize for control of European corn borer (Riba, 1984, 1985).

The requirement for controlled delivery of active ingredients is common to both agrochemicals and biologicals particularly for insecticides, and a number of systems have been developed which control release over time or in response to environmental changes (Wilkins, 1990). While controlled release of biological control agents remains in its infancy (Connick, 1990), controlled release technology is now becoming available which can be used to achieve more efficient delivery of biocontrol agents.

16.6 APPLICATION TO THE FOLIAR SURFACE

Crop protection products are applied to the foliar surface in order to control weeds, insects and plant pathogens. A variety of application techniques have been devised for this purpose (Haskell, 1985). The most technically simple is application from the ground of hydraulic sprays, the active ingredient being formulated in **liquid** or **solid** form before mixing with water. **Dusts** may also be applied from the ground in some cases, an application technique which is favoured in Japan. An alternative, where large areas are to be treated or the crop canopy makes entry difficult, is to apply the active ingredient by air. This is usually accomplished by low-volume spraying, but dusts can also be applied in this way. **Granules** may also be applied to the foliar surface, for example to control borers feeding in leaf whorls.

16.6.1 Technical difficulties

Application and maintenance of an active ingredient on the foliar surface presents a number of difficulties regardless of the type of product being applied, but the problems are especially severe in the case of microbial control agents as a consequence of the biological nature and properties of the product.

Firstly, microorganisms are **particulate** in nature. This may cause settling to occur during application. An even deposit is important since little redistribution can be expected. Unlike many chemical compounds, microbial cells cannot be absorbed by the leaf, and therefore will be exposed for extended periods to variable, and frequently hostile, conditions on the leaf surface. Microorganisms on the leaf surface are easily removed by mechanical abrasion within the canopy. **Rainfastness** is also a concern in regions of high rainfall, or where overhead irrigation is applied. **Temperature** at the foliar surface may vary widely, as may **relative humidity**. These changes will affect not only the probability of a droplet reaching the foliar surface before evaporating, but also the ability of microbial cells to remain viable and multiply.

Probably the most significant stress imposed on microbial cells in the foliar environment is **ultraviolet (UV) radiation**. Organisms used for biological control differ in their sensitivity to UV light (Couch and Ignoffo, 1981), but sensitivity to UV limits their persistence in the foliar environment in almost all cases. The use of biological control agents in the foliar environment is governed by a number of environmental constraints, and formulation technology has been used to address these.

16.6.2 Weed control

Some success has been achieved in controlling weeds by inundative foliar application of **plant-pathogenic fungi** (mycoherbicides) although the commercial impact of such products to date has been limited. Formulation aspects of mycoherbicides for foliar, soil and aquatic application has been reviewed by Connick *et al.* (1990) and Daigle and Connick (1990). Mycoherbicides have an advantage over many other biological control agents in that the requirement for foliar persistence is limited; the pathogen rapidly penetrates plant tissues, after which the mycelium is no longer exposed to the foliar environment and growth is limited only by nutrition and host defences.

In order for successful infection to take place, however, favourable environmental conditions during the period of germination and penetration are essential. **Relative humidity** is particularly important, since turgor pressure is required for fungi to penetrate plant tissues. Since high relative humidity can rarely be guaranteed over the 6- to 24-hour dew period which is required, attempts have been made to favour the infection process through formulation. Quimby *et al.* (1989) developed an invert emulsion (water in oil) which replaced the dew period requirement of *Alternaria cassiae* spores applied to control sicklepod (*Cassia obtusifolia*). This formulation was also used successfully to apply spores of *Ascochyta pteridis* to bracken (Munyaradzi *et al.*, 1990). Connick *et al.* (1991) designed an **invert emulsion** with improved physical properties, the oil phase of which contained paraffin wax, a paraffinic spray oil and an unsaturated monoglyceride emulsifier. Invert emulsions of this type retard evaporation, retaining a film of water around the spores for a sufficient period to allow germination and penetration to take place.

Secondary effects have also been noted when fungal spores have been applied in invert emulsions. A dramatic decrease in the **inoculum threshold** required for successful infection to take place was observed by Amsellem *et al.* (1990) and Munyaradzi *et al.* (1990). Amsellem *et al.* (1991) also noted that the host range of *Alternaria cassiae* and *A. crassa* was broadened by application of spores in an invert emulsion, and hypothesized that this may have been due to cuticular damage which facilitated leaf penetration, and to suppression of elicited resistance responses.

An alternative means of maintaining high moisture potentials is to incorporate **humectants** or film-forming compounds in order to retain moisture around the spore. Munyaradzi *et al.* (1990) reported that the incorporation of 2% Manucol into an aqueous spore suspension improved the germination of *Ascochyta pteridis* spores as well as significantly enhancing foliar adhesion. The commercial product Collego™, based on *Colletotrichum gloeosporoides* f. sp. *aeschynomene* is applied in a sugar solution which may have a similar function (TeBeest and Templeton, 1985; Smith, 1986).

The incorporation of nutrients into the formulation may also be advantageous in terms of maximizing **spore germination** and hence invasion of the plant. Daigle and Cotty (1991) devised an improved formulation increased spore germination and greatly enhanced efficacy as compared with spores suspended in buffer. The use of sugar in the product Collego™ has been noted above.

New developments

Collego™, which is sold for control of northern jointvetch in rice, is designed to be applied primarily by aerial spraying. The product is sold Collego™, which is sold for control of northern jointvetch in rice, is designed to be applied primarily by aerial spraying. The product is sold in two components, a sugar solution and a dry powder containing 15% by weight of spores (TeBeest and Templeton, 1985; Smith, 1986). Powdered charcoal is also provided to remove pesticide residues from aircraft spray tanks before application. Satisfactory levels of efficacy have been achieved with this product, and several other mycoherbicides are in development or under investigation. The ability to eliminate dew period requirements, alter application rates, and possibly to broaden host range through formulation, is likely to enhance the commercial attractiveness of such products.

Two products which are at an advanced stage of industrial development as foliar-applied mycoherbicides are BioMal™, produced by Philom Bios Inc. of Saskatoon, Canada, and Casst™, based on *Alternaria cassiae* which is being developed by the Mycogen Corporation of San Diego. BioMal™ is formulated as a wettable powder in which the hydrophilic spores of *Colletotrichum gloeosporoides* f. sp. *malvae* are adsorbed on to silica gel (Daigle and Connick, 1990). *Alternaria cassiae* has been evaluated in an experimental two-component formulation consisting of an emulsifiable paraffinic oil adjuvant and spores of the pathogen. The adjuvant is emulsified in water in which the spores are then dispersed (Connick *et al.*, 1990).

16.6.3 Insect control

A substantial body of literature exists on the formulation of biological control agents for control of insects, and a number of reviews are available (Angus and Luthy, 1971; Couch and Ignoffo, 1981; Soper and Ward, 1981; Young and Yearian, 1986; Devisetty, 1988). Readers are referred to these reviews for a more detailed treatment; the purpose of this section is to outline the basic principles underlying formulation requirements for biological insecticides and to review some recent technical developments.

Current strategies

At the present time, the biological control market is dominated by products based on *Bacillus thuringiensis* which account for 92% of sales (Wood Mackenzie, 1991). The major application of Bt products is for control of foliar-feeding Lepidoptera in forestry and agriculture. The use of Bt for control of mosquitoes and blackflies will be discussed in the context of formulations for application to water. Other products are likely to occupy a greater share of the market in future, however. **Baculoviruses** have been used to a small extent for insect control, particularly in forestry. Recent developments in molecular genetics are likely to enhance the efficacy of these agents and lead to renewed commercial interest (Wood and Granados, 1991). Fungi are also used for control of foliar pests in Brazil and Russia (Ferron, 1981; Filippov, 1989) and for control of glasshouse whitefly in Europe (Lisansky, 1990). A fungal product is also being developed for control of European corn borer (Rajnchapel-Messaï, 1990).

The formulation requirements for biological insecticides are similar to those for mycoherbicides except that the active ingredient is unable to escape unfavourable conditions on the foliar surface by penetrating the plant. **Persistence** on foliage is a prerequisite in most cases, since insects may continue to hatch or to invade a treated area over time. In the case of Bt and baculovirus products, the active ingredient must be ingested by the insect, and the product must therefore remain in an active state on the foliar surface until a lethal dose has been consumed. Although frequent applications of insecticides may take place during the growing season, (Haskell, 1985), the ability to persist for several days is likely to be a requirement of most formulations. During this period, formulated products will, in most cases, be exposed to the adverse environmental conditions discussed previously, and the formulation must therefore be optimized with respect to photostability, rainfastness and adhesion of deposit. In the case of biocontrol agents which are active through ingestion, the inclusion of compounds which stimulate feeding may be beneficial (Couch and Ignoffo, 1981). Agents which are sensitive to high pH may be inactivated by alkaline dew (Tuan *et al.*, 1989), so buffering of the formulation may be necessary. As with other foliar-applied biocontrol agents, formulations of microbial insecticides must achieve efficient foliar deposition using conventional application equipment.

Formulation

A number of formulation techniques have been developed to address these problems, resulting in considerable improvements over the early formulations of *Bacillus thuringiensis*. The major factor which limits the persistence of biological control agents on the foliar surface appears to be ultraviolet radiation, and numerous attempts have been made to overcome this limitation through formulation. Some fairly simple formulation approaches have proved successful; for example Jones (1990) reported that a freeze-dried baculovirus preparation containing only insect debris, china clay and a surfactant adsorbed to synthetic silica provided adequate UV protection when applied to the underside of cotton foliage. The fungus *Metarhizium anisopliae* is used successfully as an aerial spray for control of spittlebug in sugarcane, possibly because of the UV protection afforded by the dense canopy (Moscardi, 1989). Attempts to enhance photostability of Bt and baculovirus products by addition of photoprotectants (compounds which absorb or reflect UV light) have met with some success in both laboratory and field experiments, although the benefits in terms of enhanced crop protection have not always been evident (Couch and Ignoffo, 1981; Young and Yearian, 1986).

A limitation of this approach is that the **photoprotectant** is, in many cases, diluted in the spray volume, and may therefore provide only limited protection. Consequently, attempts have been made to achieve a more intimate association of the photoprotectant with the cells or virus particles. Cohen *et al.* (1991) adsorbed **cationic chromophores** on to *Bacillus thuringiensis* and reported that acriflavin provided good protection during 12 hours of exposure to UV, hypothesizing that the photoprotection was due to transfer of energy from excited tryptophan moieties to the chromophore. An alternative is to surround the biological control agent with a matrix which provides photoprotection and remains intact during and after application. A number of *encapsulation* processes are available (Tsuji, 1990), although not all are compatible with a biological active ingredient. The concept of encapsulation of biological control agents is not new; an encapsulated formulation of *Bacillus thuringiensis* was described over 25 years ago by Raun and Jackson (1966). Recently, however, an encapsulation system based on pre-gelatinized starch has been described (Dunkle and Shasha, 1988). This has the advantage that the starch matrix is digested in the insect gut, releasing the biocontrol agent after ingestion. The system can also be used for encapsulation of baculoviruses (Ignoffo *et al.*, 1991) and photoprotectants can be incorporated within the starch matrix (McGuire *et al.*, 1990). The use of self-encapsulating starch formulations may allow such formulations to be applied through spray equipment (McGuire and Shasha, 1990).

An alternative system for improving **photostability** was described by Herbig *et al.* (1991) who surrounded codling moth granulosis virus with a protective coating using a spray-dry coating process, resulting in improved UV stability. A polymeric encapsulation system, which was developed primarily for formulation of *Pseudomonas* cells, was shown to enhance the persistence of Bt in the field (Baker and Henis, 1990). Encapsulation may also be achieved by means of recombinant DNA technology. Gaertner (1990) described a cellular encapsulation system in which the *Bacillus thuringiensis* toxin is expressed in cells of *Pseudomonas fluorescens* which are then killed, providing a protective coating. Products based on this system are now reaching commercialization.

Distribution and application

In order to optimize both distribution of the biological control agent on the foliar surface and ease of application, a number of techniques originally developed for the application of agrochemicals have been investigated. Spray deposition, droplet size, spreadability and evaporation rate can be improved by the addition of surfactants, film-forming compounds, thickeners and humectants (Couch and Ignoffo, 1981). **Stability** during application and coverage can also be optimized through control of particle size distribution and foliar persistence may be improved by tank-mixing with stickers (Devisetty, 1988). Application method can have a profound effect on efficacy; for example Sopp *et al.* (1989) demonstrated that greater and better distributed spore populations, and more effective control of aphids, could be achieved by application of *Verticillium lecanii* using an ULV electrostatically-charged rotary atomizer than by hydraulic spraying.

The incorporation of feeding **stimulants** into formulations of orally-ingested biocontrol agents has been investigated and, in some cases, demonstrated to enhance efficacy (Couch and Ignoffo, 1981; Young and Yearian, 1986). A number of phagostimulants are now commercially available for tank-mixing with biological control agents. However, as with incorporation of UV protectants, improved rates of ingestion have not always been shown to result in enhancement of crop protection.

A variety of formulations are now available for **above-ground** application of biological insecticides (Table 16.3). The choice of a particular formulation approach depends on the pest to be controlled, the crop to which it is to be applied, the environment in which it is to be used, and the market in which it is to be sold. The majority of formulations which are available commercially are based on a dry form of the active ingredient which is typically formulated as a **dust** or **wettable powder** or

suspended in oil. Microbial control agents are usually more stable in a dry form than in aqueous suspension (Devisetty, 1988) although in some cases, for example in forestry where fermentation is usually carried out close to the time of application, aqueous-based formulations may be favoured for reasons of cost. **Granular formulations** may also be used in situations where a delay prior to sporulation is compatible with foliar pest control (Rombach *et al.*, 1986; Knudsen *et al.*, 1990) or for application of fungi or Bt to the leaf whorl, for example to control European corn borer in maize (Riba, 1984, 1985; Devisetty, 1988; McGuire *et al.*, 1990). New formulations of biological insecticides are likely to increase in number over the next few years as products now in development are introduced and technical advances are translated into practice.

16.6.4 Foliar disease control

To date, biological control of the major foliar plant diseases has seen very few practical applications (Rhodes, 1992). Foliar diseases present a particularly difficult challenge for microbial control agents since, following invasion, plant pathogens occupy a protected postion within host tissue, from which they may be difficult to dislodge. A further obstacle is that, in many cases, a well-distributed population of the biological control agent may be required to persist on the leaf surface for an extended period in order to provide protection from incoming spores of the pathogen. Pathogens which colonize the plant surface, such as bacteria which induce frost damage by nucleating ice crystal formation, may be easier to control than invasive pathogens (Spurr, 1990). One biological control agent which is used successfully for control of a plant pathogen on aerial, as well as underground plant surfaces is *Agrobacterium radiobacter*, used for control of crown gall (Kerr, 1980). The product is applied as a dip and is distributed both as gel and peat-based formulations. Since many of the problems associated with biological control of foliar pathogens are identical to those encountered in control of weeds and insects, this area is likely to benefit from advances in formulation technology for these purposes.

16.7 APPLICATION TO SOIL

Formulation of microorganisms for application to soil presents a quite different set of constraints to those experienced in foliar applications. Application to the soil avoids the problems of ultraviolet degradation, rainwash and abrasion. Optimal placement of an insoluble particulate active ingredient, such as a microorganism, in the soil is, however, much more difficult than in the foliar environment. In addition, microorganisms applied to soil are subject to competition from the native microflora; **timing** of application and **release** from the formulation must therefore be carefully controlled. Soil texture, chemistry and microbiology may vary considerably, and this may affect both applied microorganisms and the behaviour of the formulation. Soil moisture potential is also liable to vary with location and between and within seasons. Since these factors may have a major impact on the performance of biocontrol agents, formulation technology can be used to enhance reproducibility as well as efficacy.

16.7.1 Soil application of bioherbicides

DeVine™, a product based on *Phytophthora palmivora*, is claimed to provide very effective control of milkweed vine (*Morrenia odorata*) in citrus (Kenney, 1986). The product is sold as an aqueous suspension of chlamydospores which have a shelf-life of only six weeks (Daigle and Connick, 1990) and must be ordered prior to the season of use, although persistent control of the weed can be expected once the biocontrol agent becomes established (TeBeest and Templeton, 1985). The suspension is applied to wet soil as a **high-volume spray** in order to achieve efficient placement of the biocontrol agent in the vicinity of roots of the host. **Timing** of application is also important, since the mycoherbicide is ineffective if applied in the absence of the host. Controlled-release formulations may ultimately prove to be of value in similar situations.

Formulation of mycoherbicides as granules for soil application may allow a more extended shelf-life to be attained, while granule technology may confer controlled release properties (Connick, 1990). Weidemann and Templeton (1988a, b) reported that the soil-borne fungus *Fusarium solani* f. sp. *cucurbitae* provided effective control of Texas gourd (*Cucurbita texana*) when applied in alginate granules. Enhanced sporulation was obtained when the granules were amended with soya flour or oatmeal as a nutrient source (Weidemann, 1988). Alginate encapsulation has also been found to be compatible with the potential fungal biocontrol agents *Alternaria cassiae*, *A. macrospora*, *Fusarium lateritium*, *Colletotrichum malvarum* and *Phyllosticta* sp. (Connick, 1990).

16.7.2 Soil pests

Products based on both entomophilic nematodes and fungi are sold commercially or are under development for control of soil pests. **Nematode-based products** are currently available for use mainly in ornamentals, cranberries and domestic gardens. These products are sold in a variety of formulations including sponges, gels and wettable powders, all of which are reconstituted in water before application (Lisansky, 1990). One of the most widely used products, BioSafe™, produced by Biosys Inc. of Palo Alto, California, is sold as a package in which nematodes are immobilized in a gel matrix. The product has a shelf-life of up to six months when refrigerated. Maintenance of nematodes in a viable state during distribution has so far proved to be a barrier to the commercial development of these products beyond specialist markets, since many nematode strains must be maintained in a hydrated form. Nematodes, unlike most other biocontrol agents, actively seek out insects in the soil, but require a continuous moisture film in order to do so. Irrigation may be used to distribute nematodes in the water phase and to provide suitable conditions for location of the host. Granule approaches have also been investigated. For example, Kaya and Nelsen (1985) embedded nematodes in an alginate matrix for use as insect baits.

Reinecke *et al.* (1990) described a process for applying the **fungus** *Metarhizium anisopliae* to the soil as a granule. The fermentation is manipulated so that the fungus grows as mycelial pellets. These are then dried into granules and packed in vacuum-sealed plastic bags. The product is being developed for control of black vine weevil (*Otiorhynchus sulcatus*) under the name BIO1020. Knudsen *et al.* (1991) described a process for incorporating the fungus *Beauveria bassiana* into alginate pellets amended with wheat bran, and reported that sporulation from such pellets was enhanced by the addition of polyethylene glycol.

16.7.3 Soil-borne diseases

A bewildering array of microorganisms have been investigated as potential biological control agents of soil-borne plant diseases. The majority of successful attempts have used bacteria and fungi. Formulation of microorganisms for soil-borne disease control has been reviewed by Lewis (1991) and Connick *et al.* (1990). In general, the formulations used may be divided into seed treatments and soil-applied granules.

Seed treatment

Because of the small quantity of material which must initially be applied, seed treatment is potentially a very efficient means of applying biological control agents to soil, particularly if the agent has the capacity to colonize the soil and plant surface from the treated seed. Seed treatments are also favoured by growers as a convenient means of applying both chemicals and biologicals. Seeds may be treated using dry powder formulations (Kloepper and Schroth, 1981; Rhodes and Logan, 1986), methylcellulose slurries (Suslow and Schroth, 1982), liquid coatings (Taylor *et al.*, 1991) and pelleting techniques (McQuilken *et al.*, 1990). Unless protection of the seed itself is sufficent, the efficiency of seed treatment depends largely on the ability of the biocontrol agent to be released from the formulation and to colonize roots and soil.

Consequently, seed treatment approaches have usually been employed for formulation of organisms such as *Pseudomonas*, *Bacillus subtilis* and rhizosphere-competent strains of *Trichoderma* which are effective colonists of the root surface. Formulations must be compatible with the seed treatment techniques utilized for a particular crop. Poor shelf-life is likely to limit the use of some biocontrol agents, since treated seed may be stored for considerable periods of time after treatment. Seed treatment formulations based on *Streptomyces* spp. and *Bacillus subtilis* are sold as Mycostop™ and Quantum 4000™ respectively. Both are distributed in powder form (Connick *et al.*, 1990).

Soil-applied granules

Granules or dustable powders may be used to place the biocontrol agent directly in the soil in the vicinity of the pathogen. The simplest means of introducing microorganisms, particularly fungi, into soil is to utilize solid-state fermentation to produce the biomass, for example by growing the organism directly on solid substrates such as agricultural waste products. The entire biomass can then be incorporated into soil (Connick *et al.*, 1990).

A more sophisticated alternative is the use of semi-solid fermentation, in which an inert support coated with a nutrient film is used to grow the biomass which can either be applied directly or dried (Backman and Rodriguez-Kabana, 1974; Graham-Weiss *et al.*, 1987; Paau *et al.*, 1991). In general, industrial processes do not favour the use of solid and semi-solid fermentation, and so means of formulating the products of liquid fermentation have been sought. **Alginate encapsulation**, already discussed in the context of insect and weed control, has been investigated as a means of formulating biological fungicides. Examples include the application in alginate pellets of *Talaromyces flavus* for control of verticillium wilt (Papavizas *et al.*, 1987), and *Trichoderma* spp. and *Gliocladium* spp. for control of *Rhizoctonia* damping-off (Lewis and Papavizas, 1987). Conidiation from alginate pellets may be stimulated by the addition of nutrients and polyethylene glycol (Knudsen *et al.*, 1991). Alternative methods of applying biomass produced by liquid fermentation include the polymeric encapsulation system described by Baker and Henis (1990) in which *Pseudomonas fluorescens* was applied as a pellet in furrow. Another approach is to adsorb the cells or spores on to a solid substrate which may be compressed into a granule. Dagger G™, a biofungicide based on *Pseudomonas fluorescens* and marketed by Ecogen Inc., is formulated as a peat-based granule which is sold for control of damping-off in cotton (McIntyre and Press, 1991).

16.8 POSTHARVEST AND NON-AGRICULTURAL APPLICATIONS

Biological control of postharvest diseases has been receiving increasing attention in recent years (Wilson and Chalutz, 1991), partly in response to public concern about the use of pesticides on food products. If acceptable levels of efficacy and consistency can be obtained using biological control agents, formulation approaches will undoubtedly be required to address the particular environmental constraints of postharvest applications. Biological control has also been used for a number of purposes other than crop protection, and several innovative formulations have been developed for non-crop use. Further development of formulations of biocontrol agents for specialized purposes may be expected, particularly where environmental concerns are paramount.

16.8.1 Aquatic habitats

Formulations have been developed to date for control of both aquatic-breeding insects and aquatic weeds. Products based on *Bacillus thuringiensis* var. *israelensis* (*Bti*) are commercially available as larvicides for control of mosquitoes and blackflies (Lacey and Undeen, 1986). Because of the need to place the biocontrol agent in the vicinity of the larvae, and to sustain release over time, particularly in moving water, controlled-release is desirable and a number of formulations have been designed for this purpose, including microcapsules, effervescent granules, floating oil-based suspensions, foams and buoyant granules (Lacey *et al.*, 1984; Connick, 1990). To date, however, only one controlled-release formulation, Bactimos™ Briquettes, has been commercialized. The formulation consists of floating, doughnut-shaped briquettes which release Bti near the water surface for up to 30 days. Distribution of the biocontrol agent in the water is important. Varying degrees of buoyancy are required depending on whether bottom- or surface-feeding larvae are the target.

Control of aquatic **weeds** appears to present an attractive opportunity for biological control due to public concern over water quality although to date, only experimental formulations have been available. Control of waterhyacinth (*Eichhornia crassipes*) was obtained by Charudattan (1986) using the fungus *Cercospora rodmanii*. A wettable powder formulation was developed by Abbott Laboratories which was tank-mixed with the surfactant Triton X-100 (Daigle and Connick, 1990).

16.8.2 Professional and domestic use

Development of biocontrol agents for specialist purposes such as professional pest control, disease control in amenity turf, and home and garden use appears to have considerable market potential, and will require the development of appropriate, specialized formulations. A bran-based bait formulation of the protozoan *Nosema locustae* is produced commercially for control of grasshoppers (Henry, 1990). It is probable that formulations such as cockroach traps and insect baits will become commercially available in the near future.

16.8.3 Forestry and orchard applications

One of the major uses of microbial pest control has been in forestry for control of pests such as spruce budworm and gypsy moth. Both Bt products and baculoviruses have been used commercially for this purpose. Specialized formulations are frequently required for forestry, since formulations are typically applied by air over large areas (Sanders *et al.*, 1985). Low-volume spraying is favoured, and nozzle blockage must be avoided. Deposition of the spray deposit in the canopy and rainfastness are also important considerations. A number of specialized formulations, particularly of Bt products, are commercially available for use in forestry (Devisetty, 1988).

The fungus *Phlebia gigantea* is used in forestry in Europe for biological control of white rot of pine caused by *Heterobasidion annosum* (Rishbeth, 1988). Spores are applied as a paint to freshly cut stumps. Attempts to incorporate the agent into chainsaw oil (Artman, 1976) have had limited success since the fungus will eventually degrade the cut timber as well as the stump if applied to both surfaces.

Several formulations of the fungus *Trichoderma* sp. are sold as Binab-T^{TM} for control of silver leaf (*Chondrostereum purpureum*) in orchards. The product is inserted into the trunk in the form of powder, pellets or dowels (Lisansky, 1990; Lewis, 1991).

16.9 CONCLUSIONS

Until recently, formulation technology applied to biological control agents has lagged behind developments in strain selection and genetics, the major constraints being shelf-life, quantity of inoculum required per unit area, and persistence after application. The increasing interest in this area in the past few years, however, appears to be coming to fruition, and more effective formulations are beginning to reach the marketplace. These, when coupled with genetic improvement of biological control agents, should enable biological control to provide more realistic alternative strategies for crop protection and pest control.

REFERENCES

Amsellem, Z., Sharon, A., Gressel, J. and Quimby, P.C. Jr. (1990) Complete abolition of high inoculum threshold of two mycoherbicides (*Alternaria cassiae* and *A. crassa*) when applied in invert emulsion. *Phytopathology*, **80**, 925–9.

Amsellem, Z., Sharon, A. and Gressel, J. (1991) Abolition of selectivity of two mycoherbicidal organisms and enhanced virulence of avirulent fungi by an invert emulsion. *Phytopathology*, **81**, 985–8.

Angus, T.A. and Luthy, P. (1971) Formulation of microbial insecticides, in *Microbial Control of Insects and Mites* (eds H.D. Burges and N.W. Hussey), Academic Press, London, pp.623–38.

Artman, J.D. (1976) Further tests in Virginia using chain saw-applied *Peniophora gigantea* in loblolly pine stump inoculation. *Plant Disease Reporter*, **56**, 958–60.

Backman, P.A. and Rodriguez-Kabana, R. (1974) A system for the growth and delivery of biological control agents to the soil. *Phytopathology*, **65**, 819–21.

Baker, C.A. and Henis, J.M.S. (1990) Commercial production and formulation of microbial biocontrol agents, in *New Directions in Biological Control*, (eds R.R. Baker and P.E. Dunn), Alan R. Liss, New York, pp.333–44.

Barlow, F. (1985) Chemistry and formulation, in *Pesticide Application: Principles and Practice*, (ed P.T. Haskell), Clarendon Press, Oxford, pp.1–34.

Blachère, H., Calvez, J., Ferron, P., Corrieu, G. and Peringer, P. (1973) Étude de la formulation et de la conservation d'une préparation entomopathogene à base de blastospores de *Beauveria tenella* (Delacr. Siemaszko). *Annales de Zoologie et Écologie Animale*, **5**, 69–79.

Charudattan, R. (1986) Integrated control of waterhyacinth (*Eichhornia crassipes*) with a pathogen, insects, and herbicides. *Weed Science*, **34** (Suppl. 1), 26–30.

Cohen, E., Rozen, H., Joseph, T., Braun, S. and Margulies, L. (1991) Photoprotection of *Bacillus thuringiensis kurstaki* from ultraviolet radiation. *Journal of Economic Entomology*, **57**, 343–51.

Connick, W.J. Jr. (1988) Formulation of living biological control agents with alginate, in *Pesticide Formulations: Innovations and Developments*, (eds B. Cross and H.B. Scher), American Chemical Society, Washington D.C., pp.241–50.

Connick, W.J. Jr. (1990) Microbial pesticide controlled-release formulations, in *Controlled Delivery of Crop-Protection Agents*, (ed R.M. Wilkins), Taylor and Francis, London, pp.233–43.

Connick, W.J. Jr., Daigle, D.J. and Quimby, P.C. Jr. (1991) An improved invert emulsion with high water retention for mycoherbicide delivery. *Weed Technology*, **5**, 442–4.

Connick, W.J. Jr., Lewis, J.A., Quimby, P.C. Jr. (1990) Formulation of biocontrol agents for use in plant pathology, in *New Directions in Biological Control*, (eds R.R. Baker and P.E. Dunn), Alan R. Liss, New York, pp.345–72.

Couch, T.L. and Ignoffo, C.M. (1981) Formulation of insect pathogens, in *Microbial Control of Pests and Plant Diseases 1970–1980*, (ed H.D. Burges), Academic Press, London, pp.621–34.

Daigle, D.J. and Connick, W.J. Jr. (1990) Formulation and application technology for microbial weed control, in *Microbes and Microbial Products as Herbicides*, (ed R.E. Hoagland), American Chemical Society, Washington DC, pp.288–304.

Daigle, D.J. and Cotty, P.J. (1991) Factors that influence germination and mycoherbicidal activity of *Alternaria cassiae*. *Weed Technology*, **5**, 82–6.

Devisetty, B.N. (1988) Microbial formulations – opportunities and challenges in, *Pesticide Formulations and Application Systems: 8th Volume*, ASTM STP 980 (eds

D.A. Hovde and G.B. Beestman), American Society for Testing and Materials, Philadelphia P.A., pp.46–64.

Dulmage, H.T., Yousten, A.A., Singer, S. and Lacey, L.A. (1990) *Guidelines for Production of Bacillus thuringiensis H-14 and Bacillus sphaericus*, UNDP/ WorldBank/WHO, Geneva.

Dunkle, R.L. and Shasha, B.S. (1988) Starch-encapsulated *Bacillus thuringiensis*: a potential new method for increasing environmental stability of entomopathogens. *Environmental Entomology*, **17**, 120–6.

Ferron, P. (1981) Pest control by the fungi *Beauveria* and *Metarhizium*, in *Microbial Control of Pests and Plant Diseases 1970–1980*, (ed H.D. Burges), Academic Press, London, pp.465–82.

Filippov, NA (1989) The present state and future outlook of biological control in the USSR. *Acta Entomologica Fennica*, **53**, 11–18.

Foy, C.L. (1989) Adjuvants: terminology, classification, and mode of action, in *Adjuvants and Agrochemicals*, (eds P.N.P. Chow, C.A. Grant, A.M. Hinshalwood and E. Simundsson), CRC Press, Boca Raton, pp.1–15.

Fravel, D.R., Marois, J.J., Lumsden, R.D. and Connick, W.J. Jr. (1985) Encapsulation of potential biocontrol agents in an alginate-clay matrix. *Phytopathology*, **75**, 774–7.

Gaertner, F. (1990) Cellular delivery systems for insecticidal proteins: living and non-living microorganisms, in *Controlled Delivery of Crop-Protection Agents*, (ed R.M. Wilkins), Taylor and Francis, London, pp.245–57.

GIFAP (1984) *Catalogue of Pesticide Formulation Types and International Coding System Technical Monograph No. 2*, GIFAP, Brussels.

Graham-Weiss, L., Bennett, M.L. and Paau, A.S. (1987) Production of bacterial inoculants by direct fermentation on nutrient-supplemented vermiculite. *Applied and Environmental Microbiology*, **53**, 2138–40.

Haskell, P.T. (ed) (1985) *Pesticide Application: Principles and Practice*, Clarendon Press, Oxford.

Henry, J.E. (1990) Control of insects by protozoa, in *New Directions in Biological Control*, (eds R.R. Baker and P.E. Dunn), Alan R. Liss, New York, pp.161–76.

Herbig, S.M., Thompson, K.E., Falcon, L.A. and Berlowitz, A. (1991) Protective coatings for long-lasting biological insecticide formulations. *Proceedings of the International Symposium on Controlled Release of Bioactive Materials*, **18**, 556–7.

Ignoffo, C.M., Shasha, B.S. and Shapiro, M. (1991) Sunlight ultraviolet protection of the *Heliothis* nuclear polyhedrosis virus through starch-encapsulation technology. *Journal of Invertebrate Pathology*, **57**, 134–6.

Jones, K.A. (1990) Use of a nuclear polyhedrosis virus (NPV) to control *Spodoptera littoralis* in Crete and Egypt, in *Pesticides and Alternatives*, (ed J.E. Casida), Elsevier, The Netherlands, pp.131–42.

Kaya, H.K. and Nelsen, C.E. (1985) Encapsulation of steinernematid and heterorhabditid nematodes with calcium alginate: a new approach for insect control and other applications. *Environmental Entomology*, **14**, 572–4.

Kenney, D.S. (1986) DeVine – the way it was developed – an industrialists' view. *Weed Science*, **34** (Suppl. 1), 15–16.

Kerr, A. (1980) Biological control of crown gall through production of agrocin 84. *Plant Disease*, **64**, 25–30.

Kirsop, B.E. and Doyle, A. (eds) (1991) *Maintenance of Microorganisms and Cultured Cells*, Academic Press, London.

Kloepper, J.W. and Schroth, M.N. (1981) Development of a powder formulation of rhizobacteria for inoculation of potato seed pieces. *Phytopathology*, **71**, 590–2.

Knudsen, G.R., Eschen, D.J., Dandurand, L.M. and Wang, Z.G. (1991) Method to enhance growth and sporulation of pelletized biocontrol fungi. *Applied and Environmental Microbiology*, **57**, 2864–7.

Knudsen, G.R., Johnson, J.B. and Eschen, D.J. (1990) Alginate pellet formulation of a *Beauveria bassiana* (Fungi: Hyphomycetes) isolate pathogenic to cereal aphids. *Journal of Economic Entomology*, **83**, 2225–8.

Lacey, L.A. and Undeen, A.H. (1986) Microbial control of black flies and mosquitoes. *Annual Review of Entomology*, **31**, 265–96.

Lacey, L.A., Urbina, M.J. and Heitzman, C.M. (1984) Sustained release formulations of *Bacillus sphaericus* and *Bacillus thuringiensis* (H-14) for control of container-breeding *Culex quinquefasciatus*. *Mosquito News*, **44**, 26–32.

Lapage, S.P. and Redway, K.F. (1974) *Preservation of Bacteria with Notes on other Micro-organisms*, HMSO, London.

Lewis, J.A. (1991) Formulation and delivery systems of biological control agents with emphasis on fungi, in *The Rhizosphere and Plant Growth*, (eds D.L. Keister and P.B. Cregan), Kluwer Academic Publishers, Netherlands, pp.279–87.

Lewis, J.A. and Papavizas, G.C. (1985) Characteristics of alginate pellets formulated with *Trichoderma* and *Gliocladium* and their effect on the proliferation of the fungi in soil. *Plant Pathology*, **34**, 571–7.

Lewis, J.A. and Papavizas, G.C. (1987) Application of *Trichoderma* and *Gliocladium* in alginate pellets for control of *Rhizoctonia* damping-off. *Plant Pathology*, **36**, 438–46.

Lisansky, S.G. (1990) *Green Growers Guide*, CPL Press, Newbury.

McGuire, M.R. and Shasha, B.S. (1990) Sprayable self-encapsulating starch formulations for *Bacillus thuringiensis*. *Journal of Economic Entomology*, **83**, 1813–17.

McGuire, M.R., Shasha, B.S., Lewis, L.C., Bartelt, R.J. and Kinney, K. (1990) Field evaluation of granular starch formulations of *Bacillus thuringiensis* against *Ostrinia nubilalis* (Lepidoptera: Pyralidae). *Journal of Economic Entomology*, **83**, 2207–10.

McIntyre, J.L. and Press, L.S. (1991) Formulation, delivery systems and marketing of biocontrol agents and plant growth promoting rhizobacteria (PGPR), in *The Rhizosphere and Plant Growth*, (eds D.L. Keister and P.B. Cregan), Kluwer, The Netherlands, pp.289–95.

McQuilken, M.P., Whipps, J.M. and Cooke, R.C. (1990) Control of damping-off in cress and sugar-beet by commercial seed-coating with *Pythium oligandrum*. *Plant Pathology*, **39**, 452–62.

Moscardi, F. (1989) Production and use of entomopathogens in Brazil, in *Biotechnology, Biological Pesticides and Novel Plant-Pest Resistance for Insect Pest Management*, (eds D.W. Roberts and R.R. Granados), Boyce Thompson Institute, Ithaca, pp.53–60.

Munyaradzi, S.T., Campbell, M. and Burge, M.N. (1990) The potential for bracken control with mycoherbicidal formulations. *Aspects of Applied Biology*, **24**, 169–75.

Paau, A.S., Graham, L.L. and Bennett, M. (1991) Progress in formulation research for PGPR and biocontrol inoculants, in *Plant Growth-Promoting Rhizobacteria – Progress and Prospects*, (eds C. Keel, B. Koller and G. Defago), IOBC/WPRS, pp.399–403.

Papavizas, G.C., Fravel, D.R. and Lewis, J.A. (1987) Proliferation of *Talaromyces flavus* in soil and survival in alginate pellets. *Phytopathology*, **77**, 131–6.

Quimby, P.C. Jr., Fulgham, F.E., Boyette, C.D. and Connick, W.J. Jr. (1989) An invert emulsion replaces dew in biocontrol of sicklepod – a preliminary study,

438   *Formulation of biological control agents*

in *Pesticide Formulations and Application Systems: 8th Volume*, (eds D.A. Hovde and G.B. Beestman) ASTM STP 980, American Society for Testing Materials, Philadelphia PA, pp.264–70.

Rajnchapel-Messaï, J. (1990) Les biopesticides. *Biofutur*, **July/Aug 1990**, 23–34.

Raun, E.S. and Jackson, R.D. (1966) Encapsulation of a technique for formulating microbial and chemical insecticides. *Journal of Economic Entomology*, **59**, 620 –2.

Reinecke, P., Andersch, W., Stenzel, K. and Hartwig, J. (1990) A new microbial insecticide for use in horticultural crops, in *Brighton Crop Protection Conference, Pests and Diseases – 1990 Volume 1*, British Crop Protection Council, BCPC, Farnham, pp.49–54.

Rhodes, D.J. (1992) Microbial control of plant diseases, in *Disease Management in Relation to Changing Agricultural Practice*, (eds A.R. McCracken and P.C. Mercer), SIPP/BSPP, Belfast, pp.102–8.

Rhodes, D.J. and Logan, C. (1986) Effects of fluorescent pseudomonads on the potato blackleg syndrome. *Annals of Applied Biology*, **108**, 511–18.

Riba, G. (1984) Application en essais parcellaires de plein champ d'un mutant artificiel du champignon entomopathogène *Beauveria bassiana* (Hyphomycets) contre la pyrale du maïs *Ostrinia nubilalis* [Lep.: Pyralidae]. *Entomophaga*, **29**, 41–8.

Riba, G. (1985) Perspectives offertes par le champignon entomopathogène *Beauveria bassiana* (Bals.) Vuill. dans la lutte contre les ravageurs de grandes cultures. *Les Colloques d'INRA*, **34**, 131–40.

Rishbeth, J. (1988) Biological control of airborne pathogens, in *Biological Control of Pests, Pathogens and Weeds: Developments and Prospects*, (eds R.K.S. Wood and M.J. Way) Royal Society, London, pp.265–81.

Rombach, M.C., Aguda, R.M., Shepard, B.M. and Roberts, D.W. (1986) Entomopathogenic fungi (Deuteromycotina) in the control of the black bug of rice, *Scotinophara coarctata* (Hemiptera; Pentatomidae). *Journal of Invertebrate Pathology*, **48**, 174–9.

Sanders, C.J., Stark, R.W., Mullins, E.J. and Murphy, J. (1985) *Recent Advances in Spruce Budworms Research*, Minister of Supply and Services Canada, Ottawa.

Smith, R.J. Jr. (1986) Biological control of northern jointvetch (*Aeschynomene virginica*) in rice (*Oryza sativa*) and soybeans (*Glycine max*) – a researcher's view. *Weed Science*, **34**, (Suppl. 1), 17–23.

Soper, R.S. and Ward, M.G. (1981) Production, formulation and application of fungi for insect control, in *Biological Control in Crop Protection, BARC Symposium No. 5*, (ed G.C. Papavizas), Allanheid, Totowa, pp.161–80.

Sopp, P.I., Gillespie, A.T. and Palmer, A. (1989) Application of *Verticillium lecanii* for the control of *Aphis gosypii* by a low-volume electrostatic rotary atomiser and a high volume hydraulic sprayer. *Entomophaga*, **34**, 417–28.

Spurr, H.W. Jr. (1990) The phylloplane, in *New Directions in Biological Control*, (eds R.R. Baker and P.E. Dunn), Alan R. Liss, New York, pp.271–8.

Suslow, T.V. and Schroth, M.N. (1981) Bacterial culture preservation in frozen and dry-film methylcellulose. *Applied and Environmental Microbiology*, **42**, 872–7.

Suslow, T.V. and Schroth, M.N. (1982) Rhizobacteria of sugar beets: effects of seed application and root colonization on yield. *Phytopathology*, **72**, 199–206.

Taylor, A.G., Min, T.-G., Harman, G.E. and Jin, X. (1991) Liquid coating formulation for the application of biological seed treatments of *Trichoderma harzianum*. *Biological Control*, **1**, 16–22.

TeBeest, D.O. and Templeton, G.E. (1985) Mycoherbicides: progress in the biological control of weeds. *Plant Disease*, **69**, 6–10.

Tsuji, K. (1990) Preparation of microencapsulated insecticides and their release mechanism, in *Controlled Delivery of Crop-Protection Agents,* (ed R.M. Wilkins), Taylor and Francis, London, pp.99–122.

Tuan, S.J., Tang, L.C. and Hou, R.F. (1989) Factors affecting pathogenicity of NPV preparations to the corn earworm, *Heliothis armigera. Entomophaga,* **34,** 541–9.

Ward, M.G. (1984) Formulation of biological insecticides, surfactant and diluent selection, in *Advances in Pesticide Formulation Technology,* (ed H.B. Scher), American Chemical Society, Washington DC, pp.175–84.

Weidemann, G.J. (1988) Effects of nutritional amendments on conidial production of *Fusarium solani* f. sp. *cucurbitae* on sodium alginate granules and on control of Texas gourd. *Plant Disease,* **72,** 757–9.

Weidemann, G.J. and Templeton, G.E. (1988a) Control of Texas gourd, *Cucurbita texana,* with *Fusarium solani* f. sp. *cucurbitae. Weed Technology,* **2,** 271–4.

Weidemann, G.J. and Templeton, G.E. (1988b) Efficacy and soil persistence of *Fusarium solani* f. sp. *cucurbitae* for control of Texas gourd (*Cucurbita texana*). *Plant Disease,* **72,** 36–8.

Wilkins, R.M. (ed.) (1990) *Controlled Delivery of Crop-Protection Agents,* Taylor and Francis, London.

Wilson, C.L. and Chalutz, E. (eds) (1991) *Biological Control of Postharvest Diseases of Fruits and Vegetables, Workshop Proceedings,* USDA/ARS, Springfield VA.

Wood, H.A. and Granados, R.R. (1991) Genetically engineered baculoviruses as agents for pest control. *Annual Review of Microbiology,* **45,** 69–87.

Wood Mackenzie (1991) *Agrochemical Monitor,* 9 August 1991, Wood Mackenzie, Edinburgh.

Young, S.Y. and Yearian, W.C. (1986) Formulation and application of baculoviruses, in *The Biology of Baculoviruses, Volume 2,* (eds R.R. Granados and B.A. Federici), CRC Press, Boca Raton, pp.151–79.

17

The commercial exploitation of microorganisms in agriculture

K.A Powell

17.1 INTRODUCTION

This chapter will deal with the exploitation of microbes in agriculture. In order to keep to a sensible length, the text will concentrate on inoculants which are intended for sale to the farmer in order to deliver an effect. The effect could be pest or disease control, growth promotion, protection of harvested crops or weed control. The chapter will not attempt to cover control of pests, diseases or weeds by classical biological control, i.e. the single release of a predator in order to provide long-term control.

17.2 MICROBIAL CONTROL AGENTS

17.2.1 Commercial considerations

The world agrochemical market is about $12 000 000 000, of which biological agents account for about 1%, or $120 million (Figure 17.1). Some 92% of this biological market is for a dried preparation of *Bacillus thuringiensis* (Bt), containing toxic proteins in the form of crystals with various components of lysed bacteria and growth medium. The remainder represents a mixture of small products, but no significant living biological products have emerged. Despite considerable academic and

Exploitation of Microorganisms Edited by D.G. Jones
Published in 1993 by Chapman & Hall, London ISBN 0 412 45740 7

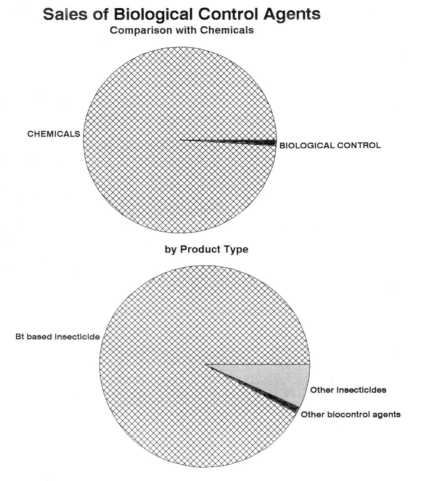

Figure 17.1 Impact of microbiological products on agrochemical markets.

commercial interest there is little real evidence of success for biological control agents. In order to examine this lack of success, some of the products or proposed products will be described, and their benefits and limitations discussed.

17.2.2 Mycoherbicides

A number of potential mycoherbicide products are on the market, have been introduced and withdrawn, or have reached development (Templeton and Heiny, 1989). A **mycoherbicide** is a fungal strain which is pathogenic for one or a few weeds. Mycoherbicides are generally applied as a spray to young weed plants within a crop. In order to compete with chemical herbicides, the target weeds are often ones which are difficult to control with chemical products. Two major target areas are perennial weeds in perennial crops and weeds closely related to crop species.

A successful product, albeit with a small market, is Collego™. This mycoherbicide is specific for northern jointvetch (*Aeschynomene virginica*) a weed of rice (*Oryza sativa*) in the USA. The weed seeds can reduce the price of the rice crop so that control is of value to the farmer. Careful formulation work has enabled the use of the fungus *Colletotrichum gloeosporoides* f. sp. *aeschynomene* in spray treatments to control the weed (Templeton, 1987). Unfortunately, the product has not achieved good penetration of the market and it remains a minor product.

DeVine™ is another mycoherbicide with a small market (Ridings, 1986). This product, which is applied to the soil, contains *Phytophthora palmivora* and is used in Florida to control strangler vine (*Morrenia osorata*) in citrus groves. The product gives excellent control but resale is limited by the lack of requirement for reuse due to the persistence of the fungus; hence the product cannot be said to be a commercial success.

Two other products have been introduced: Velgo, for the control of velvetleaf (*Abutilon theophrasti*) (Wymore *et al.*, 1987) and Casst, for the control of sicklepod (*Cassia obtusifolia*) (Walker and Riley, 1982). These products contain *Colletotrichum coccoides* and *Alternaria cassiae* respectively. Both products appear to have been withdrawn, and are reported to suffer from lack of efficacy in low humidity conditions.

These case histories demonstrate some key criteria which have to be considered if mycoherbicides are ever to provide effective weed control in large scale agriculture:

1. The target weed must be a problem to large areas of farm land with no effective control by agrochemicals.
2. The pathogen must be effective only on **one weed** or a **limited spectrum** of plants.
3. The pathogen must not survive in the environment beyond the season of use. The formulated pathogen must be effective under the conditions in which the crop is grown.
4. The formulation must protect from rainwash and UV irradiation.
5. The product must be suitable for mixing with agrochemicals in an integrated programme.

The fact that Collego and DeVine are technically successful may illustrate that the foliar environment in dry crops represents the major constraint for mycoherbicide development. Survival on the leaf surface while infection occurs is essential. Screening programmes normally use young plants under greenhouse or growth chamber conditions where survival in the natural environment is not selected. Some encouragement has been obtained by the use of invert (water in oil) emulsions to protect the fungus (Quimby *et al.*, 1988), but there is some way to go before effective commercial mycoherbicides can be said to be a reality.

17.2.3 Insecticides

Fungal insecticides

Like mycoherbicides, there have been a number of products which have appeared and then been withdrawn, but entomogenous fungi have been suggested for insect control for over 100 years (Krassilstschik, 1888). More recently, the products Mycotal and Vertelac, which contained *Verticillium lecanii* (Hall, 1981), were sold for some years.

Again, the inherent difficulties in development of fungal insecticides must be examined. There are potential markets for microbial insecticides, mainly due to the need for alternatives to control chemical-resistant species. Some target pests exist which are significant as stand-alone targets for microbial control.

Foliar-feeding pests represent the largest group of pests in most crops. It is the foliar location which causes the greatest difficulty to the potential fungal insecticides. Growth at low humidity, at potentially low temperature and under continuous ultra-violet irradiation is essential to achieve effective control. Relative humidity of lower than 95% is sufficient to prevent the progress of infection for most fungal pathogens (McCoy *et al.*, 1988). Figures 17.2 and 17.3 illustrate the effect of *Metarhizium arisophae*, and the effect of relative humidity on the infection. Again, like the mycoherbicides there is some potential for formulation to achieve an improvement in initial infection, but evidence has yet to be produced to demonstrate effective control.

Fungi which are obligate insect pathogens (e.g. *Entomophthora* spp., Dietrick, 1981), have been proposed as biological control agents but the production difficulties associated with such pathogens make commercial utilization a very remote prospect. Use of fungal pathogens as insecticides is therefore likely to be limited (by constraints of environmental limitations) to plant production systems which can be controlled to achieve the required environment. More effective products will await significant advances in fungal genetics and formulation research.

Viral insecticides

The use of insect viruses as insecticides has been practised for some time. Elcar, a commercial formulation of the *nuclearpolyhedrosis* virus of

Figure 17.2 Effect of *Metarhizium anisopliae* on insects at 100% relative humidity.

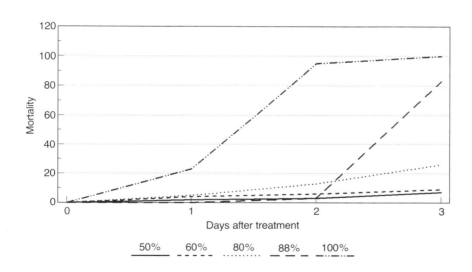

Figure 17.3 Effect of relative humidity on performance of *Metarhizium anisopliae*. Efficacy against *Heliothis virescens*.

Heliothis spp. was sold (Ignoffo and Couch, 1981) and has been recently re-launched. The practical use of virus products has been limited by inefficiency and slow speed of kill relative to chemical insecticides. The latter characteristic has limited the market to areas where speed of kill is not important, e.g. forestry. For food crops, slow kill can mean unacceptable damage to the crop which significantly reduces its value to the farmer. Formulation of virus, particularly to protect against damage by UV-irradiation is also very important (Ignoffo and Batzer, 1971). It should also be realized that the virus normally enters the host via feeding, and hence insects feeding on foliage represent the primary target.

Production of viruses is also an infant technology (Ignoffo and Couch, 1981). Most 'commercial' systems rely on production using living insects, a process with evident problems for increasing the scale of production. Use of insect tissue culture has been demonstrated on a pilot scale (Hink, 1981), but increases to commercial scale will require considerable further development.

The use of **genetic engineering** to improve the activity of baculovirus insecticides has recently been demonstrated. The genes coding for protein toxins from crustaceans or from insects have been cloned into *Autographica californica* baculovirus and the resulting recombinant viruses have been shown to be more active and to give faster speed of kill than the equivalent wild-type viruses (Stuart *et al.*, 1991; Tomalski and Miller, 1991). While this is an exciting development, the production and formulation issues must still be addressed before it will be possible for wide-scale use of these viruses in agriculture to successfully compete with agrochemicals.

Bacterial insecticides

Bacillus thuringiensis can be regarded as the major success of biological control research. In recent years, the development of resistance to chemical insecticides and the withdrawal of some products from some areas, e.g. Canadian forestry and Californian vegetables (Luthy *et al.*, 1982) has created a market for *Bt* in excess of $100 million (Wood MacKenzie, 1991). Unfortunately, in some of these markets reports are appearing of resistance to *Bt* (McGaughey, 1985). *Bt* is produced by standard fermentation techniques (Ignoffo and Anderson, 1979), but requires effective formulation and delivery in order to be effective (Bulla and Yousten, 1979). Inefficient formulation or poor spray coverage can allow the target pest to escape.

The long history of use of *Bt* in agriculture and forestry has enabled considerable study of the molecular basis of pathogenicity. The production of a single protein toxin which varies in different strains has made the genes which code for these proteins of considerable interest for further application (Li *et al.*, 1991). This has provided the basis for cloning *Bt* genes into every potential host including bacteria, viruses, fungi and plants (Vaeck *et al.*, 1987), many systems having been shown to produce effective toxin *in situ*. It is difficult to envisage the use of all these systems without the development of widespread resistance to the toxin.

Bt is not dependent on the living organism for activity. The activity resides in the protein crystals produced at the end of the fermentation. The mode of action is complex (Li *et al.*, 1991) and involves solubilization of the protein in the midgut of the insect followed by proteolytic cleavage to yield the active toxin. The active toxin then binds to a specific receptor on the epithelium. Activity is caused by the formation of a pore of 10–20 Å diameter which eventually leads to swelling and lysis of the epithelium. There is clearly opportunity for a number of routes to resistance, some of which can be specific to individual toxins but others may give more general resistance to the spectrum of *Bt* molecules. Much elegant work has been done with *Bt* leading to novel toxins and novel combinations of toxins which have a significant effect on the host range of the final product (Karamata and Pilot, 1987). Some of this work may be useful in aiding strategies to manage resistance in the future, with the next few years being the most interesting in the development of *Bt* products.

Bt production is at first sight extremely simple, especially when compared with more complex living agents. However, it remains a process which requires considerable knowledge and experience in order to produce the optimum activity and to retain that activity within the final formulation. Even this relatively simple product is more complex to produce and distribute than the chemical products with which it competes.

17.2.4 Biological control of plant pathogens

Two of the most technically successful truly biological control agents are *Agrobacterium* and *Peniophora*. The market for each is very small, because they are specific to minor pathogens of minor crops but it is worthwhile examining these products in some detail because they do illustrate some of the issues which have to be addressed in order to develop effective biological products.

Agrobacterium

Crown gall is a disease of stone fruits, grapes and some other dicotyle-donous plants caused by the bacterial pathogen *Agrobacterium tumefa-*

ciens. Kerr (1972) showed that it was possible to control the disease with a non-pathogenic variety of *Agrobacterium radiobacter*, named strain K84. If inoculated before the disease is present, *Agrobacterium radiobacter* will prevent the infection of plant tissue, although it cannot control the disease once present. K84 is specific to two of the three biovars of *A. tumefaciens* and will not control the disease on grapes caused by the third biovar.

The control method involves dipping the plant root system or seeds in a suspension of the bacteria of 10^7 or 10^8 cells per ml, immediately before sowing or planting. Activity depends on the survival of the living bacteria in soil, colonization of the root and also on the production of a toxin called **agrocin 84** (Kerr, 1980). Breakdown of control can occur if the gene for resistance to agrocin 84 is transferred to the pathogen (Panagopoulos *et al.*, 1979). For this reason a strain unable to transfer the gene was produced by genetic manipulation and is now sold for crown gall control (Ryder and Jones, 1990).

Key messages from the *Agrobacterium* work are that the pathogen and the biological control agent (BCA) must share an ecological niche where the pathogen has its effect on the host. Having a closely related species is a method of achieving this. The biological agent must be delivered to the site of action in an active state; dipping in an active suspension achieves this aim. However, there is still a requirement for a strain with the properties to give control, and many other *Agrobacterium* strains have been tested, though K84 remains the most effective (Ryder and Jones, 1990).

Peniophora

Peniophora gigantea (=*Phlebia gigantea*) was shown by Rishbeth (1963) to give effective control of *Heterobasidion annosum* (=*Fomes annosus*) on the cut stumps of pine. The purpose of this control is to prevent the colonization of nearby living trees via infection of the cut stump and transfer through root systems.

The *P. gigantea* strain effectively colonizes and destroys the stump, preventing infection by the pathogen by removal of the substrate. This BCA demonstrates the benefit of having virgin substrate to inoculate; no colonization of the substrate by indigenous flora has occurred, hence the BCA has a chance to colonize in the absence of effective competition by 'third party' microbes. A living, active, high population of the BCA is required to 'swamp' the available substrate with a large inoculum capable of domination of the infection court.

Mycostop (Lahdenperä *et al.*, 1991), a preparation based on the use of *Streptomyces griseoviridis*, has recently been registered in a limited number of countries for control of *Fusarium* spp., and is sold as a a dry formulation of spores and mycelium of the bacterium. It is too early to judge the commercial success of this product.

Other BCAs for soil-borne, foliar or post-harvest disease have been proposed or even introduced but to date no successful product has appeared.

17.3 MYCORRHIZAL INOCULANTS

Inoculation of crop plants with symbiotic fungi has been suggested for a number of years (Mosse *et al.*, 1969). Vesicular-arbuscular mycorrhizas (VAM), named after the structures formed by the mycobiont, are formed in all crop plants except beets and brassicas. VAMs are found in all soil types (Stribley, 1990). The major difficulty with these fungi is that they can only be cultured on the roots of plants or in plant tissue culture. This difficulty has prevented study of inoculation with isolated strains from pure cultures. Many studies have been done on the inoculation of crop plants with species cultured on the roots of pot-grown plants (Gianinazzi *et al.*, 1989), some of which claim benefit for yield enhancement due to inoculation. However, the overwhelming evidence is that the majority of these effects are due to enhanced uptake of phosphate from the soil (Stribley, 1989). Products from pure culture or from root culture are likely to be expensive, and it is most unlikely that this cost can be recovered, especially if the benefit from inoculation is also obtainable by the use of inexpensive phosphate fertilizer.

There is some evidence that ectomycorrhizal inoculation of trees may yield a benefit for forest trees on land which has not previously been used for forestry production (Grogan and Mitchell, 1990). This remains a small market which is unlikely to yield major return for a company producing an inoculant.

17.4 *RHIZOBIUM* INOCULANTS

Rhizobium spp. form a specific association with legumes. Much study has revealed the intricate detail of the complex association between *Rhizobium* and the root of the host plant (refer to Chapter 8). Many companies have produced and sold cultures of *Rhizobium* for crop plants, the largest one being for soya beans. For many crops, there is already an indigenous population in the soil so that inoculation does not improve yield. In many soils, a single inoculation will provide a population available to subsequent crops. *Rhizobium* inoculants tend to be available at a high price on land where the crop is introduced, for example, on the introduction of soya to Italy, though this price tends to fall as the treatment becomes an insurance against low inoculum potential.

Rhizobium inoculants have been developed on a number of substrates, among which peat remains the most popular material. *Rhizobium* inoculants are grown in a fermenter and inoculated into neutralized peat where growth occurs; in this form they can be stored and retain activity, at least under temperate conditions. The inoculum is usually applied as an on-farm dressing to seed. It is a simple and normally effective method, though problems can occur when samples are stored under unfavourable conditions which lead to loss of stability. Alternative methods of storage such as lyophilization, storage in oil or on alternative inorganic carriers have been attempted, but peat remains the preferred carrier.

Rhizobium inoculants do have some lessons for other microbial agents. The organism is relatively easy to grow, and in virgin soils can effectively reach the target site. In addition, it can be formulated for survival and for delivery to the active site. Indeed, survival of the inoculum in succeeding seasons has reduced demand.

In recent years, much effort has gone into improving *Rhizobium* strains in order to increase efficiency of **nitrogen fixation** and reduce the energy required from the host. It is instructive that genetically manipulated strains have so far failed to make a major impression. This is because the strain so constructed may be more efficient, but often fails to be competitive with the indigenous *Rhizobium* strain it is intended to replace. This is a lesson that must be learned, that much is yet to be understood about the population dynamics in the rhizosphere, about competition between microbes in the soil and about factors which affect root colonization.

17.5 HISTORICAL ASPECTS

A number of facts have become clear from experience gained with biological control agents (BCAs):

1. Biological control agents have a narrow spectrum, not only in terms of host but also in terms of environment including soil type, pH, temperature, moisture potential and crop.
2. The combination of production cost and formulation for survival is key to the economic viability of a BCA.
3. BCA may effectively control disease or pest under laboratory conditions, but under field condition the control may be negated by competitive microbes.
4. It is unrealistic to expect BCA survival under conditions where exposure to UV light, or to repeated cycles of desiccation, will affect the organism.
5. Delivery of the BCA to the site of action is critical to achieve the

desired effect. It is unrealistic to expect major movement or redistribution of the BCA in the absence of water movement.

6. The most effective microbial inoculants are those which have a niche which they can colonize before, or without competition from, indigenous microbes.

7. Delivery of a large actively growing population to the desired site of action appears to give better control of the target pest.

17.6 FUTURE PROSPECTS

Given the poor commercial history of biological control, is there a prospect for better results in the future? If there is a chance of larger markets in the future, then the process of selection, production, formulation, and delivery of biological control agents will need to be carefully planned and well resourced in order to deliver products which can be accepted by the farmer.

17.6.1 Markets

Biological control agents will have to compete with chemicals if they are to reach a larger proportion of the market. The first step for any biocontrol exercise must be to consider the ecological niche in which the putative BCA is expected to be effective. If this niche is suitable for the BCA then there is a chance of success, but further data will be needed before a suitable agent can be selected. For example, if we take the case of soil-borne disease, there is a complex of diseases which must be controlled if a significant market is to be reached. This complex of diseases can be defined by crop and by pathogen in order to fully describe the control which is required in order to develop an effective product. The environmental parameters can then be examined in order to enable an effective research strategy to be produced.

17.7 DEVELOPMENTAL INFLUENCES FOR BIOLOGICAL CONTROL AGENTS

17.7.1 The influence of soil type

It would be extremely difficult to develop as a commercial product a BCA which would be effective in a range of soil types. Clear evidence of this limitation was obtained by Defago *et al.* (1990), who isolated an effective BCA for black rot of tobacco, but later discovered that the bacterium would not give control in some soil types, especially those in which the disease is a major problem. Many BCAs have been claimed to work by production of iron-binding compounds called **siderophores**

(Schippers *et al.*, 1987), yet little consideration has been given to the effect of soil type on the availability of iron. If a BCA is to be suitable for a range of soil types, then this is an essential part of any screening programme.

17.7.2 The influence of temperature

Many papers have been published where the BCA has been grown in the laboratory at 30°C in rich, nutrient media. The BCA may have to be effective as a seed treatment under spring field conditions, in which temperature at sowing can be 8–10°C. It is at this temperature that the organism must be able to produce an effect; hence culturing and screening should be done at lower temperatures to match those of the crop and the pathogen (Merriman and Russell, 1990).

17.7.3 The influence of crop

Different plants may have different rhizosphere populations. The plant may select for the ability of the BCA to survive in the rhizosphere. **Proxy species** should be avoided to ensure the satisfactory performance of the BCA on the target crop. The chosen organism should not have a negative effect on crop plants which may be grown in the same area, and this will form an essential part of a screening process. Even cultivar variation can influence this ability of BCA to colonize the root-system. Atkinson *et al.* (1975) showed that the indigenous flora of wheat roots is altered by cultivar specificity. Weller (1988) demonstrated that the putative BCA *Pseudomonas fluorescens* 1–79 was present at 100 times greater numbers on cultivar Wampum than on cultivar Brevor.

17.7.4 The influence of available moisture

The BCA must be able to survive and recover from periods of drought, and also to grow and achieve an effect under the conditions in which the crop and disease can flourish. To rely on artificial irrigation to maintain the effectiveness of the BCA is to impose an automatic barrier to further market penetration. It is vital that screening be performed under the range of conditions which apply to the crop and disease complex.

Two examples illustrate this point quite clearly. Whipps and Magan (1987) demonstrated the effect of **water availability** on fungal growth and antagonist:pathogen interaction. Results showed clearly that limiting of water availability reduced the inhibition both by limiting growth and by reducing production of the inhibitory factor.

The second example is the poor performance of putative mycoherbicides under field conditions. The mycoherbicide will be effective under moist warm growth conditions but will fail to kill the host plant under more realistic field conditions (Greaves and MacQueen, 1990).

17.7.5 The spectrum of the BCA

Often the BCA is selected against a specific pathogen on a specific crop. Although this may offer more chance of effective control the market is likely to be limited, since farmers often treat a crop for a variety of problems in one treatment. However, since every time the crop has to be treated there is a cost in terms of labour, machinery and potential crop damage, there is a natural drive to limit the number of separate treatments. Careful targeting of the crop, disease, weed or pathogen complex will be required in order to define the niche, if a number of crops can be treated for the same problem then the market will increase. Compatibility (with both chemicals and other biological agents) will have to be considered.

17.7.6 Effect of pH

The pH range in which the BCA has to operate must also be considered. Leaf surfaces can be quite extreme in pH, while soil pH may vary between 5 and 8. The BCA should be screened across the range of pH appropriate to the target.

17.7.7 Development of the BCA

Production

It is often assumed that a BCA, once selected, can be easily produced, packaged and sold to the farmer. It is essential to be able to produce the BCA on a large scale in order to deliver the product to the farmer at a price which is competitive with chemical control. It has often been assumed that the cost of the medium ingredients is the cost of production, but there are other issues such as the capital cost of the production system which need to be considered. Table 17.1 compares the advantages and disadvantages of the main methods of production, i.e. using liquid or solid substrates. Most fermentation products are produced in **liquid** media for the reason that it is cheaper and easier to use an easily mixed system: mixing and cooling of solid substrate systems can be difficult.

The **cost** of the medium ingredients has to be considered, since expensive nutrients will cause the final product to be expensive. Less obvious is the effect of **growth rate** on cost; this is illustrated simply in

Table 17.1. Advantages and disadvantages of solid liquid and substrate fermentation

	Advantages	Disadvantages
Solid	Low capital cost	Scale up difficult
	Small scale	Control is complex
	Formulation simple	Difficult to make additions during process
Liquid	High density	Need to harvest and formulate
	Simple to control pH, oxygen, temperature, etc.	Can be difficult to get correct form, e.g. spores
	Simple to add nutrients	Capital cost may be high

Table 17.2. Effect of growth rates on fermentation cost–an illustration*

Fermentation period (days)	Time for emptying, cleaning, sterilizing, etc. (days)	Total run time (days)	Runs per year	Cost per batch (£)
2	2	4	75	73 000
4	2	6	50	100 000
6	2	8	37	128 110
8	2	10	30	153 340
10	2	12	25	180 000
12	2	14	21	210 500
14	2	16	18	242 220

*The following crude assumptions are made. The fermenter cost is £4 million per annum, comprising capital and labour costs. The materials per batch cost is £20 000. The fermenter runs for 300 days per year.

Table 17.2. The effect of slower growth rate is to lengthen the time of occupation of the fermenter vessel. Slow growing organisms can have a major disadvantage when cost is considered.

Harvesting

Once growth is complete, or is at a stage at which maximum active biomass is accumulated, this biomass must be harvested. Fungi are relatively easy to harvest because of their greater size; harvesting of bacteria may require expensive centrifugation. The harvesting process must retain as much active biomass as possible, but must also provide the biomass in a form where it can be preserved for storage before use.

Preservation

In order for the product to be sold to large numbers of farmers in different countries, it will have to be stable for sufficient time for its processing through a distribution system. Special distribution systems are possible, e.g. using cooled distribution or delivering to order, but these will add to the eventual cost of the final product.

Formulation of BCAs for survival has been based on the knowledge accumulated with *Rhizobium* inoculants. A copious literature exists but this has not provided a simple system for every strain. Laboratory methods to preserve microbial cultures have existed for many years, but these are designed to keep a small proportion of the original culture alive for as long as possible. For commercial production, the preservation of as close to 100% of the original population becomes very important, since every log decrease in viability will effectively increase the production cost of the final product ten-fold. The ideal objective is the survival of the BCA at a temperature range of −5°C to + 30°C for two years, but there is some flexibility in the practical limits.

There is evidence that microbes can be preserved in a number of ways to give some survival. Most methods have been based on the use of carriers, organic or inorganic matrices, with or without added nutrients (Table 17.3). The best known method is the use of peat, neutralized with chalk, since under the right conditions peat can support microbial growth from a low inoculum and also maintain the resulting population. This type of preservation is used for *Rhizobium* inoculants on a large scale. Inorganic carriers with added nutrients, e.g. vermiculite, have been used to achieve a similar product (Rhodes, 1990).

Table 17.3. Matrices used for formulation of BCA

Bran	Flour
Peat	Wood flour
Filtered mud	Manure
Compost	Milk powder
Lucerne powder	Soybean oil
Sugar cane bagasse	Vermiculite
Wheat straw	Coal dust
Rice straw	Mineral soil
Corn cob	Charcoal
Corn oil	Bentonite
Gum arabic	Pulverized lignite
Gum guar	Kaolinite
Methyl cellulose	PVP
Xanthan gum	

Delivery

While preservation of the active biomass is important, it is also important to consider the effect of the formulation on the delivery of the BCA in an active state to the site where it is required. If a method of preservation gives 100% survival, but does not permit application it will be of no practical use. Microbes formulated to be effective on the leaf surface will be formulated as **liquids** or **wettable powders** in order that they can be applied through conventional equipment (Figure 17.4). Further complexity may be introduced if there is an interaction with the ingredients needed to ensure activity on delivery, e.g. UV protectants, desiccation protectants or rainfastness improvers. For soil-based treatment, granules or seed treatments will be needed which will need to release the BCA at the right time and place relative to root development, and to the site of pest or disease attack. There is little point in the development of an active product which requires large amounts of formulated material in order to be effective (Merriman and Russell, 1990).

There is clearly a need for further research work on microbial preservation and formulation. The data obtained will tend to be specific to one organism, but there is no reason why such research should not be

Figure 17.4 A typical spray nozzle for aerial application of crop-protection agents.

successful. Encouragement is given by the success of microbial products such as *Rhizobium* and silage additives which are effectively formulated and sold.

17.8 CONCLUSION

Although this commercial view of the use of microorganisms in agriculture may, at first sight, seem rather negative, there is reason to believe that provided the niche is well targeted and the research planned to deliver a product which will meet the requirements of that niche, then a successful product may emerge.

REFERENCES

Atkinson, T.G., Neal, L.J., and Larson, R.I. (1975) Genetic control of the rhizosphere microflora of wheat, in *Biology and Control of Soilborne Plant Pathogens*, (ed G.W. Bruehl), American Phytopathological Society, Saint Paul, Minnesota, USA, pp.116–22.

Bulla, L.A. and Yousten, A.A. (1979) Bacterial insecticides, in *Microbial Biomass, Economic Microbiology*, vol. 4, (ed A.H. Rose), Academic Press, New York, USA, pp.91–114.

Défago, G., Berling, C.H., Berger, U., *et al.* (1990) Suppression of black root rot of tobacco and other root diseases by strains of *Pseudomonas fluorescens*: potential applications and mechanisms, in *Biological Control of Soil-borne Plant Pathogens*, (ed D. Hornby), CAB International, Wallingford, UK, pp.93–109.

Dietrick, E.J. (1981) Commercial production of entomophagons insects and their successful use in agriculture, in *Biological Control in Crop Production*, (ed G.C. Papavizas), Allenheld, Osmun, Totowa, New Jersey, USA, pp.151–80.

Gianinazzi, S., Gianinazzi-Pearson, V. and Trouvelot, A. (1989) Potentialities and procedures for the use of endomycorrhizas with special emphasis on high-value crops, in *Biotechnology of Fungi for Improving Plant Growth*, (eds J.M. Whipps and R.D. Lumsden), Cambridge University Press, Cambridge, UK, pp.41–54.

Greaves, M.P. and MacQueen, M.D. (1990) The use of mycoherbicides in the field. *Aspects of Biology*, **24**, 163–8.

Grogan, H., and Mitchell, D.T. (1990) The mycorrhizal status of some forest sites and the propagation of ectomycorrhizal Sitka spruce seedlings in Ireland. *Aspects of Appled Biology*, **24**, 123–30.

Hall, R.A. (1981) *Verticillium lecanii* as a microbial insecticide against aphids and scales, in *Microbial Control of Pests and Plant Diseases, 1970–1980*, (ed H.D. Burges), Academic Press, NY, USA, pp.483–98.

Hink, W.F. (1981) Production of *Autographica california* nuclear polyhedrosis virus in cell from large-scale suspension cultures, in *Microbial and Viral Pesticides*, (ed E. Kurstak), Marcel Dekker, New York, USA, pp.493–507.

Ignoffo, C.M. and Anderson, R.F. (1979). Bioinsecticides, in *Microbial Technology*, vol. 1, (eds H.J. Peppler and D. Perlman), Academic Press, New York, USA, pp.1–28.

Ignoffo, C.M. and Batzer, O.F. (1971) Microencapsulation and ultraviolet protectants to increase sunlight stability. *Journal of Economic Entomology*, **64**, 850–3.

Ignoffo, C.M. and Couch, T.L. (1981) The Nucleopolyhedrosis virus of *Heliothis* species, in *Microbial Control of Pests and Plant Diseases, (1970–1980)*, (ed H.D. Burges), Academic Press, New York, USA, pp.329–62.

Karamata, D. and Pilot, J.C. (1987). Hybrid *Bacillus thuringiensis* cells useful for biological control of pests with synergistic insecticidal activity and wider activity than the parent strains. *European Patent*, No.221024.

Kerr, A. (1972) Biological control of crown gall: Seed innoculation. *Journal of Applied Bacteriology*, **35**, 493–7.

Kerr, A. (1980) Biological control of crown gall through production of agocin 84. *Plant Disease*, **64**, 25–30.

Krassilstschik, I.M. (1888) La Production Industrielle des Parasites vegetaux par la destruction des Insects nuisibles. *Bulletin des Sciences France et Belgique*, **19**, 461.

Lahdenperä, M-L., Simon, E. and Uoti, J. (1991) Mycostop, a novel biofungicide based on *Streptomyces* bacteria, in *Biotic Interactions and Soil-borne Diseases*, (eds A.A.R. Beemster, G.R. Bollen, M. Gerlach, M.A. Ruissen, B. Schippers and A. Tempel), Elsevier, Amsterdam, Netherlands, pp.258–63.

Li, J., Carroll, J. and Ellar, D.J. (1991) Crystal structure of insecticidal δ-endotoxin from *Bacillus thuringiensis* at 2.5Å resolution. *Nature*, **353**, 815–19.

Luthy, P., Cordier, J-L and Fisher, M-M (1982) *Bacillus thuringiensis* as a bacterial insecticide, in *Microbial and Viral Pesticides*, (ed E. Kurstak), Marcel Dekker, New York, USA, pp.35–74.

McCaughey, E.A. (1985) Insect resistance to the biological insecticide *Bacillus thuringiensis*. *Science*, **229**, 193–5.

McCoy, C.W., Samson, R.A. and Boucias, D.G. (1988) Entomogenous fungi, in *CRC Handbook of Natural Pesticides, V (A)*, (ed C.M. Ignoffo), CRC Press, Boca Raton, Florida USA, pp.151–237.

Merriman, P. and Russell, K. (1990) Screening strategies for biological control, in *Biological Control of Soil-borne Plant Pathogens*, (ed D. Hornby), CAB International, Wallingford, UK, pp.427–35.

Mosse, B., Hayman, D.S. and Ide, G.J. (1969) Growth responses of plants in unsterilised soil to inoculation with vesicular-arbuscular mycorrhiza. *Nature*, **224**, 1031–2.

Panagopoulos, C., Psallidas, P.G. and Alivizatos, A.S. (1979) Evidence of a breakdown in effectiveness of biological control of crown gall, in *Soil-Borne Plant Pathogens*, (eds B. Schippers and W. Gams) Academic Press, New York, USA, pp.569–78.

Quimby, P.C., Fulgham, F.E., Boyette, C.D. and Connick, W.J. (1988) An invert emulsion replaces dew in biocontrol of sicklepod – a preliminary study, in *Pesticide Formulations and Application Systems*, Volume 8, ASTM STP 980, (eds D.A. Movde and G.B. Beestman), American Society for Testing and Materials: Philadelphia USA, pp.264–70.

Rhodes, D.J. (1990) Formulation requirements for biological control agents. *Aspects of Applied Biology*, **24**, 145–51.

Ridings, W.H. (1986) Biological control of strangler vine in citrus – a researcher's view. *Weed Science*, **34**, (suppl. 1), 31–2.

Rishbeth, J. (1963) Stump protection against *Fomes annosus*, III. Inoculation with *Peniophora gigantea*. *Annals of Applied Biology*, **52**, 63–77.

Ryder, M.K. and Jones, D.A. (1990) Biological control of crown gall, in *Biological Control of Soil-Borne Plant Pathogens*, (ed D. Hornby), CAB International, Wallingford, UK, pp.45–63.

Schippers, B., Bakker, A.W., and Bakker, P.A.H.M. (1987) Interactions of deleterious and beneficial rhizosphere microorganisms and the effect of cropping practices. *Annual Review of Phytopathology*, **25**, 339–58.

Stribley, D.P. (1989) Present and future value of mycorrhizal inoculants, in *Microbial Inoculation of Crop Plants*, (eds R.M. Macdonald and R.E. Campbell), IRL Press, Oxford, UK, pp.49–65.

Stribley, D.P. (1990) Do vesicular-arbuscular mycorrhizal fungi have a role in plant husbandry? *Aspects of Applied Biology*, **24**, 117–21.

Stuart, L.M.D., Hirst, M., Ferber, M.L., *et al.* (1991) Construction of an improved baculovirus insecticide containing an insect-specific toxin gene. *Nature*, **352**, 85–8.

Templeton, G.E. (1987) Mycoherbicides – achievements, developments and prospects, in *Proceedings of the Eighth Australian Weeds Conference*, Weed Society of New South Wales: Sydney, Australia, pp.489–97.

Templeton, G.E. and Heiny, D.K. (1989) Improvement of fungi to enhance mycoherbicide potential, in *Biotechnology of Fungi for Improving Plant Growth*, (eds J.W. Whipps and L.D. Lumsden) Cambridge University Press, Cambridge, UK, pp.127–52.

Tomalski, M.D. and Miller, L.K. (1991) Insect paralysis by baculovirus-mediated expression of a mite neurotoxin gene. *Nature*, **352**, 82–5.

Vaeck, M., Raynaerts, A., Hotte, H., *et al.* (1987) Transgenic plants protected from insect attacks. *Nature*, **238**, 33–7.

Walker, H.L. and Riley, J.A. (1982) Evaluation of *Alternaria cassiae* for the biocontrol of sicklepod (*Cassia obtusifolia*). *Weed Science*, **30**, 651–4.

Weller, W.M. (1988) Biological control of soil-borne plant pathogens in the rhizosphere with bacteria. *Annual Review of Phytopathology*, **26**, 379–407.

Whipps, J.M. and Magan, N. (1987) Effects of nutrient status and water potential of media on fungal growth and antagonist-pathogen interaction. *EPPO Bulletin*, **17**, 581–91.

Wood Mackenzie, (1991) *Agrochemical Service*, 1991, Wood Mackenzie, Edinburgh.

Wymore, L.A., Watson, A.K. and Gotlieb, A.R. (1987) Interaction between *Colletotrichum coccodes* and thidazuron for control of velvetleaf (*Abutilon theophrasti*). *Weed Science*, **35**, 377–83.

Index

Page numbers in *italics* refer to tables; those in **bold** refer to figures.

Index